CHEMISTRY RESEARCH AND APPLICATIONS

FOCUS ON CATALYSIS RESEARCH

NEW DEVELOPMENTS

Chemistry Research and Applications

Additional books in this series can be found on Nova's website
under the Series tab.

Additional E-books in this series can be found on Nova's website
under the E-books tab.

Chemical Engineering Methods and Technology

Additional books in this series can be found on Nova's website
under the Series tab.

Additional E-books in this series can be found on Nova's website
under the E-books tab.

CHEMISTRY RESEARCH AND APPLICATIONS

FOCUS ON CATALYSIS RESEARCH

NEW DEVELOPMENTS

MINJAE GHANG
AND
BJØRN RAMEL
EDITORS

Nova Science Publishers, Inc.
New York

Library of Congress Cataloging-in-Publication Data

Focus on catalysis research : new developments / editors, Minjae Ghang and Bjxrn Ramel.
p. cm.
Includes index.
ISBN 978-1-62100-444-8 (hardcover)
1. Catalysis--Research. I. Ghang, Minjae. II. Ramel, Bjxrn.
QD505.F64 2011
541'.395--dc23

2011033992

Published by Nova Science Publishers, Inc. †New York

CONTENTS

PREFACE

This book presents current research in the study of the principles, types and applications of catalysis. Topics discussed in this compilation include macroporous materials; a new kind of eco-compatible hybrid biocatalyst for selective reactions; new research in poylmer supported catalysts; ionic liquids as a catalyst or solvent for various organic transformations; hybrid organic-inorganic materials; preparation and characterization of coated microchannels for the selective catalytic reduction of NOx and emerging catalysts for wet air oxidation process.

Chapter 1 - Porous materials can be divided into three types according to the aperture sizes. It is defined respectively as microporous materials when the aperture sizes are lower than 2 nm, mesoporous materials when the aperture sizes fall in range of 2-50 nm, and macroporous materials when the aperture sizes become larger than 50 nm. In some cases, microporous and mesoporous materials can result in an unacceptably slow diffusion rate of reactants considering their small aperture sizes, which hinders their practical application. In order to overcome this limitation, one promising method is to prepare macroporous materials with the aperture sizes in range of 50 nm-1.5 μm to increase the chemical diffusion rate. Thus, macroporous materials have attracted considerable interest due to their important applications in catalysis, and become a research hotspot in recent years. This paper is intended to review the recent advance in macroporous materials, especially three dimensionally ordered macroporous (3DOM) materials, and their catalytic application on the basis of the research achievement on 3DOM catalysts in the authors' group. The paper is composed of four sections: (1) Overview section — a concise overview on macroporous materials will be given; (2) Controlled Synthesis: — various kinds of synthesis techniques and principles will be introduced; (3) Characterization — A variety of characterization methods for the macroporous materials will be introduced; (4) Catalytic property — the catalytic property of macroporous materials will be reviewed.

Chapter 2 - Biocatalysts and chemical catalysts are on many aspects complementary: range of substrates and reactions, operating conditions or enantioselectivity. Nowadays, the ecological factor has to be taken into account, which urges industrials to develop "green chemistry" procedures. To deserve the "green chemistry" label, an industrial production must not only resort to catalytic procedures to lower the wastes, but also use harmless solvents. The ideal one would be water, with low temperature and pressure, to lower the power consumption. From this point of view, biocatalysts seem better fitted than chemical catalysts, which are often used in organic solvents. Given their relative complementarities, it is very interesting to conceive catalysts that would combine the robustness and wide range of reactions of synthetic catalysts with the ability of enzymes to work under mild conditions in aqueous medium and with high selectivity. As a result, many teams have been working recently on the conception of hybrid biocatalysts, obtained by association of a protein with a synthetic molecule.

This chapter will first begin with a review of the strategies explored to obtain the kind of hybrid biocatalysts that are artificial metalloenzymes. They will be divided in three main categories: (i) direct insertion of an inorganic ion into a protein, (ii) covalent linking of a metallic cofactor to a protein, and (iii) supramolecular anchoring of a metallic cofactor. The results obtained with two particular systems will then be detailed. In these studies, artificial metalloenzymes have been built by non-covalent association of metalloporphyrins with monoclonal anti-steroid antibodies, or with Xylanase A from Streptomyces lividans, to catalyse oxidation reactions.

Chapter 3 - A polymer supported catalyst is a reactive system in which the catalyst or more correctly the catalytic centre is attached to a polymeric molecule. The catalytic centre is either an inherent part of the polymer, which may be a part of the monomer and which will be added to the polymer during polymerization, or attached to a previously formed polymer by some kind of chemical (covalent bonds) or physical (encapsulation) methods. These two situations can be represented as shown in figure 1.

The polymer species of these types of catalysts may be a linear or a crosslinked entity, and again the latter have proved particularly useful. Polymeric catalysts are generally used in catalytic quantities relative to reaction substrates, and can often be reused many times. The attachment of a catalyst to a support may improve its stability and selectivity. On the other hand, increased experimental convenience arising with a polymeric catalyst may be offset by a significant reduction in reactivity associated, for example, with diffusional limitations imposed by resin supports. Recent interest in the development of environmentally benign synthesis has evoked a renewed interest in developing polymer-bound catalysts and reagents for organic synthesis that maintain high activity and selectivity. [1, 2] A wide variety of catalysts have been supported in this way, ranging from strong acids and bases (ion-exchange resins), transition-metal complexes, organic catalysts and photosensitizers right through to the highly specific enzyme catalysts.

Chapter 4 - The room temperature ionic liquids (RTILs) are liquids which consist of ions and melts at or below 100 °C. They have typical properties like negligible vapor pressure, high thermal stability, and nonflammable nature. Moreover, the physicochemical properties of ionic liquids, such as their melting temperature and hydrophilicity/hydrophobicity, can be fine-tuned by changing the structure of the cations and anions. Ionic liquids were initially introduced as alternative green reaction media; but today they have marched far beyond this border, showing their significant role in controlling the reaction as a catalyst. The Brønsted and Lewis acidic ionic liquids have exhibited a great potential in replacement of conventional homogeneous and heterogeneous acidic catalysts, and have been successfully applied to a variety of reactions including the Diels-Alder reaction, Mannich reaction, Prins reaction, Acetalization of carbonyl compounds, Friedel–Crafts reaction, Friedlander Reaction, Biginelli reaction, Esterification and transesterification with enhanced selectivity and activity. Ionic liquids are also extensively used as a green reaction media for various transition metals catalyzed cross coupling reaction such as Suzuki, Heck, and Negishi. It is also used for catalyst immobilization, which not only allow recovery and recycling of costly transition metal catalyst but also has positive effect on catalysis. Catalysts often become more reactive for a reaction such as Hydroformylation, Hydrogenation and hydroamination reactions.

Chapter 5 - The influenza infection pandemic has spread on a world scale and become a major threat to human health. Neuraminidase (NA), a major surface glycoprotein of influenza virus with well-conserved active sites, offers an ideal target for the development of antiviral drugs. In this chapter, we will review the recent results on the NA active sites, NA-inhibitor interactions, proton transfers of NA inhibitors as well as respective roles of the core templates, functional groups and synergistic effects of NA inhibitors through the fragment approach. On such basis, the *de novo* designs were carried out in order to discover novel NA inhibitors as potential antiviral drugs. The calcium ion rather than conserved water molecules was found to be crucial to stabilize the NA active site; nonetheless, the conserved water molecules can greatly alter the NA-inhibitor interactions. The interaction mechanisms of NA were studied with two current antiviral drugs oseltamivir and peramivir as well as two potent lead compounds 4-(N-acetylamino)-5-guanidino-3-(3-pentyloxy) benzoic acid (BA) and 5-[(1R,2S)-1-(acetylamino)-2-methoxy-2-methylpentyl]-4-[(1Z)-1-propen-yl)-(4S,5R]-D-proline (BL). It indicated that for some NA inhibitors, for the proton transfers to form the zwitterions are not facile in aqueous solutions, which explains the low oral bioavailability of current antiviral drugs. With the fragment approach, it was revealed that the core templates rather than functional groups play a larger role during the inhibitor interactions with the NA proteins; moreover, the binding qualities are largely determined by the synergistic effects of the core templates and functional groups. Based on the understanding of NA's catalytic mechanisms and structure-activity relationships, seven tripeptides were successively designed as NA inhibitors. It was found that some of them show much higher inhibiting activities than the lead compound BA. As the tripeptides have the advantages such as low toxicities and side effects, we believe that this work will guide synthetic and medicinal chemists to the tripeptide-based antiviral drugs. It may throw light on the treatment and prevention of influenza infections, since the resistances have ever been reported for the current commercial drugs.

Chapter 6 - Oxidative catalysis has been and will be playing an important role in the production of large quantities of intermediates. This technology has a great potential for improvement and this has led to a series of better processes, including the development of a great diversity of suitable catalysts. Within these compounds, metal phosphates and phosphonates possessing layered and pillared structures were found to constitute a very good alternative as catalysts to be used in oxidative catalysis.

Metal phosphates and phosphonates can be considered as strong inorganic solid acids and much of their catalytic activity has been attributed to the Brønsted acidity of the interlayered hydroxyl groups and to the Lewis acidity of the metal center. Metal phosphates and phosphonates can be obtained at low temperatures, in aqueous media, using soft chemical routes and their preparation is quite accessible if the correct phosphonic acids and phophates are available. These heterogeneous materials are very interesting from the economical as well as from the environmental points of view because of the high yields and short reaction times, easy recovery from the reaction mixtures and the possibility to perform reactions in solvent-free conditions.

Due to their particular physical and chemical properties and high versatility, in the last forty years, this class of organic-inorganic materials has been captivating the attention of many researchers involved in heterogeneous catalysis. The main research effort in the metal phosphate and phosphonate field was initially directed towards compounds with tetravalent metal cations, but a wide variety of divalent and trivalent metals have also been reported. Nowadays, a large number of metal phosphates and phosphonates of the α- and γ-type are known.

The general interest in the chemistry of metal phosphates and phosphonates is mainly due to their unusual compositional and structural diversity varying from one-dimensional arrangements to three-dimensional microporous frameworks, passing by the most common layered networks.

Good results have been obtained with a large diversity of metal phosphates and phosphonates in a variety of organic reactions, and particularly in oxidative catalysis. The importance of these systems has also been recognized in research areas such as electrochemistry, microelectronics, biological membranes and photochemical mechanisms.

Chapter 7 - Binuclear bridged and non-bridged transition metal catalysts with definite electronic and steric modulations currently attract significant attention in the catalysis community, mainly because cooperative effects of adjacent metal ions are expected to lead to unique activation modes towards olefin polymerizations and to novel reactivity patterns. Cooperative effects between adjacent catalytic centers in binucleating ligands with aliphatic/aromatic spacers were shown to induce significant rate and selectivity enhancements in ethylene and α-olefin oligo/polymerizations. The binuclear metallocene catalysts are fascinating because of their beneficial effects in the formation of polymers for various explicit applications. In this chapter, we essentially made an attempt to focus on the catalytic role of bimetallic core towards olefin polymerizations. A comprehensive overview of this topic, which the present review is aimed at giving, seems appropriate and timely. In the main stream of this chapter, we tried to cover the recent applications of atom/bond bridged bimetallic coordination and organometallic catalysts with well defined architectures over their monometallic counter parts and an account on the usage and precise applications of supported bimetallic metallocene catalysts.

Chapter 8 - Shallow microchannels (Ø= 460 μm) were successfully coated with different catalytically active phases, e.g., Cu-ZSM-5, Cu/(ZSM-5+Al_2O_3), Au/Al_2O_3, Ag/(Al_2O_3+Ionic liquid) and Ag/Al_2O_3, and tested on the hydrocarbon-assisted selective catalytic reduction of NOx (HC-SCR) with different model bio-derived fuels, i.e., methyl- and ethyl laurate produced by transesterification and hexadecane, a paraffinic component that can be produced by decarboxylation and/or decarbonylation of natural oils and fats. Characterization of the washcoats was done by means of X-ray photoelectron spectroscopy (XPS), scanning electron microscopy with energy dispersive X-ray analysis (SEM-EDXA) and laser ablation inductively coupled plasma mass spectrometry (LA-ICP-MS), showing a dependence of the metal loading with the impregnation time and the precursor concentration. The Ag/Al_2O_3 catalysts exhibited, in general, the highest activities towards the NOx reduction. Optima in impregnation time and concentration of $AgNO_3$ solution displaying the highest activity in HC-SCR among the prepared Ag/Al_2O_3 washcoats were established. A combination of Cu-ZSM-5 or Cu/(ZSM-5+Al_2O_3) and the optimum Ag/Al_2O_3 catalyst were tested in order to improve the low temperature reduction in SCR with hexadecane as a reducing agent. The enhancement of the activity at low temperatures (< 350 °C) was up to seven-fold compared to the case when only Ag/alumina was used. The effect of the hydrocarbon concentration (hexadecane) and the presence of water in the feed were also investigated.

The reduction results over silver/alumina with bio-derived fuels were compared to those obtained when using fossil-derived fuels, e.g., octane. Octane offered better NOx reduction efficiency compared to hexadecane at temperatures higher than 350 °C. However, for temperatures lower than 350 °C, in which most of the diesel engines operate, the reduction activity of hexadecane was higher compared to octane. The obtained results in the microreactor were compared to those obtained by a conventional Ag/Al_2O_3 catalyst powder in a minireactor (Ø= 10 mm). Similar NOx reduction results were attained when working under similar conditions in both reactor systems, i.e., the same ratio between the total flow rate and the catalyst mass.

Chapter 9 - Wastewater treatment has reached a maturity state but the growing industrialization along with the more stringent environmental regulations demand an increasing dynamism in short and medium term. Accordingly, the existing technologies should improve in both versatility and efficiency and, in this sense, catalysts can play a prominent role. In this chapter, the authors offer a critical review of the catalysts currently investigated for the industrial wastewater decontamination by wet air oxidation. A survey of catalysts industrially implemented and academically investigated is presented and their nature and competitive features in activity and durability remarked. The current trend in the exploration of nanomaterials in the wet air oxidation field is highlighted. Updated research on nanotechnology-based catalysts, specially, carbon nanostructures and gold nanoparticles, is summarized and thorough discussed. Our recent results involving nanoscale gold particles are also included.

Chapter 10 - The highly efficient and selective environmentally benign bleaching approach using polyoxometalate (POMs) catalyzed ozonation of chemical pulps in organic solvent reaction media has been developed. A number of low-boiling polar aprotic and protic organic solvents showed a well-defined capacity for ozonation improvement in the presence of α-Keggin-type mixed-addenda heteropolyanions (HPAs), particularly molybdo-vanado-phosphate HPAs of series $[PMo_{(12-n)}V_nO_{40}]^{(3+n)-}$. Aqueous acetone solution was found to be the best suited reaction media for ozone delignification catalyzed by $[PMo_7V_5O_{40}]^{8-}$ (HPA-5) polyanion. The effect of solvent and catalyst concentration, pH and ionic strength on ozone bleaching of commercial eucalypt kraft pulp has been examined. The solvent proportion in solution and medium acidity were the principal factors affecting catalytic efficiency of HPA during ozonation. Under optimized conditions, the brightness improvement of bleached pulp by ca. 15% ISO, with additional lignin removal by ca. 39% and increase in intrinsic viscosity by ca. 3% was observed in comparison with conventional (solvent/catalyst free) ozonation. The elimination of the Donnan effect by increase in ionic strength (cation concentration) of pulp suspension substantially improved delignification capacity of HPA/O_3 bleaching system. The catalyst recovery/re-oxidation by ozone has been examined using partially reduced polyanion species ($HPA\text{-}5_{red}$ or heteropoly blue). It was shown that the bulk portion (ca. 90%) of $HPA\text{-}5_{red}$ is readily re-oxidized by ozone during first reaction minutes thereby confirming the closure of redox catalyst cycle, as a required condition for successful HPA-catalyzed oxidative delignification. Addition of organic solvent had favorable effect on $HPA\text{-}5_{red}$ re-oxidation by ozone (k of 0.29 min^{-1} vs. 0.43 min^{-1} for ozonation, respectively, in water and 10% v/v acetone solution; pH 2), being evidently the principal reason for high efficiency of solvent-assisted HPA/O_3 bleaching. The bleaching effect of HPA/O_3 system was substantially improved by enzymatic (xylanase) pulp pre-treatment before ozonation.

Chapter 11 - Environmental and economical considerations in past decades have urged scientists to redesign commercially important processes towards the use of more environmentally friendly substances avoiding the use of toxic compounds and generation of waste. In this way, the substitution of conventional heating by microwave irradiation as an alternative source of energy is increasingly attractive. Microwave-assisted heating is particularly interesting because it offers improved conversions (and often seletivities to target products) under milder conditions at remarkably reduced times of reaction for a wide range of conventional reactions.

In this chapter we aim to explore several asymmetric syntheses carried out under microwave irradiation, giving a general overview of recently developments in this field. Significant rate enhancements, a decrease of chiral catalyst loading and higher enantioselectivities are the key achievement of microwave activation protocols.

Chapter 12 - Since the discoveries of facile methodologies for the synthesis of triarylcorroles and the corresponding metal complexes, metallocorroles have attracted the interest of chemists immensely because of their ability to catalyze diverse reactions. The huge number of studies performed during the last decade on metallocorrole catalyzed reactions has led to several important catalytic systems encompassing oxidation catalysis, reduction catalysis, group transfer catalysis, etc. The aim of the present chapter is to provide a comprehensive account on the catalytic properties of metallocorroles with an emphasis on recent advances in the area and future directions.

Chapter 13 - Dendrimers are highly monodisperse organic nanoparticles prepared by an iterative method of synthesis. They are considered as the fourth generation polymer architecture. Due to their particular structure and the properties arising due to this structure, dendrimers have aroused the curiosity of scientists from all disciplines of research ranging from chemistry to medicine. Catalysis is an area of science really blessed by dendrimers. Contrary to their high molecular weight, many dendrimers have shown well defined solubility properties and because of that dendrimers are considered to have filled up the gap between homogeneous and heterogeneous catalysts. Dendrimers offer a number of possibilities in fine tuning the activity and selectivity of many catalysts by properly attaching them to the dendrimers. The present chapter deals with the developments in catalysis after the materialization of dendrimers.

Chapter 14 - To start with, let us recall the definition of transition metals. According to the modern definition, given by the International Union of Pure and Applied Chemistry (IUPAC) a transition metal is 'an element whose atom has an incomplete *d* sub-shell, or which can give rise to cations with an incomplete *d* sub-shell.'[1] Electrons are fed into the *d*-orbitals starting from Group 3. For the first and second transition series, the Group 3 elements, scandium (Sc^{3+}) and yttrium (Y^{3+}), have a single *d* electron in their outermost shell but usually they are not considered as transition metals. In all of their compounds they exist as Sc^{3+} and Y^{3+} ions where there are no *d* electrons. Other elements with d^1 configuration are lanthanum and actinium, but they are classified under *lanthanoid* and *actinoid* series of elements, respectively. On the other hand, the Group 12 elements, namely zinc, cadmium, and mercury have an outer shell electronic configuration $d^{10}s^2$ with no incomplete *d* shell and hence they are not transition metals according to the above definition. In their +2 oxidation state they have a d^{10} electronic configuration while in the +1 oxidation state; there are no unpaired electrons because of the formation of dimer with a covalent bond between the two atoms. An interesting property of these transition metals is that they can exhibit two or more oxidation states which usually differ by one in their compounds. They have electrons of similar energy in both the *3d* and *4s* levels and thus a particular element can form ions of nearly the same stability by losing different numbers of electrons. The first row transition metal catalysts are of great utility in the oxidation chemistry because of their high reactivity and general utility [2]. This is the reason for which the transition metals can be utilized as catalysts for redox reactions. In biological system, the transition metals play a very crucial role to perform different redox reactions which are otherwise difficult to carry other under normal conditions.

This basic knowledge about the transition metals motivated the scientists to design novel materials which contain these elements. These type of materials can be used as catalysts in various eco-friendly, selective and industrially important organic transformations. A catalyst is a compound or material that can affect the rate of a chemical reaction by providing an alternative and lower energy profile or pathway. That is, it only changes the cost of the activation energy. It is not related to the thermodynamics of the process and hence, the final product distribution. Complexation by transition metals affords access to a wide variety of oxidation states for the metal. This has the property of providing electrons or withdrawing electrons from the transition state of the reaction. Most industrially used catalysts are the transition metal in a bed, as a metal or bound structure.

In this context the development in the field of microporous materials started attracting widespread attention because of their exceptional surface areas and well-defined pore sizes. These are extremely desirable for the diffusion of bulky adsorbate or reactant molecules, which is one of the key requisites for being a good catalyst support. As early as 1983, Taramasso *et al.* invented a porous crystalline synthetic material, named as titanium silicalite 1 or TS-1 [3], where Ti atoms partially substitute Si of ZSM-5 structure. TS-1 found to be a very efficient and selective catalyst in a number of industrially important organic transformations involving small sixe molecules which could panitrate its medium size micropores of dimensions 5.4–5.6 Å. Later in 1990, Huybrechts *et al.* reported the oxidation of alkanes on a microporous crystalline titanium silicalite with hydrogen peroxide as oxidant. [4] But the major breakthrough came around two years later, when Kresge *et al.* first prepared the M41S family of mesoporous solids with regular arrays of uniform channels [5]. One of the members of this family, MCM-41 exhibited a hexagonal arrangement of uniform pores with dimensions tuneable from *ca.* 15 Å to more than 100 Å. The usefulness of these materials is attributed to their microstructure, which allows the molecules to access large surface area which enhances their catalytic activity and adsorption capacity. Followed by this discovery, Pinnavaia *et al.* reported a hexagonal mesoporous titanium containing silica, Ti-HMS [6], which worked as an efficient catalyst for the selective oxidation of alkanes, hydroxylation of phenol and epoxidation of alkenes in the presence of hydrogen peroxide. Titanium-containing mesoporous silica gradually became a substance of great interest both in academia and industry due to their potential to oxidize very bulky organic substrates which are otherwise difficult to oxidize over microporous TS-1 [3] under liquid-phase reaction conditions. This reflected the high potential of titanium containing catalysts in different industrially important reactions. Following this, numerous heterogeneous catalysts based on other transition metals started developing rapidly and they were also found to be very important catalysts for different chemical transformations [7].

Catalytic reactions can be performed either in homogeneous medium or heterogeneous. Both the media involved in the process have advantages and disadvantages over each other.

In most of the cases, the heterogenized catalytic species worked well in the immobilized state, often better than expected. In homogeneous catalysis, all catalyst, substrate(s) and the reactant are in a single phase, mainly in liquid phase. In heterogeneous catalysis, the catalysts are insoluble in the liquid phase where reactants and substrates remain. Generally, homogeneous reaction requires low temperature, whereas heterogeneous catalytic reactions are carried out at high temperature. Immobilization or heterogenizing of transition metals on different solid supports, both organic and inorganic, offer the advantages of high catalytic activity and stability, and several other benefits like easy separation. In homogeneous catalysis, separation of catalysts from reaction mixture is a cumbersome task. Heterogeneous catalysts can be used for several times without appreciable loss of its activity in its next use.

There are many strategies for the design and the preparation of heterogeneous catalysts. Various types of supports can be used viz. encapsulation in zeolites [8], immobilization in porous alumina [9], immobilization in mesoporous silica [10-13], Y-zeolite [14], resin [15], grafting on polymers [16], dendritic [17, 18] and polymeric organic support [19] have been developed in terms of heterogenization of homogeneous catalysts. In this review article we shall confine our discussion particularly on three main types of solid supports: a) microporous and mesoporous silicas with ordered pore systems, b) non-porous silicas and c) polymers or resins. A general overview of the different types of catalysts based on active transition metals that can be used as catalyst liquid phase partial oxidation reactions will be discussed. These transition metals shall include titanium, vanadium, niobium, chromium, molybdenum, manganese, iron, cobalt, nickel and copper. In this book review we will limit our discussion up to those transition metal containing catalysts, which have been synthesized and studied by our group.

In: Focus on Catalysis Research: New Developments
Editors: Minjae Ghang and Bjørn Ramel
ISBN: 978-1-62100-455-4

Chapter 1

MACROPOROUS MATERIALS: CONTROLLED SYNTHESIS, CHARACTERIZATION AND CATALYTIC PROPERTIES

Changyan Li[1,2], Baocang Liu[1], Ying Jin[1], Jun Zhang[1,2,*]
[1]College of Chemistry and Chemical Engineering, Inner Mongolia University, Hohhot 010021, P. R. China
[2]College of Life Sciences, Inner Mongolia University, Hohhot 010021, P. R. China

ABSTRACT

Porous materials can be divided into three types according to the aperture sizes. It is defined respectively as microporous materials when the aperture sizes are lower than 2 nm, mesoporous materials when the aperture sizes fall in range of 2-50 nm, and macroporous materials when the aperture sizes become larger than 50 nm. In some cases, microporous and mesoporous materials can result in an unacceptably slow diffusion rate of reactants considering their small aperture sizes, which hinders their practical application. In order to overcome this limitation, one promising method is to prepare macroporous materials with the aperture sizes in range of 50 nm-1.5 μm to increase the chemical diffusion rate. Thus, macroporous materials have attracted considerable interest due to their important applications in catalysis, and become a research hotspot in recent years. This paper is intended to review the recent advance in macroporous materials, especially three dimensionally ordered macroporous (3DOM) materials, and their catalytic application on the basis of the research achievement on 3DOM catalysts in the authors' group. The paper is composed of four sections: (1) Overview section — a concise overview on macroporous materials will be given; (2) Controlled Synthesis: — various kinds of synthesis techniques and principles will be introduced; (3) Characterization — A variety of characterization methods for the macroporous materials will be introduced; (4) Catalytic property — the catalytic property of macroporous materials will be reviewed.

[*] Corresponding Author: Professor Dr. Jun Zhang, Tel.:0086 471 4992175; Email: cejzhang@imu.edu.cn

1. INTRODUCTION

Porous materials can be divided into three types according to the aperture sizes. It is defined respectively as microporous materials when the aperture sizes are lower than 2 nm, mesoporous materials when the aperture sizes fall in range of 2-50 nm, and macroporous materials when the aperture sizes become larger than 50 nm. Microporous materials with the aperture size lower than 2nm have some advantages in a variety of applications, such as shape-selective adsorption, separation, and catalysis. However, there are some limitations when the microporous materials are used to treat heavy oil and poor quality oil. In 1992, Mobil Company has developed a series of ordered mesoporous materials-M41S [1], which represents a new family of mesoporous molecular sieve materials. MCM-41, one member of this family, exhibits a hexagonal arrangement of uniform mesopores with engineered dimensions in a range of 1.5-10nm. Tanev's group [2] have prepared hexagonal mesoporous silica (HMS), and its titanium-substituted derivatives of Ti-HMS and Ti-MCM-41, which showed selective catalytic activity on the oxidation of 2,6-di-tert-butyl phenol(2,6-DTBP). Ti-HMS is more active than Ti-MCM-41for 2,6-DTBP selective oxidation due to its larger pore structured framework and exceptionally high textural mesoporosity. Obviously, the ordered porous structure and big aperture size are beneficial for diffusional effects in some oxidation reaction.

Compared with the researches on microporous and mesoporous materials, the studies on macroporous materials was relatively late. Traditional macroporous materials are mainly prepared by porogen decomposition and foaming methods, and their aperture sizes sometime can access to micron, or even millimeter level. Moreover, the aperture size distribution and spatial arrangement are very uneven. In the late 20th century, three-dimensionally ordered macroporous (3DOM) materials with uniform aperture sizes were successfully prepared using emulsion droplets or opal structured polymer microsphere colloidal crystals as templates [3] by several groups, which has attracted great research concerns on macroporous materials for the majority of scientists.

3DOM materials not only possess general characteristics of a large surface area and high porosity of porous materials, but also have their own features, such as ordered pore structure and periodicity, narrow pore size distribution, uniform and adjustable pore size, and long-range order in the overall structure. Meanwhile, the wall of the pore structure of 3DOM materials is usually composed of nanoparticles. When integrating novel properties of nanoparticle materials into ordered macroporous structure, it may generate more new functions or features, which make 3DOM materials usable in new types of applications of catalysts or carriers, macromolecules catalysis, bio-separation and refinement, electrode materials, and photosensitive materials etc. At the same time, 3DOM materials also have great potential in the use as a second template. On the other hand, the three dimensionally ordered lattice structure of 3DOM materials has obvious effect on their photonic band gap, resulting in a brilliant light viewable by naked eyes, thus creating a new kind of photonic crystal material with complete band gap structure. This leads to the attractive prospect for 3DOM materials in applications of optical integrated circuits, light scattering diodes, photonic switches and other light-controlled fields. This paper will comprehensively review the research on the controlled synthesis, characterization and catalytic properties of 3DOM materials on the basis of the research achievements on 3DOM Au/CeO_2 material for formaldehyde catalytic oxidation in the authors' group.

2. SYNTHESIS OF MACROPOROUS MATERIALS

In the past years, macroporous materials were synthesized mainly by means of porogen decomposition and foaming methods, which usually use adding foaming agent or decompressing technology to generate large amounts of holes in the target raw materials. However, the aperture size of the resulting materials created by the above methods is greatly uneven, and the holes are basically independent. Even more, the aperture shapes are hard to be finely controlled. These shortcomings seriously limit the practical application of macroporous materials.

The Template method then emerged as a new technology for the preparation of 3DOM materials. By filling the target material precursor in the gap of the ordered structure of the template following the removal of the template with high temperature treatment or other methods, 3DOM materials with uniform and ordered skeleton can be created. To this end, many special materials can be used as templates for the synthesis of 3DOM materials. It is found that there were a lot of symmetry structures in animals and plants after a long natural evolutionary development. These symmetry structures with long-range order can be used as templates for the creation of 3DOM materials, as Padhi [4] , Krishnaro [5] , and Greil's [6] groups have demonstrated the synthesis of SiC materials with ordered macroporous structure by using natural fiber, shell and wood grains as bio-templates. However, the bio-template seldom shows well-ordered arrangement that can be used to duplicate for preparation of 3DOM materials, limiting their wide applications.

In addition to the bio-template, colloidal crystals organized by an ordered arrangement of inorganic or polymer microspheres can also be used as templates for the synthesis of macroporous materials, and the colloidal crystal template method gradually emerged as the widest and the most mature technology in the field of synthesis of ordered macroporous materials. Due to its lower cost and lower demanding on equipment than other methods, the colloidal crystal template method is very suitable for general laboratory and industry. So far, many kinds of macroporous materials have already synthesized using this method, such as silicon oxide [7] , metal oxides [8] , carbon [9] , metal [10] , polymers [11] and composite materials [12]. In this paper, the studies on 3DOM materials synthesized by the colloidal crystal template method will be largely emphasized and the related synthetic methodology will be reviewed in more detail.

2.1. Colloidal Crystal Template Method for Preparation of 3DOM Materials

The preparation of 3DOM materials by the colloidal crystal template method generally consists of three steps: (1) synthesizing monodisperse colloidal microspheres and the assembling of the colloidal crystal template with an appropriate method into an ordered opal structure; (2) filling and solidifying of appropriate precursor into voids of opal structure; (3) removal of the template with high temperature treatment or organic solvent dissolution, as the synthetic procedures were schematically displayed in Figure 1. Through these procedures, the aperture size and structure of 3DOM materials could be well-controlled. Moreover, the composition and structure of the skeleton of 3DOM are largely affected by the colloidal crystal template size, precursor type, filling method, and template removing technique.

2.1.1. Synthesis of Colloidal Microspheres

Basically, monodispersity and uniformity of colloidal microspheres are the foundation and prerequisite for the assembly of colloidal crystal templates with long-range order. Currently, three kinds of colloidal microspheres could act as templates, including silica (SiO_2), polystyrene (PS), and polymethyl methacrylate (PMMA) microspheres.

Usually, SiO_2 microspheres can be synthesized by the Stober method. Namely, monodispersed SiO_2 microspheres can be prepared by hydrolysis of TEOS or condensation of silicate when using ammonia as catalyst. This method is available for the control of the size of SiO_2 microspheres from tens of nanometer to several microns. Even more, the hydroxyl functional groups on the surfaces of SiO_2 microspheres will facilitate the further functionalization. Meanwhile, SiO_2 microspheres obtained by this method usually possess high monodispersity, and the synthesis process is simple.

Polymer colloidal microspheres could be synthesized by a soap-free emulsion polymerization method. Compared with the conventional emulsion polymerization, soap-free emulsion polymerization overcomes the shortcomings of a large consumption of an emulsifier of emulsion polymerization. Moreover, the polymerization process easily occurs with the addition of a small amount of initiator residual base or monomer. At the same time, the polar groups on the microsphere surface could form a charged layer, which could stabilize the emulsion, and the surface of monodispersed colloidal crystal microspheres is very clean.

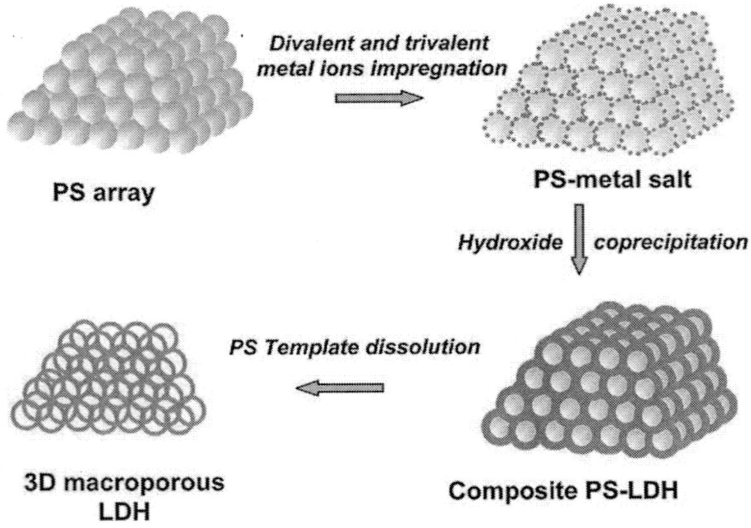

Figure 1. Schematic illustration of the preparation of 3DOM materials via colloidal crystal template method [13].

2.1.2. Assembly of Colloidal Crystal Template

The formation of 3DOM materials can be regarded as the replication process of an inverse colloidal crystal template. Namely, the void between the colloidal crystal microsphere templates is filled by precursor solution, which then converts into the corresponding solid. By removing the templates, the solid framework is retained, while the location of the original template microspheres is replaced by air balls, thus the final 3DOM structure forms. During the preparation, if colloidal microspheres are not well-assembled, the obtained 3DOM materials would be chaotic [14]. Therefore, the assembly of the colloidal crystal template is the basis and prerequisite for the follow-up experiment and the ordered homogeneity degree of the template will largely affect the final structure of 3DOM materials. So far, there are mainly six methods that have been developed for the assembly of the colloidal crystal template including sedimentation method [15] , vertical deposition method [16] , electrophoretic deposition method [17] , physical limit method [18] , electromagnetism fluid technology, and evaporative self-assembly method [19].

2.1.2.1. Sedimentation Method

Colloidal microspheres can be assembled into the colloidal crystal template under the gravity. Lopez et al [13] have studied the gravitational sedimentation of SiO_2 microspheres in detail. In this method, the successful assembly of microspheres into the colloidal crystal template mainly depends on the relevant parameters of the size, density, sedimentation velocity, and dispersion medium viscosity of colloidal microspheres. If the density and size of colloidal microspheres are suitable, it would be easy to deposit the microspheres at the bottom of the container. When the colloidal microspheres deposit, they will go through the conversion from disorder to order. In particular, when the deposition rate of colloidal microspheres is slow, they can form ordered colloidal crystal template with a long range, three dimensionally-ordered structure. As the density discrepancy between the PS or PMMA colloidal microsphere and solvent is small, the sedimentation rate is too slow to subside. Thus, filter deposition is a method that could accelerate the sedimentation rate. By adopting vacuum filtration to remove the liquid solution of colloidal microspheres, the colloidal microspheres could be rapidly accumulated. Compared with the natural sedimentation, this method is easy to control the speed of deposition and the thickness of opal structure, and easy to wash away the different media filled in the opal structure gaps. However, the natural and filter sedimentation methods are somehow time consuming, and it is hard to control the layer and morphology of colloidal crystal structure. Besides, it is very difficult to unify the order in the gravity direction.

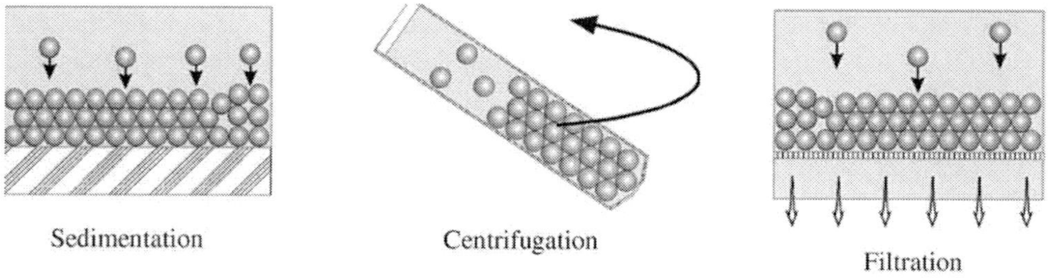

Figure 2. Schematic illustration of sedimentation , centrifugation and filtration methods [20].

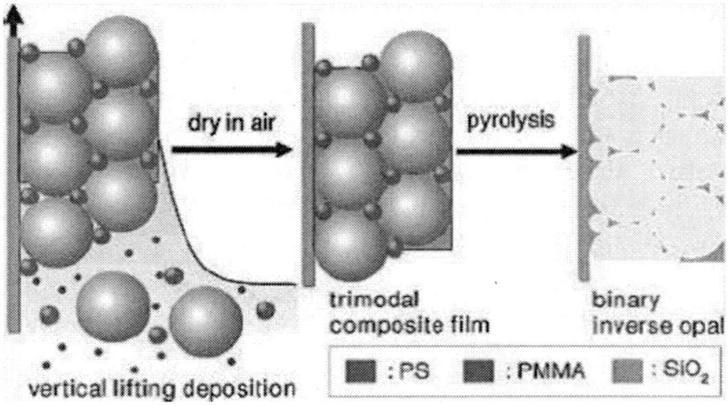

Figure 3. Schematic illustration of the assembly process of the colloidal crystal template via the vertical deposition method. [16, 21].

The centrifugal sedimentation method allows the selection of suitable speed according to the size of colloidal microspheres by modulating the centrifuge force to accelerate the assembly of the colloidal crystal template. Usually it takes a relatively short time to obtain a colloidal crystal template. The structure of a colloidal crystal template obtained with centrifuge sedimentation method has more defects than that prepared with natural and filter sedimentation methods, and the order degree is reduced as well. However, the operation of the centrifuge sedimentation method is simple and easy to get a long-range ordered structure. Thus, the centrifuge sedimentation method is versatile and important for the preparation of 3DOM materials.

2.1.2.2. Vertical Deposition Method

The vertical deposition method mainly uses the curved liquid surface to complete the assembly process. The assembly process is schematically illustrated in Figure 3. When a substrate contacts with a colloidal microsphere dispersion medium, it will be gradually wetted, and then a curved liquid surface will then be formed between the microspheres and substrate surface. With the evaporation of the solution in a curved liquid surface, colloidal microspheres fall into the crystal growth area to complete the assembly of the colloidal crystal template.

The key of the vertical deposition method is to control the balance between sedimentation velocity of colloidal microspheres in suspension and the evaporation rate of the curved liquid surface. Too fast deposition and too rapid evaporation may undermine the balance of the assembly, resulting in sharp concentration variation.

Vlasov and co-workers [22] have studied the formation of the three-dimensional ordered arrangement of SiO_2 colloidal microspheres through the convection generated by temperature difference via the vertical deposition method. But this method needs high demand on temperature controlling. If the large temperature deviation exists, it would be vulnerable to generate colloidal crystal stacking or other defects. Gu et al [23] have successfully attained the purpose of controlling the thickness of the colloidal crystal template by pulling the substrate with speed more quickly than the evaporation rate to offset the impact on the concentration gradient caused by the rapid sedimentation of large microspheres, Zhao et al [16] realized the controlling of the declining rate of the liquid level to weaken the microsphere sedimentation speed through extracting the bottom liquid via an inverse process of the lift-pulling method. The advantage is that it not only could control the flowing velocity of the solution, but also avoid the wobbling question when lifting and pulling the glass substrate. At the same time, it raises the quality of the colloidal crystal template. Therefore, slow speed is a key factor for achieving three dimensionally ordered arrangements. Recently, Ma et al [24] improved the vertical deposition process by dissolving SiO_2 colloidal microspheres obtained via by the stober method into acetone, so that SiO_2 colloidal microspheres could be rapidly assembled onto the quartz substrate at room temperature. Thus, a well-organized, three dimensionally ordered structure can be obtained in one or two days.

2.1.2.3. Electrophoretic Deposition Method

The electrophoretic deposition method utilizes the DC field to drive the electrically charged colloidal microspheres moving and depositing onto the oppositely charged electrode. After the microspheres are deposited, a gas-liquid-crystal phase transition, which can be controlled by the current, takes place. In order to prevent the generation of bubbles by water electrolysis affecting the crystallization of ordered colloidal microspheres, a water and alcohol mixed solvent is selected as medium. Compared with the gravitational sedimentation method, the electrophoretic deposition method can greatly shorten the assembling time of colloidal microspheres. However, this method needs high requirements on the instrument, and limits its further development and promotion.

2.1.2.4. Physical Limit Method

The physical limit method was first created by Xia and co-workers. [11a,19a]. A reactor consisting of two parallel slides, which are adhered by photosensitive material on the edge, is used in this method. At the edge of slide, there is a small rectangular channel. In order to separate the solvent and colloidal microspheres, the channel height is less than the diameter of colloidal microspheres. Driven by nitrogen pressure and ultrasonic oscillation, colloidal microspheres form a cubic, close-packed structure at the bottom of the reactor. This method has several advantages, for instance, the obtained sample is regularly ordered, and shows good compactness; the accumulation layer and surface topography could be strictly controlled; the assembling period is short and less affected by temperature and colloidal microsphere charges; the cubic, close-packed structure in a large area can be easily obtained; the assembled colloidal crystal layers could be adjusted by modifying the thickness of photosensitive materials. However, the disadvantage is that the method demands strictly on instrument and equipment.

2.1.2.5. Electromagnetism Fluid Technology

When high dielectric constant colloidal microspheres are dispersed into fluid-insulating oil, it can form electric magnetic rheological suspension. Under the role of electrical and magnetic fields, the suspended microspheres will form a three dimensional mesh, chain or columnar structure. If the suspended colloidal microspheres are monodispersed, they would form a single fcc and bct structure. This technology is called electromagnetism fluid technology that provides a unique method for the synthesis of 3DOM materials.

2.1.2.6. Evaporative Self-assembly Method

If the microsphere dispersed medium is completely evaporated at certain conditions, it would form a gas-liquid phase interface, on which three dimensionally ordered colloidal crystals could be assembled. During the assembly process, two physical processes of the evaporation of dispersion medium and the assembly of colloidal microspheres occur on the gas-liquid phase interface. With the assembling of the colloidal microspheres, the evaporation speed of dispersion medium may be hindered. Therefore, through adjusting the assembling area and evaporation temperature, a high quality colloidal crystal template could be achieved. The method features the high efficiency, good quality, large area, and few defects for the obtained three dimensionally ordered assemblies.

2.1.3. Template Thermal Treatment

The thermal treatment of a colloidal crystal template is an important step before the precursor filling. Before the sintering of the colloidal crystal template, the arrangement of microspheres is very loose and independent on each other. The filling of the reactant precursor into the voids of the colloidal crystal template mainly depends on capillarity. However, the assembled microsphere templates are often not of enough mechanical strength , and the loosening or shifting may occur. Thus, the colloidal crystal template may split into pieces, and it is difficult to obtain large area 3DOM materials with high mechanical strength.

When a colloidal crystal template is heat treated, it may form a neck connection with each other to some extent. This can improve the mechanical strength of the colloidal crystal template, and create interconnected small windows among macro-pores when the template is removed. Gates et al [25] have studied the treatment of the PS colloidal crystal microsphere template at low temperature in detail, and found that when the treatment temperature was higher than glass state temperature, the deformation of PS colloidal microspheres may occur and lead to the great contact area of adjacent and contact microspheres. If the thermal treatment temperature is long enough at 100°C, the PS colloidal crystal template would melt and become a transparent caked mass, so that loses the role of the template. Studies showed that the SiO_2 colloidal crystal template has a high glass state temperature at 700-950°C. If the temperature is below 700°C, it would not be conducive to melt and form neck connection, while if the temperature exceeded 900°C, SiO_2 microspheres would be excessively melted, which will eventually cause the voids of SiO_2 microspheres to be too narrow to be conducive to filling.

2.1.4. Template Filling

After the colloidal crystal template is assembled, the filling technology and liquid-solid or gas-solid conversion reaction in the gap of the colloidal crystal template are the key factors for the synthesis of 3DOM materials. Great attention should be paid to the following conditions for the precursor selection and template filling: First, the templates should be saturated by the precursor dissolved solvent, which could make the precursors flow through the entire template and could eliminate the crystal defects generated by the infiltration area. Second, the precursor should have good solubility in the solvent in order to ensure that the more precursors could enter into the gap of the template and prevent pore wall collapsing. In addition, the melting point of the organic precursor should be higher than the glass state temperature of the template; otherwise the precursor will lose the role due to the melting before the template removal. According to the filling ways of precursor, the methods for synthesis of 3DOM materials can be divided into the following categories.

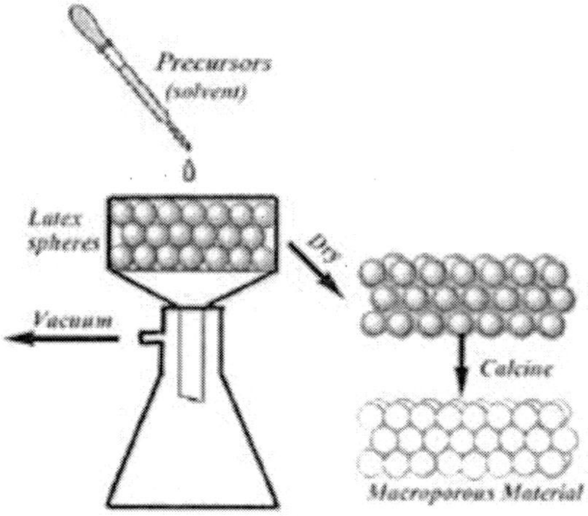

Figure 4. Schematic illustration of preparation 3DOM materials sol-gel chemistry method [3c].

(1) In situ sol - gel method

Metal alkoxide or other metal salts solution could form sol-gel at a lower temperature through solution hydrolysis and polymerization reactions, which can penetrate into the gap of template microspheres by means of capillary force through impregnation or dropping pattern. Figure 4 schematically illustrates the procedures for synthesis of the macroporous metal oxide networks. Firstly, millimeter-thick layers of latex spheres were deposited on filter paper in a Buchner funnel under vacuum, and soaked with ethanol. Then, metal alkoxide solution was added dropwise to cover the latex spheres completely while suction was applied. Lastly, after drying the composite in a vacuum desiccator for 3 to 24 hours, the latex spheres were removed by calcination in flowing air at a certain temperature for 7 to 12 hours, achieving hard and brittle powder particles. At present, many uniform 3DOM oxide, such as SiO_2, ZrO_2, TiO_2, Al_2O_3, WO_3, Fe_2O_3, Sb_4O_6 [26] , have been synthesized by the in situ sol-gol method. Although this method is simple and easy to operate, it is hard to the aperture size shrinkage. Therefore, in situ sol-gel technology is only suitable for preparing high valence 3DOM oxide materials.

The low state alkoxide is quite reactive, and their hydrolysis reaction is completely acute. Besides, its solubility is low. Therefore, it is very difficult to enter into the gap of the template. In addition, its price is high and shows low commercialization levels. This greatly limits the development of low cost 3DOM oxide materials. Transition metal nitrate, citric acid and ethanol solution could form sol at a certain temperature. This sol could act as a precursor with reduced cost and improved thermal stability. Moreover, it could also prevent the pore structure from deforming, which favors the preparation of low cost state 3DOM oxide materials.

(2) Salt precipitation and chemical conversion method

Yan et al [27] have studied the synthesis of 3DOM materials by using water, acetic acid, ethanol or their mixture as solvent, and transition metal nitrate or acetate as precursors. Due to the low decomposition temperature and melting point, metal nitrate or acetate could deposit in the gap of the template by means of capillary force after the solvent was evaporated. Besides, Yan et al [28] have also studied the availability of using oxalate salts as precursors for the preparation of the macroporous framework, and MgO, Cr_2O_3, Mn_2O_3, and NiO 3DOM oxide materials were successfully obtained. Although this method is relatively complicated and not suitable for multi-composites, it still provides a new way to synthesize 3DOM oxide materials.

It is well-known that ethylenediaminetetraacetic acid is a bidentate ligand and could strongly coordinate with rare earth ions. Zhang et al [29] have synthesized Ln^{3+}-EDTA chelate, which could easily enter into the template. When the template was removed, Ln^{3+}-EDTA complex would be decomposed at high temperatures to obtain 3DOM rare earth oxide materials. In addition, Li et al [8c, d] have also synthesized 3DOM CeO_2 and ZrO_2 materials by adopting cheap nitrate and chloride as precursors. Various inorganic salts as precursors not only can reduce the production cost of 3DOM oxide materials, but also greatly enrich the 3DOM oxide species.

(3) Electrochemical deposition method

Electrochemical deposition is a relatively new method that uses a conductive carrier deposited with microsphere template as the electrolytic cell cathode. Metal ions or semiconductor materials can deposit in the gap of the template through the reduction of the electrode, as the synthetic procedures are schematically shown in Figure 5. Wang [30] et al had already selected SiO_2 as the template, in which the $HAuCl_4$ was reduced into Au via the constant potential method.

Although this method can control the deposition of the layer, it is difficult to completely fill the template. Therefore, the mechanical strength of 3DOM material is not high. Gu et al [7b] have synthesized the macroporous SiO_2 and TiO_2 using the electrophoresis, 10% aqueous suspensions of SiO_2 (6 nm) and TiO_2 (15 nm). In this study, Gu et al resolved the key question of complete infiltration of opal template with a low-shrinkage material without destroying the ordering of the opal. High quality, inverse opals 3DOM structure with flat surfaces and large domains can be prepared. The area of a single crack-free domain of the 3DOM structure can exceed 10 000μm^2.

(4) Chemical vapor deposition method

The chemical vapor deposition method employs the gas phase precursor to penetrate into the gap of the colloidal microsphere template under high temperature and pressure. The gas phase precursor will decompose and deposit into the gap of the template, forming the macroporous materials. Carbon and silicon macroporous materials could be acquired by this method, while other methods are not available for the synthesis of such materials. It is important that the chemical vapor deposition method could greatly control and adjust the precursor deposition quantity even up to 100%. Zakhidov et al [9b] had first synthesized 3DOM diamond using the chemical vapor deposition method. Subsequently, Blanco et al [31] have obtained the fcc structured SiO_2, and treated it in silane gas at 200 torr and 250-350°C in order to enhance the mechanical strength. When the reaction has continuously reacted for 24h, Si and SiO_2 compounds could be obtained. At last, SiO_2 was removed by dissolution in HF. This example showed that thechemical vapor deposition was an effective method for the synthesis of 3DOM materials.

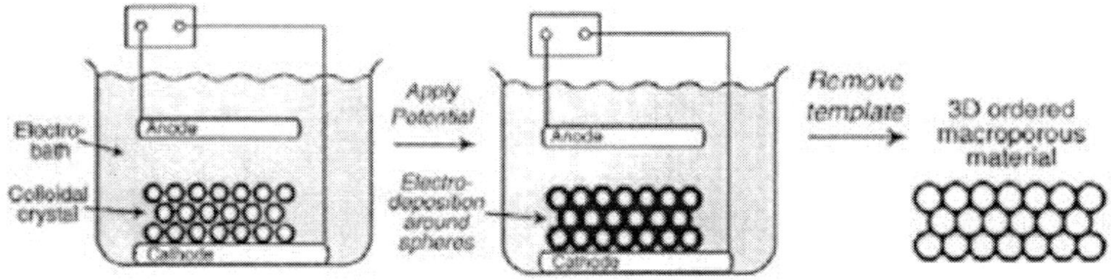

Figure 5. Schematic illustration of synthesis of 3DOM macroporous materials by electrodeposition method [7b].

(5) Pulsed laser sputtering and ion sputtering method

In addition to liquid or gas form, the precursor could enter into the template interior in solid form. Generally, a solid precursor could be broken into nanoparticle pieces under laser or current impacting. Then, these nanoparticle pieces could be deposited into the gap of template. In general, an Si chip soaked in HF solution is placed on the target location, while an Si chip covered with colloidal crystals is placed on the deposition location. When the target slice is irradiated with a certain pulse laser, silicon nanoparticles are soon sputtered on the crystal surface of the deposition position. With the deposition tower rotating, silicon nanoparticles will deposit uniformly on the surface of the colloidal crystal template. The microsphere template can then be removed to form 3DOM materials by high temperature treatment or other methods.

(6) Co-deposition method

Co-deposition method is a technology used to form 3DOM materials via the process of natural sedimentation of template microspheres and nanoparticles simultaneously. When the microspheres self-assemble into a three dimensionally ordered structure, nanoparticles also accumulate in the gap of the template. With the evaporation of the solvent, the deposition quantity of nanoparticles gradually increases until the gap of the template is completely filled. By removing the template, 3DOM materials could be obtained. The preparation process is schematically illustrated in Figure 6. As the nanoparticles are used as raw materials directly, chemical conversion processes are not necessary for the preparation. Unlike the sol-gel method that the target materials need chemical conversion through the process of condensation, gelation and calcinations, the co-deposition method is simple, and the nanoparticles of the target materials are prepared in advance before the assembly. In addition, the nanoparticles are introduced during the assembly process rather than after the assembly. Due to the less shrinkage of nanoparticles and the formation of the compact nanoparticle structure after the assembly, the co-deposition method could avoid some defects resulting from the shrinkage of the template during the removing process.

Xia et al [32] have demonstrated the synthesis of 3DOM ferromagnets using nanoparticles with the diameter 10 times less than the PS microspheres. Yoshio Sakka et al [26a] have modified the PMMA microspheres with PEI, and then dispersed them into monodispersed nanosized TiO_2 solution. Following the flocculation, suction filtration, drying, and calcination processes, 3DOM TiO_2 materials can be achieved.

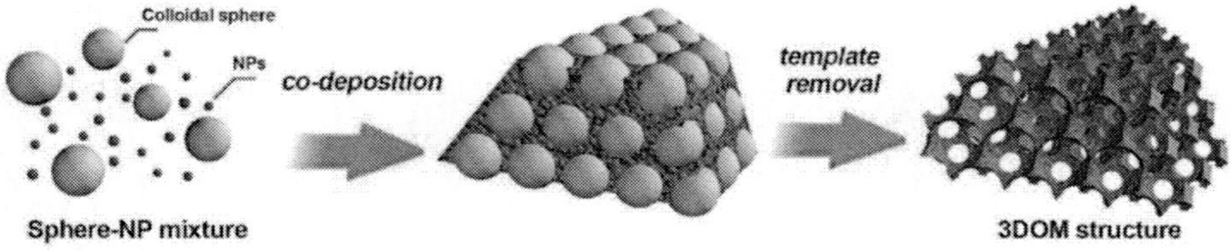

Figure 6. Schematic illustration of preparation of 3DOM materials via co-deposition method [33].

(7) Polymerization method

Choosing an organic monomer as a precursor, 3DOM materials can be synthesized by filling the gap of the template with the monomer precursor followed with the polymerization initiated by heat radiation, light radiation and catalysis, and then removing the template with organic solvent., as the example of organic macroporous materials was demonstrated by Xia [11b]. Compared with inorganic macroporous materials, organic macroporous materials are not liable to absorb water and have low surface tension, and have optical transparency with strong toughness, thus may find a wide range of applications.

2.1.5. Removal of Template

The calcination method is usually used to remove the organic template. However, rapidly rising temperature is liable to cause many bubbles, unfavorable for the formation of the ordered macroporous structure [26c, 31]. Therefore, the appropriate heating rate and calcination temperature are crucial to the formation of the ordered macroporous structure and framework. In addition, the template can also be removed by dissolving with an organic solvent, for instance, the PS microsphere template can be dissolved by toluene, tetrahydrofuran and carbon tetrachloride, while; SiO_2 microsphere template can be removed with HF.

3. Characterization of 3DOM Materials

In fact, the characterization of 3DOM materials is the same as that of the general inorganic materials. 3DOM materials can be characterized by means of scanning electron microscopy (SEM), transmission electron microscopy (TEM), X-ray powder diffraction (XRD), BET surface area analysis, FT-IR, thermal analysis (TG-DTA), and energy dispersive X-ray (EDX) analysis etc.

3.1. FT-IR Analysis

Uniformity of templates is the foundation and prerequisite for the assembly of an ordered colloidal crystal template. It will largely affect the ordering level and structural integrity of the final structure of 3DOM materials. Currently, polystyrene (PS), polymethyl methacrylate (PMMA) and silica microspheres are three common templates that are used for the preparation of 3DOM materials. Xu et al [34] have prepared the PMMA microspheres with diameters of 300-500nm via a modified emulsifier-free emulsion polymerization technique using water-oil biphase double initiators. A series of 3DOM $LaCo_xFe_{1-x}O_3$ (x=0-0.5) perovskite-type materials were synthesized by the colloid crystal template method. Samples have been analyzed by means of the FT-IR method, as shown in Figure 7.

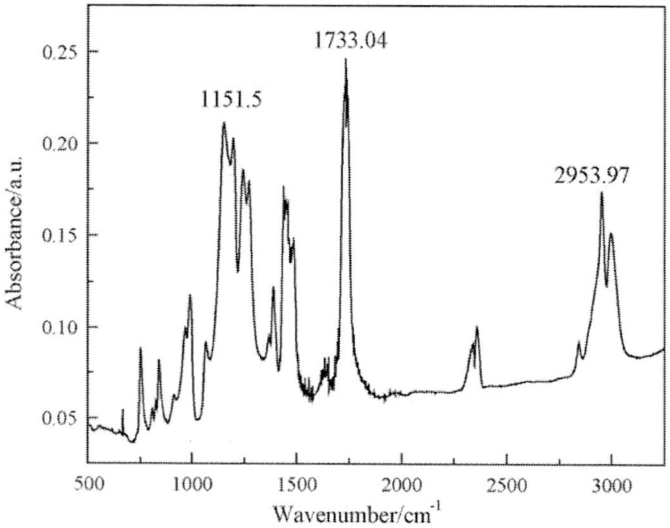

Figure 7. The FT-IR spectrum of PMMA microspheres. [34]

From Figure 7, we could learn that the sample has three strong peaks at 1152, 1733, and 2954cm^{-1}, which corresponded to the stretching vibration of a carbon oxygen single bond of the carbonyl group, and carbon hydrogen single bond of the methyl group, respectively. All these three peaks are the characteristic peaks of PMMA, indicating the presence of PMMA in microspheres. Moreover, other functional groups could not be seen from the FT-IR spectrum, implying the surface of the PMMA microspheres obtained by emulsifier-free emulsion polymerization is clean. Because of this , we could use same method to measure the other template such as PS.

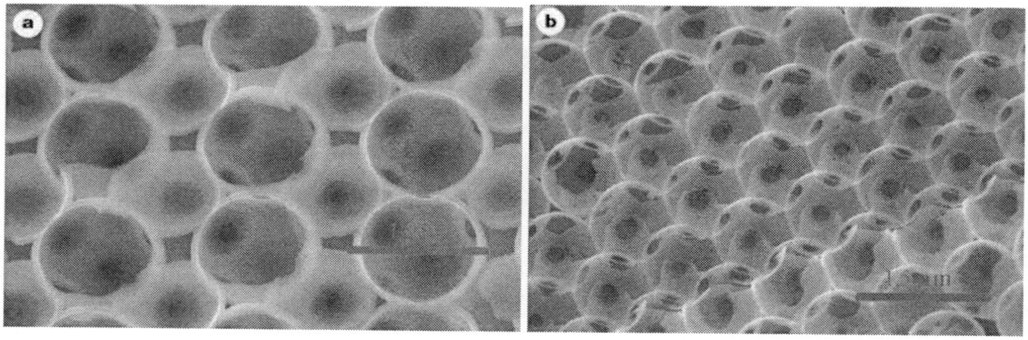

Figure 8. SEM images of internal facets of an Si inverse opal: a, [110] facet. b, [111] facet [31].

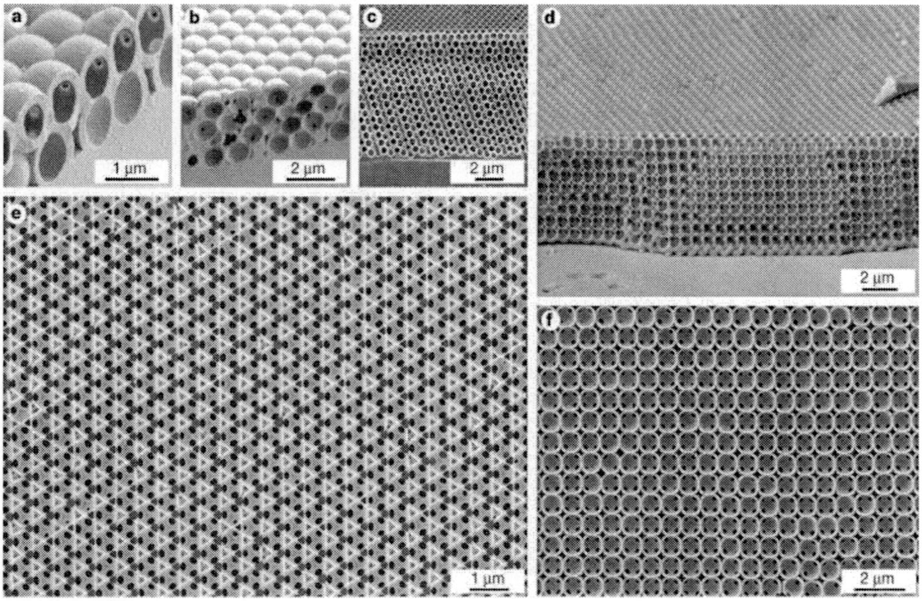

Figure 9. SEM images of planar Si photonic crystals.

3.2. SEM and TEM Characterization of 3DOM Materials

Blanco et al [31] have obtained the fcc structured SiO_2 via CVD method. SEM images of a typical inverse Si opal taken after etching the infiltrated structure is shown in Figure 8. The images reveal an interconnected network of air spheres surrounded by Si shells, inheriting the fcc order of the opal template. Adjacent air spheres are connected by windows.

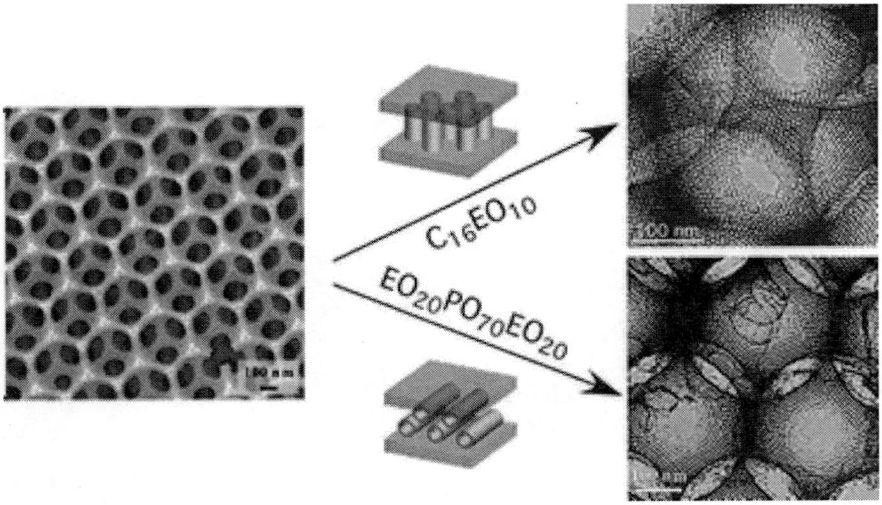

Figure 10. Photograph of a 3DOM/m silica monolith, SEM image showing the macropore structure, and TEM images of two different mesopore orientations [35].

Vlasov [22] et al have used an alternative method to form synthetic opals. SEM images indicating planar and cross-sectional Si photonic crystals are shown as a function of the thickness of the initial opal template for 2 [Figure 9 (a)], 4 [Figure 9 (b)], and 16 [Figure 9 (c)] layers. The Si substrate is snapped, and the fracture propagates up through the photonic crystal. Close examination of the lowest layer reveals that the photonic crystals are completely integrated into the wafer. Apparent defects in [Figure 9 (c)] are due to a ripple in the fracture surface that occurs in thicker photonic crystals. Sample edge showing the (100) surface confirms that the crystals are fcc [Figure 9 (e and f)]. The sphere diameters are 670 nm [Figure 9 (c and e)], 855 nm [Figure 9 (d and f)], and 1mm [Figure 9 (e and f)].

In the synthesis of 3DOM SiO_2 with surfactant-templated mesoporous walls (3DOM/m silica), different mesostructures may be present, including disordered, worm-like structures, 2D-hexagonal channels, and cubic mesopore arrays. [33] When the surfactant Brij 56 ($C_{16}H_{33}(OCH_2CH_2)_nOH$, $n\sim10$) is used together with a co-surfactant, they run parallel to the latex spheres when the polymeric surfactant P123 ($EO_{20}PO_{70}EO_{20}$) is employed (Figure 10).

This orientation can affect the connection of adjacent macropores and mass transport through the pore systems. When multiple types of surfactants are used in one system (e.g., block copolymers and smaller surfactants), the complex interactions between them provide the means to influence the hierarchical structure, but they can also result in phase separation that may prevent hierarchical templating (shown in Figure 10).

Tonti et al [36] synthesized 3DOM $LiMn_2O_4$ spinel by a colloidal templating process. The SEM image of Figure 11(a) shows the ordered hollow spheres that form the macroporous fcc lattice replica of the latex opal and the large circular interconnecting windows at the positions corresponding to the contact points of contiguous latex particles of the template (darker areas). The space between the nearest macropores that corresponds to tetrahedral (Th) and octahedral (Oh) holes in the template is occupied by the oxide and appears as clearer areas. Figure 11a corresponds to a commonly found [111] surface, but occasionally different surfaces, such as the [100] [Figure 11(b)], can also be observed. The opal assembling process usually starts by packing spheres at the surface of the latex suspension. The hexagonal symmetry is the most compact 2D arrangement; however, other arrangements can also kinetically form.

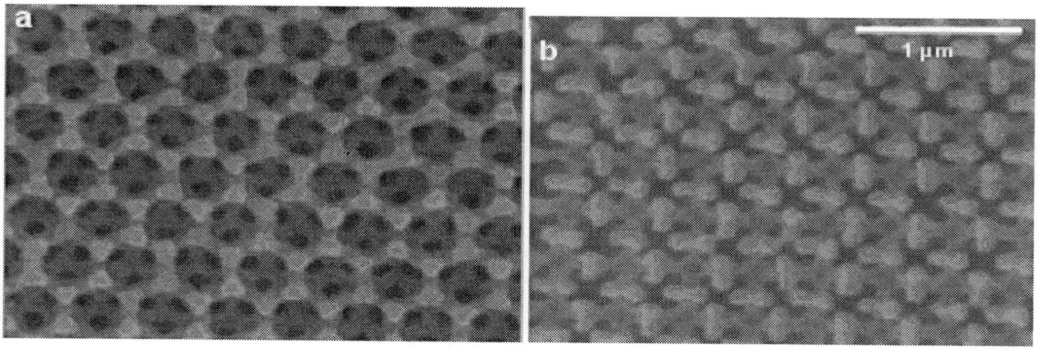

Figure 11. SEM micrographs of 3DOM $LiMn_2O_4$. Images (a) and (b) correspond to [111] and [100] surfaces [36].

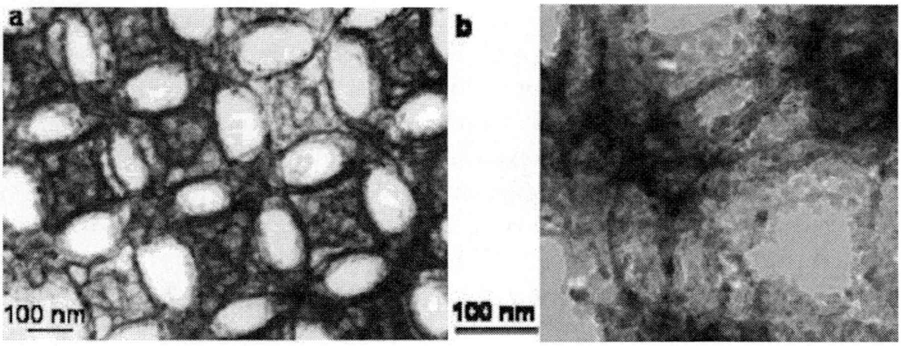

Figure 12. TEM micrographs showing the macropore walls formed by $LiMn_2O_4$ nanocrystals. The images correspond to projections of 3DOM $LiMn_2O_4$ inverse opals close to [100] (a) and [110] (b) directions [36].

In Figure 12a, the TEM image shows the octahedral and tetrahedral holes projected in the [100] direction (hence the same symmetry is observed as in Figure 11). These cavities appear partially mineralized, with the inorganic material forming a wall around the hollow spheres and creating {missing text?}

Zhang et al [37] have reported the synthesis of 3DOM Au/CeO_2 catalyst for formaldehyde oxidation using colloidal crystal template methods. The formaldehyde oxidation tests revealed that the 3DOM Au/CeO_2 catalyst exhibited superior catalytic activity with 100% formaldehyde conversion at 75°C. The 3DOM Au/CeO_2 catalyst was characterized by means of XRD, SEM, and TEM in detail.

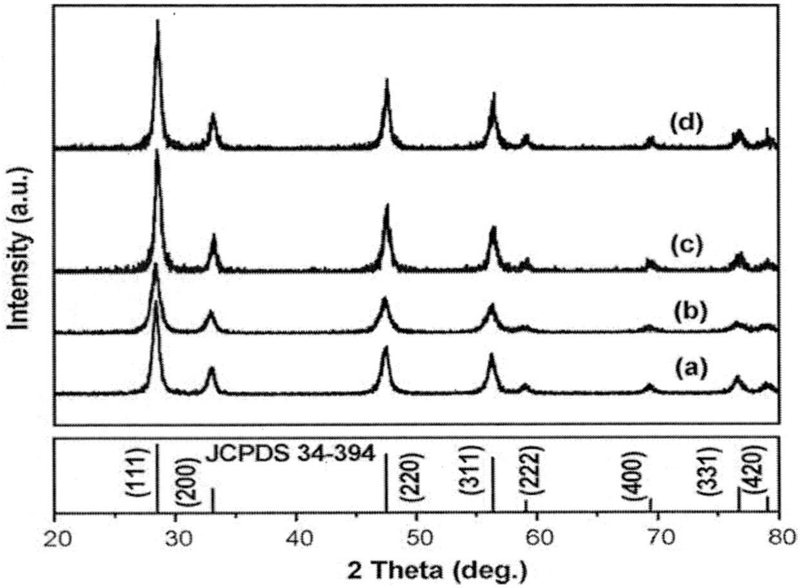

Figure 13. XRD patterns of (a) 3DOM CeO_2 with 130 nm pore size, (b) 3DOM Au-1 wt%/CeO_2 with 130 nm pore size (60°C drying), (c) 3DOM CeO_2 with 80 nm pore size, (d) 3DOM Au -1 wt%/CeO_2 with 80 nm pore size (60ºC drying).

Figure 14. SEM images showing the assembly of (a)200 nm and (b)400 nm PS colloidal spheres to form three-dimensional ordered PS colloidal crystal arrays, and (c)the 3DOM Au/CeO_2 materials.

Figure 13 showed XRD patterns of 3DOM Au/CeO_2 catalyst. It is clearly illustrated from XRD characterization that the phase structure of the 3DOM catalyst can be indexed to cubic phased CeO_2 (JCPDS: 34-394) with the diffraction peaks of 2θ at 28.50, 33.08,47.49, 56.34, and 76.70 ° corresponding to (111), (200), (220), (311), and (331) lattices of cubic structured CeO_2,, respectively.

Figure 14 indicated SEM images for the observation of the size distribution of PS template microspheres. From the SEM images, it could be seen that the microspheres are uniform with narrow size distribution, and the average size is 200 [Figure 14 (a)] and 400nm [Figure 14 (b)]. The microspheres are spherical without any aggregation. By using the uniform microspheres as templates, 3DOM Au/CeO_2 materials with long-range ordered periodicity were successfully prepared [Figure 14 (c)].

To verify the existence of Au in a 3DOM Au/CeO_2 catalyst, EDX analysis of a 3DOM Au/CeO_2 catalyst was carried out. It was found that the representative peaks corresponding to Ce, Au, and O elements are detected, confirming the formation of CeO_2 and the existence of Au element in 3DOM Au/CeO_2 catalyst. To further explain the existence of Au nanoparticles, TEM was conducted. The 3DOM structure of Au/CeO_2 catalyst with the overlapped pores can be clearly observed in Figure 15. Besides, the Au nanoparticles precipitated on the surface of macroporous CeO_2 support can be observed. The size of Au nanoparticles formed on the surface of 3DOM supports is estimated to be around 5 nm. The HRTEM observation of the lattice fringes of d=0.236 and 0.312 nm corresponding to Au (111) and CeO_2 (111) shown in the insert of Figure 15, further confirms the existence of small Au nanoparticles on 3DOM CeO_2 support. Moreover, due to the uniform macroporous structures leading to a good distribution of catalytic species of Au nanoparticles loaded on the 3DOM structures with less aggregation, the 3DOM Au/CeO_2 catalysts showed enhanced capability for formaldehyde catalytic oxidation. On the other hand, the ordered macroporous structures benefit the mass transfer of formaldehyde.

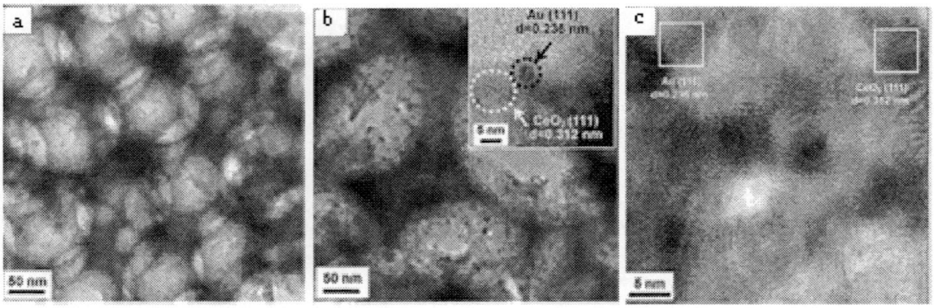

Figure 15. TEM (a) and HRTEM (b and c) images of 3DOM Au /CeO_2 synthesized via a gas bubbling-assisted deposition precipitation method The black and white circles in the insert of (e) and the white rectangles in (f) clearly showing the lattice fringes of Au (111) and CeO_2 (111) firmly suggest the formation of small Au nanoparticles on 3DOM CeO_2 support.

3.4. Thermogravimetric Analysis

Thermogravimetric analysis is another valuable technique that can be used to characterize the moisture content, reaction temperature and organic species stability of 3DOM materials. Lianget al [38] had reported the studies of thermal properties of 3DOM TiO_2 materials prepared by template and impregnation methods using PS colloidal crystals as templates. Figure 16 displays the TG and DTA diagrams of 3DOM TiO_2 materials. It is shown that there is a slow mass loss when the materials are treated from room temperature to 350 °C due to the evaporation of solvent. The sharp mass loss of an exothermic process in range of 350 to 400 °C is owing to the decomposition and combustion of PS template. The phase transformation of 3DOM TiO_2 materials from anatase to rutile occurs at about 450°C. By virtue of thermogravimetric analysis, we could determine the calcination temperature at 550°C in order to get 3DOM TiO_2. Because of this, , during the characterization process, thermogravimetric analysis needs to combine with other characterization techniques such as FT-IR, XRD, SEM and TEM to give overall structural, morphological and thermal properties of 3DOM materials

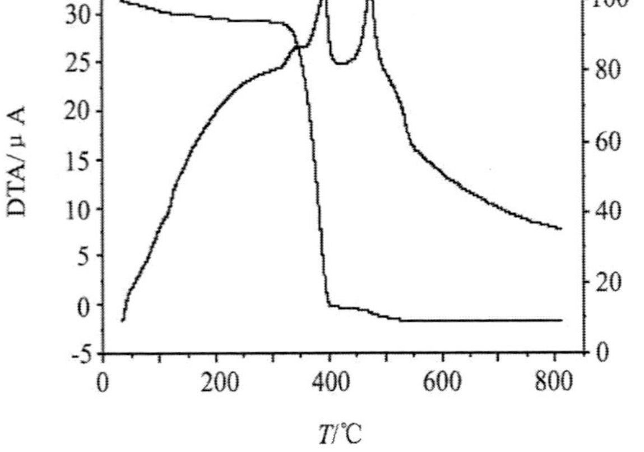

Figure16. TG and DTA diagrams of 3DOM TiO_2 materials [38].

Table 1. Specific Surface Areas of 3DOM WO_3, WC, and Pt-Modified WC As Determined by N_2 Adsorption and BET Analysis

PMMA template sphere diameter	Specific surface area (m^2 g^{-1})		
	WO_3	WC(indirect)	WC(direct)
180nm	17.1	16.0	54.2
260nm	16.9	15.8	55.8
370nm	13.4	17.8	41.0
490nm	8.9	18.6	47.7
1wt%Pt, 490nm	10.4	24.1	28.7
3wt%Pt, 490nm	12.3	26.4	34.0
5wt% Pt, 490nm	13.6	28.2	38.0

3.5. BET Analysis

Three-dimensionally ordered macroporous tungsten carbide (3DOM WC) with varying pore sizes was synthesized using the "inverse-opal" method. Poly{methyl methacrylate} (PMMA) colloidal crystal template with sphere diameters ranging from 180 to 490 nm was used. The effect of Pt modification of the template prior to carburization on the resulting material composition was also investigated. [39] The material properties were evaluated using surface and bulk characterization techniques. Specific surface areas of the unmodified and Pt-modified 3DOMWO3 and WC materials for both indirect and direct carburization procedures are displayed in Table 1. The BET results demonstrated a substantially higher surface area for 3DOMWC prepared using the direct method (40-56 m^2 g^{-1}) compared to those using the indirect method (16-19 m^2 g^{-1}) and 3DOMWO$_3$ (9-17 m^2 g^{-1}) samples. The WO3 samples were observed to increase in surface area with decreasing pore size due to crystallite formation around the 3DOM skeleton as previously reported.

In contrast, 3DOMWC samples produced using the direct and indirect carburization methods had surface areas roughly independent of the observed pore size. Furthermore, Pt modification of WO_3 samples and WC made using the indirect method caused an increase in surface area with increasing Pt metal loading.

Nitrogen adsorption isotherms were analyzed using BJH pore size distribution analysis. The 3DOM WC material synthesized using the direct carburization procedure showed uniform mesoporosity, whereas the 3DOM WC produced using indirect carburization and 3DOM WO3 samples demonstrated no porosity. A uniform pore size of approximately 4 nm was observed for the directly carburized 490 nm 3DOM sample (Figure 17).This feature was determined to be independent of the PMMA template size used during synthesis.

4. Application of 3DOM Material in Catalysis

3DOM materials emerge as a new kind of material with features of large surface area, uniform and adjustable pore size, narrow pore size distribution, high porosity, and long range ordered porous periodicity etc. The wall of a 3DOM structure is commonly composed of nanoparticles, which could endow 3DOM materials novel functionality. Thus, 3DOM materials show great potential in the application of heterogeneous catalysis, photocatalysis and waste treatment. Due to the excellent permeability of 3DOM materials, the process of adsorption and desorption can be completed in a 3DOM structure quickly. Compared with the conventional microporous and mesoporous materials, 3DOM materials have many advantages, including the large aperture size, quick mass transfer rate, less blockage and high transmission efficiency etc.

4.1. Selective Oxidation of Hydrocarbons

Selective oxidation of hydrocarbons is an important chemical reaction in chemical industry that is applied to produce bulk chemical raw materials and intermediates. 3DOM materials are believed to be the important catalysts for effectiveness in selective oxidation of hydrocarbons due to their novel structural properties.

Johnson et al [40] have anchored two kinds of $[Co(II)(H_2O)\text{-}PW_{11}O_{39}]^{5-}$, $\{SiW_9O_{37}[Co(II)(H_2O)]_3\}^{10-}$ species into mesoporous (3-6 nm) and macroporous (350-450 nm) SiO_2 surface. The results showed that although the mesoporous SiO_2 has a higher surface area, the pore channel of mesoporous SiO_2 restricted the transmission of macromolecules to some extent. On the contrary, the macroporous SiO_2 allowed the connection of more groups, leading to higher activity and selectivity in an epoxidation catalytic reaction.

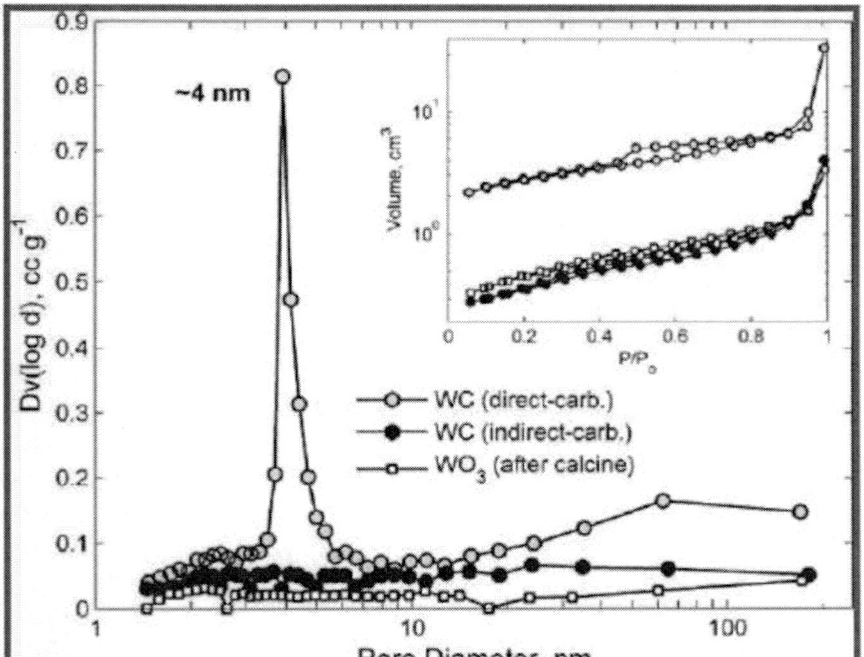

Figure 17. BJH pore-size distribution for 3DOM WO_3 and WC synthesized using both direct and indirect carburization methods with 490 nm PMMA [39].

In industry, inert α-Al_2O_3 is usually used as a carrier of catalysts for producing ethylene oxide. In general, the α-Al_2O_3 microsphere carrier has good mobility. However, its surface area is small and is non-porous, so it is not easy to diffusion, which is easy to cause deep oxidation of ethylene oxide, thus reducing the product selectivity. While Stein et al [41] have adopted 3DOM α-Al_2O_3 carrier as a catalyst for the same reaction. The results indicated that the selectivity and conversion rate was largely improved.

For the selective oxidation of n-butane, the production rate of maleic anhydride would be only 40%, when the traditional organic vanadium-phosphorus-oxide (VPO) [42] catalyst was used. However, the production of maleic anhydride can be increased to 54% under the same reaction conditions when using 3DOM VPO as a catalyst, which further verified the effectiveness of using a 3DOM catalyst with an ordered macroporous structure and high surface area as the catalyst for selective oxidation of n-butane.

4.2. Fuel Cells Hydropower

3DOM materials are also valuable candidate materials for a methanol fuel cell. Chai et al [43] have found that Pt/Au alloy supported 3DOM carbon showed a better catalytic property for methanol oxidation compared with the commercial carbon black. In addition, 3DOM carbon with good physical and chemical stabilities can be applied to acid, alkali and other harsh environments. As a carrier of lithium-ion batteries [44] , the reaction efficiency was significantly improved as well.

3DOM materials are also used as catalysts for ethanol reforming reaction to obtain hydrogen energy. [45] Compared with the conventional hydrogen production methods of water electrolysis, methanol reforming, and hydrocarbon reforming, the ethanol reforming reaction method has broad prospects. In general, the ethanol reforming method uses noble metals as catalysts, but its cost is too high to widely use. Although the existing Ni-based catalyst has a high conversion rate, its anti-carbon and anti-sintering ability is poor, which seriously affects the application of ethanol steam reforming. Li et al [72] have designed a 3DOM Ni/CeO_2 catalyst for ethanol steam reforming reaction. The catalyst not only leads to the high conversion rate, but also fully makes use of the hydrogen storage function of CeO_2 and the advantage of a 3DOM structure.

3DOM materials have been explored as effective catalysts for water-gas shift reaction and CO preferential oxidation as well. Hao [38] and Guan et al [46] have prepared 3DOM Pt/TiO_2 catalyst and Pt-Rh/Al_2O_3, respectively, for water-gas shift reaction and CO preferential oxidation. The results showed that a 3DOM Pt/TiO_2 and Pt-Rh/Al_2O_3 microchannel reactor has more activity than ordinary powder Pt/TiO_2 and Pt-Rh/Al_2O_3 materials.

4.3. Petrochemical Complex

Fluid catalytic cracking is the core technology of modern oil refining industry. At present, it is facing a major challenge for the production of lighter and better products with heavier and low-quality oils. [47]

The average diameter of residual oil such as heavy distillate oil is about 1.0-2.5 nm, while the aperture size of Y-type zeolite is only around 0.74nm. Thus, if using the Y-type zeolite for oil refining, there will be a major obstacle for the heavy oil molecules to approach to the catalytic activity center. Dong and Chen et al [48] have prepared a macroporous catalyst with the pore size in range of 5-2000 nm by template method, and attempted to improve the accessibility of heavy distillate oil to the active center. O'Connor [49] and Yanik's groups [50] proved that a good residual oil catalyst should have a better gradient pore structure and acid distribution. It is worth mentioning that 3DOM materials could provide smooth channels for heavy oil cracking in order.

Conventional hydrodesulfurization usually adopts a single component carrier for catalytic hydrogenation to remove sulfur at a certain pressure and temperature. In general, the single component carrier mainly focuses on Al_2O_3, TiO_2, ZrO_2, activated carbon, and $BaTiO_3$, but these catalytic carriers are not effective on removing thiophene and its derivatives. Chen et al [51] have studied the use of macroporous catalysts for fluid catalytic cracking and hydrodesulfurization. Due to their good mass transfer property, macroporous catalysts have high activity than conventional catalysts. Tan and co-workers [52] have also found that MY/kaolin composite catalyst with a macroporous-mesoporous-microporous structure displayed excellent catalytic activity in the heavy oil cracking reactions. Macroporous structure, especially 3DOM materials, has particular superiority to macromolecule catalytic reaction due to their good mass transfer and heat transfer properties.

4.4. Diesel Soot Combustion

Diesel engines are widely used in manufacture and transportation industries. However, the diesel exhaust from diesel engines may cause serious environmental problems. Studies on the treatment of diesel exhaust have been carried out for many years, but have yet to find some valid way to completely solve the problems. Thus, a highly efficient catalytic filter is urgently needed for capturing the soot particles, reducing the soot ignition temperature and making the soot complete combustion at the exhaust pipe temperature. Sadakane et al [12] have synthesized 3DOM $La_{1-x}Sr_xFeO_3$ (x=0-0.4) perovskite oxide materials with greatly improved catalytic combustion activity by the colloidal crystal template method for catalytic combustion of carbon particles. The main reason for the improvement of the catalytic activity is that the diffusion rate of carbon particles on the 3DOM $LaFeO_3$ catalyst is faster than that on non-porous perovskite $LaFeO_3$, catalyst, and the surface contact area is also larger . [53]

4.5. Catalytic Oxidation of Formaldehyde

Formaldehyde is recognized as an indoor air pollutant. It is well-known that prolonged exposure to concentrations of formaldehyde that exceed safe limitations is greatly harmful to human health. If humans were exposed to an indoor environment polluted with elevated formaldehyde levels for an extended length of time, serious health problems including burning sensations in the eyes and throat, nausea, difficulty in breathing, and even lethiferous diseases can be caused [54]. In recent years, due to its health effects, formaldehyde pollution is currently of increasing concern. Supported nanoscaled Au catalysts have been demonstrated as an important series of catalysts for formaldehyde catalytic oxidation. When formaldehyde oxidation occurs, the formic acid is produced. For the ordinary catalyst, formic acid could be deposited on the catalyst and adsorbed on an active center, causing the loss of the catalytic activity. Meanwhile, the generated CO_2 during formaldehyde catalytic oxidation is hard to outflow, which seriously affects the catalyst activity and life, thus restricting its practical application. Zhang et al [37] have prepared a 3DOM Au/CeO_2 catalyst by the colloidal crystal template method. Compared to the powder CeO_2 catalyst, 3DOM Au/CeO_2 catalyst showed the improved formaldehyde catalytic activity, since the 3DOM structure of the catalyst facilitates the uniform distribution of Au nanoparticle species, improves the mass transfer of formaldehyde molecules, and releases the generated CO_2. Thus, the development of 3DOM materials with feature of macroporous structure provides a new thinking for elimination of formaldehyde pollution via catalytic oxidation process.

4.6. Photocatalytic Reaction

TiO_2 photocatalytic oxidation technology is paid more and more attention because of its mild reaction condition and wide application without secondary pollution. TiO_2 photocatalyst is mainly used in sewage treatment and air purification. Traditional power TiO_2 photocatalyst has many shortcomings in practical use, including low surface area, easy inactivation, separation difficulty, large air resistance, and unsuitability in flow system etc, which seriously limits its application for waste water and gas treatment. Although, powder nanoscaled TiO_2 shows the improved catalytic efficiency, the contract surface areas of catalyst with reactants are still limited due to the sever aggregation of the catalyst when the reaction is carried out in organic solution. Thus, this kind of supported TiO_2 photocatalysts still have a certain distance to practical application. Madhvi et al have prepared 3DOM Au/TiO_2 catalyst for degradation of methyl orange and methyl blue. Results showed that the larger aperture size of 3DOM materials makes the reactants well-dispersed in the pore structure, leading to the effective adsorption of methyl orange and methyl blue, which eventually improves the catalytic property. Therefore, 3DOM materials might be available for solving the problem of the low contact surface of catalyst with reactants, and thus, improve the efficiency of photocatalytic reaction. [55]

Conclusion

3DOM materials have attracted much research attention in recent years due to their special structural and functional properties. Currently, 3DOM materials is becoming a hot research area in materials science, and the research in this area is mainly focused on: (1) development of synthesis technology; (2)modulation of porous structure and pore size; (3) construction of various materials systems with 3DOM structure; (4) functionalization of 3DOM structure surface; (5) exploration of novel functionality of 3DOM structures. However, all these research aspects are somehow difficult to achieve. Applications of 3DOM materials in various areas still face challenges such as improving the pore wall compactness and mechanical stability, forming ordered structure in large scale, and finding simple synthesis route to decrease synthesis cost. The research on the application of 3DOM materials still needs to be further pushed forward. It is expected that the advance and development of research on 3DOM materials will greatly benefit the applications of such materials in photonic crystals, catalysis, adsorption and separation, electrode materials, thermal resistance materials, communication, energy conversion, energy storage, and environmental health, etc.

ACKNOWLEDGMENT

We gratefully acknowledge the financial supports from the National Natural Science Foundation of China (209610052), Key Project of Ministry of Education (327011), and Oversea Returned Talent Start-up Funding for Research of Ministry of Education of China (208138), Department of Science and Technology of Inner Mongolia (Public Security Foundation208096), Inner Mongolia University Funds (10013-121008, 206077, 206043).

REFERENCES

[1] (a) Kresge, C. T.; Leonowicz, M. E.; Roth, W. J.; Vartuli, J. C.; Beck, J. S., Ordered mesoporous molecular sieves synthesized by a liquid-crystal template mechanism. *Nature* 1992, *359* (6397), 710-712; (b) Casey, C. P.; Underiner, T. L.; Vosejpka, P. C.; Gavney, J. A.; Kiprof, P., Rearrangement of a propargyl vinyl rhenium complex to a rhenium allenyl vinyl ketone complex. *J Am Chem Soc* 1992, *114* (27), 10826-10834.

[2] Tanev, P. T.; Chibwe, M.; Pinnavaia, T. J., Titanium-containing mesoporous molecular sieves for catalytic oxidation of aromatic compounds. *Nature* 1994, *368* (6469), 321-323.

[3] (a) Imhof, A.; Pine, D. J., Uniform Macroporous Ceramics and Plastics by Emulsion Templating. *Adv Mater* 1998, *10* (9), 697-700; (b) Velev, O. D.; Jede, T. A.; Lobo, R. F.; Lenhoff, A. M., Porous silica via colloidal crystallization. *Nature* 1997, *389* (6650), 447-448; (c) Holland, B. T.; Blanford, C. F.; Stein, A., Synthesis of Macroporous Minerals with Highly Ordered Three-Dimensional Arrays of Spheroidal Voids. *Science* 1998, *281* (5376), 538-540; (d) Wijnhoven, J. E.; nbsp; G; J; Vos, W. L., Preparation of Photonic Crystals Made of Air Spheres in Titania. *Science* 1998, *281* (5378), 802-804.

[4] Patel, M.; Padhi, B. K., Production of alumina fiber through jute fiber substrate. *J Mater Sci* 1990, *25* (2), 1335-1343.

[5] Krishnarao, R. V.; Mahajan, Y. R.; Kumar, T. J., Conversion of raw rice husks to SiC by pyrolysis in nitrogen atmosphere. *J Eur Ceram Soc* 1998, *18* (2), 147-152.

[6] Greil, P.; Lifka, T.; Kaindl, A., Biomorphic Cellular Silicon Carbide Ceramics from Wood: I. Processing and Microstructure. *J Eur Ceram Soc* 1998, *18* (14), 1961-1973.

[7] (a) Greil, P.; Lifka, T.; Kaindl, A., Biomorphic Cellular Silicon Carbide Ceramics from Wood: II. Mechanical Properties. *J Eur Ceram Soc* 1998, *18* (14), 1975-1983; (b) Gu, Z. Z.; Hayami, S.; Kubo, S.; Meng, Q. B.; Einaga, Y.; Tryk, D. A.; Fujishima, A.; Sato, O., Fabrication of structured porous film by electrophoresis. *J Am Chem Soc* 2001, *123* (1), 175-176.

[8] (a) Subramanian, G.; Manoharan, V. N.; Thorne, J. D.; Pine, D. J., Ordered Macroporous Materials by Colloidal Assembly: A Possible Route to Photonic Bandgap Materials. *Adv Mater* 1999, *11* (15), 1261-1265; (b) Subramania, G.; Constant, K.; Biswas, R.; Sigalas, M. M.; Ho, K. M., Optical photonic crystals fabricated from colloidal systems. *Appl Phys Lett* 1999, *74* (26), 3933-3935; (c) Li, S.; Zheng, J.; Yang, W.; Zhao, Y., Preparation of Three-dimensionally Ordered Macroporous Oxides by Combining Templating Method with Sol–Gel Technique. *Chem Lett* 2007, *36* (4), 542-543; (d) Li, S.; Zheng, J.; Yang, W.; Zhao, Y.; Liu, Y., Preparation and characterization of three-dimensional ordered macroporous rare earth oxide—CeO₂. *J Porous Mat* 2008, *15* (5), 589-592.

[9] (a) Holland, B. T.; Blanford, C. F.; Do, T.; Stein, A., Synthesis of Highly Ordered, Three-Dimensional, Macroporous Structures of Amorphous or Crystalline Inorganic Oxides, Phosphates, and Hybrid Composites. *Chem Mater* 1999, *11* (3), 795-805; (b) Zakhidov, A. A.; Baughman, R. H.; Iqbal, Z.; Cui, C.; Khayrullin, I.; Dantas, S. O.; Marti, J.; Ralchenko, V. G., Carbon Structures with Three-Dimensional Periodicity at Optical Wavelengths. *Science* 1998, *282* (5390), 897-901.

[10] (a) Jiang, P.; Cizeron, J.; Bertone, J. F.; Colvin, V. L., Preparation of Macroporous Metal Films from Colloidal Crystals. *J Am Chem Soc* 1999, *121* (34), 7957-7958; (b) Velev, O. D.; Tessier, P. M.; Lenhoff, A. M.; Kaler, E. W., Materials: A class of porous metallic nanostructures. *Nature* 1999, *401* (6753), 548-548.

[11] (a) Park, S. H.; Xia, Y., Fabrication of Three-Dimensional Macroporous Membranes with Assemblies of Microspheres as Templates. *Chem Mater* 1998, *10* (7), 1745-1747; (b) Park, S. H.; Xia, Y., Macroporous Membranes with Highly Ordered and Three-Dimensionally Interconnected Spherical Pores. *Adv Mater* 1998, *10* (13), 1045-1048.

[12] Sadakane, M.; Asanuma, T.; Kubo, J.; Ueda, W., Facile Procedure To Prepare Three-Dimensionally Ordered Macroporous (3DOM) Perovskite-type Mixed Metal Oxides by the Colloidal Crystal Templating Method. *Chem Mater* 2005, *17* (13), 3546-3551.

[13] Géraud, E.; Rafqah, S.; Sarakha, M.; Forano, C.; Prevot, V.; Leroux, F., Three Dimensionally Ordered Macroporous Layered Double Hydroxides: Preparation by Templated Impregnation/Co-precipitation and Pattern Stability upon Calcination†. *Chem Mater* 2007, *20* (3), 1116-1125.

[14] Antonietti, M.; Berton, B.; Göltner, C.; Hentze, H. P., Synthesis of Mesoporous Silica with Large Pores and Bimodal Pore Size Distribution by Templating of Polymer Lattices. *Adv Mater* 1998, *10* (2), 154-159.

[15] (a) Miguez, H.; Meseguer, F.; Lopez, C.; Blanco, A.; Moya, J. S.; Requena, J.; Mifsud, A.; Fornes, V., Control of the photonic crystal properties of fcc-packed submicrometer SiO_2 spheres by sintering. *Adv Mater* 1998, *10* (6), 480-483; (b) Zhu, J.; Li, M.; Rogers, R.; Meyer, W.; Ottewill, R. H.; Crew, S. T. S. S. S.; Russel, W. B.; Chaikin, P. M., Crystallization of hard-sphere colloids in microgravity. *Nature* 1997, *387* (6636), 883-885.

[16] Zhou, Z. C.; Zhao, X. S., Flow-controlled vertical deposition method for the fabrication of photonic crystals. *Langmuir* 2004, *20* (4), 1524-1526.

[17] Schroden, R. C.; Al-Daous, M.; Blanford, C. F.; Stein, A., Optical Properties of Inverse Opal Photonic Crystals. *Chem Mater* 2002, *14* (8), 3305-3315.

[18] (a) Holgado, M.; García-Santamaría, F.; Blanco, A.; Ibisate, M.; Cintas, A.; Míguez, H.; Serna, C. J.; Molpeceres, C.; Requena, J.; Mifsud, A.; Meseguer, F.; López, C.,

Electrophoretic Deposition To Control Artificial Opal Growth. *Langmuir* 1999, *15* (14), 4701-4704; (b) Park, S. H.; Qin, D.; Xia, Y., Crystallization of Mesoscale Particles over Large Areas. *Adv Mater* 1998, *10* (13), 1028-1032; (c) Li, S.; Zheng, J.; Yang, W.; Zhao, Y., A new synthesis process and characterization of three-dimensionally ordered macroporous ZrO2. *Mater Lett* 2007, *61* (26), 4784-4786.

[19] (a) Gates, B.; Qin, D.; Xia, Y., Assembly of Nanoparticles into Opaline Structures over Large Areas. *Adv Mater* 1999, *11* (6), 466-469; (b) Park, S. H.; Xia, Y. N., Assembly of mesoscale particles over large areas and its application in fabricating tunable optical filters. *Langmuir* 1999, *15* (1), 266-273.

[20] Velev, O. D.; Lenhoff, A. M., Colloidal crystals as templates for porous materials. *Current Opinion in Colloid & Interface Science* 2000, *5* (1-2), 56-63.

[21] Wang, J.; Li, Q.; Knoll, W.; Jonas, U., Preparation of Multilayered Trimodal Colloid Crystals and Binary Inverse Opals. *J Am Chem Soc* 2006, *128* (49), 15606-15607.

[22] Vlasov, Y. A.; Bo, X.-Z.; Sturm, J. C.; Norris, D. J., On-chip natural assembly of silicon photonic bandgap crystals. *Nature* 2001, *414* (6861), 289-293.

[23] Gu, Z.-Z.; Fujishima, A.; Sato, O., Fabrication of High-Quality Opal Films with Controllable Thickness. *Chem Mater* 2002, *14* (2), 760-765.

[24] Ma, X.; Li, B.; Chaudhari, B. S., Fabrication and annealing analysis of three-dimensional photonic crystals. *Appl Surf Sci* 2007, *253* (8), 3933-3936.

[25] Gates, B.; Park, S. H.; Xia, Y., Tuning the Photonic Bandgap Properties of Crystalline Arrays of Polystyrene Beads by Annealing at Elevated Temperatures. *Adv Mater* 2000, *12* (9), 653-656.

[26] (a) Tang, F.; Uchikoshi, T.; Sakka, Y., A practical technique for the fabrication of highly ordered macroporous structures of inorganic oxides. *Mater Res Bull* 2006, *41* (2), 268-273; (b) Gundiah, G.; Rao, C. N. R., Macroporous oxide materials with three-dimensionally interconnected pores. *Solid State Sci* 2000, *2* (8), 877-882; (c) Kuai, S.; Badilescu, S.; Bader, G.; Brüning, R.; Hu, X.; Truong, V. V., Preparation of Large-Area 3D Ordered Macroporous Titania Films by Silica Colloidal Crystal Templating. *Adv Mater* 2003, *15* (1), 73-75; (d) Wijnhoven, J. E. G. J.; Bechger, L.; Vos, W. L., Fabrication and Characterization of Large Macroporous Photonic Crystals in Titania. *Chem Mater* 2001, *13* (12), 4486-4499; (e) Wang, D.; Caruso, R. A.; Caruso, F., Synthesis of Macroporous Titania and Inorganic Composite Materials from Coated Colloidal SpheresA Novel Route to Tune Pore Morphology. *Chem Mater* 2001, *13* (2), 364-371; (f) Dionigi, C.; Calestani, G.; Ferraroni, T.; Ruani, G.; Liotta, L. F.; Migliori, A.; Nozar, P.; Palles, D., Template evaporation method for controlling anatase nanocrystal size in ordered macroporous TiO2. *J Colloid Interf Sci* 2005, *290* (1), 201-207; (g) Yi, G.-R.; Moon, J. H.; Yang, S.-M., Ordered Macroporous Particles by Colloidal Templating. *Chem Mater* 2001, *13* (8), 2613-2618; (h) Velev, O. D.; Jede, T. A.; Lobo, R. F.; Lenhoff, A. M., Microstructured Porous Silica Obtained via Colloidal Crystal Templates. *Chem Mater* 1998, *10* (11), 3597-3602.

[27] Yan, H.; Blanford, C. F.; Holland, B. T.; Smyrl, W. H.; Stein, A., General Synthesis of Periodic Macroporous Solids by Templated Salt Precipitation and Chemical Conversion. *Chem Mater* 2000, *12* (4), 1134-1141.

[28] (a) Yan, H.; Blanford, C. F.; Holland, B. T.; Parent, M.; Smyrl, W. H.; Stein, A., A Chemical Synthesis of Periodic Macroporous NiO and Metallic Ni. *Adv Mater* 1999, *11* (12), 1003-1006; (b) Yan, H.; Blanford, C. F.; Smyrl, W. H.; Stein, A., Preparation

and structure of 3D ordered macroporous alloys by PMMA colloidal crystal templating. *Chem Commun* 2000, (16), 1477-1478.

[29] Zhang, Y.; Lei, Z.; Li, J.; Lu, S., A new route to three-dimensionally well-ordered macroporous rare-earth oxides. *New Journal of Chemistry* 2001, *25* (9), 1118-1120.

[30] Wang, C. H.; Yang, C.; Song, Y. Y.; Gao, W.; Xia, X. H., Adsorption and direct electron transfer from hemoglobin into a three-dimensionally ordered. macroporous gold film. *Adv Funct Mater* 2005, *15* (8), 1267-1275.

[31] Blanco, A.; Chomski, E.; Grabtchak, S.; Ibisate, M.; John, S.; Leonard, S. W.; Lopez, C.; Meseguer, F.; Miguez, H.; Mondia, J. P.; Ozin, G. A.; Toader, O.; van Driel, H. M., Large-scale synthesis of a silicon photonic crystal with a complete three-dimensional bandgap near 1.5 micrometres. *Nature* 2000, *405* (6785), 437-440.

[32] Xia, Y. N.; Gates, B.; Li, Z. Y., Self-assembly approaches to three-dimensional photonic crystals. *Adv Mater* 2001, *13* (6), 409-413.

[33] Stein, A.; Li, F.; Denny, N. R., Morphological Control in Colloidal Crystal Templating of Inverse Opals, Hierarchical Structures, and Shaped Particles†. *Chem Mater* 2007, *20* (3), 649-666.

[34] Xu, J.; Liu, J.; Zhao, Z.; Zheng, J.; Zhang, G.; Duan, A.; Jiang, G., Three-dimensionally ordered macroporous LaCoxFe1-xO3 perovskite-type complex oxide catalysts for diesel soot combustion. *Catal Today* 2010, *153* (3-4), 136-142.

[35] Li, F.; Wang, Z.; Ergang, N. S.; Fyfe, C. A.; Stein, A., Controlling the Shape and Alignment of Mesopores by Confinement in Colloidal Crystals: Designer Pathways to Silica Monoliths with Hierarchical Porosity. *Langmuir* 2007, *23* (7), 3996-4004.

[36] Tonti, D.; Torralvo, M. J.; Enciso, E.; Sobrados, I.; Sanz, J., Three-dimensionally ordered macroporous lithium manganese oxide for rechargeable lithium batteries. *Chem Mater* 2008, *20* (14), 4783-4790.

[37] Zhang, J.; Jin, Y.; Li, C. Y.; Shen, Y. N.; Han, L.; Hu, Z. X.; Di, X. W.; Liu, Z. L., Creation of three-dimensionally ordered macroporous Au/CeO2 catalysts with controlled pore sizes and their enhanced catalytic performance for formaldehyde oxidation. *Appl Catal B-Environ* 2009, *91* (1-2), 11-20.

[38] Liang, H.; Zhang, Y.; Liu, Y., Three-dimensionally ordered macro-porous Pt/TiO2 catalyst used for water-gas shift reaction. *J Nat Gas Chem* 2008, *17* (4), 403-408.

[39] Bosco, J. P.; Sasaki, K.; Sadakane, M.; Ueda, W.; Chen, J. G. G., Synthesis and Characterization of Three-Dimensionally Ordered Macroporous (3DOM) Tungsten Carbide: Application to Direct Methanol Fuel Cells. *Chem Mater* 2010, *22* (3), 966-973.

[40] Johnson, B. J. S.; Stein, A., Surface Modification of Mesoporous, Macroporous, and Amorphous Silica with Catalytically Active Polyoxometalate Clusters. *Inorganic Chemistry* 2001, *40* (4), 801-808.

[41] Stein, A.; Schroden, R. C., Colloidal crystal templating of three-dimensionally ordered macroporous solids: materials for photonics and beyond. *Current Opinion in Solid State and Materials Science* 2001, *5* (6), 553-564.

[42] Carreon, M. A.; Guliants, V. V., Synthesis of catalytic materials on multiple length scales: from mesoporous to macroporous bulk mixed metal oxides for selective oxidation of hydrocarbons. *Catal Today* 2005, *99* (1-2), 137-142.

[43] (a) Chai, G. S.; Shin, I. S.; Yu, J. S., Synthesis of ordered, uniform, macroporous carbons with mesoporous walls templated by aggregates of polystyrene spheres and

silica particles for use as catalyst supports in direct methanol fuel cells. *Adv Mater* 2004, *16* (22), 2057-+; (b) Chai, G. S.; Yoon, S. B.; Yu, J.-S.; Choi, J.-H.; Sung, Y.-E., Ordered Porous Carbons with Tunable Pore Sizes as Catalyst Supports in Direct Methanol Fuel Cell. *The Journal of Physical Chemistry B* 2004, *108* (22), 7074-7079.

[44] (a) Su, F.; Zhao, X. S.; Wang, Y.; Zeng, J.; Zhou, Z.; Lee, J. Y., Synthesis of Graphitic Ordered Macroporous Carbon with a Three-Dimensional Interconnected Pore Structure for Electrochemical Applications. *The Journal of Physical Chemistry B* 2005, *109* (43), 20200-20206; (b) Kim, P.; Joo, J. B.; Kim, W.; Kang, S. K.; Song, I. K.; Yi, J., A novel method for the fabrication of ordered and three dimensionally interconnected macroporous carbon with mesoporosity. *Carbon* 2006, *44* (2), 389-392; (c) Yu, J.-S.; Kang, S.; Yoon, S. B.; Chai, G., Fabrication of Ordered Uniform Porous Carbon Networks and Their Application to a Catalyst Supporter. *J Am Chem Soc* 2002, *124* (32), 9382-9383; (d) Sakamoto, J. S.; Dunn, B., Hierarchical battery electrodes based on inverted opal structures. *J Mater Chem* 2002, *12* (10), 2859-2861; (e) Kim, P.; Joo, J. B.; Kim, W.; Kim, H.; Song, I. K.; Yi, J., Direct fabrication of Pt-supported macroporous carbon with nanoporous walls. *Carbon* 2005, *43* (11), 2409-2412; (f) Lee, K.; Lytle, J.; Ergang, N.; Oh, S.; Stein, A., Synthesis and Rate Performance of Monolithic Macroporous Carbon Electrodes for Lithium-Ion Secondary Batteries. *Adv Funct Mater* 2005, *15* (4), 547-556.

[45] Soboleva, T.; Zhao, X. S.; Mallek, K.; Xie, Z.; Navessin, T.; Holdcroft, S., On the Micro-, Meso- and Macroporous Structures of Polymer Electrolyte Membrane Fuel Cell Catalyst Layers. *Acs Appl Mater Inter* 2010, *2* (2), 375-384.

[46] Guan, G.; Zapf, R.; Kolb, G.; Hessel, V.; Löwe, H.; Ye, J.; Zentel, R., Preferential CO oxidation over catalysts with well-defined inverse opal structure in microchannels. *Int J Hydrogen Energ* 2008, *33* (2), 797-801.

[47] Míguez, H.; Meseguer, F.; López, C.; Mifsud, A.; Moya, J. S.; Vázquez, L., Evidence of FCC Crystallization of SiO2 Nanospheres. *Langmuir* 1997, *13* (23), 6009-6011.

[48] (a) Dong, D.; Chen, X. H.; Xiao, W. T.; Yang, G. B.; Zhang, P. Y., Preparation and properties of electroless Ni-P-SiO2 composite coatings. *Appl Surf Sci* 2009, *255* (15), 7051-7055; (b) Chen, S. L.; Chen, L. Z.; Pan, E., Three-dimensional time-harmonic Green's functions of saturated soil under buried loading. *Soil Dynamics and Earthquake Engineering* 2007, *27* (5), 448-462.

[49] (a) O'Connor, P., Chapter 15 Catalytic cracking: The Future of an Evolving Process. In *Studies in Surface Science and Catalysis*, Ocelli, M. L., Ed. Elsevier: 2007; Vol. Volume 166, pp 227-251; (b) O'Connor, P.; Verlaan, J. P. J.; Yanik, S. J., Challenges, catalyst technology and catalytic solutions in resid FCC. *Catal Today* 1998, *43* (3-4), 305-313.

[50] Kuehler, C. W.; Jonker, R.; Imhof, P.; Yanik, S. J.; O'Connor, P., Catalyst assembly technology in FCC. Part II: The influence of fresh and contaminant-affected catalyst structure on FCC performance. In *Studies in Surface Science and Catalysis*, Dr, M. L. G.; Dr, P. O. C., Eds. Elsevier: 2001; Vol. Volume 134, pp 311-332.

[51] Chen, S.-L.; Dong, P.; Xu, K.; Qi, Y.; Wang, D., Large pore heavy oil processing catalysts prepared using colloidal particles as templates. *Catal Today* 2007, *125* (3-4), 143-148.

[52] Tan, Q.; Bao, X.; Song, T.; Fan, Y.; Shi, G.; Shen, B.; Liu, C.; Gao, X., Synthesis, characterization, and catalytic properties of hydrothermally stable macro-meso-micro-

porous composite materials synthesized via in situ assembly of pre-formed zeolite Y nanoclusters on kaolin. *Journal of Catalysis* 2007, *251* (1), 69-79.

[53] (a) Xu, J. F.; Liu, J. A.; Zhao, Z.; Zheng, J. X.; Zhang, G. Z.; Duan, A. J.; Jiang, G. Y., Three-dimensionally ordered macroporous LaCoxFe1-xO3 perovskite-type complex oxide catalysts for diesel soot combustion. *Catal Today* 2010, *153* (3-4), 136-142; (b) Zhang, G. Z.; Zhao, Z.; Liu, J.; Jiang, G. Y.; Duan, A. J.; Zheng, J. X.; Chen, S. L.; Zhou, R. X., Three-dimensionally ordered macroporous Ce1-xZrxO2 solid solutions for diesel soot combustion. *Chem Commun* 2010, *46* (3), 457-459.

[54] (a) Li, C.; Shen, Y.; Jia, M.; Sheng, S.; Adebajo, M. O.; Zhu, H., Catalytic combustion of formaldehyde on gold/iron-oxide catalysts. *Catalysis Communications* 2008, *9* (3), 355-361; (b) Shen, Y.; Yang, X.; Wang, Y.; Zhang, Y.; Zhu, H.; Gao, L.; Jia, M., The states of gold species in CeO2-supported gold catalyst for formaldehyde oxidation. *Applied Catalysis B: Environmental* 2008, *79* (2), 142-148; (c) Zhang, Y.; Shen, Y.; Yang, X.; Sheng, S.; Wang, T.; Adebajo, M. F.; Zhu, H., Gold catalysts supported on the mesoporous nanoparticles composited of zirconia and silicate for oxidation of formaldehyde. *Journal of Molecular Catalysis A: Chemical* 2010, *316* (1-2), 100-105.

[55] (a) Fan, G. L.; Xiang, X.; Fan, J.; Li, F., Template-assisted fabrication of macroporous NiFe2O4 films with tunable microstructural, magnetic and interfacial properties. *J Mater Chem* 2010, *20* (35), 7378-7385; (b) Srinivasan, M.; White, T., Degradation of methylene blue by three-dimensionally ordered macroporous titania. *Environ Sci Technol* 2007, *41* (12), 4405-4409; (c) Sadakane, M.; Sasaki, K.; Kunioku, H.; Ohtani, B.; Abe, R.; Ueda, W., Preparation of 3-D ordered macroporous tungsten oxides and nano-crystalline particulate tungsten oxides using a colloidal crystal template method, and their structural characterization and application as photocatalysts under visible light irradiation. *J Mater Chem* 2010, *20* (9), 1811-1818.

In: Focus on Catalysis Research: New Developments
Editors: Minjae Ghang and Bjørn Ramel
ISBN: 978-1-62100-455-4

Chapter 2

A NEW KIND OF ECO-COMPATIBLE HYBRID BIOCATALYSTS FOR SELECTIVE REACTIONS: ARTIFICIAL METALLOENZYMES

Quentin Raffy[1], Rémy Ricoux[2], Jean-Pierre Mahy[2,*]

[1]Université de Strasbourg – CNRS,Institut Pluridisciplinaire Hubert Curien IPHC, Chimie Nucléaire, 23 Rue du Loess, 67037, France

[2]LaboratoiredeChimie Bioorganiqueet Bioinorganique, Institut deChimieMoléculaireet des materiauxd'Orsay (ICMMO), UMR 8182 CNRS, Universitéde Paris 11, 91405 Orsay Cedex, France

ABSTRACT

Biocatalysts and chemical catalysts are on many aspects complementary: range of substrates and reactions, operating conditions or enantioselectivity. Nowadays, the ecological factor has to be taken into account, which urges industrials to develop "green chemistry" procedures. To deserve the "green chemistry" label, an industrial production must not only resort to catalytic procedures to lower the wastes, but also use harmless solvents. The ideal one would be water, with low temperature and pressure, to lower the power consumption. From this point of view, biocatalysts seem better fitted than chemical catalysts, which are often used in organic solvents. Given their relative complementarities, it is very interesting to conceive catalysts that would combine the robustness and wide range of reactions of synthetic catalysts with the ability of enzymes to work under mild conditions in aqueous medium and with high selectivity. As a result, many teams have been working recently on the conception of hybrid biocatalysts, obtained by association of a protein with a synthetic molecule.

* Corresponding author : Prof. Jean-Pierre Mahy, Laboratoire de Chimie Bioorganique et Bioinorganique, Institut de Chimie Moléculaire et des materiaux d'Orsay (ICMMO), UMR 8182 CNRS, Université de Paris 11, 91405 Orsay Cedex, France. Ph. (33) 1 69 15 74 21; Fax. (33) 1 69 15 72 81; E-mail. jean-pierre.mahy@u-psud.fr

This chapter will first begin with a review of the strategies explored to obtain the kind of hybrid biocatalysts that are artificial metalloenzymes. They will be divided in three main categories: (i) direct insertion of an inorganic ion into a protein, (ii) covalent linking of a metallic cofactor to a protein, and (iii) supramolecular anchoring of a metallic cofactor. The results obtained with two particular systems will then be detailed. In these studies, artificial metalloenzymes have been built by non-covalent association of metalloporphyrins with monoclonal anti-steroid antibodies, or with Xylanase A from Streptomyces lividans, to catalyse oxidation reactions.

INTRODUCTION

Nowadays, many examples are known of chiral products the two enantiomers of which possess different biological activities. As a result, chemical industry, and particularly pharmaceutical industry, has to develop processes to isolate enantiopure compounds, in order to commercialize them, or at least to study the biological properties of each isomer. Indeed, since 1992, the American Food and Drug Administration (FDA) and the European Committee for Proprietary Medicinal Products (CPMP) ask for an individual study of the effects of each enantiomer of a pharmaceutical product commercialized as a racemic mixture. Moreover, the FDA has allowed, since 1997, a much faster registration for a medicine in the case of a so-called "racemic-switch", when the "new" medicine is in fact the most active enantiomer of a former product commercialized as a racemic mixture. [1] The improved activity of the product, which allows to lower the dose, as well as the opportunity to expand a lucrative patent, motivate industry to develop enantiomerically pure compounds. As for an example, in 2002, the two most sold medicines, Lipitor and Zocor (14 Billion dollars), were optically pure compounds. The same year, the global sum of all enantiopure compounds reached 159 billion dollars. This value should still rise, owing to the products being developed now. [2]

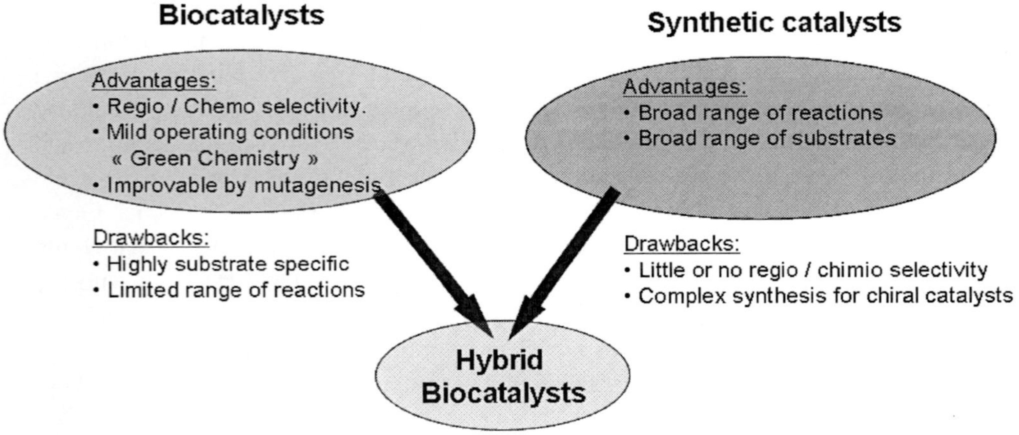

Figure 1. Advantages of the Biocatalysts and the Synthetic catalysts that Hybrid biocatalysts could combine, thus eliminating several drawbacks.

The most widely used technique in pharmaceutical industry to synthesise optically pure compounds is still chiral resolution of diastereoisomer salts. [3] Stoichiometric synthesis methods are also being used, via chiral auxiliaries, linked either to the substrates or to the reactive species. However, these procedures necessitate many chiral auxiliaries, which increase their cost. Asymmetric catalysis makes it instead possible to produce enantiomerically enriched or pure products with low quantities of chiral catalyst. Moreover, the use of catalysts lowers the quantities of waste, which is both cost-effective and eco-friendly. This makes asymmetric catalysis a very promising method, as shown by the attribution of the Nobel Prize to MM. Knowles, Sharpless and Noyori for their work in this area. [4,5,6]

There are three main ways to perform asymmetric catalysis: organocatalysis, [7] organometallic catalysis and biocatalysis. Among these, the two last are used in industry to produce enantiopure amino acids, amino alcohols, amines, alcohols and epoxides. [8] The optimal procedure will be determined by several factors: the quantity of product needed, the cost of the catalyst, its efficiency, its stability and its enantioselectivity.

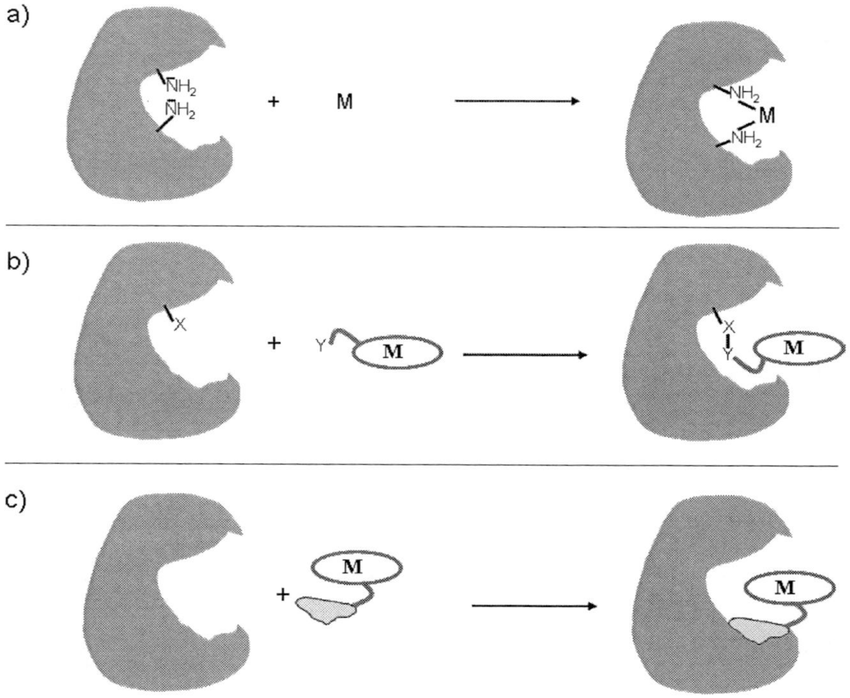

Figure 2. Schematic representation of the three main strategies used to build artificial metalloenzymes. a) direct insertion of an inorganic ion in the cavity of a protein b) Covalent linkage of a metal cofactor c) Non-covalent insertion of a metal cofactor.

Biocatalysts and chemical catalysts are on many aspects complementary: range of substrates and reactions, operating conditions or enantioselectivity (Figure 1). Nowadays, the ecological factor has to be taken into account, which urges industrials to develop "green chemistry" procedures. [9] To deserve the "green chemistry" label, an industrial production must not only resort to catalytic procedures to lower the wastes, but also use harmless solvents. The ideal one would be water, with low temperature and pressure, to lower power consumption. From this point of view, biocatalysts seem better fitted than chemical catalysts, which are often used in organic solvents. Given their relative complementarities, it is very interesting to conceive catalysts that would combine the robustness and wide range of reactions of synthetic catalysts with the ability of enzymes to work under mild conditions in aqueous medium and with high selectivity. As a result, many teams have been working recently on the conception of hybrid biocatalysts, obtained by association of a protein with a synthetic molecule.

One particular kind of hybrid biocatalyst is being especially studied, built by association of a metal cofactor to a protein, to form an artificial metalloenzyme. This chapter will first begin with a review of the strategies explored to build artificial metalloenzymes. They will be divided in three main categories: (i) direct insertion of an inorganic ion into a protein, (ii) covalent linking of a metallic cofactor to a protein, and (iii) supramolecular anchoring of a metallic cofactor. The results obtained with two particular systems will then be detailed. In these studies, artificial metalloenzymes have been built by non-covalent association of metalloporphyrins with monoclonal anti steroid antibodies or with Xylanase A from Streptomyces lividans to catalyse oxidation reactions.

Artificial Metalloenzymes

Direct Insertion of an Inorganic Ion into a Protein

Many teams around the world have been working on a way to create enzymatic activities by introducing inorganic metallic cofactors – ion or oxide – directly in a chelating pre-existent site of a protein.

Serum albumins are transport proteins that are present in great quantity in the blood, the broad cavity of which allows to accommodate as various substrates as fatty acids, bilirubine, hemin or metallic ions. [10] This makes them very interesting systems for the design of new metalloenzymes by non-covalent insertion of metal cofactors or ions. The first team reported to have used them for this purpose is Okano's team, who was the first to obtain, in 1983 an artificial metalloenzyme with bovine serum albumin, in which cavity they inserted osmium tetroxide. [11] The BSA, the substrate and osmium tetroxideare supposed to form a complex (Figure 3 a), which, after reducing hydrolysis, leads to the bis-hydroxylated product. In the presence of tert-butyl hydroperoxide (^{t}BuOOH), the system becomes catalytic, and up to 40 equivalents of bis-hydroxylated product are obtained starting from α-methylstyrene. In this case, an enantiomeric excess of 68 % in favour of S-diol is obtained(Figure 3b).

In Italy, Bertucci and coll. have also capitalized on the ability of the serum albumins to complex metal ions. They have built an artificial metalloenzyme by association of human serum albumin (HSA) with an excess of rhodium (I) salts. The complex formed catalysed quantitatively the hydroformylation of styrene and olefines under high pressure of carbon monoxide and hydrogen (1:1). Moreover, this reaction showed a good regioselectivity in favour of the branched aldehyde (Figure 4). [12, 13]

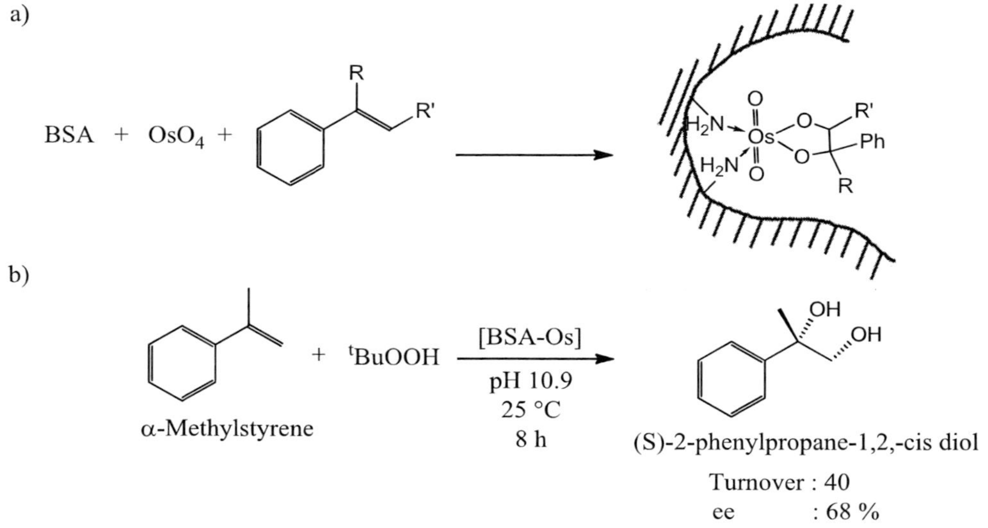

Figure 3. a) Postulated complex of BSA with the substrate and osmium oxide. b) *Cis*-hydroxylation of α-methylstyrene by tert-butyl hydroperoxide, catalysed enantioselectively.

Similar complexes obtained with chicken egg albumin and papain also catalyse styrene hydroformylation, with similar conversion rate and regioselectivity. However, they proved to be much less stable than those formed with HSA. Further studies by mass-spectrometry showed that an excess of rhodium led to a strong denaturing of the protein, most likely arising from interactions between the metal and cysteines sulphur atoms, which may even lead to a cleavage of disulfide bonds. [14]

Vanadium chloroperoxidases are non-heminic metalloenzymes that are more resistant to oxidative degradation than their heminic analogues. [15] However, due to their relatively small active site, they can only accommodate small substrates, which decreases their potential for a use in asymmetric synthesis. Their active site shows very high similarities with that of phytases, which are metal-free. [16] Using this observation, the team of Sheldon came with the idea of building a new protein by inserting vanadate in the phytase of *Aspergillus ficuum* (Figure 5). Vanadate was well accommodated in the active site of the phytase, with dissociating constants ranging between 3 and 15 μM. Moreover, the new artificial metalloprotein proved to have a catalytic activity similar to that of natural vanadium chloroperoxidase. It catalysed the sulfoxidation of thioanisole by H_2O_2, with an enantiomeric excess reaching 66 % when the reaction proceeded at 4 °C (Figure 5 c). [15] In later experiments, they tried to modulate the system by varying the host protein (acid-phophatase, phospholipase, sulfatase, apo-ferritin, BSA) and the metal moiety (Mo, Re, W, Se). [17] The first artificial metalloenzyme, built with *Aspergillus ficuum* phytase and vanadate proved to be the most efficient one, for turnover number as well as for enantioselectivity. Nonetheless, the obtained artificial peroxidase is less selective than the heminic chloroperoxidase of *C. fumago* (ee of 66 % (S), compared to 99 % (R)), and less efficient (Turnover frequency of 5.5 min^{-1}, compared to 900 min^{-1}). However, its very low cost (0,84 $\$.kg^{-1}$ compared to 14000 $\$.kg^{-1}$) and stability makes it a very interesting tool for industrial use in asymmetric catalysis.

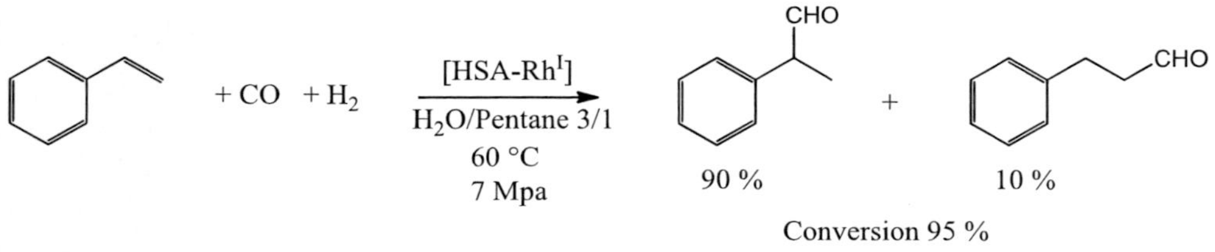

Figure 4. Regioselective hydroformylation of styrene catalysed by the HSA-Rhodium complex.

Carbonic anhydrase (CA) is a natural zinc-metalloenzyme catalysing the hydration of carbon dioxide into bicarbonate. The active-site-zinc can be substituted by other metals, among which manganese. Recently, Okrasa and Kazlauskas have built an artificial protein this manner. [18] After dialysis with 2, 6-carboxylatopyridine, to eliminate zinc, manganese was inserted by dialysis, and was found to bind to the protein with a dissociation constant K_D of about $10^{-3.4}$-$10^{-4,0}$ M. The artificial metalloenzyme obtained catalysed very efficiently the oxidation of the o-dianisidine, an usual substrate used for the measurement of the peroxidase activity, with a molar efficiency $k_{cat}/K_M = 1.4x10^6\ M^{-1}.s^{-1}$, [19] close to that of horseradish peroxidase ($k_{cat}/K_M= 57x10^6\ M^{-1}.s^{-1}$). It also catalysed the enantioselective epoxidation of olefins, with enantiomeric excesses up to 67 %, but with poor yields, due to the degradation of the enzyme occurring during the reaction.

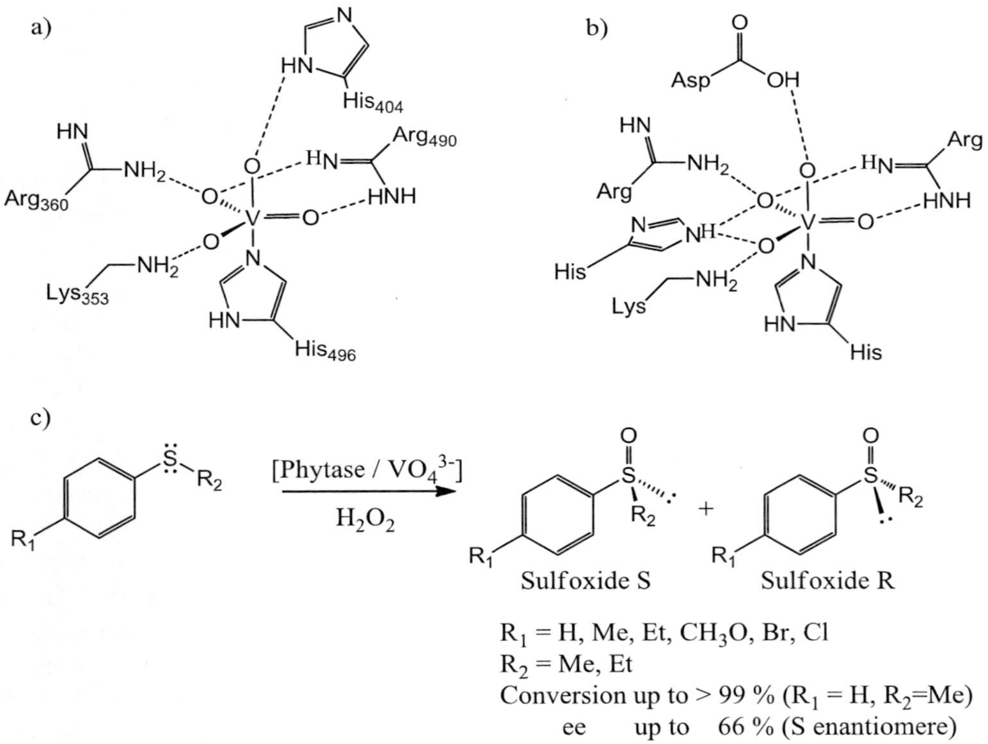

Figure 5. Scheme of the active sites of a) *C. inaequalis* vanadium chloroperoxidase b) *A. ficuum* phytase with vanadate. [17] c) Sulfoxidation of thioanisole by H_2O_2, catalysed by the artificial metalloenzyme.

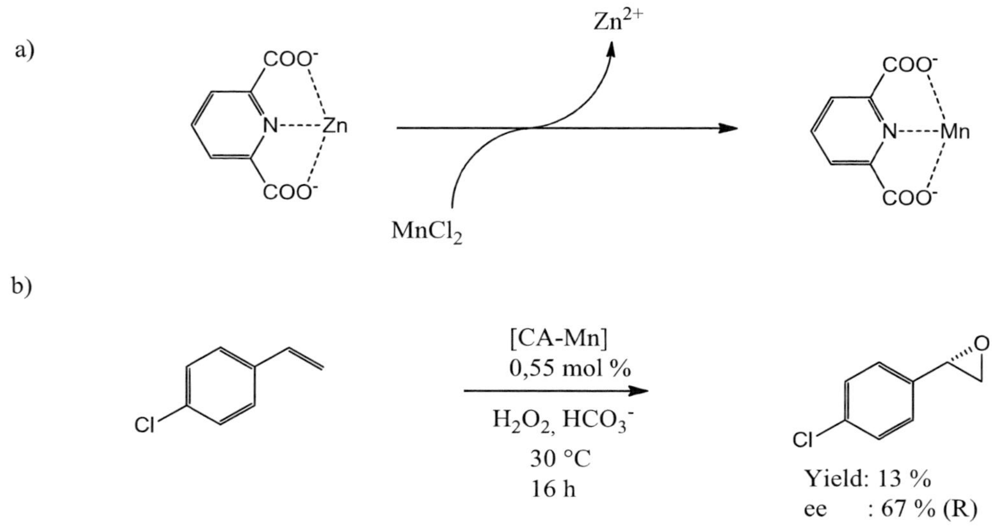

Figure 6. a) Substitution of the active-site zinc by manganese. b) Enantioselective epoxidation of *p*-chlorostyrene, catalysed by the artificial metalloenzyme.

Several teams are using the direct insertion of metal ions in proteins clefts to build artificial metalloenzymes. These modifications can be non-specific, like the insertion of rhodium by treating serum albumin with an excess of this metal, [12] or very precise, like the incorporation of vanadate in phytase active site. [15,17] These methods have the advantage that they necessitate only limited protein engineering, the metal being "spontaneously" accommodated, which can lower the cost of hybrid biocatalysts for an eventual industrial use. However, except changing the metal, there is very little room for chemical optimisations. This is not the case with hybrid biocatalysts built by the association of a protein with a synthetic inorganic catalyst, the activity and selectivity of which can also both be improved chemically. This is a reason why many groups are working on such systems.

The strategies developed in this area can be divided in two main categories: covalent linking of cofactors, and non-covalent insertion in a protein.

Covalent Linking of a Metallic Cofactor to a Protein

After the work published by Kaiser, who had conferred an oxidating activity to papain by covalently grafting a flavin on a cysteinyl residue, [20-21] Distefano and coworkers were the first to describe a method leading to an artificial metalloenzyme catalysing enantioselective hydrolysis of esters. [22-23] The artificial enzyme was obtained by using adipocyte lipid-binding protein (ALBP) as a host protein. ALBP is a small protein of 131 residues, with a quite large cavity of 600 Å^3, broad enough to allow the positioning of both cofactor and substrate. It has also a single cysteinyl residue inside the cavity, which made possible for Distefano to selectively link phenanthroline, which binds copper(II) ions with a good affinity (Figure 7 - a). The metalloprotein obtained catalysed the hydrolysis of non-activated amino acids, inducing an enantioselectivity characterized by ees up to 86 % (Figure 7 - b).

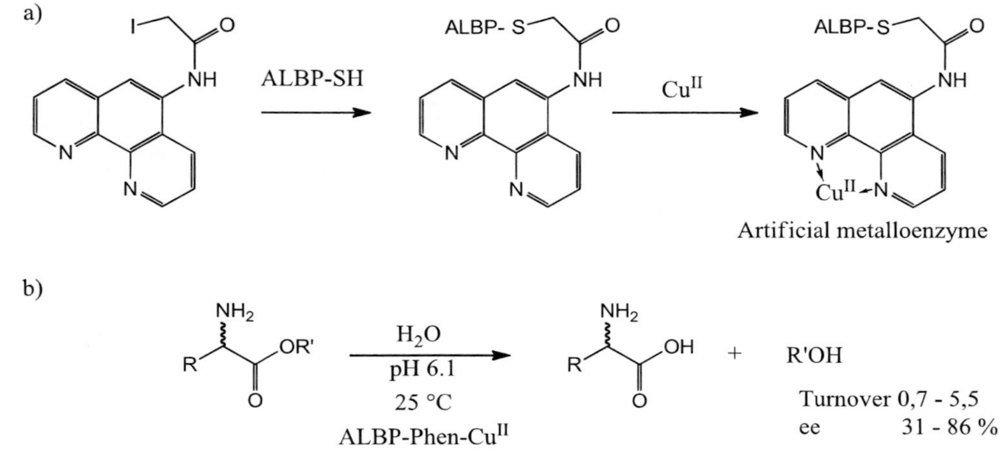

Figure 7. a) Building of the artificial metalloenzyme. b) Catalysed hydrolysis of amino-acid esters.

Papain is another very interesting protein for this kind of chemical modification. Indeed, this protease has a single cysteinyl residue located in its active site. This allows selective chemical modifications, by reaction with the thiol, which explains why several groups have been using it to build hybrid biocatalysts. As a case in point, Reetz and co-workers have managed to covalently bind to papain a manganese salen, as well as a bis-pyridiniumyl rhodium complex, by Michael addition (Figure 8). [24-25] The two artificial enzymes appeared to be able to catalyse reactions of epoxidation and hydrogenation, with weak enantiomeric excesses of about 10 %, that the authors planned to improve by directed evolution.

Panella and coll. have also covalently modified papain by coupling a phosphorylated ligand to the nucleophilic thiol of the cysteinyl residue. The linked ligand then formed a stable complex with rhodium (Figure 9 - a). The hybrid biocatalyst obtained catalysed the hydrogenation of methyl 2-acetamidoacrylate with a great efficiency (Figure 9 - b), but without any enantioselectivity. The authors assumed that this lack of selectivity was due to a too great flexibility of the catalyst or to the large size of the papain's cavity.

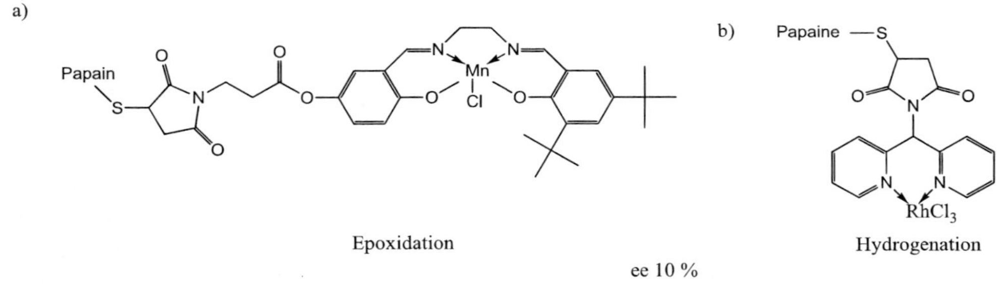

Figure 8. Catalysts covalently bound to papain, and reactions catalysed. a) Manganese salen.
b) bis-pyridinumyl rhodium complex.

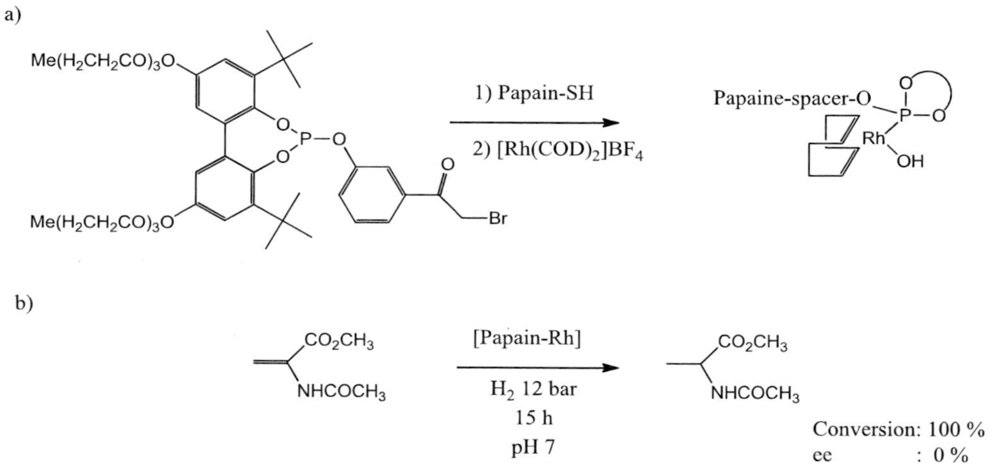

Figure 9. a) Building of the artificial metalloenzyme. b) Catalysed hydrogenation of methyl 2-acetamidoacrylate.

The catalytic antibody 38C2 has a hydrophobic cavity, which contains a lysine with a highly nucleophilic amino moiety. In a similar way than with papain, the team of Janda has taken advantage of this amine to covalently bind a bis-imidazole ligand selectively in the cavity of the protein (Figure 10 – a). [26] This ligand has a strong affinity for copper(II), with a dissociation constant of 10^{-12} M. After complexation with copper(II), the built metalloenzyme catalysed the hydrolysis of 4-Nitrophenyl picolinate with a conversion rate improved by a factor $3.5x10^4$, compared to the complex formed between the isolated ligand and copper (Figure 10 - b).

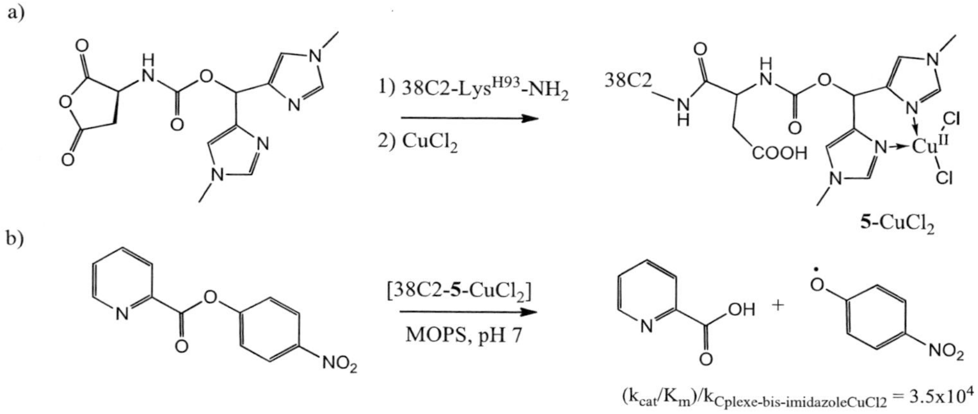

Figure 10. a) Covalent coupling of the bis-imidazole ligand to the amino moiety of the lysine, and complexation of Cu(II). b) Catalysed hydrolysis of 4-Nitrophenyl picolinate.

a)

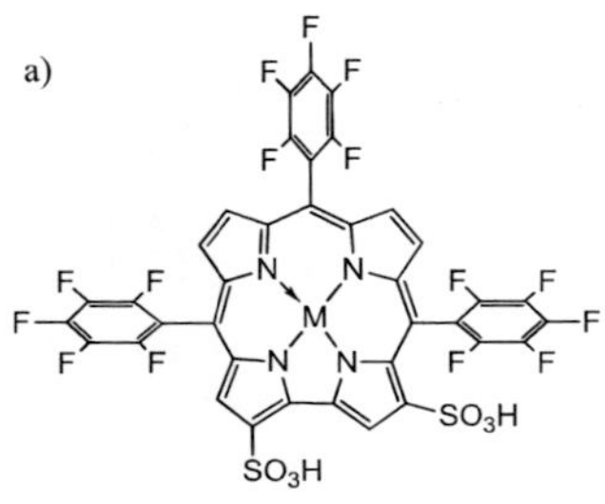

M = Fe^{III} or Mn^{III}

b)

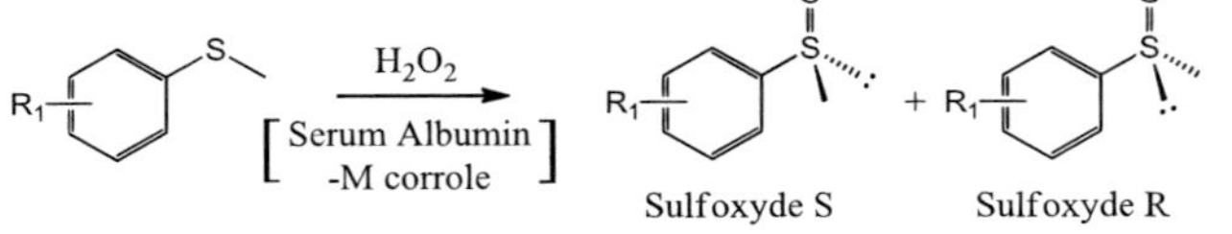

Host protein	M	R_1	Conversion	ee
BSA	Mn	2-F	76 %	68 % (S)
PSA	Mn	2-F	98 %	65 % (S)
BSA	Mn	2-Br	16 %	74 % (S)
BSA	Fe	H	87 %	38 % (S)
HSA	Fe	H	76 %	10 % (S)
HSA	Mn	H	69 %	17 % (R)

Figure 11. a) Metal-corrole cofactor. b) Sulfoxidation of thioanisole and analogues catalysed by the artificial metalloenzymes.

Supramolecular Anchoring of a Metallic Cofactor

Supramolecular anchoring (non-covalent) of a cofactor in the cavity of a protein is today the most widely explored method to obtain artificial metalloenzymes. The current works will be presented based on the host proteins used.

Serum Albumins

As stated above, serum albumins are transport proteins that accept a broad range of substrates. Human serum albumin (HSA) for example, has a strong affinity for hemin ($K_D < 10$ nM), the structure of the corresponding 1:1 complex formed by the protein with this molecule and a fatty acid was described in 2003 by Zunszain *et al.* [27] This showed that the HSA could accommodate compounds with structures close to that of a porphyrin in their cavity.

Gross and coll. have taken advantage of this property, and inserted a bis sulfonated metal corrole into HSA. This led to a stable 1:1 complex, the corrole being accommodated in a specific binding site of strong affinity (micromolar K_D). [28] This corrole, containing iron or manganese, has then been inserted into several serum albumines; human, bovin, porcin, rabbit and sheep serum albumines; to build new artificial metalloenzymes. [29] The corresponding complexes formed did actually act as artificial metalloenzymes catalyzing the sulfoxidation of thioanisole and some analogs with conversion rates up to 98 % and enantiomeric excesses up to 74 % (Figure 11 b). In these studies, the manganese corrole was always superior to its iron counterpart, in terms of activity, selectivity as well as stability.

Former studies having shown that BSA could induce enantioselectivity in Diels-Alder reaction, Reetz and coll. have also used this protein to build an artificial metalloenzyme catalysing this reaction. They choose copper phtalocyanine as a cofactor, which BSA could accommodate. The artificial metalloprotein built this way actually catalysed cycloaddition of azachalcones to cyclopentadiene, with conversion yields above 80 %, and enantiomeric excesses up to 98 %. [30] The *endo* / *exo* selectivity was due to the copper phtalocyanine itself, as showed by the results obtained without the protein. On the other side, the enantiomeric excesses measured on the main *endo* compound were caused by the protein. Similar results were obtained with other serum albumins (HSA, porcin SA, sheep SA, rabbit

SA and chicken SA), with the exception of the two last, which gave very low ees (7 % and 1 %). The association of SA with copper phtalocyanine has led to selective artificial metalloenzymes, which the authors planned to improve by directed evolution.

Myoglobin

The oxygen-binding protein, myoglobin, has also been used by several teams to build artificial metalloenzymes. The oxygen-binding site consists in a 10 Å diameter cavity containing an iron-containing heme prosthetic group. The heme is non-covalently linked to the protein, by hydrophobic interactions, electrostatic interactions via its two carboxylate moieties, and chelation of the iron by the imidazole of the histidine 93. [31] This non-covalent anchoring allows removal of the heme from the protein to yield the apo-myoglobin, with a free cavity able to accommodate another metallic cofactor.

a)

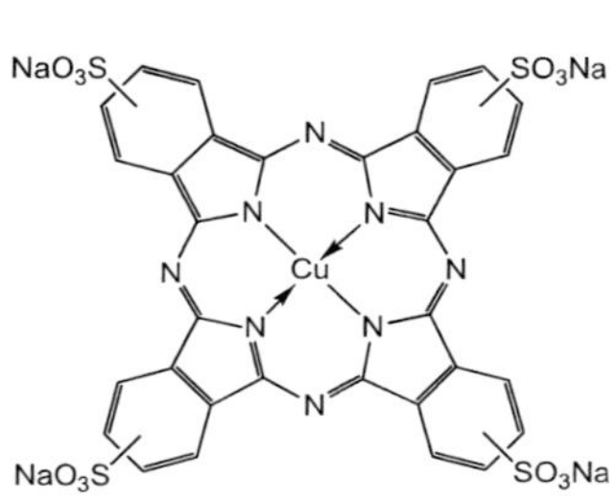

b)

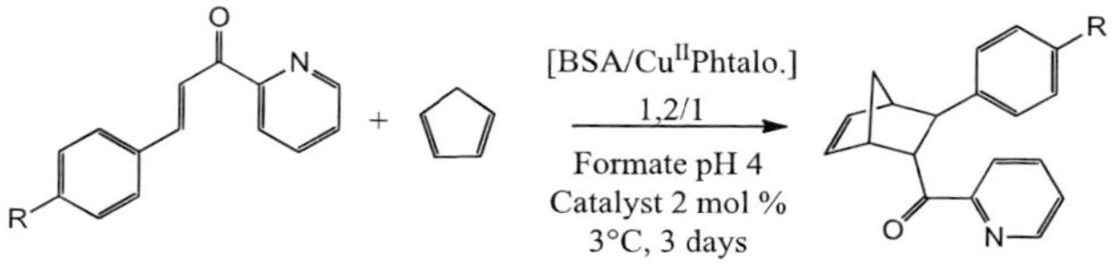

Host protein	R	Conversion	Endo:exo	ee endo
---	H	n.d.	95:5	0 %
BSA	H	80 %	96:4	93 %
BSA	CH_3	84 %	93:7	87 %
BSA	OCH_3	71 %	95:5	88 %
BSA	NO_2	91 %	91:9	98 %
BSA	Cl	89 %	95:5	85 %
HSA	H	89 %	91:9	85 %
PSA	H	87 %	89:11	68 %
SSA	H	76 %	88:12	75 %

Figure 12. a) Copper phtalocyanine used as a cofactor. b) Diels Alder reaction catalysed by several serum albumine – copper phtalocyanine complexes.

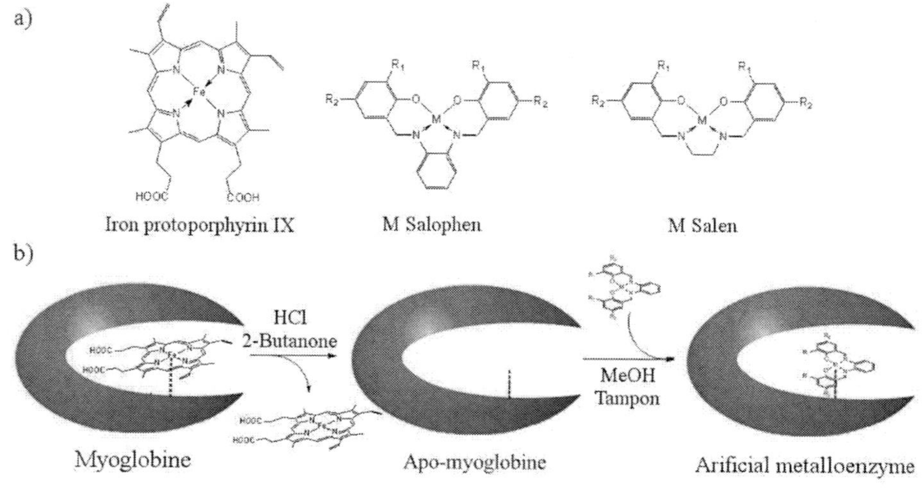

Figure 13. a) Structures of the native heme and of the synthetic metal cofactors. b) Preparation of the artificial metalloenzyme from myoglobin.

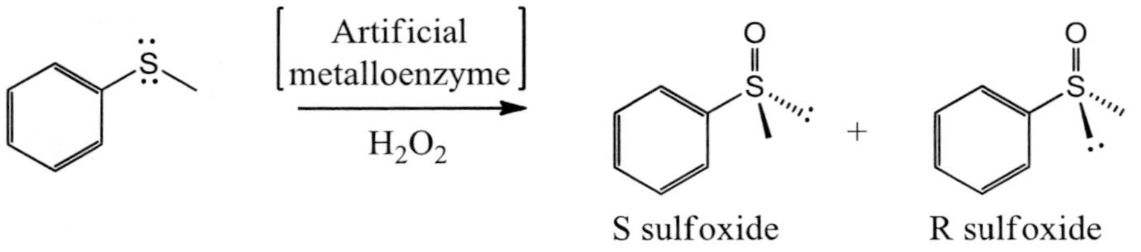

Mutant	Cofactor	R_1	R_2	TON 10^{-3} TO.min^{-1}	ee
WT	Cr-salophen	H	^{t}Bu	46	4 % (R)
H64D/A71G	Cr-salophen	H	^{t}Bu	54	13 % (S)
H64D/A71G	Cr-salophen	Me	H	130	30 % (S)
H64D/A71G	Cr-salen	Me	H	210	33 % (S)
A71G	Mn-salen	Me	H	1295	17 % (R)
H64D/A71G	Mn-salen	Me	H	464	32 % (S)
A71G	Mn-salen	n-Pr	H	2724	27 % (R)

Figure 14. Sulfoxidation of thioanisole enantioselectively catalysed by the hybrid biocatalysts (Cr/Mn)-salens/salophens-myoglobins.

Watanabe and coll. have used this strategy to insert synthetic chromium salophens into apo-myoblogin. Owing to their resemblance with the heme; size, hydrophobicity, coordination of the metal; these molecules are well fitted to the cavity of the apo protein. The first catalytic studies, performed with apo-myoglobin and mutants as host proteins, have shown that such systems stereoselectively catalysed the sulfoxidation of thioanisole, but with rather low turnover (≤ 130.10^{-3} TO.min^{-1}) and enantiomeric excesses (ee ≤ 16 % (S)). [32]

The determination of the crystal structure of two complexes formed by association of manganese and chromium salophenes with the A71G mutant of the apo-myoglobin confirmed that the molecules were bound to the protein by coordination of the metal by the imidazole moiety of the histidine 93. It also showed that the bulky phenylenediamine moiety of the cofactors lowered the accessibility to the active site of the artificial metalloenzyme for the substrate. As a result, they chose to use salens instead, which had a rather low effect on the efficiency of the chromium cofactors. Indeed, the TON increased from 130.10^{-3} TO.min^{-1} with the salophen, to 210 ± 76.10^{-3} TO.min^{-1} with the salen. The effect of the substitution was much more noticeable with the manganese cofactor, the activity of which increased up to 2724.10^{-3} TO.min^{-1}. The selectivity of the reaction was also modified by changing the hydrophobic subsituents of the salen. This led to an inversion of the enantioselectivity induced in the reaction (Figure 14). [33-34]

At the same time, Lu and coll. have also built artificial metalloenzymes by inserting a manganese salen into the cavity of apo-myoglobin mutants. [35] Their cofactor had thiosulfonate methane moieties, aimed at covalently linking the molecule to the protein after reaction with cysteinyl residues (Figure 15). The mutants needed for such a linkage were designed after modelling of the position of the manganese salen in the cavity.

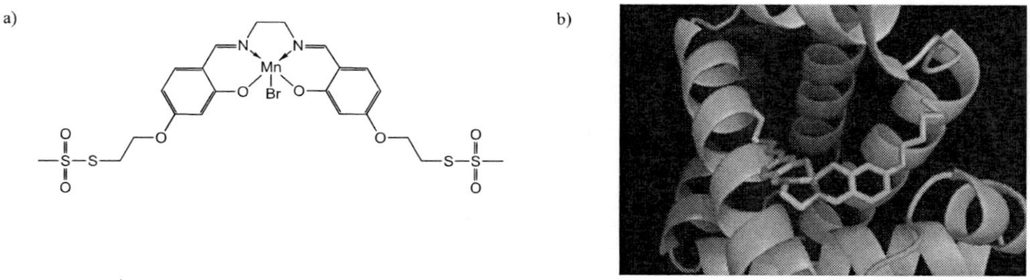

c)

Catalyst	Turnover (min-1)	ee
Cr-Salen-apo(H64D/A71G)Mb	78	13 % (S)
Mn-**1**-apo(Y103C)Mb	51	12 % (S)
Mn-**1**-apo(L72C/Y103C)Mb	390	51 % (S)

Figure 15. a) Manganese salen used as cofactor. b) Representation of the manganese salen twice covalently bound to L72C/Y103C apo-myoglobin. c) Main results obtained with sulfoxidation of thioanisole.

At first, the salen was linked by a single position to the Y103C mutant of the protein. This led to an artificial metalloenzyme catalysing sulfoxidation of thioanisole with a rather low enantiomeric excess (12 %), similar to that obtained with a non-covalently bound chromium salen (Figure 14). The twofold linkage of the cofactor was made possible by the use of the L72C/Y103C double mutant, as confirmed by mass-spectrometry. The enantiomeric excess induced in the sulfoxidation of thioanisole was significantly improved, to reach 51 %, the twofold linkage stabilizing the cofactor in the cavity.

Hayashi and Hiseada have built an artificial protein by replacing the native heme of the myoglobin by another porphyrin, with anionic moieties (Figure 16). This protein had a peroxidase activity which catalysed oxidation reactions, with rates up to 11 fold higher than native myoglobin with catechol. [36] More recently, Hiseada and coll. have also obtained an artificial metalloenzyme with the same efficiency by using Iron(III)-13,16-dicarboxyethyl-2,7-diethyl-3,6,12,17-tetramethylporphycen to replace the heme. [37]

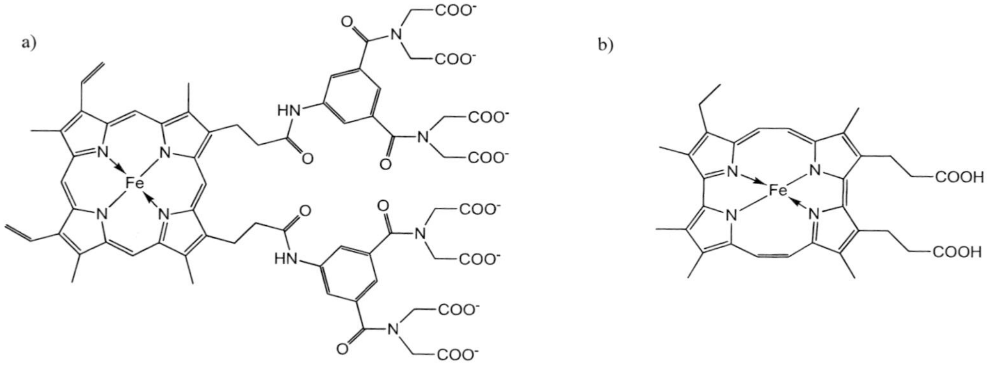

Figure 16. a) Anionic porphyrin inserted by Hayashi and Hiseada. b) Iron (III) 13,16-dicarboxyethyl-2,7-diethyl-3,6,12,17-tetramethylporphycen.

Among the promising works aiming at using apo-myoglobin to build artificial metalloenzymes, we can also name that of Hitomi and coll., who inserted a phenanthroline into the cavity of apo-myoglobin. This led to a hybrid protein, with a chelating site that the authors planned to use to obtain artificial metalloenzymes. [38]

Avidin, Streptavidin and Neutravidin

In the field of artificial metalloenzymes built by supramolecular anchoring of cofactors, avidin, neutravidin and streptavidin have been extensively studied. These proteins consist in four identical sub-units, each able to bind biotin with an extremely low dissociation constant ($K_D \approx 10^{-15}$), one of the lowest measured to date for a protein-ligand complex. This very strong affinity allows almost irreversible – although non-covalent – insertion of biotin in the cavity of the protein, which several groups have used to obtain artificial metalloenzymes.

The first published works in this area were those of Wilson and Whitesides, [39] who described in 1978 that a rhodium complex covalently bound to biotin (Figure 17 – a) quantitatively catalysed hydrogenation of α -acetamidoacrylic acid, with a great enantioselectivity when associated with avidin. When this complex was used with avidin pre-incubated with biotin, the measured efficiency drastically dropped, and the enantiomeric excess disappears. This showed that the enantioselectivity observed was induced by the fixation of the complex in the biotin recognition pocket of avidin.

a)

b)

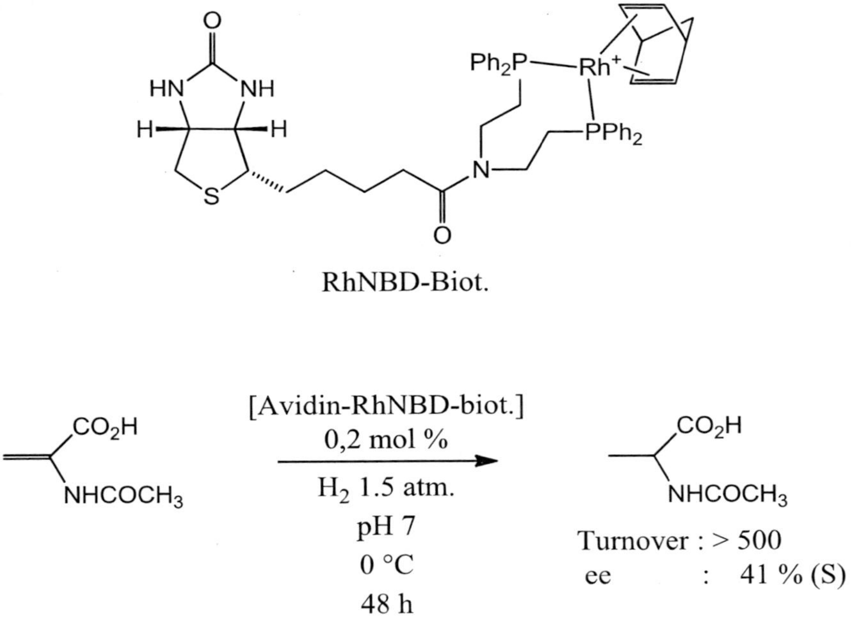

Figure 17. a) Rhodium complex used by Wilson and Whitesides. b) Hydrogenation of α-acetamidoacrylic acid catalysed by the rhodium complex-avidin system (NBD = norbornadiene).

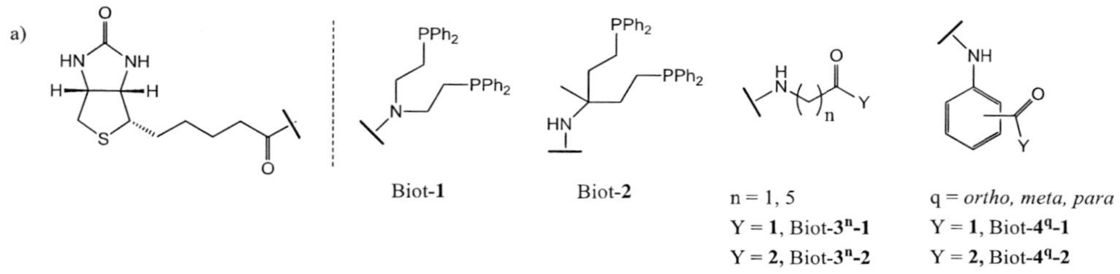

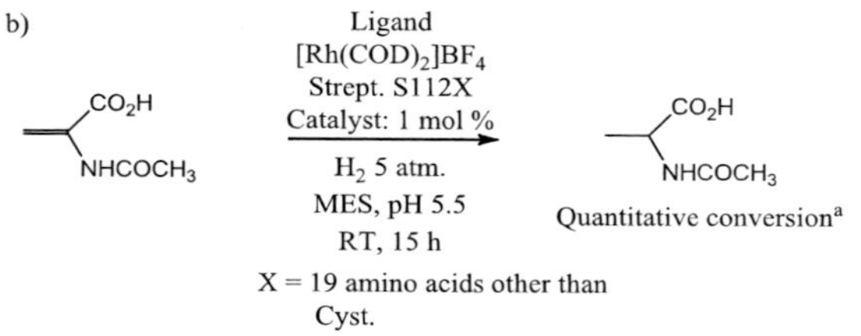

Mutant	Cofactor	ee
WT	Biot-1	94 % (R)
S112C	Biot-1	76 % (R)
S112P	Biot-1	31 % (R)
S112H	Biot-4^{meta}-1	58 % (S)
S112K	Biot-4^{meta}-1	63 % (S)
S112P	Biot-4^{meta}-1	36 % (S)
S112R	Biot-4^{meta}-1	63 % (S)
S112P	Biot-2	63 % (S)
S112Q	Biot-3^4-2	87 % (R)

Figure 18. a) Cofactors synthesized by Ward and Coll. and used with streptavidin and mutants. b) Hydrogenation of α-acetamidoacrylic acid catalysed enantioselectively. c) Quantitative conversion, except for the mutant S112C with Biot-1 (19%). [42]

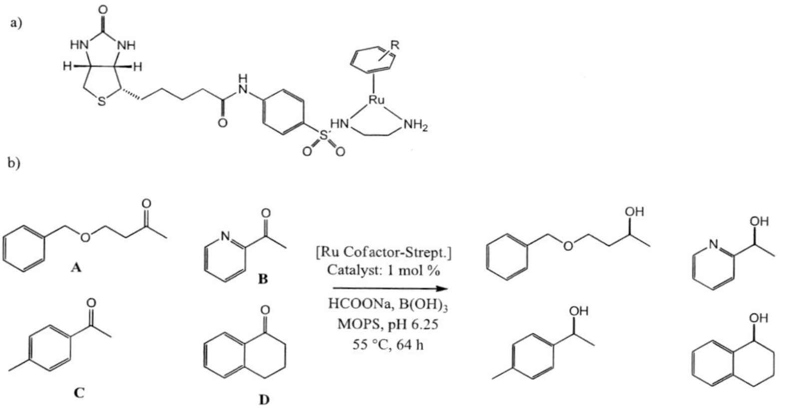

Mutant	Substrate	Conversion	ee
S112A	A	97 %	69 % (R)
S112R	B	95 %	70 % (S)
P64G	C	92 %	94 % (R)
S112Y	D	79 %	97 % (R)

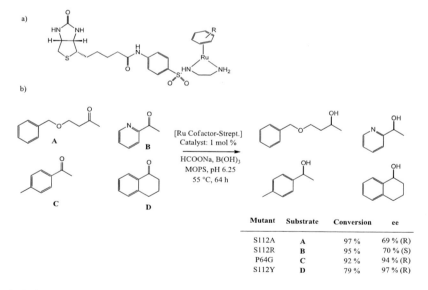

Figure 19. a) Cofactors developed for catalytic hydrogenation of ketones. b) Catalysed reactions and main results. [42]

More recently, Ward and coll. took up this strategy to develop hybrid biocatalysts associating new biotinylated rhodium complexes with avidin, streptavidin and neutravidin. They used several cofactors, modulating the spacer separating biotin from the rhodium complex, as well as streptavidin mutants, to obtain artificial metalloenzymes catalysing the quantitative hydrogenation of α-acetamidoacrylic acid with enantiomeric excesses up to 96

%. The best results were obtained with streptavidin and its mutants, the cavity of which is deeper than that of avidin. Furthermore, the selectivity of the reaction in favour of one or the other enantiomer proved to be adaptable upon changing the cofactor / protein couple. [39] Subsequent results showed that the protein did not only induce a stereoselectivity in the catalysed reaction, but that it also increased the efficiency of the catalyst. [40] Ward's team subsequently managed to improve the enantioselectivity and substrate-selectivity of these artificial metalloenzymes by mutagenesis of streptavidin. [41]

More recently, spacers containing enantiopure R-proline led to a selectivity never achieved before with such hybrid biocatalyst in favour of the S enantiomer (ee 91%) in the hydrogenation reaction. [43]

The association of ruthenium-biotinylated-cofactor complexes with streptavidin mutants yielded artificial metalloenzymes catalysing the enantioselective reduction of several ketones, with both conversion rates and enantiomeric excesses up to 97 % (Figure 19). [42] These hybrid biocatalysts also catalysed the opposite reaction, that is oxidation of alcohols to ketones or aldehydes, by peroxides. [44] Although the association of the chemical catalyst with the protein lowered its efficiency, this last had a protecting effect toward oxidative degradation. Such an artificial metalloenzyme could be used for the kinetic resolution of chiral alcohols, if it proved to be substrate-selective.

Ward and coll. bank on chemical modifications of cofactors, associated with directed mutagenesis of the host proteins to improve their artificial metalloenzymes. Meanwhile, Reetz and coll. have modulated the selectivity of the Wilson and Whitesides hybrid biocatalyst by directed evolution. By using streptavidin as host protein, they reached enantiomeric excesses up to 65 % in favour of the S-enantiomer in the reaction of α-acetamidoacrylic acid hydrogenation. This can be compared to the 23 % ee obtained with the wild-type protein. The selectivity was even inverted by directed evolution, to reach an ee of 7 % in favour of the R enantiomer.

Anti-cofactors Antibodies

Antibodies are proteins that are able to bind their antigen with very high affinity. As a result, they were host-proteins of choice to build artificial metalloenzymes by supramolecular anchoring of a cofactor structurally close to the antigen. By generating monoclonal antibodies against specifically designed antigens, several teams have obtained proteins that were able to bind non-covalently metal cofactors, sometimes with a very high affinity. Some of these results will be presented, especially those about anti-porphyrin antibodies.

The first work concerning an anti-porphyrin antibody having a catalytic activity when associated to an analogue of its antigen was published in 1990 by Cochran and Schultz. [45] The antibody was generated against N-methylmesoporphyrin IX (*N*-MMP, Figure 20), that was supposed to be analogous to the transition state in the metallation of porphyrins by ferrochelatases. Some of the isolated monoclonal antibodies were effectively able to catalyse mesoporphyrin IX metallation by several metal ions. [45] The study of the association of the antibody 7G12-A10 with iron mesoporphyrin IX (Fe-MMP) showed that this last seemed to be effectively located in the recognition pocket of the antibody. The complex formed with Fe-MMP showed a peroxidase activity, catalysing the oxidation of pyrogallol, *o*-dianisidine and 2,2'-azinobis(3-éthylbenzothiazoline-6-sulfonic acid) by hydrogen peroxide, with a

$k_{cat}/K_M(H_2O_2)$ = 274 $M^{-1}.s^{-1}$ more than four time higher than with the Fe-MMP alone ($k_{cat}/K_M(H_2O_2)$ = 64 $M^{-1}.s^{-1}$).

Thereafter, several antibodies were generated against synthetic porphyrins, which also showed a peroxidase activity when associated to the corresponding metalloporphyrin cofactor (Table 1).

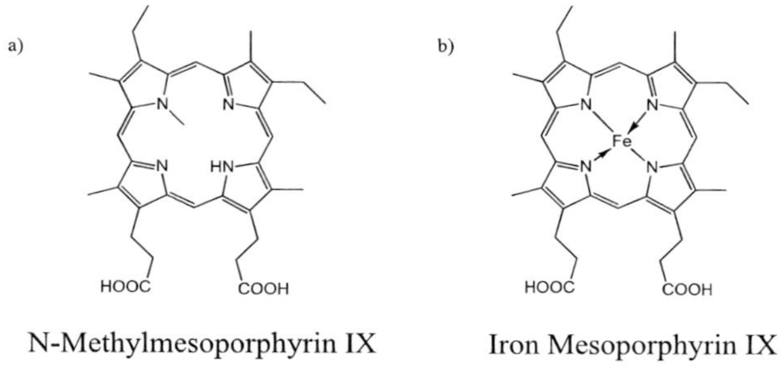

Figure 20. a) N-methylmesoporphyrin IX used to generate the antibodies. b) Iron mesoporphyrn IX used as a cofactor.

Table 1. Kinetic constants measured with H_2O_2 for anti-porphyrin antibodies showing peroxidase activity when associated to a metalloporphyrin cofactor

Antibody	Hapten used	Substrate	k_{cat} (min^{-1})	K_M (mM)	k_{cat}/K_M $M^{-1}.s^{-1}$	Reference
7G12-A10	*N*-MMP	*o*-dianisidine	394	24	274	[45]
2B4	*N*-MMP	*o*-dianisidine	330	43	128	[46]
9A5	*N*-MMP	pyrogallol	132	35	63	[47]
11D1	*N*-HMMP	pyrogallol	86	13	110	[47]
13-1 L	TCPP	pyrogallol	667	2,3	4833	[48]
03-1	TCPP	pyrogallol	50	4	300	[49]
13G10	ToCPPFe	ABTS	100	16	105	[50]
14H7	ToCPPFe	ABTS	63	9	119	[50]
12E11G	3MPy1CPP	pyrogallol	nd	nd	nd	[51]

TCPP : *meso*-Tetrakis(para-carboxyphenyl)porphyrin, ToCPP : α,α,α,β-*meso*-Tetrakis (ortho-carboxyphenyl)porphyrin, 3MPy1CPP : 5-(4-carboxyphenyl)-10,15,20-tris(4-methylpyridyl) porphyrin, ABTS: 2,2′-Azino-bis(3-ethylbenzothiazoline-6-sulfonic acid) diammonium salt.

More particularly, Mahy and coll. have obtained two antibodies generated against iron α,α,α,β-*meso*-Tetrakis(ortho-carboxyphenyl)porphyrin (ToCPPFe, Figure 21), 13G10 and 14H7, with affinities characterized by K_D values of $2.9x10^{-9}$M and $5.5x10^{-9}$M respectively for their antigen, the best found until now for anti-porphyrin antibodies. [50] The artificial metalloenzymes built by association of these antibodies with ToCPPFe catalysed the oxidation of 2,2′-Azino-bis(3-ethylbenzothiazoline-6-sulfonic acid) (ABTS) with a five time better efficiency than ToCPPFe alone (cf. Table 1, 20 $M^{-1}.s^{-1}$ for ToCPPFe). A consecutive study of the 13G10-ToCPPFe complex showed that its catalytic activity reached an optimum around pH 5, which was interpreted as a consequence of the participation of a carboxylic acid in the catalysis. [52] This hypothesis was strengthened by the loss of 50 % of the peroxidase activity of the artificial metalloenzyme after treatment of the carboxylic acid side-chains of the protein with glycinamide.

Artificial metalloenzymes catalysing mono-oxygenation reactions have also been obtained from anti-porphyrin antibodies. In these cases, the antigen porphyrin was generally N-substituted, in order to generate a cavity in the antibody, in which the substrate could be accommodated.

The team of T. Yang has obtained the two antibodies 2B4 and 7C7 using a *N*-4-bromophenyl substituted porphyrin as hapten. When associated with a metalloporphyrin, they catalysed styrene epoxidation by NaOCl, as well as aminopyrine demethylation by H_2O_2, with k_{cat}/K_M of 1389 $M^{-1}.s^{-1}$. [53]

In 1999, Nimri and Keinan have generated antibodies against tin(IV) *meso*-tetrakis(para-carboxyphenyl)porphyrin bearing an axial α-naphtoxy ligand (Figure 22 a), in order to build artificial metalloenzymes catalysing sulfoxidation reactions. [54] The α-naphtoxy ligand mimicked thioanisole in the transition state of the sulfoxidation reaction, as well as iodosylbenzene, used as the oxidant. Such an antigen was used to lead to an antibody with a cavity adapted to both molecules. The selected antibody SN37.4 retained a good affinity for the antigen ($K_D = 5.10^{-8}$M), but it seems that when associated to the Ru *meso*-tetrakis(para-carboxyphenyl) porphyrin, 25 % of the porphyrin remains in solution, outside the cavity of the protein. Nevertheless, in the presence of the antibody, the ruthenium porphyrin catalyses sulfoxidation of thioanisole and other sulphurs of similar size by iodosylbenzene, with enantiomeric excesses up to 43 % (Figure 22 c).

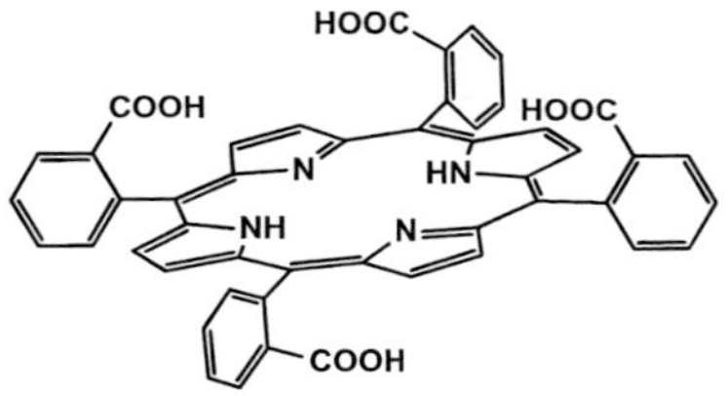

Figure 21. Iron α,α,α,β-*meso* Tetrakis((ortho carboxy)phenyl) porphyrin, used as antigen and as cofactor.

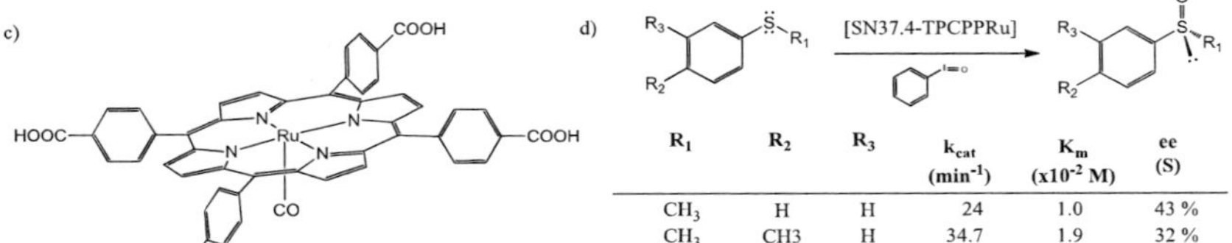

R₁	R₂	R₃	k_{cat} (min^{-1})	K_m ($x10^{-2}$ M)	ee (S)
CH_3	H	H	24	1.0	43 %
CH_3	CH3	H	34.7	1.9	32 %
CH_3	OCH3	H	174.3	2.0	27 %
CH_3	Br	H	14.1	1.9	27 %
CH_3	COCH3	H	8.8	2.3	24 %
CH_3	H	CH_3	333.0	2.6	20 %
CH_2CH_3	H	H	25.3	2.4	21 %

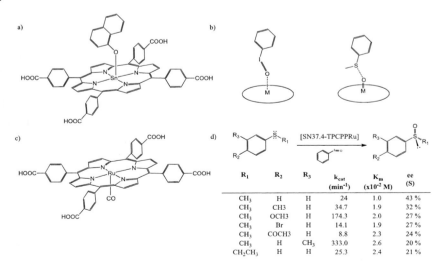

Figure 22. a) Tin(IV) *meso*-tetrakis(para-carboxyphenyl)porphyrin used as hapten, with an α-naphtoxy ligand. b) How iodosylbenzene and thioanisole are supposed to approach the metal of the heme. c) Ru

meso-tetrakis(para-carboxyphenyl) porphyrin used as cofactor. d) Reactions of sulfoxidation catalysed, and main results.

Finally, artificial metalloenzymes have been built with antibodies generated against semi-synthetic porphyrins, such as the microperoxidase 8 (MP8, Figure 23). [55] The MP8 is a heme bearing an octapeptide chain, which is obtained after pepsic and trypsic digestion of equine hearth cytochrome c. [56] The fifth residue of the peptide is the histidine 18 of cytochrome c, which imidazole side chain complexes the iron of the heme. The MP8 has a peroxidase activity, [57] as well as a mono-oxygenase activity.

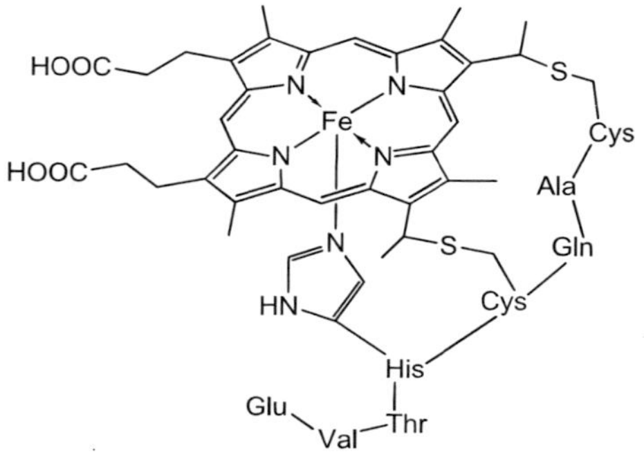

Figure 23. Structure of the microperoxidase 8 (MP8).

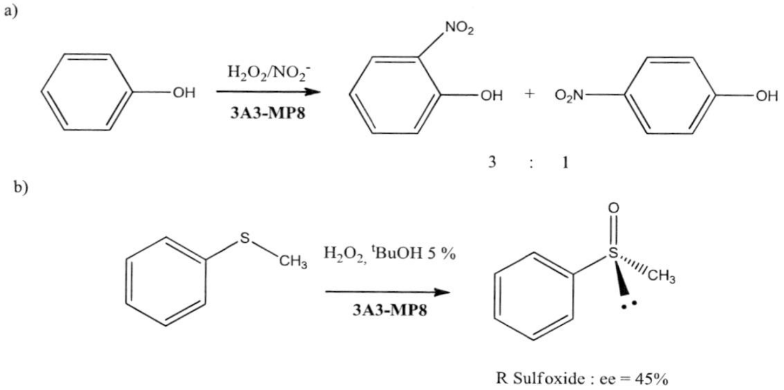

Figure 24. a) Nitration of phenol catalysed by the artificial metalloenzyme 3A3-MP8. b) Enantioselective sulfoxidation of thioanisole by 3A3-MP8.

As an antigen, the microperoxidase 8 has two main assets. First, its terminal amine allows an easy linkage to the protein responsible for the immune response. Second, the carboxylate functions of its propionate side chains and of the terminal glutamate of the peptide induce complementary functions in the antibody generated, increasing its affinity for the molecule.

The axial complexation of the iron by the imidazole moiety of the histidine, as in most natural peroxidases, which makes MP8 a very interesting cofactor..

Antibodies were generated using MP8 as a hapten, and selected on their affinity to the molecule. In particular, the monoclonal antibody 3A3 showed a good affinity for the MP8, with a K_D of 10^{-7}M. When associated to its antigen – used as cofactor – it formed an artificial metalloenzyme, with one of the best peroxidase activity ever reported for such system, characterized by a k_{cat}/K_M of 2.10^6 $M^{-1}.min^{-1}$. [55,58] Moreover, the artificial metalloenzyme also catalysed *ortho* and *para* nitration of phenols by NO_2^- with H_2O_2, with a regioselectivity of 3 to 1 in favour of the *ortho*-nitrophenol (Figure 24 a). [59]

It also enantioselectively catalysed the sulfoxidation of thioanisole by water peroxide with enantiomeric excesses up to 45 % (Figure 24 b). [60]

Artificial Metalloenzymes Built by Non-covalent Association of Porphyrin Cofactors and Antibodies

As mentioned in the previous chapter, antibodies are proteins with two cavities in which their hapten can be accommodated with a very high affinity. Thus, we took advantage of this feature to develop versatile new artificial metalloenzymes, without generating anti-porphyrin antibodies. Indeed, metallic cofactors bearing an antigen pattern can be non-covalently anchored into the cavity of the protein, which could provide an asymmetric environment to the catalyst. This "Trojan horse" strategy, in which the hapten moiety is used to drag the catalyst into the cavity of the protein, has been used to build several metalloenzymes based on anti-steroid antibodies and metalloporphyrin cofactors.

The choice of the antibody was made on its affinity for its antigen. First of all, a good affinity is important to ensure that the quantity of free cofactor is negligible, even without an excess of antibody. Moreover, the highest the affinity, the most tightly the cofactor is bound to the protein. This could reduce the possible conformations and environments of the catalyst in the cavity of the protein, which is a key feature to achieve reproducible chemo- or stereo-selective catalysis. The antibody we used was 7A3, a monoclonal anti-estradiol antibody. It had been generated by immunization of mice with an antigen obtained by covalent linkage of estradiol in 3-position to BSA. [61] This antibody had a very strong affinity for its antigen; $K_D \approx 10^{-9}M$; which made it an excellent candidate for our strategy.

Several cofactors were synthesized, following the scaffold of a metalloporphyrin, responsible for the catalytic activity, covalently linked to the estradiol. Owing to the substitution of the estradiol in the antigen used to generate *7A3*, the estradiol was linked using its phenol moiety. In the studies that we will present, the same anchor was used, which was synthesized from commercial oestrone (Scheme 1). [63] After a nucleophilic substitution of the oestrone phenolate with ethyl bromo acetate, the resulting ester was hydrolysed to yield 3-O-carboxymethyl oestrone. The ketone was then selectively reduced by sodium borohydride.

The choice of the porphyrins was based both on their stability towards the catalysis of oxidation reactions and on synthetic issues. The methine bridge can easily be hydroxylated under oxidative conditions, thus, our catalysts were all based on *meso*-aryl-substituted porphyrins. In addition, as the cofactors had to be water-soluble to interact with the antibody, we have synthesized ionic porphyrins, bearing three methyl-pyridiniumyl, or three sulfonato moieties (Scheme 2). [63,64]

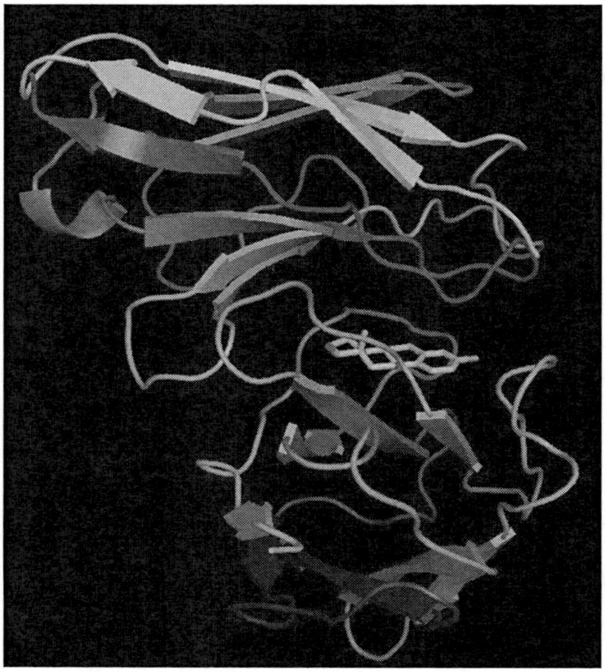

Figure 25. Model structure of 7A3 Fab, with the estradiol inside the cavity. [62]

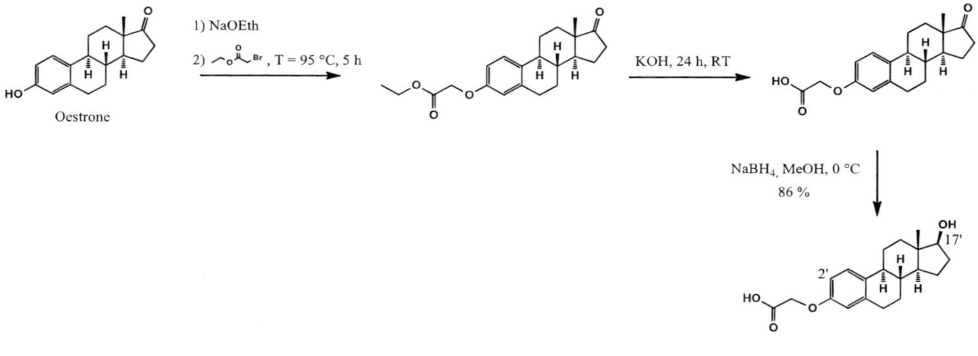

Scheme 1. Synthesis of the estradiol anchor.

After covalent linkage of the porphyrins to the modified estradiol through peptidic coupling, the porphyrins were metallated by iron (III) or manganese (III). The pyridine moieties of the first porphyrin had been methylated prior to the metallation (Scheme 2 b).

The cofactors synthesized, the first step consisted in checking that the estradiol anchor was still well accommodated in the cavity of the antibody, despite the bulky porphyrin. Increasing amounts of cofactor were added to a solution of antibody, [65] or of antibody to a solution of cofactor. [64] In each case, a stoichiometry of two cofactor molecules per antibody was measured by UV-Visible spectroscopy experiments. Moreover, the dissociation constants of the complexes formed with the protein were determined (Table 2). Despite the presence of the porphyrin, the affinity of the antibody for the cofactors remains quite good, up to $K_D = 4.10^{-7}$M for the association of *1-Fe* with *7A3*. When compared to the affinity of the antibody for estradiol ($K_D \approx 10^{-9}$M), the values measured are quite similar. This shows that the nature of the porphyrin (cationic or anionic), as well as the metal inserted have little influence on the association of the cofactor with the antibody.

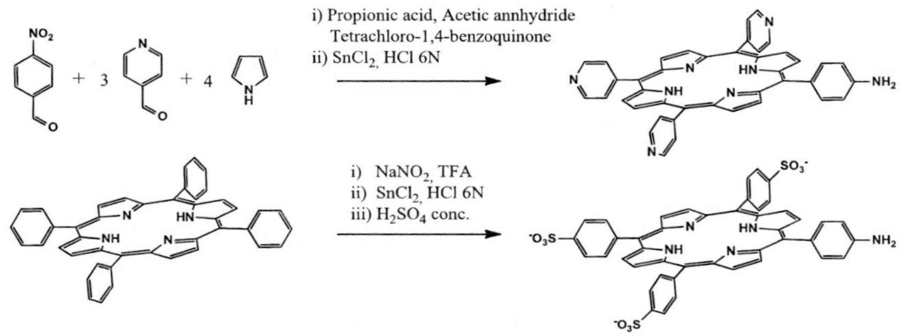

Scheme 2. a) Synthesis of the two core porphyrins. b) Synthesis of the metal cofactors *1*-M and *2*-M.

Table 2. Dissociation constants measured for the complexes metal cofactors – antibody 7A3

	1	*2*
Fe	$K_D = 4.10^{-7}$ M	$K_D = 2.10^{-6}$ M
Mn	$K_D = 2.10^{-6}$ M	$K_D = 9.10^{-6}$ M

More detailed studies were conducted on the complex formed by association of 1-*Fe* with *7A3,* in order to examine the environment of the catalytic centre and the eventual steric hindrance due to amino-acid side chains.

They showed that despite the association of *1-Fe* with the antibody, the iron of the porphyrin can be complexed by two imidazole molecules, on the axial positions. The C_{50} of imidazole was measured to be 23 mM. This value can be compared to that found under similar conditions with the *tetra*-paracarboxyphenyl porphyrin, which is seven times lower, with a value of 3.4 mM. [66] This showed that, even though the steric hindrance brought by the antibody around the porphyrin does not prevent the complexation of the iron by imidazole, it penalizes it. Moreover, the porphyrin is buried deeply enough in the cavity of the protein to prevent its dimerization, which occurs under basic conditions.

Taken altogether, the above-mentioned results are in agreement with the anchoring of the metal cofactors thanks to the specific recognition of the estradiol by the antibody 7A3. The porphyrin seems to remain very near to the entrance of the cavity, with a moderate interaction with the protein.

Catalytic Activities of the Artificial Metalloenzymes

After having verified that the cofactors synthesized were actually recognized by the antibody through their estradiol anchor, the catalytic activity of the artificial proteins formed was estimated.

The peroxidase activity of *1-Fe-7A3* was assayed and compared to that of the cofactor *1-Fe* alone, using hydrogen peroxide as the oxidant, and 2,2'-azino-bis(3-ethylbenzothiazoline-6-sulfonate) (ABTS) as co–substrate of the reaction (Figure 26). ABTS is a classical substrate of peroxidases, which upon one electron oxidation yields a radical cation strongly absorbing at 414 nm. The experiments performed showed that both *1-Fe* and *1-Fe-7A3* had a peroxidase activity, with a pseudo first-order kinetics regarding the catalyst or the hydrogen peroxide concentration. The protein had a very noticeable influence on the catalysis, the rate of the reaction with *1-Fe-7A3* being more than two times higher than with *1-Fe* alone. The kinetic model we developed, based on the one of Lente and Espenson [67,68] showed that the antibody had an accelerating effect mainly on the first step of the reaction, that is the reaction of hydrogen peroxide with the iron(III)-porphyrin.

To measure the influence of the antibody on the selectivity of the oxidation reactions catalysed by iron(III) cofactors, the reaction of sulfoxidation of thioanisole by hydrogen peroxide was studied. Sulfoxidation reactions were conduced with an excess of thioanisole, and the results obtained with the artificial metalloenzymes were compared to that of the free cofactors.

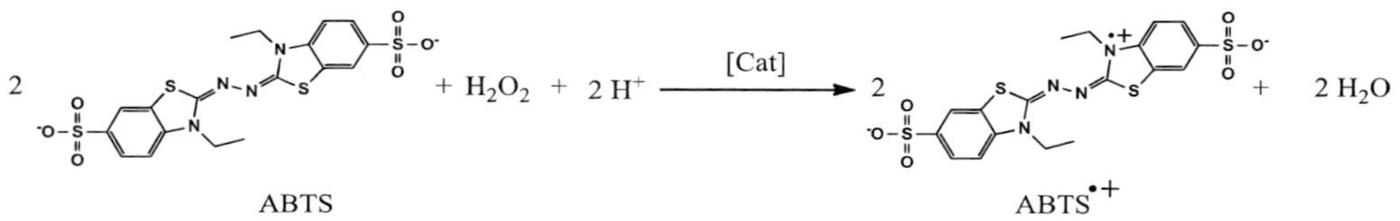

Figure 26. Catalysed oxidation of ABTS by H_2O_2. [Cat] = *1*-Fe or*1*-Fe-7A3.

First of all, the antibody increases the efficiency of the reaction, with final amounts of sulfoxide up to 5 times higher than with the cofactor alone. This shows a protecting effect of the antibody against cofactor's self-oxidation in the presence of H_2O_2. Moreover, kinetic studies showed that the initial velocity of the sulfoxidation reaction in the presence of the artificial metalloenzyme *1-Fe-7A3* was about twice higher than with *1-Fe* alone, which is consistent with the results obtained with ABTS. [63]

Whereas the product of the reaction was found completely racemic with any of the cofactors alone, enantiomeric excesses of about 10 % were measured with the artificial metalloenzymes. These enantiomeric excesses are rather low, but not very different from that obtained by other teams in their first attempts to build artificial metalloenzymes by non-covalent insertion of a cofactor, such as Watanabe and coll. (4 %) or Lu and coll. (12 %).

The manganese cofactors did not catalyse the sulfoxidation of thioanisole by hydrogen peroxide, which was certainly due to their catalase activity, which led to decomposition of H_2O_2. However, it is known that manganese porphyrins can catalyse epoxidation reactions, especially when imidazole is used as a co-catalyst. Therefore, the oxidation of styrene by oxone was chosen to test the influence of the antibody on the selectivity of a reaction catalyzed by manganese cofactors.

Under the conditions of the catalysed oxidation reactions, epoxidation of styrene occurs, to give 2-phenyloxirane. However, other products are also formed, the main one being 2-phenylacetaldehyde, and traces of benzaldehyde (Figure 28). It is likely that 2-phenylacetaldehyde is formed by over-oxidation of the epoxide, which has been reported to occur with ruthenium porphyrins. [69] First, one can clearly see a huge difference in the efficiencies of the two cofactors. Whereas with *1-Mn* the conversionis almost quantitative, the yields appear much lower with *2-Mn*. The protein has apparently little influence on the conversion, whatever the cofactor, as the global yields of oxidation products remain unchanged with the artificial metalloenzymes (Figure 28). No stereoselectivity of the catalysis could be observed, whether with the cofactors alone or with the artificial metalloenzymes.

Conclusion

A new kind of artificial metalloenzymes has been developed, based on the Trojan-horse strategy, with an anti-estradiol antibody and antigen-based cofactors. The stoichiometry of 2:1 and the quite good affinities measured have shown that the cofactors were recognized by the antibody through their estradiol anchor, and so that their association with it led to a real artificial metalloprotein. According to the structural studies, it seems that the porphyrin is located at the entrance of the cavity of the protein.

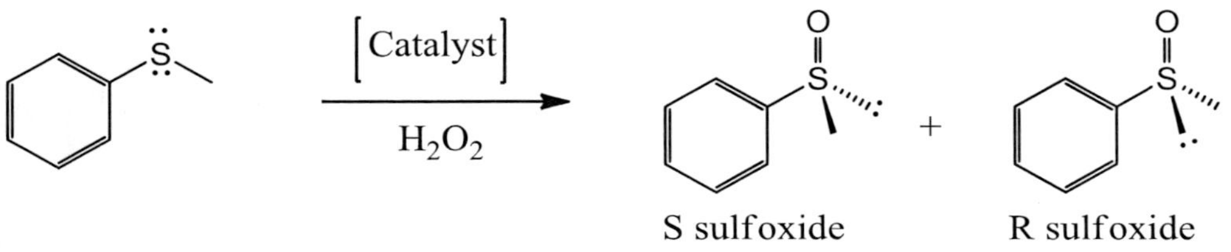

Catalyst	Final [Sulfoxide]	ee
1-Fe	27 μM	0 %
1-Fe-7A3	60 μM	8 % (S)
2-Fe	8 μM	0 %
2-Fe-7A3	44 μM	10 % (S)

Figure 27. Sulfoxidation of thioanisole catalysed by the iron cofactors and the artificial metalloenzymes formed by association with 7A3.

When built with an iron cofactor, the artificial metalloprotein had a peroxidase activity, as showed by the experiments made with ABTS. Even more importantly, the protein has a clear accelerating effect on the first step of the catalysis that is oxygen activation by the porphyrin. The same accelerating effect was observed with the sulfoxidation of thioanisole, together with a twofold raise of the final yield, showing a protective effect of the protein against oxidative degradation of the catalyst. Moreover, the protein actually induced stereoselectivity in the sulfoxidation reaction, with moderate enantiomeric excesses of about 10 %.

The artificial metalloenzymes obtained with manganese cofactors catalysed styrene oxidation by oxone with medium to very good efficiency, up to almost total conversion. In fact, among all the metalloenzymes described until now, *1-Mn-7A3* catalyses the epoxidation of styrene with the best yields ever reported.

This was the first attempt to build artificial metalloenzymes with such associations, and the selectivity induced in the sulfoxidation reaction, however rather weak, is very encouraging. With the cofactors synthesized, it appears that the porphyrins are located at the edge of the antibody's cavity. Therefore, there is high hope that with a shorter spacer between the estradiol anchor and the metalloporphyrin, this last will be buried enough in the cavity for the amino-acid side chains to generate important steric hindrance, which will induce higher enantioselectivity in the reactions catalysed.

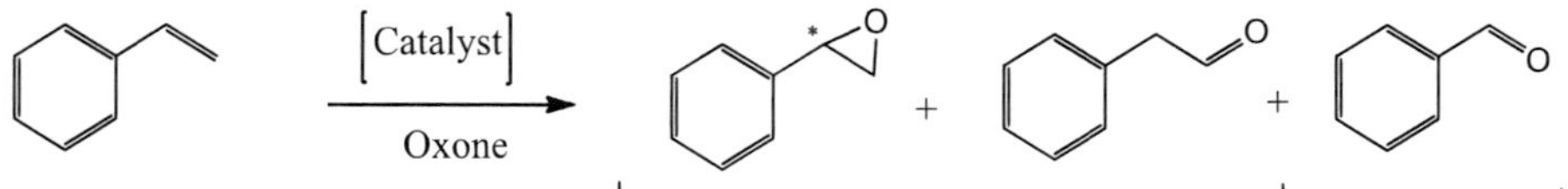

Catalyst	Yield	Yield	Yield	Total
1-Mn	22 %	66 %	2 %	90 %
1-Mn-7A3	21 %	63 %	1 %	85 %
2-Mn	12 %	6 %	2 %	20 %
2-Mn-7A3	9 %	9 %	2 %	20 %

Figure 28. Epoxidation and oxidation of styrene catalysed by the manganese cofactors and the artificial metalloenzymes formed by association with 7A3.

Artificial Metalloenzymes Built by the "Host-Guest" Strategy Involving the Non-covalent Insertion of Fe(III)-porphyrins into Xylanase A (Xyl A) from *Streptomyces Lividans*

Finally another strategy, the so-called "host–guest" strategy can be envisioned. It involves the covalent or non-covalent insertion of a guest metal cofactor inside the cavity of a non-related host protein chosen not only to bring a chiral environment around the cofactor, but also, in some occasions, for its particular thermodynamic properties: thermoresistance, work under a wide range of pH. Such a strategy has been already used for the non-covalent insertion of a rhodium complex inside human serum albumin (HSA), to build up a hybrid metalloprotein that is able to catalyze the regioselective hydroformylation of styrene with CO and H_2 (see first chapter). [12, 13] We have thus chosen, to build up new hemoproteins, a simple strategy which involves the non-covalent incorporation of Fe(III)-tetra-*p*-carboxyphenylporphyrin (Fe(TpCPP), Figure 29 a) ***3-Fe***into thermostable β-1,4-endoxylanase or xylanase A (Xyl A) from *Streptomyces lividans*, an enzyme available at low cost and in large quantities. Xylanases are glycoside hydrolases that hydrolyze β-1,4 bonds in the main chain of xylan.

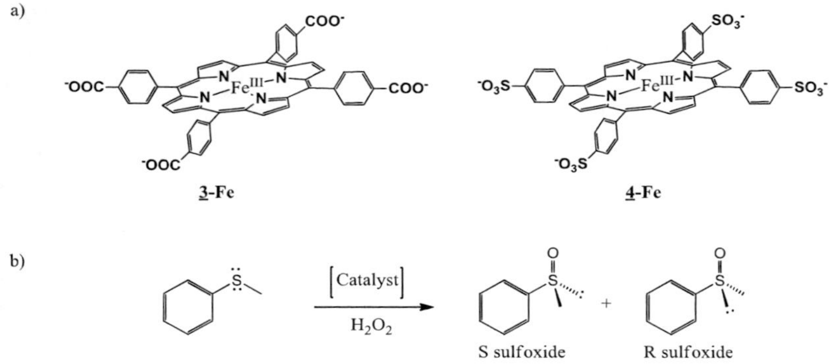

Catalyst	Yield	ee
3-Fe	33 %	0 %
3-Fe-XylA	24 %	24 % (S)
4-Fe	45 %	0 %
4-Fe-XylA	24 %	36 % (S)

Figure 29. a) Fe(III)-tetra-*p*-carboxyphenylporphyrin and *meso*-tetra(4-sulfonatophenyl) porphyrinatoiron(III) b) Catalysed sulfoxidation of thioanisole ans main results.

Table 3. Enantiomeric excess and configuration of the sulfoxide obtained by oxidation of thioanisole by hydrogen peroxide in the presence of various hybrid metalloprotein catalysts

Protein	Ligand	Type of binding	Enantiomeric excess	Configuration	Reference
Xln10A	FeTpSPP	Non-covalent	10%	S	[72]
Xln10A	FeTpCPP	Non-covalent	36%	S	[72]
Apo-myoglobine and mutants	Cr-Salen	Non-covalent	4–13%	S	[73]
	Mn-Salen	Covalent	51%	S	[74]
Anti-MP8-antibody, 3A3	MP8	Non-covalent	46%	R	[60]
Anti-estradiol antibody, 7A3	Fe-porphyrin-estradiol	Non-covalent	8%	S	[63]

Fe(T*p*CPP)-Xln10A was found to have peroxidase activity. [70] Enzymatic kinetic studies, as well as coordination chemistry studies suggested that steric hindrance and an increased hydrophobicity was brought by the protein around the iron atom of the porphyrin. In addition, a very interesting protecting effect against oxidative degradation of the porphyrin was provided by the protein. [70] These two properties made Fe(TpCPP)-Xln10A an excellent candidate as catalyst for the enantioselective oxidation of substrates such as sulfur containing compounds that play an important role in medicine and agriculture. The oxidation of thioanisole by H_2O_2 was thus performed in the presence of either ***3-Fe*** or *meso*-tetra(4-sulfonatophenyl)porphyrinatoiron(III) (FeTpSPP) ***4-Fe*** alone or of the corresponding Fe–porphyrin–Xln10A acting as a catalyst (Figure 29 a). [71, 72]

Hemozymes were obtained by the addition of 1 eq. ***3-Fe*** or ***4-Fe*** to 1.5 eq to Xln10A 2 µM in 50 mM citrate/phosphate buffer pH 3. Both ***3-*** and ***4-Fe***–Xln10A complexes were characterized by UV-visible spectroscopy in 50 mM citrate buffer pH 3. The spectrum of the ***3-Fe***–Xln10A complex was very similar to that of ***4-Fe*** alone, with maxima at 411, 565 and

609 nm, suggesting that no amino acid side-chain of Xln10A was coordinating the iron. On the contrary, whereas ***4-Fe*** exhibited a UV-visible spectrum characteristic of an high-spin pentacoordinate porphyrin-iron(III) species, with maxima at 394, 530 and 666 nm, ***4-Fe***–Xln10A was characterized by a very different UV-visible spectrum with maxima at 416, 552 and 619 nm, characteristic of a low-spin hexacoordinate porphyrin-iron(III). This clearly indicated that the environment of the iron atom changed upon insertion of ***4-Fe*** into Xln10A, which could be due to the binding of one or two amino acid side-chain of Xln10A to the iron atom.

To explain these results, molecular modeling analysis was performed. Both ***3-Fe*** and ***4-Fe*** have been docked into the Xln10A binding site with the program Gold (version 3.2), following a protocol described previously. [71] The main polar contacts occur between the negatively charged substituents of the phenyl groups of the ligands and the patches {His80, Lys48, Trp266} and {Ser212, Asn208, Asn173} of the receptor. [71] Owing to the larger size of the sulfonato substituents with respect to the carboxylato ones, ***4-Fe*** is buried less deeply than ***1-Fe*** in the Xyln10A binding pocket. In addition, in the case of ***4-Fe***, two residues, Serine 86 and the Histidine 85, though lying, respectively, at 6.19 and 6.35 Å from the iron ion, could coordinate it after a structural rearrangement of the protein, which could be the origin of the change in UV-visible spectrum observed when ***4-Fe*** was added to Xln10A.

The influence of the Xln10A protein on the yield and selectivity of the oxidation of thioanisole catalyzed by the ***3-Fe*** and ***4-Fe*** was then studied (Figure 29 b). The reactions with ***4-Fe*** and ***4-Fe***-Xln10A were performed at pH 3, since at this pH, ***2-Fe*** had been shown to be able to catalyze the oxidation of aromatic sulfides by H_2O_2. In 50 mM glycine pH 3, ***4-Fe*** as catalyst led to a 45 % yield (with respect to H_2O_2) and a turnover of 0.56 min^{-1} but no enantiomeric excess was observed. Insertion of ***4-Fe*** into Xln10A led to a decrease in the yield (24 %) and in the turnover (0.30 min^{-1}) but an enantiomeric excess of 24% in favor of the S sulfoxide was observed (Table 3). Both the lower yields and turnover and the induction of an enantiomeric excess were in agreement with the insertion of ***4-Fe*** into the cleft of Xln10A. The reactions with ***3-Fe*** and ***3-Fe***-Xln10A were performed at pH 7.4, since at this pH the para substituents of the *meso*-phenyl groups of ***1*** were predominantly under the carboxylate form. Once again, under these conditions better yields were obtained with ***3-Fe*** (33 %) than with ***3-Fe***-Xln10A (24 %) as well as better turnovers with ***3-Fe*** (0.41 min^{-1}) than with ***3-Fe***-Xln10A (0.30 min^{-1}). However, with ***3-Fe***-Xln10A as catalyst, an enantiomeric excess of 36 % in favor of the S sulfoxide was observed, that was larger than that observed with ***4-Fe***-Xln10A in 50 mM glycine pH 3. In agreement with the molecular modeling studies, this could be due to a deeper insertion of ***3-Fe*** than ***4-Fe*** into the cleft of Xln10A. When compared to the enantiomeric excesses already reported with other hybrid metalloproteins (Table 3), it is larger than those obtained with antibody 7A3-Fe-porphyrin-estradiol (8%) [63] and non-covalent Cr-salen-apoMyoglobin mutant (4–13%) [73] and compares with the highest ones obtained with antibody 3A3-MP8 (45%) [60] and covalent Mn-salen-apo myoglobin mutant complexes (51%). [74]

Conclusion

In conclusion, the results described above show that the association of iron(III)-meso-tetra-paracarboxylato-phenylporphyrin ***3-Fe*** with Xylanase A leads to a new "Hemozyme" that is able to catalyze the selective oxidation of sulfides by H_2O_2. Xylanase A not only protects the cofactor against self-oxidation, but also induces an enantiomeric excess in favor of the S sulfoxide (36%) that is one of the highest ever obtained for the oxidation of sulfides by hybrid metalloproteins.

GENERAL CONCLUSION

As we have seen in this chapter, the construction of artificial metalloenzymes is a growing field in bioinorganic chemistry. Many teams are working on building new artificial metalloenzymes or improving existing systems.

The aim of such association of a metallic cofactor with a protein is generally to take advantage of the natural asymmetry of the protein to induce a stereoselectivity in the reaction catalysed. From this point of view, several teams have succeeded using any of the three strategies. Enantiomeric excesses up to 66%were obtained after insertion of vanadate in the phytase of *Aspergillus ficuum,* [17] up to 94% by association of a biotinylated cofactor with streptavidin, [42] and up to 51% by covalent double linkage of a manganese salen into the cavity of an apo-myoglobin mutant. [35] The work done by Lu and coll. in this last example has clearly pointed out one essential condition to induce reproducible and noticeable stereoselectivity: the cofactor has to be firmly attached to the protein, so that the second coordination sphere of the metal is well defined. All the aforementioned results were obtained with such systems.

Another very important condition to obtain an artificial metalloenzyme catalysing reactions stereoselectively is that the cofactor has to be located in an encumbered enough environment. This has been well shown by the results obtained by Panella and coll. with papain. [3] The cofactor was covalently bound in the cavity of the protein, thus complying with the first condition, however, no stereoselectivity was observed in the reaction catalysed. This was explained by the largeness of the papain's cavity, the substrates having little or no interaction with the protein during catalysis. To our knowledge, no artificial metalloenzyme built with papain led to high enantioselectivities.

With these two rules in mind, our team has designed two kinds of new artificial metalloenzymes associating anti-steroid antibodies to steroid-based cofactors (high antigen-antibody affinity, narrow cavity)or XylanaseA to metalloporphyrins (size-fitted cavity).

As regards to the first system, the affinity of the antibody for the cofactors synthesized remained still quite good, but, owing to the length of the spacer between the steroid and the catalytic centre, this last seemed to stay on the edge of the cavity. This explained the rather low enantiomeric excesses measured. Better enantioselectivities could be obtained with a shorter spacer driving the porphyrin further into the cavity, providing the affinity remains good.

Good enantioselectivities were measured in the sulfoxidation reaction of thioanisole with the xylanase-porphyrin artificial metalloenzymes. The computer models showed that the porphyrins were indeed buried into the cavity of the protein. Moreover, the highest enantioselectivity was obtained with Fe(III)-tetra-*p*-carboxyphenylporphyrin, which was deeper inserted into the cleft of xylanase A than meso-tetra(4-sulfonatophenyl) porphyrinatoiron(III).

Once artificial metalloenzymes following the two above mentioned conditions have been built, they can be improved by two ways: chemically, by varying the cofactor and the metal, and biochemically, by selective mutagenesis of the host protein, to design the accommodating cavity.Several teams have modified their systems both ways (Ward and coll. [43,42], Watanabe and coll. [32,34,75], Lu and coll. [35]), always achieving better activities and enantioselectivities.

There is high hope that artificial metalloenzymes catalysing reactions with high efficiency and high stereoselectivity will be obtained after optimization, which could be used industrially to replace non environment-friendly catalysts.

REFERENCES

[1] Breuer, M.; Ditrich, K.; Habicher, T.; Hauer, B.; Kesseler, M.; Sturmer, R.; Zelinski, T. *Angew Chem Int Edit*2004, *43*, 788.

[2] Rouhi, A. M. *Chem Eng News* 2003, *81*, 45.

[3] Panella, L.; Broos, J.; Jin, J.; Fraaije, M. W.; Janssen, D. B.; Jeronimus-Stratingh, M.; Feringa, B. L.; Minnaard, A. J.; de Vries, J. G. *Chem Commun (Camb)*2005, 5656.

[4] Knowles, W. S. *Angew Chem Int Edit* 2002, *41*, 1999.

[5] Sharpless, K. B. *Angew Chem Int Edit* 2002, *41*, 2024.

[6] Noyori, R. *Angew Chem Int Edit* 2002, *41*, 2008.

[7] Lelais, G.; MacMillan, D. W. C. *Aldrichim Acta* 2006, *39*, 79.

[8] Letondor, C.; Pordea, A.; Humbert, N.; Ivanova, A.; Mazurek, S.; Novic, M.; Ward, T. R. *J Am Chem Soc* 2006, *128*, 8320.

[9] Anastas, P. T.; Zimmerman, J. B. *Environ Sci Technol* 2003, *37*, 94a.

[10] Peters, T. *All about albumin : biochemistry, genetics, and medical applications*; Academic Press: San Diego, 1996.

[11] Kokubo, T.; Sugimoto, T.; Uchida, T.; Tanimoto, S.; Okano, M. *J Chem Soc Chem Comm* 1983, 769.

[12] Bertucci, C.; Botteghi, C.; Giunta, D.; Marchetti, M.; Paganelli, S. *Adv Synth Catal* 2002, *344*, 556.

[13] Marchetti, M.; Mangano, G.; Paganelli, S.; Botteghi, C. *Tetrahedron Lett.* 2000, *41*, 3717.

[14] Crobu, S.; Marchetti, M.; Sanna, G. *J Inorg Biochem* 2006, *100*, 1514.

[15] van de Velde, F.; Konemann, L.; van Rantwijk, F.; Sheldon, R. A. *Chem Commun* 1998, 1891.

[16] Hemrika, W.; Renirie, R.; Dekker, H. L.; Barnett, P.; Wever, R. *P Natl Acad Sci USA* 1997, *94*, 2145.

[17] van de Velde, F.; Arends, I. W. C. E.; Sheldon, R. A. *J Inorg Biochem* 2000, *80*, 81.

[18] Okrasa, K.; Kazlauskas, R. J. *Chemistry* 2006, *12*, 1587.

[19] Ugarova, N. N.; Lebedeva, O. V.; Berezin, I. V. *J Mol Catal* 1981, *13*, 215.

[20] Levine, H. L.; Kaiser, E. T. *J Am Chem Soc* 1980, *102*, 343.

[21] Kaiser, E. T.; Lawrence, D. S. *Science* 1984, *226*, 505.

[22] Davies, R. R.; Distefano, M. D. *J Am Chem Soc* 1997, *119*, 11643.

[23] Qi, D. F.; Tann, C. M.; Haring, D.; Distefano, M. D. *Chem Rev* 2001, *101*, 3081.

[24] Reetz, M. T. *Tetrahedron* 2002, *58*, 6595.
[25] Reetz, M. T.; Rentzsch, M.; Pletsch, A.; Maywald, M.; Maiwald, P.; Peyralans, J. J. P.; Maichele, A.; Fu, Y.; Jiao, N.; Hollmann, F.; Mondiere, R.; Taglieber, A. *Tetrahedron* 2007, *63*, 6404.
[26] Nicholas, K. M.; Wentworth, P., Jr.; Harwig, C. W.; Wentworth, A. D.; Shafton, A.; Janda, K. D. *Proc Natl Acad Sci U S A* 2002, *99*, 2648.
[27] Zunszain, P. A.; Ghuman, J.; Komatsu, T.; Tsuchida, E.; Curry, S. *BMC Struct Biol* 2003, *3*, 6.
[28] Mahammed, A.; Gray, H. B.; Weaver, J. J.; Sorasaenee, K.; Gross, Z. *Bioconjugate Chem* 2004, *15*, 738.
[29] Mahammed, A.; Gross, Z. *J Am Chem Soc* 2005, *127*, 2883.
[30] Reetz, M. T.; Jiao, N. *Angew Chem Int Ed Engl* 2006, *45*, 2416.
[31] Hunter, C. L.; Lloyd, E.; Eltis, L. D.; Rafferty, S. P.; Lee, H.; Smith, M.; Mauk, A. G. *Biochemistry-Us* 1997, *36*, 1010.
[32] Ohashi, M.; Koshiyama, T.; Ueno, T.; Yanase, M.; Fujii, H.; Watanabe, Y. *Angew Chem Int Ed Engl* 2003, *42*, 1005.
[33] Ueno, T.; Koshiyama, T.; Ohashi, M.; Kondo, K.; Kono, M.; Suzuki, A.; Yamane, T.; Watanabe, Y. *J Am Chem Soc* 2005, *127*, 6556.
[34] Ueno, T.; Koshiyama, T.; Abe, S.; Yokoi, N.; Ohashi, M.; Nakajima, H.; Watanabe, Y. *J Organomet Chem* 2007, *692*, 142.
[35] Carey, J. R.; Ma, S. K.; Pfister, T. D.; Garner, D. K.; Kim, H. K.; Abramite, J. A.; Wang, Z.; Guo, Z.; Lu, Y. *J Am Chem Soc* 2004, *126*, 10812.
[36] Hayashi, T.; Hisaeda, Y. *Accounts Chem Res* 2002, *35*, 35.
[37] Hayashi, T.; Murata, D.; Makino, M.; Sugimoto, H.; Matsuo, T.; Sato, H.; Shiro, Y.; Hisaeda, Y. *Inorganic Chemistry* 2006, *45*, 10530.
[38] Hitomi, Y.; Mukai, H.; Yoshimura, H.; Tanaka, T.; Funabiki, T. *Bioorg Med Chem Lett* 2006, *16*, 248.
[39] Collot, J.; Gradinaru, J.; Humbert, N.; Skander, M.; Zocchi, A.; Ward, T. R. *J Am Chem Soc* 2003, *125*, 9030.
[40] Collot, J.; Humbert, N.; Skander, M.; Klein, G.; Ward, T. R. *J Organomet Chem.* 2004, *689*, 4868.
[41] Klein, G.; Humbert, N.; Gradinaru, J.; Ivanova, A.; Gilardoni, F.; Rusbandi, U. E.; Ward, T. R. *Angew Chem Int Ed Engl* 2005, *44*, 7764.
[42] Letondor, C.; Pordea, A.; Humbert, N.; Ivanova, A.; Mazurek, S.; Novic, M.; Ward, T. R. *J Am Chem Soc* 2006, *128*, 8320.
[43] Rusbandi, U. E.; Skander, M.; Ivanova, A.; Malan, C.; Ward, T. R. *Comptes Rendus Chimie* 2007, *10*, 678.
[44] Thomas, C. M.; Letondor, C.; Humbert, N.; Ward, T. R. *J Organomet Chem*2005, *690*, 4488.
[45] Cochran, A. G.; Schultz, P. G. *J Am Chem Soc* 1990, *112*, 9414.
[46] Kawamura-Konishi, Y.; Asano, A.; Yamazaki, M.; Tashiro, H.; Suzuki, H. *J Mol Catal B-Enzym* 1998, *4*, 181.
[47] Feng, Y.; Liu, Z.; Gao, G.; Gao, S. J.; Liu, X. Y.; Yang, T. S. *Enzyme Engineering Xii* 1995, *750*, 271.
[48] Takagi, M.; Kohda, K.; Hamuro, T.; Harada, A.; Yamaguchi, H.; Kamachi, M.; Imanaka, T. *Febs Lett* 1995, *375*, 273.

[49] Harada, A.; Fukushima, H.; Shiotsuki, K.; Yamaguchi, H.; Oka, F.; Kamachi, M. *Inorganic Chemistry* 1997, *36*, 6099.

[50] Quilez, R.; deLauzon, S.; Desfosses, B.; Mansuy, D.; Mahy, J. P. *Febs Lett* 1996, *395*, 73.

[51] Yamaguchi, H.; Tsubouchi, K.; Kawaguchi, K.; Horita, E.; Harada, A. *Chem-Eur J* 2004, *10*, 6179.

[52] De Lauzon, S.; Mansuy, D.; Mahy, J. P. *Eur J Biochem* 2002, *269*, 470.

[53] Liu, X. M.; Chen, S. G.; Feng, Y.; Gao, G.; Yang, T. S. *Enzyme Engineering Xiv*1998, *864*, 273.

[54] Nimri, S.; Keinan, E. *J Am Chem Soc* 1999, *121*, 8978.

[55] Ricoux, R.; Sauriat-Dorizon, H.; Girgenti, E.; Blanchard, D.; Mahy, J. P. *J Immunol Methods* 2002, *269*, 39.

[56] Aron, J.; Baldwin, D. A.; Marques, H. M.; Pratt, J. M.; Adams, P. A. *J Inorg Biochem* 1986, *27*, 227.

[57] Adams, P. A. *J Chem Soc Perk T 2* 1990, 1407.

[58] Ricoux, R.; Raffy, Q.; Mahy, J. P. *Comptes Rendus Chimie* 2007, *10*, 684.

[59] Ricoux, R.; Girgenti, E.; Sauriat-Dorizon, H.; Blanchard, D.; Mahy, J. P. *J Protein Chem* 2002, *21*, 473.

[60] Ricoux, R.; Lukowska, E.; Pezzotti, F.; Mahy, J. P. *Eur J Biochem* 2004, *271*, 1277.

[61] Delauzon, S.; Desfosses, B.; Moreau, M. F.; Trang, N. L.; Rajkowski, K.; Cittanova, N. *Hybridoma* 1990, *9*, 481.

[62] Girgenti, E. *PhD Thesis* 2003.

[63] Raffy, Q.; Ricoux, R.; Mahy, J. P. *Tetrahedron Lett.* 2008, *49*, 1865.

[64] Sansiaume, E.; Ricoux, R.; Gori, D.; Mahy, J. P. *Tetrahedron-Asymmetr* 2010, *21*, 1593.

[65] Raffy, Q.; Ricoux, R.; Sansiaume, E.; Pethe, S.; Mahy, J. P. *J Mol Catal A-Chem.* 2010, *317*, 19.

[66] DeLauzon, S.; Mansuy, D.; Mahy, J. P. *Eur J Biochem* 2002, *269*, 470.

[67] Lente, G.; Espenson, J. H. *New J Chem* 2004, *28*, 847.

[68] Lente, G.; Espenson, J. H. *Int J Chem Kinet* 2004, *36*, 449.

[69] Jiang, G.; Chen, J.; Thu, H. Y.; Huang, J. S.; Zhu, N.; Che, C. M. *Angew Chem Int Edit* 2008, *47*, 6638.

[70] Ricoux R.; Dubuc R.; Dupont C.; Marechal, J.-D; Martin, A.; Sellier, M.; Mahy, J.-P. *Bioconjugate Chem* 2008, *19*, 899

[71] Mahy, J.-P., Raffy, Q., Allard, M., Ricoux, R. *Biochimie* 2009, *91*, 1321.

[72] Ricoux, R.; Allard, M.; Dubuc, R., Dupont, C., Marechal, J.-D.; Mahy, J.-P. *Org Biomol Chem* 2009, *7*, 3208.

[73] Ohashi, M.; Koshiyama, T.; Ueno, T.; Yanase, M.; Fujii, H.; Watanabe, Y. *Angew Chem Int Ed* 2003, *42*, 1005.

[74] Carey, J. R.; Ma, S. K.; Pfister, T. D.; Garner, D. K.; Kim, H. K.; Abramite, J. A.; Wang, Z.; Guo Z.; Lu, Y. *J. Am Chem Soc* 2004, *126*, 10812.

[75] Ozaki, S.; Yang, H. J.; Matsui, T.; Goto, Y.; Watanabe, Y. *Tetrahedron-Asymmetry* 2006, *17*, 3192.

In: Focus on Catalysis Research: New Developments
Editors: Minjae Ghang and Bjørn Ramel

ISBN: 978-1-62100-455-4

Chapter 3

RECENT TRENDS IN POLYMER SUPPORTED CATALYSTS

G. Rajesh Krishnan and Krishnapillai Sreekumar*

Department of Applied Chemistry, Cochin University of Science and Technology, Cochin, Kerala, India

1. INTRODUCTION

A polymer supported catalyst is a reactive system in which the catalyst or more correctly the catalytic centre is attached to a polymeric molecule. The catalytic centre is either an inherent part of the polymer, which may be a part of the monomer and which will be added to the polymer during polymerization, or attached to a previously formed polymer by some kind of chemical (covalent bonds) or physical (encapsulation) methods. These two situations can be represented as shown in figure 1.

The polymer species of these types of catalysts may be a linear or a crosslinked entity, and again the latter have proved particularly useful. Polymeric catalysts are generally used in catalytic quantities relative to reaction substrates, and can often be reused many times. The attachment of a catalyst to a support may improve its stability and selectivity. On the other hand, increased experimental convenience arising with a polymeric catalyst may be offset by a significant reduction in reactivity associated, for example, with diffusional limitations imposed by resin supports. Recent interest in the development of environmentally benign synthesis has evoked a renewed interest in developing polymer-bound catalysts and reagents for organic synthesis that maintain high activity and selectivity. [1, 2] A wide variety of catalysts have been supported in this way, ranging from strong acids and bases (ion-exchange resins), transition-metal complexes, organic catalysts and photosensitizers right through to the highly specific enzyme catalysts.

* ksk@cusat.ac.in

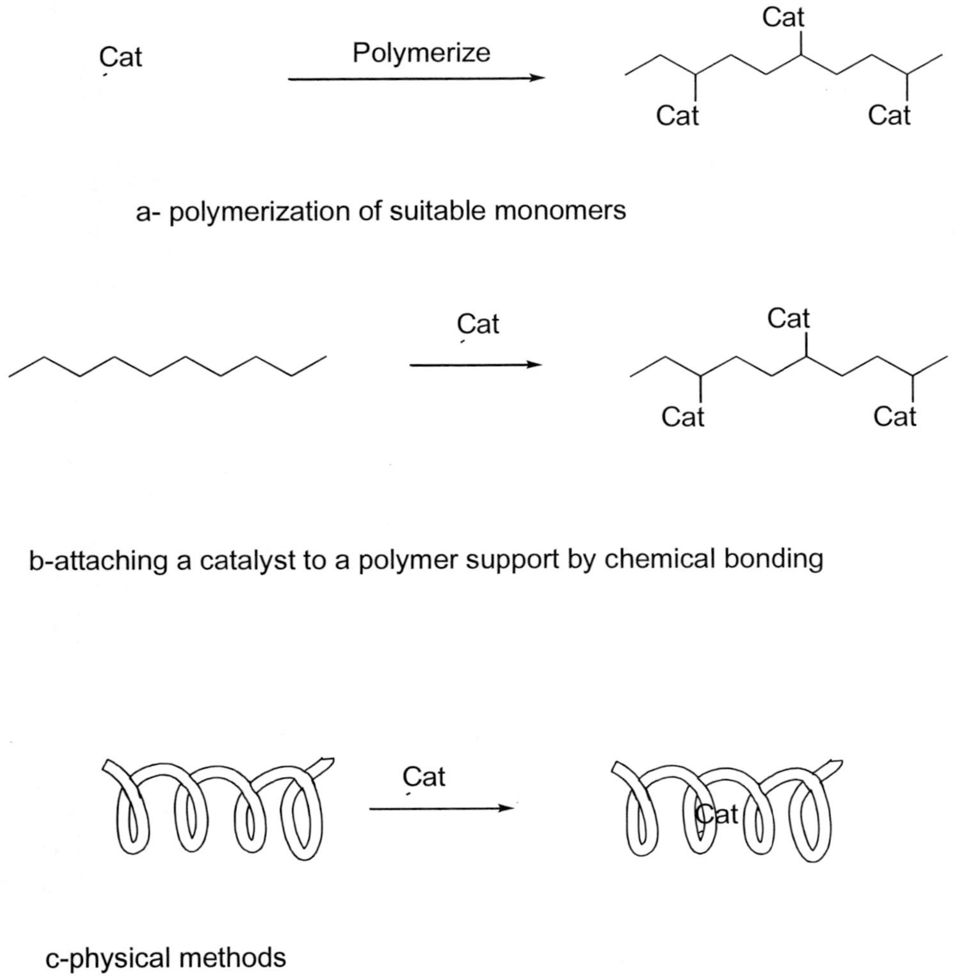

Figure 1. General types of polymer supported catalysts.

2. Advantages and Disadvantages of Polymer Supported Catalysts

Probably, the most important advantage in using a functionalized polymer as support of a catalyst is the simplification of product work-up, separation, and isolation. In the case of crosslinked polymer resins, simple filtration procedures can be used for isolation and washing, and the need for complex chromatographic techniques can be eliminated. Scarce and/or expensive materials can be efficiently retained when attached to a polymer and, if appropriate chemistry is available, they can in principle be recycled many times and this in metal-based catalysis means using precious metal catalysts several times. Mal-odorous and harmful substances can be handled easily by attaching them to polymers and they can be easily separated from the final product easily. In addition, polymer supported catalysts can be manipulated easily and large number of diversity can be introduced by techniques of combinatorial chemistry like mix and split method. [3] Due to these advantages, polymer supported catalysts are widely used in combinatorial library preparation and pharmaceutical industry. [4,5] Resins, in addition, provide the possibility of automation in the case of repetitive stepwise synthesis and the facility of carrying out reactions in flow reactors on a commercial scale. Moreover, immobilization of catalysts on an insoluble polymer has significant influence on the activity and selectivity. The reactivity of an unstable reagent or catalyst may be attenuated when supported on a resin, and the corrosive action of, for example, protonic acids can also be minimized by effective immobilization.

In addition to these factors, a number of potentially important reactivity changes may be induced by the use of a functionalized polymer. When the latter is crosslinked, restricted interaction of functional groups may be achieved. A high degree of cross-linking, a low level of functionalization, low reaction temperatures, and the development of electronic charges near the polymer backbone tend to encourage this situation, which may be regarded as mimicking the solution condition of "infinite dilution". In these circumstances, intermolecular reaction of bound molecules is prevented, and such attached residues can be made either to react intramolecularly or to react selectively with an added soluble reagent. Polymer supported metal complexes with vacant coordination sites can be regarded as fulfilling this description, with the resin inhibiting the normal solution oligomerization processes of such species. Under certain circumstances, it is also possible to achieve the complementary state of "high concentration" by heavily loading a flexible polymer matrix with one particular moiety in an attempt to force its reaction with a second polymer-bound species. Another advantage is due to the unique microenvironment created for the reactants within the polymer support. Improved catalyst stability within the polymer matrix, increased selectivity for intramolecular reactions, enhanced regioselectivity due to steric hindrance, and the superior activity of some supported chiral catalysts due to site cooperation have all been reported.

Balancing the above advantages, there are also a number of important disadvantages. Probably the most important of these is the likely additional time and cost in synthesizing a supported reagent or catalyst. This may well be offset by the potential advantages, and certainly in the case of regenerable and recyclable species, this objection essentially disappears. The occurrence of slow reactions and poor yields, however, can seldom be accommodated, and this can be a problem. Appropriate choice of support and reaction conditions can overcome these difficulties.

Eventhough, a large number of polymer supports were developed in the last fifty years, the most widely used support for the preparation of polymer-supported catalysts is polystyrene. The synthesis and applications of polymer-supported catalysts were widely reviewed. [6-20]

3. General Methods of Synthesis of Polymer Supported Catalysts

General methods of synthesis of polymer supported catalyst can be classified in to two. In the first method a ligand is attached to a polymer having a suitable functional group using suitable and efficient chemical reactions (figure 2). The polymer supported ligand is used for the preparation of catalytically active metal complexes. Polymer supported organocatalysts can also be prepared similarly by attaching an already prepared organocatalyst to a polymer support using ideal chemical reactions. Instead of attaching the preformed ligand to a polymer support, it can also be possible to do the solid phase synthesis of the ligand in many cases. In this case, the ligand is synthesized on the polymer support having suitable functional groups using solid phase synthesis strategies.

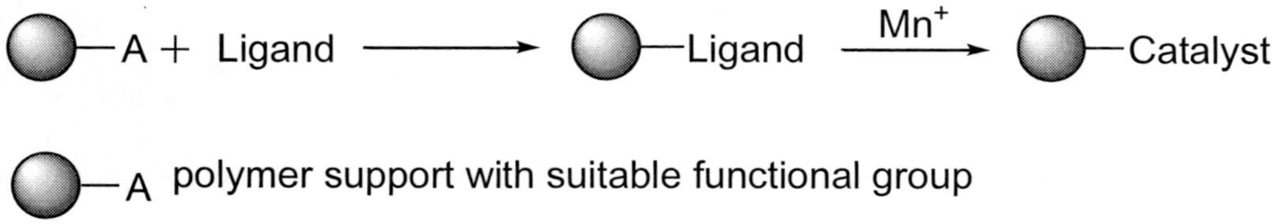

Figure 2. Attaching ligands to the polymer supports

Second method of preparation of polymer supported catalyst involves preparation of a ligand having polymerizable functional groups followed by its polymerization either alone or in the presence of other monomers (figure 3). In many cases, using a second monomer become very important for introduction of crosslinking as well as for reduction of the number of catalytic species on the polymer (crowding of catalytsts may reduce its activity). The polymer carrying the ligand so prepared can be used in the preparation of metal catalysts. Polymer supported organocatalysts can also be prepared using this method and it requires the design and synthesis of suitable monomers that carry both polymerizable functionality and organocatalytic functionality.

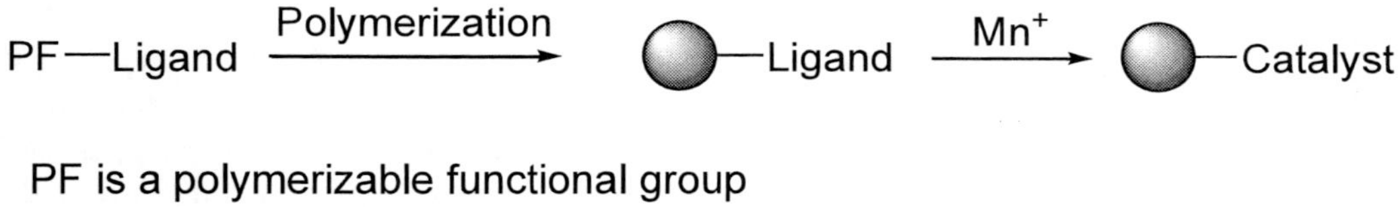

Figure 3. Polymerization of ligand to obtain polymer supported ligand

Both these methods can be used for the synthesis of polymer-nanoparticle composites for catalytic applications. For the preparation of such materials it requires, in many cases, preparation of polymer supported metal complexes followed by their reduction using suitable methods.

Both these methods have their own advantages and disadvantages. The first method is more common, because, there are a number of chemical reactions that can be used for efficient immobilization of ligands to the supports. Due to the rapid progress in combinatorial chemistry, a large number of polymer supports has became commercially available and almost all functional groups can be generated on polymer supports by suitable chemical transformation. So the number of possibilities is much higher in the first method. Added advantages are that, with the help of modern day analytical techniques, it becomes more easy to follow the progress of chemical reactions on polymer supports and there by optimize the conditions for maximum yield in the synthesis of polymer supported catalysts. [21] The disadvantages include finding a suitable reaction for immobilization is not easy always and in some cases poor yield and complex chemistry make the process tedious.

The second method is not a very common one because it is not always easy to prepare the ligand having polymerizable functional groups due to the highly reactive nature of such groups. On the other hand, due to the rapid development of controlled polymerization techniques, this method has found growing interest in preparing polymer supported catalyst having well defined structure and loading.

4. Classification of Polymer Supported Catalysts

Since the number of polymer supported catalysts is huge and it goes on increasing every year it is a very difficult task to classify polymer supported catalyst in one manner or other. Here these catalysts are classified according to the type of reaction they catalyze.

4. 1. Polymer-Supported Catalysts for Coupling Reactions

One of the most studied areas in polymer-supported catalysts is the development of novel catalysts for coupling reactions. Colacot et al. reported a novel polymer supported catalyst for Suzuki coupling reaction in which palladium complexes of commercially available polymer (FibreCat^{TM}-1001) containing triphenyl phosphine ligand was treated with palladium salts and the resulting catalyst was used for catalyzing Suzuki coupling. The reaction proceeded well within a short period of time compared to many previously reported catalysts. [22] There was no metal leaching observed, but the catalyst turned black after the first run and so no recycling studies were described. In the same year, Shieh *et al* developed a catalyst for Suzuki coupling reaction from triphenyl phosphine ligand supported on DVB crosslinked polystyrene and palladium chloride. The catalysts performed well under a given set of conditions and could be recycled many times with out considerable leaching. [23] Phan et al. used polystyrene supported salen type palladium catalyst for Suzuki coupling. The catalyst was so efficient that all the substrates attempted gave 100% yield and the catalyst was recyclable. [24] Another interesting example to mention is polystyrene supported palladium –N-heterocyclic carbene reported by Lee and co-workers. [25, 26] A short and versatile synthesis of reusable diarylphosphinopolystyrene-supported palladium catalyst for Suzuki coupling was described by Schweizer *et al.* [27] In this method, nucleophilic substitution of the chlorine atoms of a Merrifield resin by diarylphosphinolithium followed by introduction of palladium with a soluble palladium salt gave the catalyst. Ten different catalysts were prepared by varying the substitution on the aryl groups on the phosphino groups and all these catalysts were screened in the cross coupling between 4-bromoacetophenone and benzene boronic acid. All these catalysts gave quantitative yield with considerably lower amount of catalyst. A study by Betch and co-workers using various newly synthesized polymer supported Pd catalysts showed that the (*tert*-butylphenylphosphinomethyl) polystyrene-supported Pd catalyst was highly efficient for versatile Suzuki reactions from aryl chlorides (figure 4). These couplings were performed in the presence of low amounts (4 mequiv) of supported Pd and the catalyst can be reused more than seven times without loss of efficiency and the Pd leaching was found to be extremely low (<0.1% of the initial amount). [28]

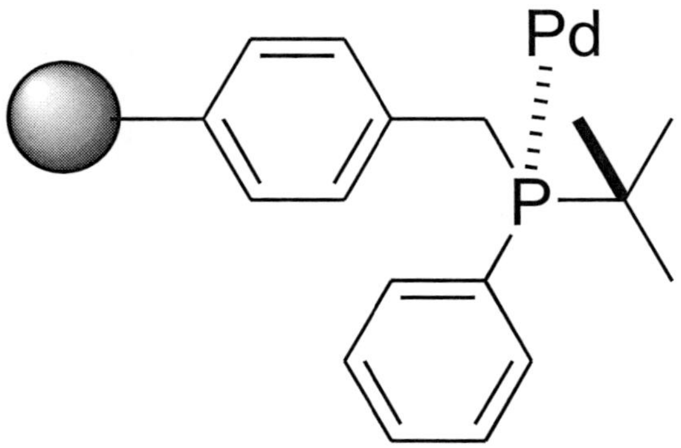

Figure 4. Polystyrene-DVB co-polymer supported Pd catalyst for Suzuki coupling reactions

Polymer supported Pd nanoparticles were also found to show prominent catalytic activity in many reactions especially Suzuki coupling and hydrogenation. Krishnan and Sreekumar showed that DVB crosslinked polystyrene supported poly(amidoamine) dendrimer stabilized Pd nanoparticles were efficient catalysts in Suzuki cross couplings between aryl halides and aryl boronic acids. The catalyst could be removed by simple filtration and could be reused a number of times without loss of activity. [29] Astruc and co-workers used polystyrene with suitable functional groups as ligands for the synthesis Pd nanoparticles. Complexation of $Pd(OAc)_2$ or K_2PdCl_4 to "click" polymers functionalized with phenyl, ferrocenyl and sodium sulfonate groups gave polymeric palladium(II)-triazolyl complexes that were reduced to "click" polymer-stabilized palladium nanoparticles (PdNPs). Water soluble poly(sodium sulfonate-triazolylmethyl)styrene was also used for the preparation of nanoparticles. These nanoparticles showed excellent catalytic activity, with relatively low loading of Pd, in both Suzuki coupling and hydrogenation. The structure of the catalyst is shown in figure 5. [30]

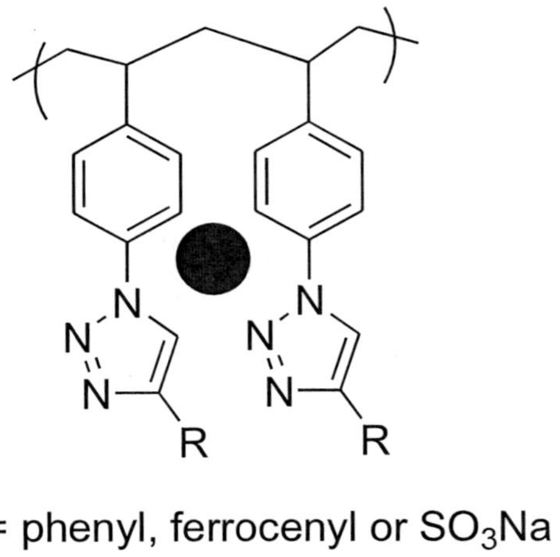

Figure 5. Polymer stabilized Pd nanoparticles for Suszuki couplings

A β-ketoester complex of palladium was prepared and used as a recyclable catalyst for Heck reaction between aryl iodides and bromides and alkenes by Dell'Anna *et.al.* [31] The catalyst was stable in air and water and gave good to excellent yields. Polymer supported N-methylimidazolium-palladium complex and polymer supported N-heterocyclic carbene-palladium complex were developed by Altaya *et.al.* [32] and Shokouhimehr *et.al.* [33] respectively for Heck reaction. The latter catalyst gave good yield with all the substrates attempted while the yield varied from 10%-100% in the case of the former. Polymer supported palladium catalyst for Stille coupling was reported by Dell'Anna *et.al.* [34] The coupling between aryl iodides and bromides with organostannanes has been investigated in the presence of a polymer supported palladium catalyst. The reaction could be performed in air without any activating ligand and with non-dried solvents. The catalyst, which acted by releasing controlled amounts of soluble active species, could be recycled several times in the coupling between $Sn(CH_3)_4$ or nBu_3SnPh with iodoarenes or activated bromoarenes. In a recent report, Bai and Wang used polymer-supported triphenyl phosphine-Pd complex as reusable catalyst for atom efficient coupling reaction of aryl halides with sodium tetraphenylborate in water under microwave irradiation to give polyfunctional biaryls in excellent yields (figure 6). [35] Polymer supported $PPh_2Pd(0)$ and Schiff base-Pd complexes also showed notable catalytic activity in Heck coupling and the catalyst could be recovered and reused with out loss of activity. [36,37]

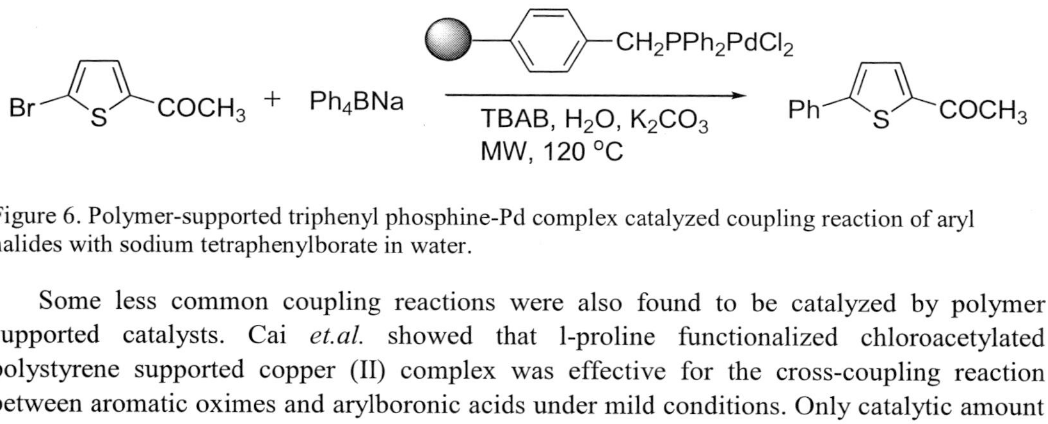

Figure 6. Polymer-supported triphenyl phosphine-Pd complex catalyzed coupling reaction of aryl halides with sodium tetraphenylborate in water.

Some less common coupling reactions were also found to be catalyzed by polymer supported catalysts. Cai *et.al.* showed that l-proline functionalized chloroacetylated polystyrene supported copper (II) complex was effective for the cross-coupling reaction between aromatic oximes and arylboronic acids under mild conditions. Only catalytic amount of catalyst was required and no deoximation occurred. The catalyst could be recovered by simple filtration and reused several times without significant loss of activity. [38]

4.2. Polymer Supported Catalysts for Other Carbon-Carbon Bond Forming Reactions

There are few examples of polymer-supported catalysts for aldol reaction in literature. Itsuno *et.al.* developed a chiral catalyst for asymmetric Mukaiyama aldol reaction. [39] Chiral *N*-sulfonylated α-amino acid monomer derived from (*S*)-tryptophan was copolymerized with styrene and divinylbenzene under radical polymerization conditions to give a polymer-supported *N*-sulfonyl-(*S*)-tryptophan. Treatment of the polymer-supported chiral ligand with 3,5-bis(trifluoromethyl)phenyl boron dichloride afforded a polymeric Lewis acid catalyst effective for asymmetric Mukaiyama aldol reaction of silyl enol ethers and aldehydes. Various aldehydes were allowed to react with silyl enol ethers in the presence of the polymeric chiral Lewis acid to give the corresponding aldol adducts in high yield with high levels of enantioselectivity. Catalytic efficiency of polymer-supported sulphonic acid in cross aldol condensation between arylaldehydes and cycloketones to afford α,α'-bis(substituted benzylidene)cycloalkanones was shown by An *et.al.* [40]

Madhavan and Weck described polymer-supported (*R,R*)-(salen)AlCl complexes that were immobilized on poly(norbornene)s that displayed excellent activities and enantioselectivies as catalysts for the 1,4-conjugate addition of cyanide to α.β-unsaturated imides (figure 7). [41] These supported catalysts could be recycled up to 5 times without compromising catalyst activities or selectivities. Furthermore, the catalyst loadings could be reduced from 10-15 mol%, the common catalyst loadings for non-supported (salen)Al catalysts, to 5 mol%, a decrease of metal content by 50-66%, without lowering product yields or enantioselectivities. Kinetic studies indicated that the polymer-supported catalysts were significantly more active than their corresponding unsupported analogues.

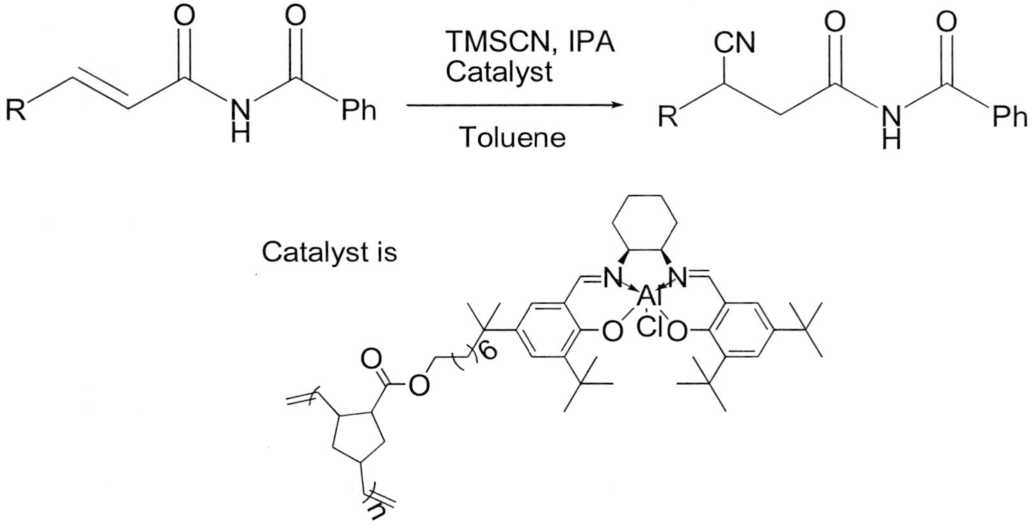

Figure 7. Polymer supported Al catalysts for the 1,4-conjugate addition of cyanide to α,β-unsaturated imides.

Liu and Jiang showed that *Janda Jel* supported tertiaryphosphine was an efficient green organocatalyst for α-addition of carbon nucleophile to α,β-unsaturated compounds under mild conditions. [42] Trost and co-workers reported polymer supported C_2-symmetric ligands for palladium catalyzed asymmetric allylic alkylation reactions. [43] In the presence of 25 mol% of ligand and 5 to 10 mol% palladium, the reaction proceeded with up to 97% yield and 68-99% ee.

A novel polymer supported BisBinol ligand was prepared by Sekiguti *et.al.* for the preparation of bimetallic catalysts for various organic transformations. [44] The polymer was prepared by co-polymerization of MMA and a monomer containing the BisBinol moiety in the presence of AIBN. A bimetallic complex of this polymeric ligand with Al and Li was found to be efficient catalyst in the formation of Michael adduct between 2-cyclohexen-1-one with dibenzyl malonate with up to 96% ee. The same ligand was used in the preparation of a bimetallic titanium catalyst, which promoted the asymmetric carbonyl-ene reaction of methyl glyoxylate with α-methyl styrene in high yield with excellent enantioselectivity.

A novel polymer-supported *N*-heterocylic carbene (NHC)–rhodium complex was prepared from chloromethyl polystyrene resin using a simple procedure by Yan *et.al.* [45] This polymer-supported NHC–rhodium complex was used as a catalyst for the addition of arylboronic acids to aldehydes affording arylmethanols in excellent yields.

Rajagopal *et.al.* used poly(ethylene glycol) monomethyl ether as a soluble support while *J*anda*J*el and Merrifield resins as insoluble supports for attaching chiral Cu(salen) complex through a glutarate spacer (figure 8). Various ketones underwent asymmetric trimethylsilylcyanation at room temperature with $(CH_3)_3SiCN$ in the presence of these supported Cu(salen) complexes and Ph_3PO as the catalysts. Aromatic, aliphatic, and heterocyclic ketones have been converted into the corresponding cyanohydrin trimethylsilyl ethers in 83−96% yields with 52−84% ee. A double activation where Cu(salen) plays the role of Lewis acid and Ph_3PO acts as a Lewis base was assumed to be the driving factor for catalysis. The soluble catalysts were recovered by precipitation with a suitable solvent while the insoluble catalysts were simply filtered from the reaction mixture and were reusable. Comparative studies showed that *J*anda*J*el-attached catalyst could be used for five cycles with the retention of efficiency while the Merrifield supported one lost activity with each use. [46]

Polymer-supported chiral β-hydroxy amides and C_2-symmetric β-hydroxy amides were found to be highly efficient catalysts for enantioselective addition of phenylacetylene to aldehydes. The products were obtained in high yields (up to 93%) and enantioselectivities (up to 92% ee). Among the catalysts, polymer-supported chiral β-hydroxy amide showed better activity and selectivity. The catalyst can be recovered and reused at least four times, without considerable loss of enantioselectivity. The catalyst was suitable not only for aromatic aldehydes but also for aliphatic aldehyde. [47]

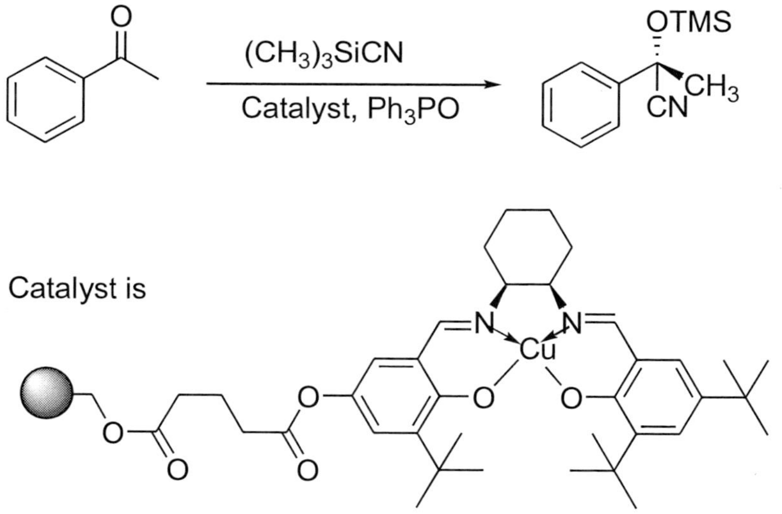

Figure 8. Polymer supported Cu(salen) catalyst for trimethylsilyl cyanation.

Alza and Pericas prepared a polymer-supported α,α-diarylprolinol silyl ether that showed catalytic activity and enantioselectivity comparable to the best homogeneous catalysts in the Michael addition of aldehydes to nitroolefins (figure 9). The peculiarity of the catalyst is that it showed unprecedented substrate selectivity in favor of linear short-chain aldehydes and this selectivity was assumed to be due to the combinational effects of polymer backbone, triazole linker, and the catalytic unit. The catalyst is shown in figure 9. [48]

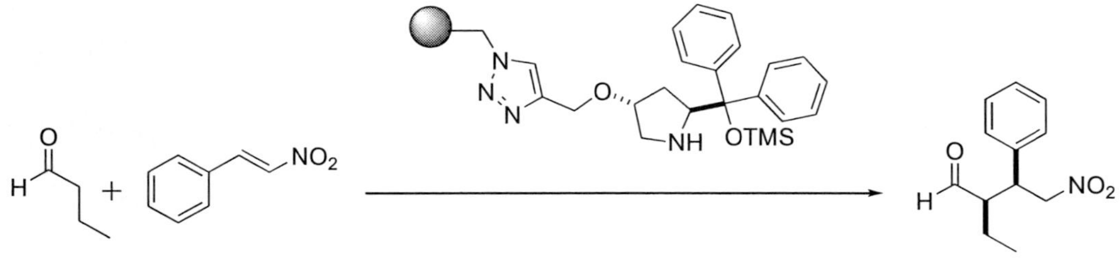

Figure 9. Polymer supported silyl ether catalyst for Michael addition.

Krishnan and Sreekumar reported the organocatalytic property of poly(amidoamine) dendrimers supported on DVB crosslinked polystyrene in Knoevenagel condensations. The polymer supported dendrimer was able to catalyze the synthesis of substituted olefins by the condensation of carbonyl compounds and active hydrogen compounds. The beauty of the catalytic system was that the products were obtained in high yield (100% in many cases). The catalyst could be easily recovered and reused at least for ten times without loss of activity. The catalyst was compatible to a number of substrates and the reaction could be performed in water or ethanol contrary to many homogeneous catalysts, where the reaction was usually performed in aromatic solvents. [49]

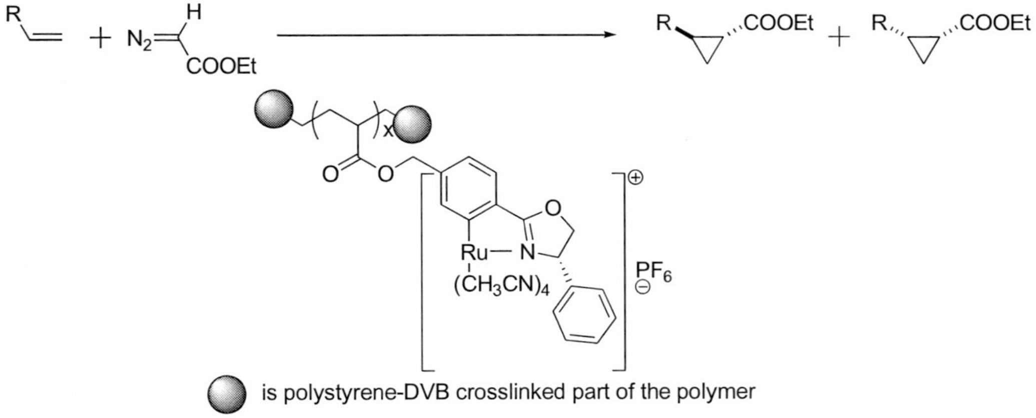

Figure 10. Cycloporpanation catalyzed by polymer supported Ru complex.

Iwasa and co-workers prepared a macroporous polymer by AIBN catalyzed copolymerization of styrene, DVB and 4-((S)-4,5-dihydro-4-phenyloxazol-2-yl)benzyl acrylate. This copolymer was used as the ligand in the preparation of a novel polymer supported Ru complex capable to catalyze enantioselective inter and intra molecular cyclopropanation (figure 10). The catalyst was active to a broad range of substrates. It could be recovered by filtration and could be reused for eleven cycles without considerable loss of activity and selectivity. The authors also prepared the same catalyst by copolymerization of styrene, DVB and Ru complex of 4-((S)-4,5-dihydro-4-phenyloxazol-2-yl)benzyl acrylate. This offers additional advantage of preparing the catalyst with tunable catalyst loading by changing the amount of metal complex used in polymerization. [50]

4.3. Polymer-Supported Catalysts for Oxidation and Epoxidation

An important type of polymer-supported catalyst developed to maturity is catalyst for oxidation and epoxidation. Polymer supported oxidation catalysts attain great attention because of their resemblance with enzymatic catalysts in action. [51, 52]

Divinyl benzene crosslinked polystyrene supported β-diketone linked complexes of Mn(II) have been prepared, characterized and used as heterogeneous catalyst in the oxidation of secondary alcohols to ketones in the presence of $K_2Cr_2O_7$ as oxidant. [53]

The preparation of a series of polymer supported chiral Schiff bases derived from salicylaldehyde and optically active amino alcohols were reported by Barbarini *et.al.* [54] These heterogeneous ligands have been complexed with $VO(acac)_2$ and employed to catalyze the enantioselective oxidation of sulfides to sulfoxides with hydrogen peroxide as an environmentally acceptable oxidant (figure 11). The procedure afforded the sulfoxides in good yield and selectivity and with enantiomeric excess values comparable to those obtained with the homogeneous counterparts. The catalyst can be used for at least four cycles without any significant decrease in both efficiency and enantioselectivity.

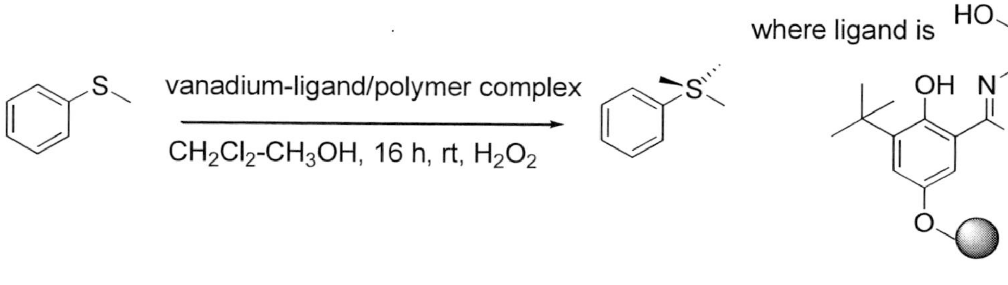

Figure 11. Polymer supported chiral Schiff base-vanadium complex catalyzed oxidation of sulphides.

Anchoring of the amino acid L-valine on chloromethyl polystyrene in the presence of a base gave a fresh ligand for polymer-supported catalyst as described by Valodkar and co-workers. [55] Reaction of cupric acetate with this polymeric ligand resulted in chelate formation with Cu(II) ion. The supported Cu(II) complexes behaved as versatile catalysts in the oxidation of various substrates such as benzyl alcohol, cyclohexanol and styrene in the presence of *t*-butyl hydroperoxide as oxidant. The effect of reaction conditions on conversion and selectivity to products has been studied in detail. Preliminary kinetic experiments revealed that the Cu(II) complexes attached to polymer matrix can be recycled about four times with no major loss in activity. Kang et al. used polystyrene supported 1,10-phenanthroline as the ligand for the development of a new ruthenium based oxidation catalyst. [56] 5-amino-1,10-phenanthroline prepared according to a standard procedure was attached to DVB crosslinked chloromethyl polystyrene. The polymer-supported metal complex prepared from this ligand and $RuCl_3$ was used as heterogeneous catalyst in the oxidation of primary and secondary alcohols in the presence of iodosyl benzene as oxidizing agent. The conversion was from moderate to excellent. The catalyst was recycled five times, but after the third cycle the efficiency was lost considerably. Benaglia and co-workers prepared a new oxidation catalyst by anchoring TEMPO on PEG chains. [57]This system was used as catalyst in the aerobic oxidation of alcohols in the presence of $Co(NO_3)_2$ or $Mn(NO_3)_2$ as co-catalyst. Primary and benzylic alcohols were converted to carbonyl compounds with an yield up to 99%. The authors showed that introduction of a spacer between the catalyst and the polymer considerably increased the activity of the catalyst. The catalyst was recycled six times with out loss of activity. Lie *et al* used polystyrene supported 2-iodobenzamide as an efficient organic catalyst for the oxidation of alcohols. [58] The catalyst prepared from aminomethyl polystyrene and 2-iodobenzoic acid in a single step was used in the oxidation of both primary and secondary alcohols in CH_3CN-H_2O mixture and the oxidizing agent selected was Oxone ($2KHSO_5 \cdot KHSO_4 \cdot K_2SO_4$). The conversion was moderate to excellent and the catalyst was reused four times without any loss of activity.

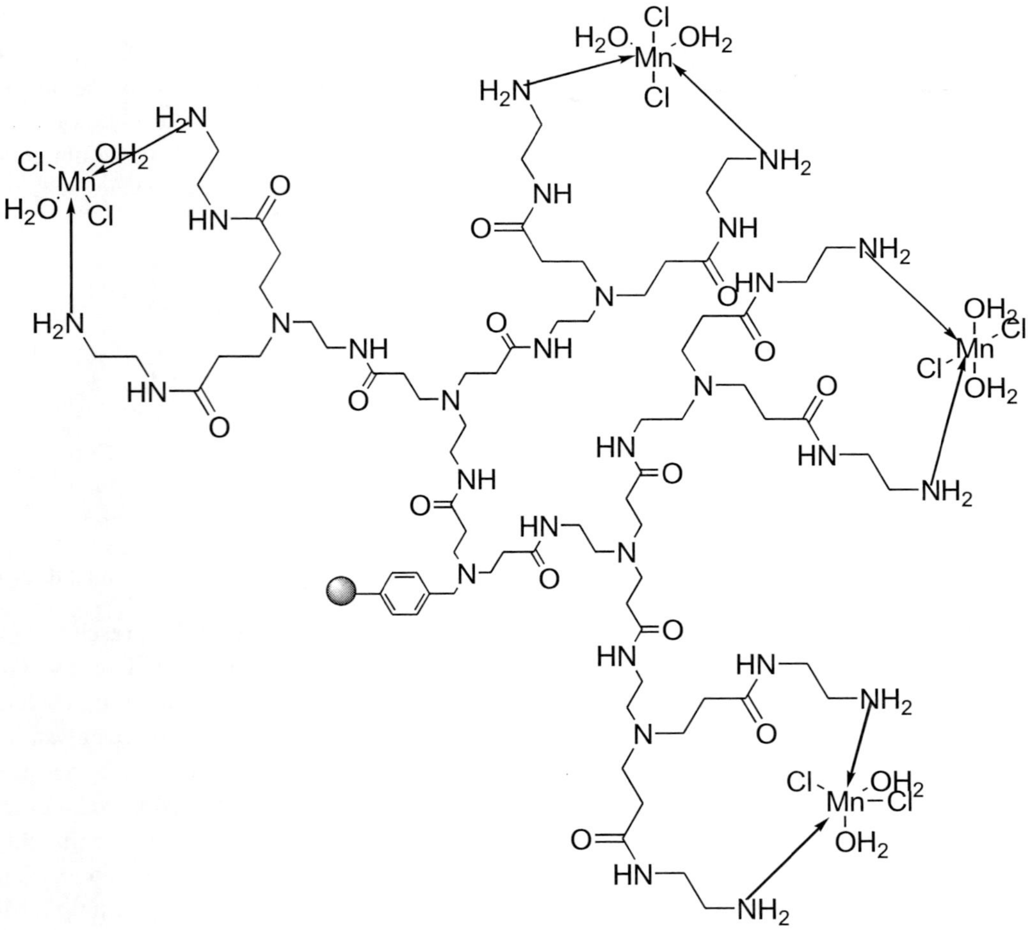

Figure 12. Polymer supported PAMAM-Mn complex used as catalyst for oxidation of alcohols.

Krishnan and Sreekumar used Mn complex of poly(amidoamine) supported on lightly crosslinked polystyrene as a reusable catalyst for oxidation of alcohols (figure 12). The catalyst was compatible to both inorganic oxidants like $K_2Cr_2O_7$ and $KMnO_4$ and green oxidant urea-hydrogen peroxide complex. Since the ligand used was a dendrimer, the catalyst showed some selectivity for smaller substrates. So it effectively catalyzed oxidation of small alcohols like cyclohexanol to corresponding ketones while failed in the case of larger molecules like cholesterol. The catalyst could be easily recovered and reused six times without loss of efficiency. [59]

Ruthenium and cobalt complexes of polymer supported Schiff base were prepared by Trakampruk and Kanjina. Schiff base ligand was prepared by reacting chloromethylated poly(styrene-divinyl-benzene) with 4-hydroxybenzaldehyde, followed by condensation with 2-aminopyridine. The metal complexes were prepared by the reaction of the polymer supported ligand and corresponding metal salt in methanol. The metal complexes acted as efficient catalysts in the oxidation of alcohols and ethylbenzene in the presence of *tert*-butyl hydroperoxide (*t*-BuOOH) as the oxidant. The catalyst was active with other oxidants like H_2O_2 and iodosylbenzene, but the yield was low under the same reaction conditions. Among the complexes reported the Ru complex showed better activity than the Co complex. [60]

Another polymer supported Schiff base metal complex as oxidation catalyst was reported by Gupta and Sutar. Iron(III), copper(II) and zinc(II) complexes were prepared from corresponding metal salts and polymer supported *N,N'*-bis (*o*-hydroxy acetophenone) hydrazine Schiff base. The catalytic activity of unsupported and polymer supported Schiff base complexes of metal ions was evaluated by studying the oxidation of phenol and epoxidation of cyclohexene. The polymer supported metal complexes showed better catalytic activity than unsupported metal complexes. The catalytic activity of metal complexes was optimum at a molar ratio of 1:1:1 of substrate to oxidant and catalyst. The selectivity for catechol and epoxy cyclohexane in oxidation of phenol and epoxidation of cyclohexene was better with polymer supported metal complexes in comparison to unsupported metal complexes. The polymer supported iron(III) complexe was found to be a better catalyst than the supported of copper(II) and zinc(II) complexes. [61]

Polymer-supported copper complexes were used as catalysts for the catalytic wet hydrogen peroxide oxidation (CWPO) of phenol by Bengoa *et.al.* Both polybenzimidazole resin and poly(styrene-divinylbenzene) resins were used for the preparation of supported Cu(II) complexes. Comparison of the catalytic activity of these catalysts in oxidation of phenol in batch stirred tank reactor showed that the highest phenol conversion was obtained for polystyrene supported catalysts. But metal leaching was also high for the same. This study has dealt with important factors that are useful in transforming polymer supported catalysts from laboratory scale to industrial scale. [62]

Jian *et.al.* used poly(phthalazinone ether sulfone ketone) supported heteropoly anion $[PW_9O_{34}]^{9-}$ in liquid-phase oxidation of benzyl alcohol to benzaldehyde with hydrogen peroxide as oxidant. The supported catalyst exhibited high catalytic activity under mild reaction conditions and is reusable. [63]

Sreekumar and co-workers have reported a number of polymer supported catalysts for epoxidation of olefins. [64-67] Metal complexes of polymer supported β-diketones were used effectively as heterogeneous catalysts in the epoxidation of various olefins by H_2O_2. Both DVB crosslinked polystyrene and poly(methyl methacrylate) were used as supports. Better results were obtained with poly(methyl methacrylate) supported catalyst under similar conditions.

Saldino *et.al.* reported a novel catalyst for the epoxidation of alkenes by H_2O_2. Methylrhenium trioxide (CH_3ReO_3, MTO) was supported on poly(4-vinyl pyridine) 25% crosslinked with divinylbenzene and on poly(4-vinyl pyridine-N-oxide) 2% crosslinked with DVB. In addition, microencapsulated MTO with polystyrene 2% crosslinked with DVB or a mixture of PS and PVP (both 2% crosslinked with DVB) were also studied. A detailed characterization of the catalyst was done with FTIR, SEM and WAXS which revealed a distorted, octahedral rhenium coordination geometry in the supported catalyst and the reticulation grade of polymer was shown to be linked to surface morphology. [68] All these catalysts showed good activity and the yield of the epoxide was above 90% in many cases. Moreover, the formation of the diol as a side product from epoxide was diminished considerably. The same group also reported a convenient and efficient synthesis of monoterpene epoxides by application of heterogeneous poly(4-vinyl pyridine)/methylrhenium trioxide (PVP/MTO) and polystyrene/methylrhenium trioxide (PS/MTO) systems. [69] Even, highly sensitive terpenic epoxides were obtained in excellent yield. The epoxidation was carried out with H_2O_2 and the catalysts were stable systems for at least five recycling experiments.

Lu and co-workers attached dimeric cinchonine, cinchonidine, and quinine via nitrogen to long linear PEG chains to afford soluble polymer-supported chiral ammonium salts, which were employed as phase-transfer catalysts in the asymmetric epoxidation of chalcones. [70] The highest enantiomeric excess obtained was 86%. The structures of the catalysts are shown in figure 13

Polymer-supported $Co^{II}LCl_2$ {L = 2-(alkylthio)-3-phenyl-5-(pyridine-2-ylmethylene)-3,5-dihydro-4*H*-imidazole-4-one} complex has been synthesized and employed as a catalyst for the epoxidation of alkenes using iodosylbenzene and hydrogen peroxide as oxidants by Beloglazkina *et.al.* [71] The epoxides were obtained in 10 to 85% yield.

4.4. Polymer-Supported Catalysts for Ring Opening Reactions

β-amino alcohols have found wide-spread applications in pharmaceutical synthesis and fine chemical industry. The most successful method for their synthesis is the nucleophilic ring opening of epoxides by amines in the presence of a catalyst. Bandini et al. developed a polymer-supported Indium Lewis acid catalyst for the ring opening of epoxides to give β-amino alcohols. [72] This new heterogeneous Amberlyst 15/Indium complex effectively catalyzed (20 mol % based on indium) the formation of new C-C as well as C-S bonds through the highly regio and stereoselective ring-opening reaction of enantiomerically pure epoxides. The easily prepared Amberlyst 15/Indium Lewis acid did not require inert atmosphere conditions or anhydrous media and could be easily recovered and recycled for several times without loss of activity.

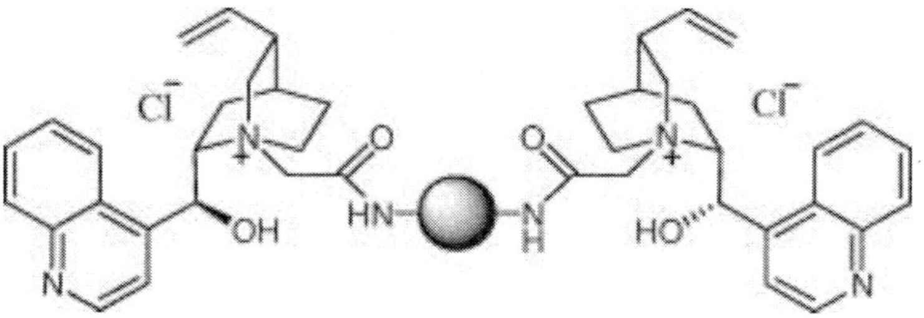

Diacetamido-PEG2000 N-bound cinchoninium chloride

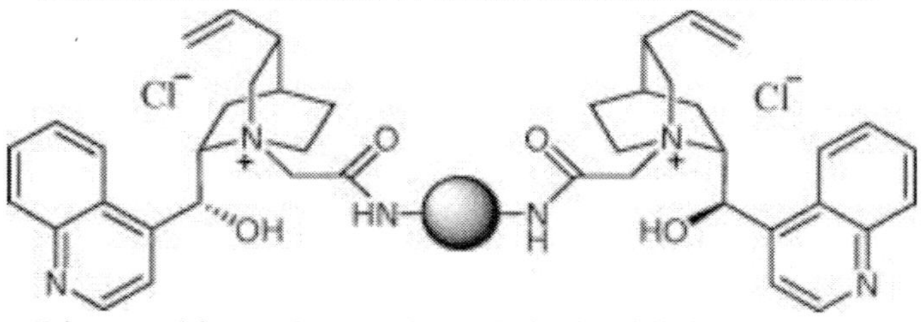

Diacetamido-PEG_{2000} N-bound cinchonidinium chloride

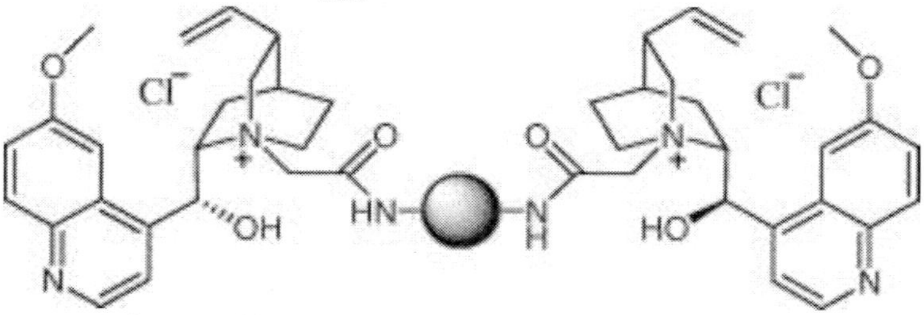

Diacetamido-PEG_{2000} N-bound quininium chloride

Figure 13. Polymer supported chiral ammonium salts used as phase transfer catalysts.

Polymer supported copper sulphate was used as heterogeneous catalyst for the ring opening of epoxides with amines. An alkylamine functional, insoluble and crosslinked polymer support was prepared by the reaction of a glycidyl group containing polymer with ethylenediamine. Copper complex of this polymer was prepared. This complex could be used as a heterogeneous catalyst to afford a rapid, efficient and mild method for synthesis of β-amino alcohols by aminolysis of epoxides. The reaction proceeded in a short time and the yield of the amino alcohols varied from 60 to 100%. [73] Lee and co-workers reported another novel catalyst for ring opening of epoxides. Polymer-supported metal (Fe or Ru) complexes for epoxide ring opening reactions were successfully prepared by anchoring the bis(2-picolyl)amine ligand onto the polymer, poly(chloromethylstyrene- co-divinylbenzene); they showed heterogeneous catalytic activity and easy recyclability in the ring opening reaction of various epoxide substrates with methanol or H_2O at room temperature under mild and neutral conditions. The catalyst was recycled ten times with out loss of activity. [74]

Divinyl benzene crosslinked polystyrene supported first to third generation poly(amidoamine) dendrimers were used as heterogeneous organocatalysts by Krishnan and Sreekumar in the synthesis of β-amino alcohols by ring opening of epoxides using anilines. Study on the influence of the crosslinking of support and the generation of dendrimers showed that third generation dendrimer attached to lightly crosslinked resin showed better activity. The β-amino alcohols were obtained in high yield and the catalyst can be recovered by simple filtration and reused four times without loss of activity. [75]

4.5. Polymer-Supported Catalysts for Reduction

A number of polymer-supported catalysts for reduction and hydrogenation of functional groups can be found in the literature. Saluzzo *et.al.* reported polymer supported catalysts for asymmetric heterogeneous reduction of a C=O bond by hydrogen transfer or molecular hydrogen. [76] Hydrogen transfer reduction was performed with amino alcohol derivatives of enantiopure poly((S)- glycidyl methacrylate-co-ethyleneglycol dimethacrylate) (poly(GMA-co-EGDMA)) and poly((S)-glycidyl methacrylate- co-divinyl benzene) (poly(GMA-co-DVB)) for solid–liquid catalysis. Finally, hydrogenation with molecular hydrogen was employed both in solid–liquid and in liquid–liquid biphasic catalysis. The first one was performed with BINAP grafted onto a polyethylene glycol and the second one with polyureas containing the BINAP structure. All the catalysts gave excellent yield with more than 99% enantiomeric excess. *N*-Methyl-α,α-diphenyl- L -prolinol derivatives with *para*-bromo substituents in one or both of the phenyl rings are easily bound to crosslinked polystyrene beads containing phenylboronic acid residues using Suzuki reaction. When the products were used as catalysts for the reaction of aldehydes with diethylzinc in toluene at 20 °C, the alcohols were produced in chemical yields >90% and with ees of up to 94%. The best of the two supported catalysts gave ees only 0–9% lower than those obtained with the corresponding soluble catalyst. One of the supported catalysts was recycled successfully nine times. [77] The same catalysts were used to catalyze reductions of several prochiral ketones with borane in tetrahydrofuran at 22 °C. The expected secondary alcohols were obtained in high chemical yields and ees were generally in the range 79–97 %. The catalyst was recycled 14 times without loss of stereochemical performance. [78]

Zarka *et.al.* reported the synthesis of new amphiphilic block copolymers with (2S, 4S)-4-diphenylphosphino-2- (diphenylphosphinomethyl) pyrrolidine (PPM) units in the side chain and their application in the asymmetric hydrogenation. [79] The polymers prepared were used as macroligands for the rhodium catalyzed hydrogenation of two prochiral enamides, acetamido cinnamic acid and its methyl ester. Additionally, catalyst recovery and reuse was possible by simple extraction of the substrate/product from the aqueous polymer phase after each cycle.

Attaching a homogeneous rhodium catalyst to a fluoroacrylate copolymer backbone developed a novel hydrogenation catalyst soluble in supercritical carbon dioxide. [80] The polymer was synthesized by the polymerization of 1*H*,1*H*,2*H*,2*H*-heptadecafluorodecyl acrylate monomer (zonyl TAN) and *N*-acrylosuccinimide (NASI), the former increasing the solubility in supercritical carbon dioxide and the latter providing attachment sites for the catalyst. Diphenylphosphinopropylamine, $NH_2(CH_2)_3PPh_2$ (DPPA), was used to exchange the NASI groups in the polymer, which was then reacted with $[RhCl(COD)]_2$ to obtain the catalyst. The catalyst was soluble in supercritical carbon dioxide and its hydrogenation activity was evaluated using 1-octene and cyclohexene hydrogenation as model reactions. The synthesis route for the catalyst was reproducible, as shown by reaction activity studies on different batches of catalyst. The catalyst was evaluated at different substrate-to-rhodium molar ratios and at different temperatures.

Two polymer-supported chiral ligands were prepared based on Noyori's (1*S*,2*S*)- or (1*R*,2*R*)-*N*-(*p*-tolylsulfonyl)-1,2-diphenylethylenediamine by Li et al. [81] The ligand prepared by the standard procedure was coupled to aminomethyl polystyrene in the presence of DCC. The combination of this supported ligand with $[RuCl_2(p\text{-cymene})]_2$ has shown to exhibit high activities and enantioselectivities for heterogeneous asymmetric transfer hydrogenation of aromatic ketones with formic acid–triethylamine azeotrope as the hydrogen donor, whereby affording the respective optically active alcohols, the key precursors of chiral fluoxetine, a drug used for the treatment of depression and anxiety. The catalysts could be recovered and reused in three consecutive runs with no significant decline in enantioselectivity. The procedure avoided the plausible contamination of fluoxetine by the toxic transition metal species.

Amphiphilic PS-PEG resin dispersed palladium nanoparticles were used as catalysts for the hydrogenation of olefins and hydrodechlorination of chloroarenes by Nakao and co-workers. [82] In many cases of the substrates attempted, up to 99% yield was obtained in both reactions. The catalyst was recycled ten times with out considerable loss of activity. Polymer supported Ni-B nanoparticles prepared by ion-exchange-chemical reduction method showed good activity in catalytic transfer hydrogenation of aromatic nitrocompounds with hydrazine hydrate as hydrogen donor. The catalyst could be recovered and reused. The reusability experiments showed that the catalyst was stable and could be used three times with no decrease in activity. [83]

Gayathri and co-workers prepared a polystyrene (DVB crosslinked) supported palladium-imidazole complex, which was used as catalyst in the hydrogenation of benzylideneaniline and a few of its para substituted derivatives at ambient conditions (figure 14). The influence of variation in temperature, concentration of the catalyst as well as the substrate on the rate of reaction was studied. The catalyst showed an excellent recycling efficiency over six cycles without leaching of metal from the polymer support. [84]

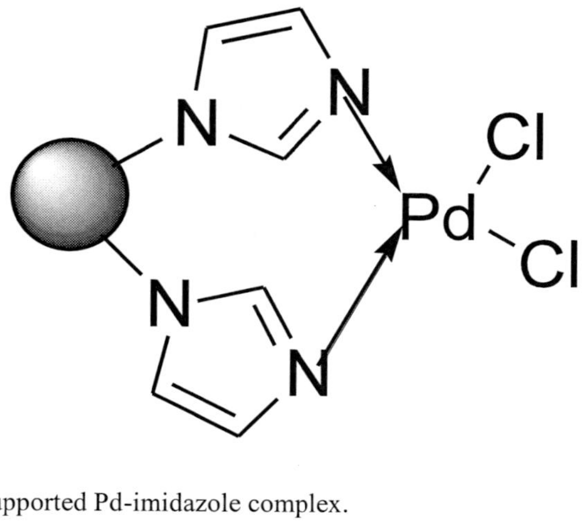

Figure 14. Polystyrene supported Pd-imidazole complex.

4.6. Polymer-Supported Catalysts for Olefin Metathesis

In the last thirty years, olefin metathesis have become one of the indispensable synthetic tools. [85] Because of this importance of metathesis reaction, many groups have developed polymer-supported catalysts for metathesis reaction. Schrock and Hoveyda reported the first recyclable chiral polymer supported catalyst for olefin metathesis. [86] A chiral bis(styrene) derivative was co-polymerized with styrene under the conventional suspension polymerization conditions and this polymer-supported ligand was complexed with a molybdenum triflate to get the respective catalyst. The catalyst performed well in asymmetric ring closing metathesis and an ee as high as 98% was obtained. Even if the catalyst exhibited less activity than the homogeneous counter part, the stability of the polymer-supported system was very high as observed from the low metal leaching. Akiyama and Kobayashi reported a linear polystyrene supported arene-ruthenium catalyst for ring closing olefin metathesis for the synthesis of various heterocycles (figure 15). [87] The products were obtained from 50 to 100% yield. The authors have also described the catalyst recycling.

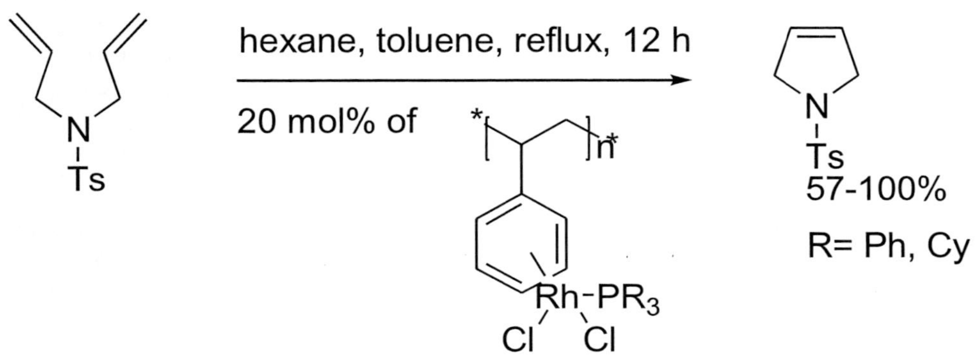

Figure 15. Polystyrene supported arene-ruthenium catalyst for ring closing olefin metathesis.

Another interesting example of polymer supported metathesis catalyst was reported by Jafarpour et al. [88] They have attached a number of metal complexes to the macroporous poly(DVB). Under the reaction conditions, the metal complexes underwent cross metathesis with poly-DVB and exchanged their carbene moiety for the unbound vinyl groups of the supporting polymer, giving rise to polymer supported catalysts.

4.7. Polymer-Supported Catalysts for Other Types of Reactions

There are a number of polymer-supported catalysts for which atleast one reference can be found in the literature. These miscellaneous catalysts are reviewed in the following section.

Li et al. used the polymer-supported bimetallic catalyst system PVP–$PdCl_2$–$NiCl_2$/TPPTS/PPh_3 in the hydroxycarbonylation of styrene under aqueous–organic two-phase condition. [89] The reaction proceeded to 100% conversion and with a selectivity of up to 64% to the branched acid. The catalyst was recycled three times with out loss of activity. Zhang reported a silica-supported chitosan (CS)–palladium complex CS–$PdCl_2$/SiO_2, which showed good conversion and higher regioselectivity in carbonylation of 6-methoxy-2-

vinylnaphthalene. [90] The high selectivity of the catalyst was achieved by the synergic effect of Pd–Ni bimetallic system and by polymer protection. Effects of reaction variables have been studied to optimize the reaction conditions. The hydroesterification of various substrates were also investigated. XPS and TEM analysis showed that the catalytically active species was composed of particles of nanometric size and the polymer could serve as a ligand. Recycling of the catalyst was also studied.

Homochiral 2,2′-bis(oxazolin-2-yl)-1,1′-binaphthyl (boxax) ligands were anchored on various polymer supports including PS-PEG, PS, PEGA, and MeO-PEG via selective monofunctionalization at the 6-position of the binaphthyl backbone by Hocke and Uozumi. Palladium(II) complexes of these supported boxax ligands catalyzed Wacker-type cyclization of 2-(2,3-dimethyl-2-butenyl)phenol to give 2-methyl-2-isopropenyl-2,3-dihydrobenzofuran with up to 96% ee. [91] The scheme of the reaction and structure of the catalyst are given in figure 16.

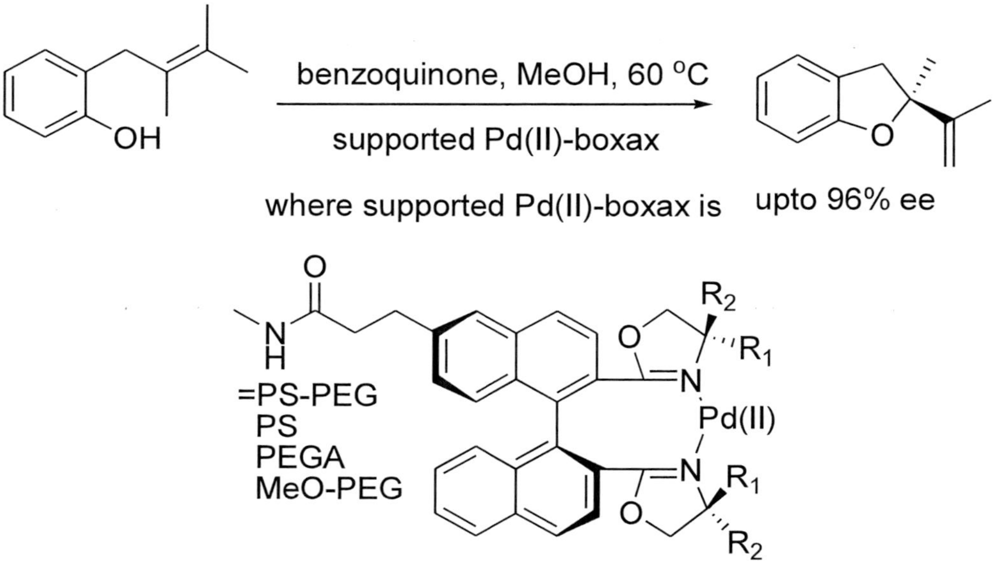

Figure 16. Polymer supported palladium(II) complexes catalyzed Wacker-type cyclization.

Iimura *et al.* showed that polystyrene sulphonic acid containing a long alkyl chain on the aromatic ring was a highly efficient catalyst for many organic transformations in water. [92] The authors showed that, with the increase in the length of the alkyl chain, the efficiency of the catalyst was increased and very low loading of the sulphonyl group was required to get excellent results. Only 1 mol% of the catalyst was required for getting excellent yield in reactions like hydrolysis of thioesters, deprotection of ketals, transthioacetalization of acetal and hydration of epoxides with in a very short reaction time under mild reaction conditions in water as the solvent.

Montchamp and co-workers have described a reusable polymer-supported hydrophosphinylation catalyst for the preparation of H-phosphinic acids. [93] Reaction of excess commercially available polystyryl isocyanate with nixantphos in refluxing toluene directly produced the desired urea-linked ligand with a loading of 0.1 to 0.3 mmol of ligand per gram of resin. The active catalyst was prepared by treating the ligand with Pd_2dba_3. The catalyst was successfully employed in the hydrophosphinylation of alkenes with H_3PO_2. The same catalyst was used for allylation of H_3PO_2 with allylic alcohols by the same group. [94] It was observed that the polymer supported catalyst showed less efficiency than the homogeneous counter part in both reactions.

A series of polar group functionalized polystyrene-supported phosphine reagents were examined as catalysts in the aza-Morita–Baylis–Hillman reaction of *N*-tosyl arylimines and a variety of Michael acceptors with the aim of identifying the optimal polymer/solvent combination by Toy *et.al.* [95] For these reactions *J*anda*J*el-PPh_3 (1 mmol PPh_3/g loading) resin containing methoxy groups (*JJ*-OMe-PPh_3) on the polystyrene backbone in THF solvent provided the highest yield of all the catalyst/solvent combinations examined. The methyl ether groups were incorporated into *JJ*-OMe-PPh_3 using commercially available 4-methoxystyrene, and thus such polar polystyrene resins were easily accessible and found utility as nucleophilic catalyst supports. Up to 81% yield was reported by the authors.

In another interesting report, polymer-supported palladium catalysts prepared from commercially available phosphine-functionalised polymers (PS-PR_2), $Pd_2(dba)_3$ and P(*t*-Bu)$_3$ were used as heterogeneous catalysts in the amination of aryl halides. [96] Four commercially available resins were used and the Pd species generated insitu from $Pd_2(dba)_2$ and P(t-Bu)$_3$ were attached to the resin to get the corresponding catalyst. Catalyst stability was investigated using ^{31}P NMR spectroscopy. One of the catalysts was reused in the amination of bromobenzene and chlorotoluene, up to three times, without loss in yield. Recyclability of the catalyst was dependent on the method of preparation and the nature of the polymer-bound phosphine.

Commercially available polystyrene supported triphenyl phosphines were found to be highly active catalyst in the Trost's γ-addition of various pro-nucleophiles with methyl 2-butynoate in water-toluene mixture. [97] The catalyst was recyclable up to three times with gradual loss in activity. Eventhough, this was considered as a green route for Trost's addition, the requirement of large amount of catalyst and a large variation in the yield of products with various substrates limited its practical application.

A poly(vinyl pyrrolidone) supported catalyst was reported by Chari and Syamasundar. [98] Condensation of o-phenylenediamine with ketones under solvent free conditions to afford the corresponding 1,5-benzodiazepine derivatives in high yield was catalyzed by this PVP supported ferric chloride catalyst.

Huisgen's [3+2] cycloaddition reaction between azides and alkynes was catalyzed by a polymer-supported catalyst prepared from CuI and Amberlyst A-21. [99] The catalyst prepared by simple stirring of polymer and copper salt followed by filtration and washing showed high efficiency and the triazole derivatives were obtained in excellent yield.

Zhao and Li showed that polymer supported sulfonamide of N-glycene was a highly efficient and selective catalyst for allylation of aldehydes and imines (generated in situ from aldehydes and amines) with allytributyltin. [100] Commercially available polystyrene-DVB resin was chlorosulphonylated followed by grafting of glycene ethyl ester. Saponification of this resin gave the active catalyst. The allylation proceeded with small amount of catalyst with in a short time interval. The allylic alcohols and amines were produced in good to excellent yield.

Polymer-bound *p*-toluenesulfonic acid was shown to catalyze efficiently the direct nucleophilic substitution of the hydroxy group of allylic and benzylic alcohols with a large variety of carbon- and heteroatom-centered nucleophiles. [101] The reaction conditions were mild, the process was conducted under an atmosphere of air without the need for dried solvents, and water was the only side product of the reaction and up to 90% conversion was observed.

Sulfenylphosphinoferrocene (Fesulphos) ligand carrying an alcohol linker was attached to Wang resin or Merrifield resin via an ether linkage and used in heterogeneous catalysis by Martin-Matute *et.al.* [102 This ligand was used in Cu(I) catalyzed asymmetric 1,3-dipolar cycloaddition of imines and N-phenylmaleimide. Among the catalysts, the Merrifield resin supported one showed better activity and up to 99% ee was obtained in many cases. The same ligand was also used in the enantioselective Pd catalyzed allylic substitution. It was observed that, even if the yield was little lower compared to the homogeneous counterparts, the enantioselectivity was higher for the polymer-supported ligand. The recycling experiments failed because of the oxidation of the ligand. The structure of the ligand is shown in figure 17.

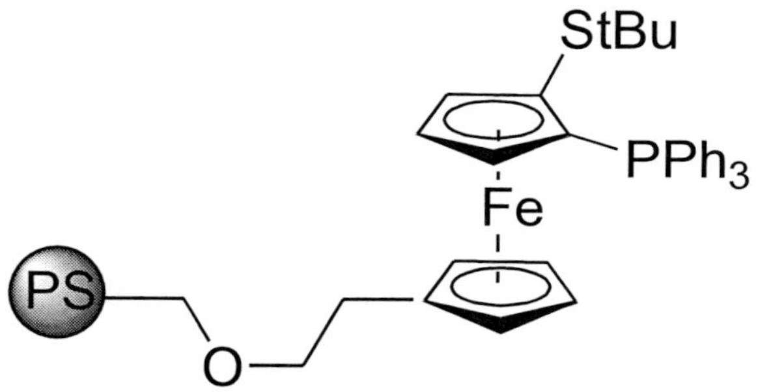

Figure 17. Polymer supported sulfenylphosphinoferrocene ligand.

Oxidative conversion of ketones and alcohols using a polymer supported catalyst was studied by Yamamoto *et.al.* [103] Various ketones were converted to the corresponding α-tosyloxyketones with *m*CPBA and *p*-toluenesulfonic acid in the presence of a catalytic amount of poly(4-iodostyrene). Moreover, secondary alcohols were directly converted to the corresponding α-tosyloxyketones using *m*CPBA and catalytic amounts of iodobenzene and potassium bromide, followed by treatment with *p*-toluenesulfonic acid in a one-pot manner. Both linear poly(4-iodostyrene) and macroporous crosslinked poly(-iodostyrene) were used and better results were obtained by the linear one. The catalysts were compared with the simple iodobenzene and the results showed that the polymer supported versions were less efficient and the efficiency was very low in the case of latter reaction.

A polymer supported triazole [104] prepared by Cu (I) catalyzed Huisgen 1,3-dipolar cycloaddition between trans-4-hydroxyproline and Merrifield type resin was used as highly efficient catalyst for α-aminoxylation of aldehydes and ketones by Font *et.al.* [105] The enantioselectivity of the catalyst was so high that the product obtained in up to 99% ee. The catalyst was recycled three times with out considerable loss of yield and enantioselectivity.

Linear polystyrene supported sulphonamide [10-(4-perfluorobutylsulfonamino sulfonylphenyl)decyl polystyrene] was developed by Zhang *et.al.* [106] A long alkyl chain separated the polymer support and the catalytically active species. This strongly acidic catalyst performed well in esterfication of carboxylic acids in water. In contrast to other catalysts, no excess of alcohol was required and the reaction proceeded well with equimolar quantities of acids and alcohols. The authors compared the catalyst with Nafion NR50 and NKC-9, two conventional solid acid catalysts and found that the reported catalyst was much more active than the former in esterification under the same condition. The catalyst was recycled five times with out appreciable loss of activity. Deleuze *et.al.* showed that polymer-supported titanium alkoxide effectively catalyzed transesterification of (meth)acrylic esters, an important industrial reaction. The catalyst showed good stability with low metal leaching. [107]

A polymer-supported gadolinium triflate catalyst was prepared from chloromethyl polystyrene resin by Lee *et.al.* Chloromethyl polystyrene was functionalized using an NHC moiety and this supported ligand was used in the synthesis of polymer supported metal complex. This polymeric catalyst was used as an efficient Lewis acid catalyst for the acetylation of various alcohols and phenols with acetic anhydride, affording high yields under mild conditions. The reaction was completed in a short period of time with small amounts of the catalyst. The catalyst was reused over 10 times without any significant loss of its catalytic activity. [108]

12-Tungstophosphoric acid impregnated on polyaniline was found to be an efficient catalyst for Friedel–Crafts benzylation of aromatics with benzyl alcohol as benzylating agent. A comparative study of the same catalyst supported on inorganic materials like silica and zirconia showed that the catalyst supported on polyaniline was selective towards monobenzylation and no dibenzylation was observed. This increased catalytic activity and selectivity was attributed to the dispersion of tungstophosphoric acid in the polyaniline matrix. The catalyst was found to be reusable for five cycles without appreciable loss in activity. [109]

Merrifield resin supported cinchona ammonium salts bearing 2′-fluorobenzene, 2′-cyanobenzene and 2′-*N*-oxypyridine groups were prepared and applied to the phase-transfer catalytic alkylation of *N*-(diphenylmethylene)glycine *tert*-butyl ester for the enantioselective synthesis of α-amino acids by Shi *et.al.* [110] Various α-amino acids were obtained in good yield (60-87%) and good enantiomeric excess (76–96%).

(Dimethylamino)methyl polystyrene supported CuI was used as an efficient and environmentally benign heterogeneous catalyst for the cyclization of propargyl alcohols with CO_2 to alkylidene cyclic carbonates under supercritical conditions by Jiang *et.al.* [111]

Polymer-supported sulfonic acid (NKC-9) was found to be a reusable catalyst in the synthesis of diaminotriphenylmethanes (DTMs) in high yield through one-step condensation of arylaldehydes and *N,N*-dimethylaniline under mild reaction condition. [112]

Kim and co-workers have found out a highly efficient method for nucleophilic fluorination using an alkali metal fluoride through the synergistic effect of the polymer-supported ionic liquid as a catalyst and *tert*-alcohol as an alternative reaction media (figure 18). This system not only enhances the reactivity of alkali metal fluorides and reduces the formation of by-products but also allows the use of a polymer-supported catalyst protocol. The fluorinated products were obtained in high yield. [113]

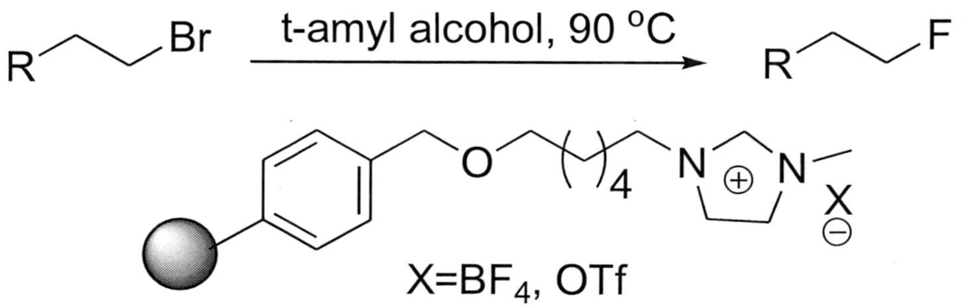

Figure 18. Nucleophilic fluorination catalyzed by polymer supported ionic liquid.

Yi and Cai treated Amberlyst A-21, a kind of well-known and cheap polymeric material, with ytterbium perfluorooctanesulfonate [Yb(OPf)$_3$] giving a catayst with a ytterbium loading of 1.34 (wt%) (figure 19). The polymer-supported fluorous ytterbium catalyst was highly efficient in nitration, esterification, Fridel-Crafts acylation, and aldol condensation. The catalyst can be recovered by simple filtration and used again without a significant loss of catalytic activity. The advantages of this catalyst are that the method of preparation was simple and this protocol avoided the use of fluorous solvents during the reaction or workup, which are expensive and can leach in small amounts. [114]

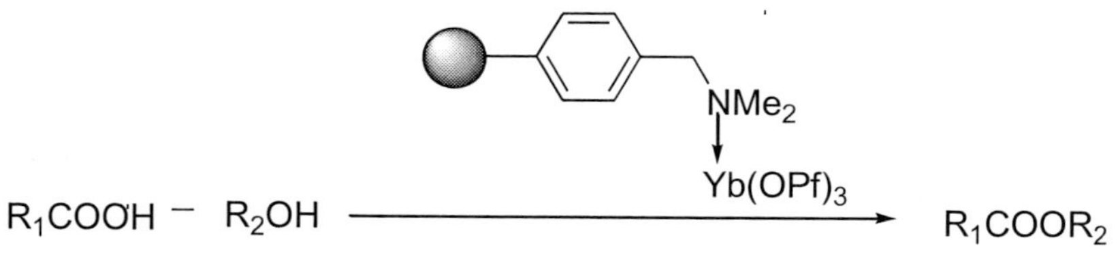

Figure 19. Polymer supported ytterbium catalyst

Ring opening polymerization of ε-caprolactone was catalyzed under mild conditions by scandium trifluoromethanesulfonate attached to a polystyrene support (figure 20). Using this catalyst it was possible to prepare poly(ε-caprolactone) with low polydispersity. Comparison of supported catalyst with unsupported one showed that there was no reduction in activity by attaching the catalyst to a polymer support. After polymerization the catalyst could easily be recovered and no loss of Sc was observed. [115,116]

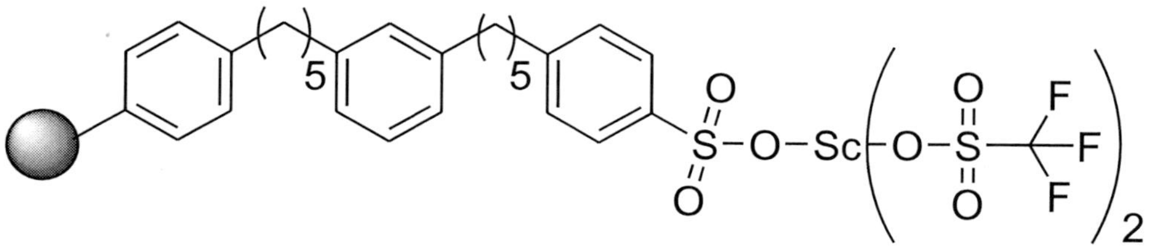

Figure 20. Polymer supported catalyst used for ring opening polymerization of caprolactone.

Sreekumar and co-workers had used polystyrene supported β-diketone metal complexes for the efficient oxidation of olefins to carbonyl compounds [117] and for aromatic coupling and substitution reactions. In both cases, the yields of the products were excellent and the catalysts could be recycled at least half a dozen times with out loss of activity.

Conclusion

A large number of supported ligands and corresponding metal complexes have been prepared and used as catalysts for synthetic organic chemistry along with many polymer supported organocatalysts and nanoparticles. Almost all kinds of reactions ranging from oxidations to polymerization could be catalyzed using polymer supported catalysts. Polymer supported catalysts outweigh common heterogeneous catalysts, used in many industrial processes, in aspects of activity and selectivity. One main disadvantage associated with many early stage polymer supported catalysts was metal leaching. Due to the continuous efforts from various research groups this problem could be solved to a good extend. The rapid growth of organocatalysis has also helped to develop new polymer supported catalysts without the fear of metal leaching. Recent developments in nanoscience has also helped to develop highly efficient nanocatalysts attached to polymer supports. Due to the growing concern in favor of environmental friendly and cost effective processes, it has become necessary to change from conventional processes to new ones and polymer supported catalysts are good alternatives. In short, polymer supported catalyst is an area of growing interest due to the pressure from pharmaceutical and fine chemical industry because of the potential advantages the supported catalysts have.

References

[1] *Solid-Phase Synthesis and Combinatorial Techniques*; Seneci, P., Ed.; John Wiley: New York, 2001.

[2] *Catalysis by Polymer-Immobilized Metal Complexes;* Pomogailo, A. D., Ed.; Gordon and Breach: Australia, 1998.

[3] Frechet, J. M. J. *Tetrahedron*, 1981, 37, 663

[4] Kobayashi, S. *Chem. Soc. Rev*. 1999, 28, 1.

[5] Kobayashi, S. *Curr. Opin. Chem. Biol*. 2000, 4, 338.

[6] Akelah, A.; Sherrington, D. C. *Chem. Rev*. 1981, 81, 557.

[7] Shuttleworth, S. J.; Allin, S. M.; Sharma, P. K. *Synthesis* 1997, 11, 1217.

[8] Shuttleworth, S. J.; Allin, S. M.; Wilson, R. D.; Nasturica, D. *Synthesis* 2000, 8, 1035.

[9] Ley, S. V.; Baxendale, I. R.; Bream, R. N.; Jackson, P. S.; Leach,A. G.; Longbottom, 62. D. A.; Nesi, M.; Scott, J. S.; Storer, I.; Taylor,S. J. *J. Chem. Soc., Perkin Trans.* 1 2000, 3815.

[10] de Miguel, Y. R. *J. Chem. Soc., Perkin Trans*. 1 2000, 4213.

[11] de Miguel, Y. R.; Brule′, E.; Margue, R. G. *J. Chem. Soc., Perkin Trans*. 1 2001, 3085.

[12] Clapham, B.; Reger, T. S.; Janda, K. D. *Tetrahedron* 2001, 57,4637.

[13] Eames, J.; Watkinson, M. *Eur. J. Org. Chem*. 2001, 7, 1213.

[14] Leadbeater, N. E.; Marco, M. *Chem. Rev.* 2002, 102, 3217.
[15] McNamara, C. A.; Dixon, M. J.; Bradley, M. *Chem. Rev.* 2002, 102, 3275.
[16] Dickerson, T. J.; Reed, N. N.; Janda, K. D. *Chem. Rev.* 2002, 102, 3325.
[17] Bergbreiter, D. E. *Chem. Rev.* 2002, 102, 3345.
[18] Benaglia, M.; Puglisi, A.; Cozzi, F. *Chem. Rev.* 2003, 103, 3401.
[19] Benaglia, M.; *New. J. Chem.* 2006, 11, 1525.
[20] Cozzi, F. *Adv. Synth. Catal.* 2006, 348, 1367.
[21] *Analytical methods in combinatorial chemistry*, Yan, B., Zhang, B., Eds.; CRC: Pennsylvania, 2000
[22] Colacot, T. J.; Gore, E. S.; Kuber, A. *Organometallics*, 2002, 21, 16.
[23] Shieh, W. C.; Shekhar, R.; Blacklock, T.; Tedesco, A. *Synth. Commun.* 2002, 32, 1059.
[24] Phan, N. T. S.; Brown, D. H.; Styring P. *Tetrahedron Lett.,* 2004, 45, 7915.
[25] Kim, J. H.; Jun, B. H.; Byun, J. W.; Lee, Y. S. *Tetrahedron Lett.*, 2004, 45, 5827.
[26] Byun, J. W.; Lee, Y. S. *Tetrahedron Lett.*, 2004, 45,1837.
[27] Schweizer, S.; Becht, J. M.; Drian, C. L. *Adv. Synth. Catal.* 2007, 349, 1150.
[28] Schweizer, S.; Becht, J. M.; Drian, C. L.; *Tetrahedron*, 2010, *66,* , 765-772
[29] Krishnan, G. R.; Sreekumar, K. *Soft Mater.* 2010, 8, 114
[30] Ornelas, C.; Diallo, A. K.; Ruiz, J.; Astruc, D. *Adv. Synth. Catal.* 2009, 351, 2147
[31] Dell'Anna, M. M.; Mastrorilli, P.; Muscio, F.; Nobile, C. F.; Suranna, G. P. *Eur. J. Inorg. Chem.*, 2002, 5, 1094.
[32] Altaya, B.; Burguete, M. I.; Verdugo, E. G.; Karbass, N.; Luis, S. V.; Sans, P. V. *Tetrahedron Lett.* 2006, 47, 2311.
[33] Shokouhimehr, M.; Kim, J. H.; Lee, Y. S. *Synlett*, 2006, 618.
[34] Dell'Anna, M. M.; Lofu, A.; Mastrorilli, P.; Mucciante, V.; Nobile, C. F. *J. Organomet. Chem.* 2006, 691, 131.
[35] Bai, L.; Wang, J. *Adv. Synth. Catal.* 2008, 350, 315.
[36] Yao, C.; Li, H.; Wu, H.; Liu, Y.; Wu, P. *Catal. Commun.* 2009, *10*, 1099,
[37] Liu, Y.; Jia, J.; Tan, H.; Sun, Y.; Tao, J. Chin. J. *Chem.* 2010, 28, 967
[38] Wang, L.; Huang, C.; Cai, C. *Catal. Commun.* 2010, 11, 532
[39] Itsuno, S.; Arima, S.; Haraguchi, N. *Tetrahedron*, 2005, 61,12074.
[40] An, L. T.; Zou, J. P.; Zhang, L. L. *Catal. Commun.* 2008, 9, 349.
[41] Madhavan, N.; Weck, M. *Adv. Synth. Catal.* 2008, 350, 419.
[42] Liu, H. L.; Jiang, H. F. *Tetrahedron* 2008, 64, 2120.
[43] Trost, B. M.; Pan, Z.; Zambrano, J.; Kuiat, C. *Angew. Chem. Int. Ed. Engl.* 2002, 41, 4691.
[44] Sekiguti, T.; Lizuka, Y.; Takizawa, S.; Jayaprakash, D.; Arai, T.; Sasai, H. Org. Lett. 2003, 5, 2647.
[45] Yan, C.; Zeng, X.; Zhang, W.; Luo, M. *J. Organomet. Chem.* 2006, 691, 3391.
[46] Rajagopal, G.; Selvaraj, S.; Dhahagani, K. *Tetr. Asym.* 2010, *21*, 2265
[47] Hui, X. P.; Huang, L. N.; Li, Y. M.; Wang, R. L.; Xu, P. F. *Chirality*, 2010, *22*, 347
[48] Alza, E.; Pericas, M. A. *Adv. Synth. Catal.* 2009, *351*, 3051
[49] Krishnan, G. R.; Sreekumar, K. *Eur. J. Org. Chem.* 2008, 4763
[50] Abu-Elfotoh, A. M.; Phomkeona, K.; Shibatomi, K.; Iwasa, S. *Angew. Chem. Int. Ed.* 2010, 49, 8439
[51] Sherrington, D. C. *Pure & Appl. Chem.* 1988, 60, 401.
[52] Vinodu, M. V.; Padmanabhan, M. *Proc. Indian Acad. Sci.* (Chem. Sci.), 2001, 113, 1.

[53] Nair, V. A.; Mustafa, S. M.; Sreekumar, K, *J. Polym. Res*. 2003, 10, 267.
[54] Barbarini, A.; Maggi, R.; Muratori, M.; Sartori, G.; Sartorio, R. *Tetrahedron Asymmetry,* 2004, 15, 2467.
[55] Valodkar, V. B.; Tembe, G. L.; Ravindranathan, M.; Ram, R. N.; Rama, H. S. *J. Mol. Catal, A. Chem.* 2004, 208, 21.
[56] Kang, Q.; Luo, J.; Bai, Y.; Yang, Z.; Lei, Z. *J. Organomet. Chem*. 2005, 690, 6309.
[57] Benaglia, M.; Puglisi, A.; Holczknecht, O.; Quici, S.; Pozzi, G. *Tetrahedron*, 2005, 61, 12058.
[58] Lei, Z.; Yan, P.; Yang, Y. *Catal. Lett*. 2007, 118, 69.
[59] Krishnan G. R.; Sreekumar, K. *Appl. Catal. A. Gen*. 2009, 353, 80
[60] Trakampruk, W.; Kanjina, W. *Ind. Eng. Chem. Res*. 2008, 47, 964
[61] Gupta, K. C.; Sutar, A. K. *J. Mol. Catal. A. Chem*. 2008, 280, 173
[62] Castro, I. U.; Sherrington, D. C.; Fortuny, A.; Fabregat, A.; Stuber, F.; Font, J.; Bengoa, C. *Catal. Today*, 2010, 157, 66
[63] Weng, Z.; Wang, J.; Zhang, S.; Yan, C.; Jian, X. *Catal. Commun*. 2008, 10, 125
[64] Nair, V. A.; Sreekumar, K. *Curr. Sci*. 2001, 81, 194.
[65] Nair, V. A.; Sreekumar, K. *J. Polym. Mater*. 2002, 19, 155 & 265; 2003, 20,267
[66] Nair, V. A.; Suni, M. M.; Sreekumar, K. *Proc. Indian Acad. Sci.* (Chem. Sci.) 2002, 114, 481.
[67] Nair, V. A.; Suni, M. M.; Sreekumar, K. *Designed Monomers and Polymers* 2003, 6, 81.
[68] Saladino, R.; Neri, V.; Pelliccia, A. R.; Caminiti, R.; Sadun, C. *J. Org. Chem*. 2002, 67, 1323.
[69] Saladino, R.; Neri, V.; Pelliccia, A. R. Mincione, E. *Tetrahedron*, 2003, 59, 7403.
[70] Lu, J.; Wang, X.; Liu, J.; Zhang, L.; Wang, Y. *Tetrahedron Asymmetry*, 2006, 17, 330.
[71] Beloglazkina, E. K.; Majouga, A. G.; Romashkina, R, B.; Zyk, N. V. *Tetrahedron Lett.* 2006, 47, 2957.
[72] Bandini, M.; Fagioli, M.; Melloni, A.; Umani-Ronchi, A. *Adv. Synth. Catal*. 2004, 346, 573.
[73] Yarapathy, V. R.; Mekala, S.; Rao, B. V.; Tammishetti, S. *Catal. Commun*. 2006, 7, 466.
[74] Lee, S. H.; Lee, E. Y.; Yoo, D. W.; Hong, S. J.; Lee, J. H.; Kwak, H.; Lee, Y. M.; Kim, J.; Kim, C.; Lee, J. K. *New J. Chem*. 2007, 31, 1579.
[75] Krishnan G. R.; Sreekumar, K. *Polymer*, 2008, 49, 5233
[76] Saluzzo, C.; Lamouille, T.; Herault, D.; Lemaire, M. *Bioorg. Med. Chem.* Lett. 2002, 12, 1841.
[77] Kell, R. J.; Hodge, P.; Nisar, M.; Watson, D. *Bioorg. Med. Chem. Lett*. 2002, 12, 1803.
[78] Kell, R. J.; Hodge, P.; Snedden, P.; Watson, D. *Org. Biomol. Chem*. 2003, 1, 3238.
[79] Zarka, M. T.; Nuyken, O.; Weberskirch, R. *Chem. Eur. J*. 2003, 9, 3228.
[80] Lopez-Castillo, Z. K.; Flores, R.; Kani, I.; Fackler Jr, J. P.; Akgerman, A. *Ind. Eng. Chem. Res*. 2002, 41, 3075.
[81] Li, Y.; Li, Z.; Li, F.; Wang, Q.; Tao, F. *Org. Biomol. Chem*. 2005, 3, 2513.
[82] Nakao, R.; Rhee, H.; Uozumi, Y. *Org. Lett*. 2005, 7, 163.
[83] Wen, H.; Yao, K.; Zhang, Y.; Zhou, Z.; Kirschning, A. *Catal. Commun*. 2009, 8, 1207
[84] Udayakumar, V.; Alexander, S.; Gayathri, V.; Shivakumaraiah, Patil, K. R.; Viswanathan, B. *J. Mol. Catal. A. Chem*. 2010, 317, 111

[85] Grubbs, R. H. *Nobel Lecture*, 2005.
[86] Hultzsch, K. C.; Jernelius, J. A.; Hoveyda, A. H.; Schrock, R. R. Angew. *Chem. Int. Ed. Engl.*2002, 41, 589.
[87] Akiyama, R.; Kobayashi, S. *Angew. Chem. Int. Ed. Engl.* 2002, 41, 2602.
[88] Jafarpour, L.; Heck, M. P.; Baylon, C.; Lee, H. M.; Mioskowski, C.; Nolan, S. P. *Organometallics*, 2002, 21, 671.
[89] Li, F. W.; Xu, L. W.; Xia, C. G. *Appl. Catal. A.* 2003, 253, 509.
[90] Zhang, J.; Xia, C. G. *J. Mol. Catal. A. Chem.* 2003, 206, 59.
[91] Hocke, H.; Uozumi, Y. *Synlett*, 2002, 2049.
[92] Iimura, S.; Manabe, K.; Kobayashi, S. *Org. Biomol. Chem.* 2003, 1, 2416.
[93] Deprele, S.; Montchamp, J. L. *Org. Lett.* 2004, 6, 3805.
[94] Bravo Altamirano, K.; Montchamp, J. L. *Org. Lett.* 2006, 8, 4169.
[95] Zhao, L. J.; Kwong, C. K. W.; Shi, M.; Toy, P. H. *Tetrahedron*, 2005, 61, 12026.
[96] Guino, M.; Hii, K. K. *Tetrahedron Lett.* 2005, 46, 7363.
[97] Skouta, R.; Varma, R. S.; Li, C, *J. Green Chem.* 2005, 7, 571.
[98] Chari, M. A.; Syamasundar, K. *Catal. Commun.* 2005, 6, 67.
[99] Girard, C.; Onen, E.; Aufort, M.; Beauviere, S.; Samson, E.; Herscovici, J. *Org. Lett.* 2006, 8, 1689.
[100] Li, G. L.; Zhao, G. *Org. Lett.* 2006, 8, 633.
[101] Sanz, R.; Martinez, A.; Miguel, D.; Alvarez-Gutierrez, J. M.; Rodriguez, F. *Adv. Synth. Catal.* 2006, 348, 1841.
[102] Martin-Matute, B.; Pereira, S. I.; Pena-Cabrera, E.; Adrio, J.; Silva, A. M. S.; Carretero, J. C. *Adv. Synth. Catal.* 2007, 349, 1714.
[103] Yamamoto, Y.; Kawano, Y.; Toy, P. H.; Togo, H. *Tetrahedron*, 2007, 63, 4680.
[104] Tornoe, C. W.; Christensen, C.; Meldal, M. *J. Org. Chem.* 2002, 67, 3057.
[105] Font, D.; Bastero, A.; Sayalero, S.; Jimeno, C.; Pericas, M. A. Org. Lett. 2007, 9, 1943.
[106] Zhang, Z.; Zhou, S.; Nie, J. *J. Mol. Catal. A. Chem.* 2007, 265, 9.
[107] Alves, M. H.; Riondel, A.; Paul, J. M.; Birot, M.; Deleuze, H. *Comp. Rend. Chim.* 2010, 13, 1301
[108] Yoon, H. J.; Lee, S. M.; Kim, J. H.; Cho, H. J.; Choi, J. W.; Lee, S. H.; Lee, Y. S. *Tetr. Lett.* 2008, 49, 3165
[109] Satam, J. R.; Jayaram, R. V. *Catal. Commun.* 2008, 9, 1937
[110] Shi, Q.; Lee, Y. J. Kim, M. J.; Park, M. K.; Lee, K.; Song, H.; Cheng, M.; Jeong, B. S.; Park, H. G.; Jew, S. S. *Tetrahedron Lett.* 2008, 49, 1380.
[111] Jiang, H. F.; Wang, A. Z.; Liu, H. L. Qi, C. R. *Eur. J. Org. Chem.* 2008, 2008, 2309.
[112] An, L. T.; Ding, F. Q.; Zou, J. P. *Dyes Pigm.* 2008, 77, 478
[113] Kim, D. W.; Jeong, H. J.; Lim, S. T.; Sohn, M. H.; Chi, D. Y. *Tetrahedron*, 2008, 64, 4209
[114] Yi, W. B.; Cai, C. J. *Fluor. Chem.* 2008, 129, 524
[115] Oshimura, M.; Takasu, A.; Nagata, K. *Macromolecules*, 2009, 42, 3086
[116] Takasu, A.; Oshimura, M.; Hirabayashi, T. J. *Polym. Sc. A. Polym. Chem.* 2008, 46, 2300
[117] Nair, V. A.; Suni, M. M; Sreekumar, K. *Reactive & Functional Polymers*, 2003, 57, 33

In: Focus on Catalysis Research: New Developments
Editors: Minjae Ghang and Bjørn Ramel
ISBN: 978-1-62100-455-4

Chapter 4

IONIC LIQUIDS AS A CATALYST OR SOLVENT FOR VARIOUS ORGANIC TRANSFORMATIONS

Krishna M. Deshmukh, Ziyauddin S. Qureshi and Bhalchandra M. Bhanage**[*]**
Department of Chemistry, Institute of Chemical Technology, Matunga, Mumbai-400019, India

ABSTRACT

The room temperature ionic liquids (RTILs) are liquids which consist of ions and melts at or below 100 °C. They have typical properties like negligible vapor pressure, high thermal stability, and nonflammable nature. Moreover, the physicochemical properties of ionic liquids, such as their melting temperature and hydrophilicity/ hydrophobicity, can be fine-tuned by changing the structure of the cations and anions. Ionic liquids were initially introduced as alternative green reaction media; but today they have marched far beyond this border, showing their significant role in controlling the reaction as a catalyst. The Brønsted and Lewis acidic ionic liquids have exhibited a great potential in replacement of conventional homogeneous and heterogeneous acidic catalysts, and have been successfully applied to a variety of reactions including the Diels-Alder reaction, Mannich reaction, Prins reaction, Acetalization of carbonyl compounds, Friedel–Crafts reaction, Friedlander Reaction, Biginelli reaction, Esterification and transesterification with enhanced selectivity and activity. Ionic liquids are also extensively used as a green reaction media for various transition metals catalyzed cross coupling reaction such as Suzuki, Heck, and Negishi. It is also used for catalyst immobilization, which not only allow recovery and recycling of costly transition metal catalyst but also has positive effect on catalysis. Catalysts often become more reactive for a reaction such as Hydroformylation, Hydrogenation and hydroamination reactions.

[*] E-mail address: bhalchandra_bhanage@yahoo.com; km.deshmukh@yahoo.co.in

1. INTRODUCTION

Ionic liquids (ILs) have attracted increasing interest past 20 years in the context of green organic synthesis and catalysis. Although ionic liquids were initially introduced as alternative green reaction media because of their unique chemical and physical properties of non-volatility, thermal stability, non-flammability, and controlled miscibility, today they have marched far beyond this boundary, showing their significant role in controlling reactions as solvent or catalysts. In the last few years, there have been several reviews published in which ionic liquids considered as a central theme [1-12]. It was used as a catalyst or liquid support for replacement of conventional homogeneous and heterogeneous acidic or transition metal catalysts [13-16].

The discovery of ionic liquid went back to 1914 when Paul Walton reported ethyl ammonium nitrate (EAN), a truly room temperature ionic liquid (RTIL), with a melting point of 12 °C [17]. Ionic liquid entirely consists of ions but it is different from the classical definition of molten salts. Melting point criterion has been proposed to distinguish between molten salts and ionic liquids. Molten salts are usually defined as a highly-melting, highly viscous and highly corrosive liquid medium, while ionic liquids are defined as pure compounds, consisting of only cations and anions (i. e. salts), which melts at or below 100 °C and has lower viscosity. As both the anion and cation can be varied, these solvents can be designed for particular end use in mind or for a particular set of properties. Hence they are also known as 'Designer solvents'. They are also termed as 'neoteric solvents' as they have remarkable new properties that can break new grounds to clean up the modern chemical industry. They have many fascinating properties since both the thermodynamics and kinetics of the reactions in ionic liquids are different from those in conventional molecular solvents. Following are the some of the properties of that make ionic liquids an attractive substitute for molecular solvent as well as catalyst in chemical processes [18, 19].

1. They serve as an excellent solvent for wide range of organic, inorganic and organometallic reagents.
2. Negligible vapour pressure even at high temperature hence they can be used in high vacuum systems.
3. The hydrophilicity/hydrophobicity can be fine-tuned by changing the structure of cations and anions.
4. They are immiscible with a number of organic solvents and provides a nonaqueous, polar alternative for biphasic systems.
5. They are often composed of poorly coordinating ions, so they have the potential to be highly polar yet non coordinating solvents.
6. The Bronsted and Lewis acidity can be controlled or tuned depending upon the counter anion precursors.
7. The tetrafluoroborate (BF_4) and the hexafluorophosphate (PF_6) based ILs can be easily prepared and effectively reused.

1.1. Classification of Ionic Liquids

The ionic liquids can be divided into three categories depending upon the functionality on the cation or variety of anions.

[I] Binary Ionic Liquids or 1st Generation Ionic Liquids

Theses generally include mixture of metal halides and organic cations. The organochloroaluminate ionic liquid are the most investigated class of molten salts and can be prepared easily by mixing quaternary ammonium salts, especially *N*-alkylpyridinium, 1,3-dialkylimidiazolium halides, choline chloride with $AlCl_3$ in various proportions.

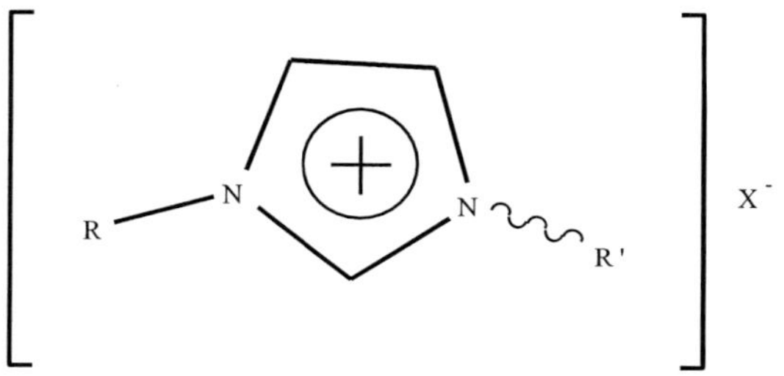

R, R' = -alky chain. X = anions like $AlCl_4^-$, $FeCl_4^-$, etc.

Figure 1. Binary ionic liquids.

These ionic liquids have been used for a variety of reactions promoted by Lewis acids considering the advantage of their controlled acidity.

[II] Simple Ionic Liquids or 2nd Generation Ionic Liquids

This includes the liquids derived from single anion and cation. Theses ambient temperature air and water-stable ionic liquids can be obtained by substitution of halide anion of *N*-alkylpyridinium and 1,3-dialkylimidiazolium cation by other weakly coordinating anions such as BF_4^- and PF_6^-.

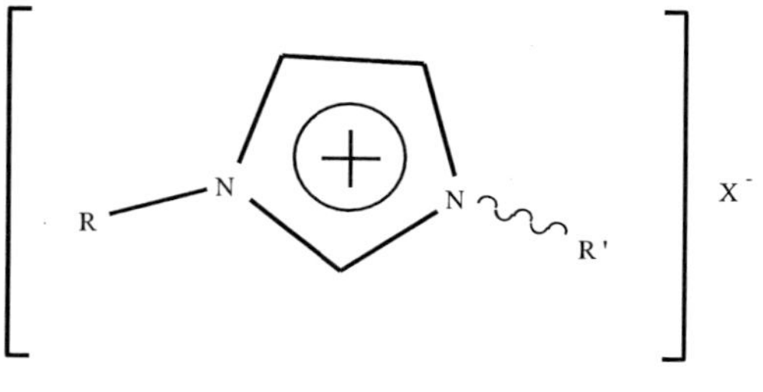

R, R' = -alky chain. X = anions like PF_6^-, BF_4^- etc.

Figure 2. Simple ionic liquids.

Unlike the chloroaluminate ionic liquids, they are stable towards air and water. The recyclability of these ionic liquids has made them extremely popular as they can be used as solvent for a wide range of reaction.

[III] Task Specific Ionic Liquids or 3rd Generation Ionic Liquids

This type of compounds includes the ionic liquids in which a functional group is covalently tethered to the cation or anion (or both) of the ILs. Which were further applied for the specific purpose application as reagents or catalysts in organic reactions. The concept of task-specific ionic liquids was first introduced by Davis [20, 21]

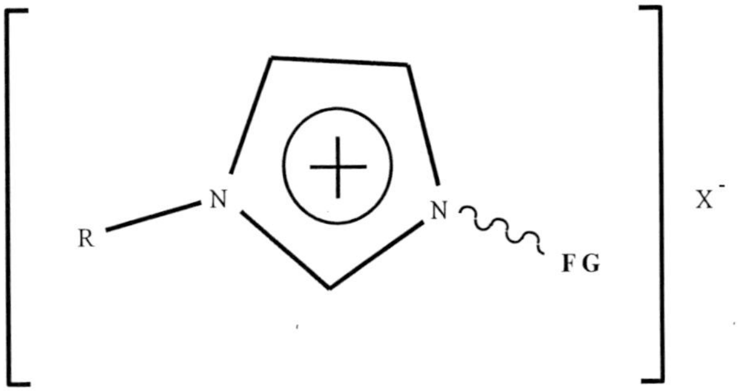

FG = $-NH_2$, -OH, -OR, -SH, $-OPPh_2$, $-PPh_2$, $-Si(OR)_3$, -Metal complex, -urea, -thiourea, etc.

Figure 3. Task specific ionic liquids (TSILs).

1.2. Synthesis of Ionic Liquids

Synthesis of ionic liquids generally consists of two steps. First step is common for both the first and second generation ionic liquids i.e. desired cation has to be generated, usually by direct alkylation/quaternization of a nitrogen or phosphorus atom. In the second step, the anion resulting from the alkylation reaction can be exchanged for a different one by metathesis reaction or by direct combination with Lewis acid [22, 23]. The general and detailed synthesis of imidazolium-based ionic liquids is represented in Figure 4 and discussed further.

The incorporation of functional groups can impart a particular capability to the ionic liquids, enhancing their ability for catalyst reusability and stability compared with the unfunctionalized ILs.

Imidazolium salts with different anions are obtained by the quaternization reaction depending upon the alkylating reagent. If it is not possible to obtain desired imidazolium salt with required anion, then Path A and B (Figure 4) can be followed. Two different paths are possible to replace anion formed resulting from initial quaternization step. First is the imidazolium salts directly treated with Lewis acids, this leads to the formation of first generation ionic liquids of the type [RR'im][MXy+1] (Path A Figure 4). Alternatively it is possible to exchange anion with desired anion by addition of metal salt $M^+[A]^-$ (with precipitation of M^+X^-), by displacement of anion by a strong acid $H^+[A]^-$ (with evaporation of

HX) or by passing over ion-exchange resin (Path B, Figure 4) this lads the formation of second generation ionic liquids.

Task specific ionic liquid in which a functional group is covalently tethered to the cation or anion (or both) of the ILs were prepared by direct alkylation/quaternization of a nitrogen or phosphorus atom (Path C, Figure 4) [20].

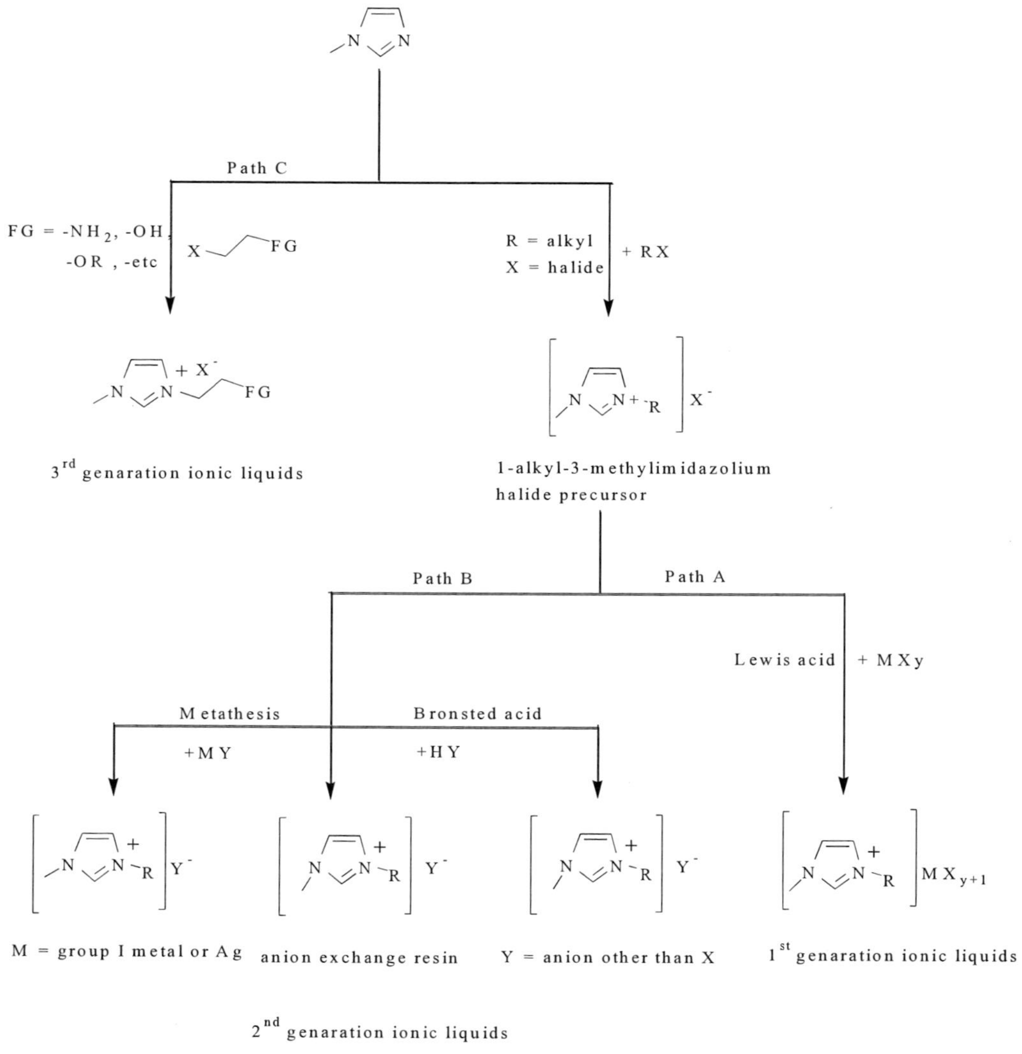

Figure 4. General method for synthesis of ionic liquids.

2. PRODUCT SELECTIVITY IN IONIC LIQUIDS

The concept of the ionic liquids as a "designer solvent" by making changes to the structure of either the anion, or the cation (or both), have been proved. Seddon and co-worker reported dramatic effect on the outcome of the chemical reaction by changing ionic liquid.

The authors demonstrated that reaction of toluene with nitric acid in HCl, gives three completely different products in three different ionic liquids(Scheme 1) [24].

When ionic liquid with trifluoromethanesulfonate (OTf) as anion was used the nitration of toluene was observed. When the anion was a halide, halogenated products obtained. Finally, when the anion is a methanesulfonate salt [OMs], the methyl group was oxidized to the organic acid, nitric acid here acts as an oxidizing agent rather than a nitrating agent

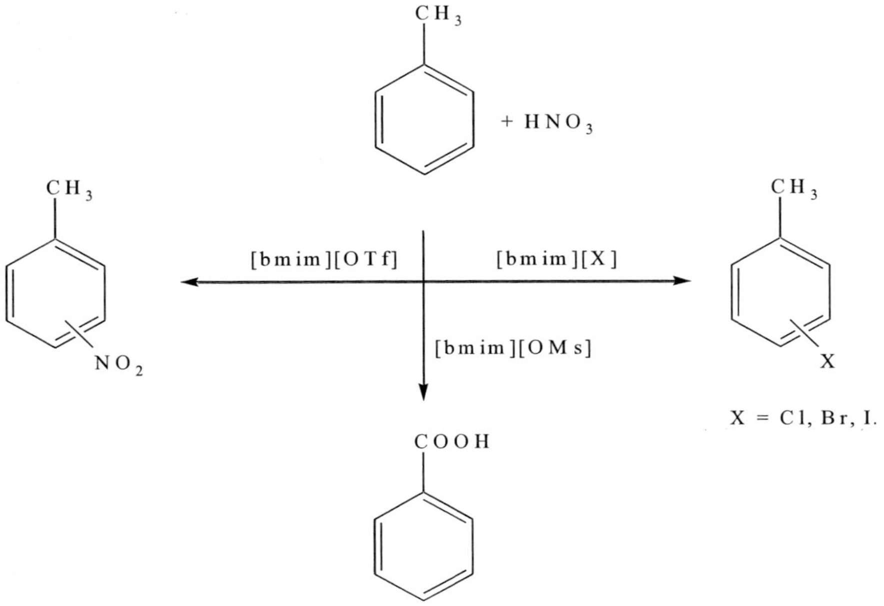

Scheme1.

3. Chemical Stability of 1, 3-dialkylimidazolium Based Ionic Liquids

The 1, 3-dialkylimidazolium cation based ionic liquid have acidic proton at the C-2 position, which may be get deprotonated under basic condition (e.g. potassium t-butoxide) to form a stable carbene (Scheme 2). The resulting carbene is strongly stabilized by the presence of the two adjacent nitrogen atoms [25].

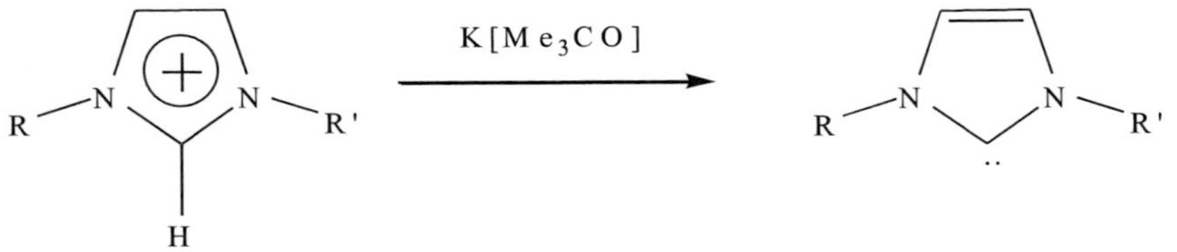

Scheme 2.

A breakthrough was achieved by Arduengo et al. by isolating and fully characterizing a stable singlet *N*-heterocyclic carbene, 1,3-di-1-adamantylimidazol-2-ylidene, by deprotonation of an imidazolium salt [26,27]. Formation of *N*-heterocyclic carbene in base-catalyzed reactions, can be prevented by 'blocking the C-2 position of the 1,3-dialkylimidazolium cation by methylation or by changing to another inert cation.

4. Ionic Liquids in Organic Synthesis

There are three major applications of ionic liquids in organic synthesis depending on the nature of the RTILs. The first one involves the use of organo-aluminate ionic liquids in reactions promoted by Lewis acids taking advantage of the controlled acidity of such melts. The second application is related to the use of RTILs as "neutral" solvents or liquid support for the transition metal catalyst. The final application takes the advantage of task specific ionic liquid as a ligand, catalyst or extractant for metal ion.

4.1. Organic Synthesis in Organoaluminate Ionic Liquids

Application of organoaluminate ionic liquids was well reviewed by Welton et al. Early report for the application of organoaluminate ionic liquids reports the Friedel-Crafts alkylation and acylation reaction.

4.1.1. Friedel-Crafts Alkylation and Acylation

Boon et al. first reported the alkylation and acylation of benzene using 1-methyl-3-ethylimidazolium-aluminium trichloride ionic liquid ([EMIM]Cl – $AlCl_3$), ionic liquid and methyl chloride and acetyl chloride as alkylating and acylating agents (Scheme 3) The authors reported that rate of the reaction depends on the Lewis acidity of the ionic liquid [28].

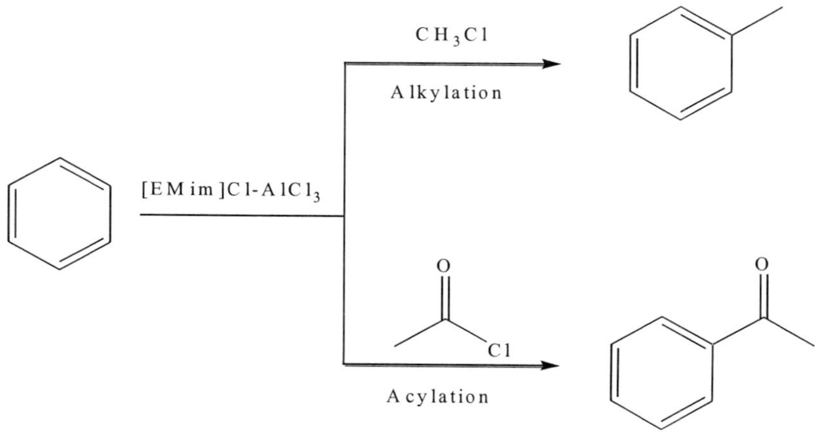

Scheme 3.

Later on Seddon and co-worker further studied the alkylation and acylation of various aromatic compounds in [EMIM]Cl – $AlCl_3$ ionic liquids and results compared with conventional molecular solvents [29]. Further, acylation of the organometallic complex

ferrocene was performed in [EMim]I-$AlCl_3$ (Scheme 4), affording solely the monoacylated product in good to excellent yields [30].

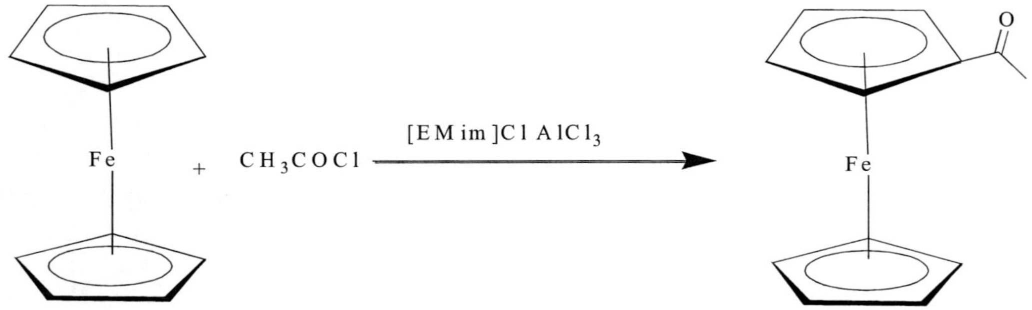

Scheme 4.

Recently, Wasserscheid and co-worker reported, supported ionic liquid phase (SILP) catalyst material using silica as a support coated with an acidic chloroaluminate ionic liquid. The catalyst proved to be very active and selective for Slurry-Phase alkylation of cumene [31].

4.1.2. Friedel-Crafts Sulfonylation and Sulfamoylation

Salunkhe and co-worker reported sulfonylation and sulfamoylation of benzene and substituted benzenes with 4-methyl benzenesulfonyl chloride and sulfamoyl chlorides (Scheme 5). Using 1-Butyl-3-methylimidazolium chloroaluminate ionic liquids as a Lewis acid catalyst.

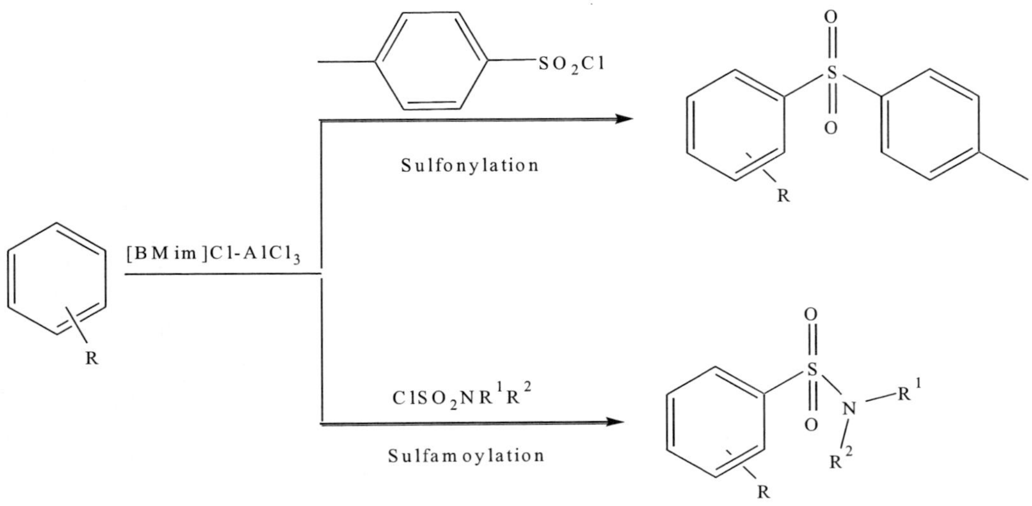

Scheme 5.

^{27}AlNMR spectroscopy has been explored as a tool to investigate the mechanistic details of the reaction. Spectral studies show the predominance of $[Al_2Cl_7]^-$ species in [BMim]Cl-$AlCl_3$, $N = 0.67$, acidic ionic liquid in the presence of 4-methyl benzenesulfonyl chloride, and after the reaction with the aromatic hydrocarbon, $[AlCl_4]^-$ species predominates. This change in speciation of aluminum can be attributed to the interaction of the Lewis acidic species $[Al_2Cl_7]^-$ of the ionic liquid with the formed HCl during the sulfonylation reaction, which was further evidenced by the control experiment [32, 33].

4.2. Diels- Alder Reaction

Carols Lee firstly reported the Diels-Alder reaction between cyclopentadiene with methyl acrylate to give endo/exo products using room temperature chloroaluminate ionic liquids as solvent/catalyst (Scheme 6). The author reported, selectivity and reactivity of reaction was strongly influenced by the Lewis acidity of the chloroaluminate IL [34, 35].

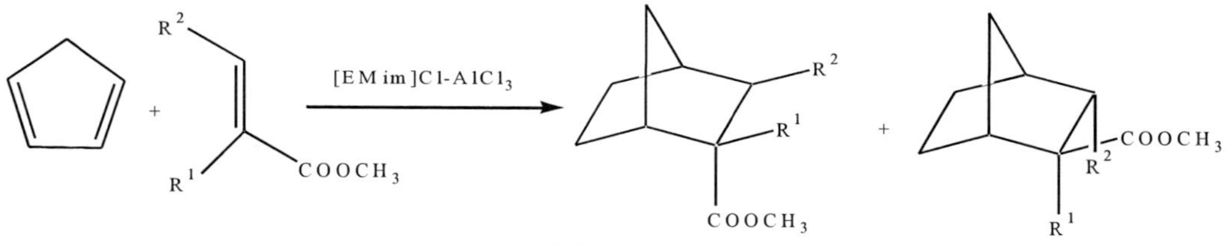

Scheme 6.

Later on, Abbott et al. reported the choline chloride based zinc- or tin-containing ionic liquids as water insensitive, recyclable catalysts for Diels –Alder reaction [36].

4.3. Henry Reaction

Kumar et al. reported Henry reactions in various chloroaluminate ionic liquids. The author reported the higher compositions of organic species of the chloroaluminates was more efficient as compared with lower compositions organic species in catalyzing reaction (Scheme 7).The order in which the organic cations influence the yields for these reactions is shown below.

[EMIM] > [MEIM] > [BPC] > [BMP] > [BMIM]

Scheme 7.

Both aliphatic and aromatic carbonyl compounds show the good yield of product further the ionic liquids were recycled five times [37].

4.4. Fischer Indole Synthesis

Khadilkar and co-worker described, Fisher Indole synthesis of different ketones using 1-butylpyridinium-$AlCl_3$ ionic liquid as a solvent as well as a catalyst (Scheme 8). The amount of $Alcl_3$ required is much lesser than other reported catalyst [38].

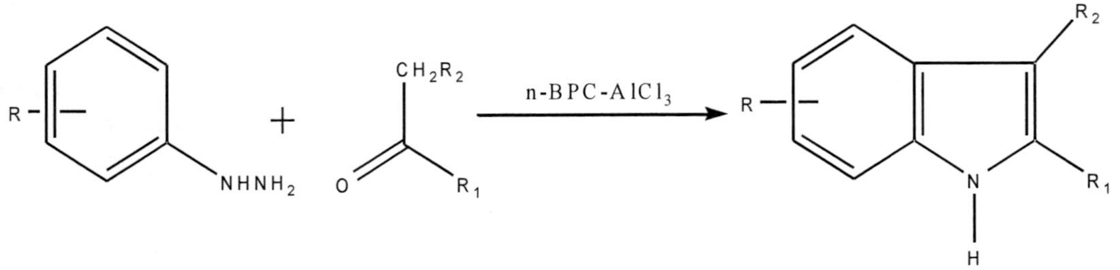

Scheme 8.

Subsequently, Abbott and co-worker reported, regiospecific Fischer indole reaction in choline chloride·$2ZnCl_2$., the reaction gives high yield with one equivalent of the ionic liquid, exclusive formation of 2,3-disubstituted indoles is observed in the reaction of alkyl methyl ketones, and the products readily sublime directly from the ionic liquid [39].

5. Transition Metal Catalyzed Reaction Using Ionic Liquid as a Media or Support

Transition metal catalyzed reactions have gained importance in modern organic chemistry, as powerful synthetic transformations can be facilitated with high selectivity (chemo-, regio-, and enantio-selectivity). In 2001, trio Sharpless, Noyori and Knowles shared Nobel Prize for their research on "homogeneously catalyzed asymmetric catalysis" and similarly, Chauvin, Grubbs and Schrock received 2005 Nobel for the "development of the metathesis method catalyzed by transition metals in organic synthesis". Recently, Heck, Negishi and Suzuki received 2010 Nobel for their work on "palladium catalyzed cross coupling in organic synthesis", Their discoveries have been made use of transition metals in the production of chiral selective drugs and materials.

5.1. Ionic Liquids as a Reaction Media

Ionic liquids can dissolve organometallic compounds and provide a polar, weakly coordinating medium for transition-metal catalysts [1, 40]. In this case, ionic liquids are used as inert solvents or co-catalysts. Application of ionic liquids as a reaction media for transition catalyzed reaction is early reviewed by Wasserscheid et al. [13] and Liu et al. [14].

Ionic liquids are attractive as potential solvents for a number of reasons:

1. Relatively low viscosities and high thermal stability.
2. Exhibit very low vapor pressures and thus are effectively non-volatile.
3. Good solvents for a wide range of both inorganic and organic materials and unusual combinations of reagents can be brought into the same phase.
4. Highly polar yet non co-ordinating solvents with the metal catalysts which can result in enhanced rate of reaction.
5. Immiscible with a number of organic solvents which provide a non-aqueous, polar alternative for two-phase systems to effectively carry out catalyst-product separation. Hydrophobic ionic liquids can be used as immiscible polar phases with water.
6. Being composed of two parts, there is a synthetic flexibility that is not available to conventional solvents. This makes the spectrum of their physical and chemical properties much larger than that of organic solvents.
7. The use of highly active phosphites, phosphonites and phosphinites as modifying ligands for two-phase catalysis comes within reach in room temperature ionic liquids because degradation reactions, such as hydrolysis are less likely to occur as compared to water.

5.2. Ionic Liquids as a Support

Ionic liquids have been demonstrated to be ideal immobilizing agents for various “classical” transition-metal catalyst precursors in reactions ranging from Ziegler-Natta type to the hydroformylation of olefins and are useful in organic synthesis [41, 42]. In all these applications the ionic liquid can play its role as either “innocent” solvent, ligand precursor, co-catalyst or the catalyst itself depending on the specific cation/anion combination and the reaction under investigation.

5.2.1. Supported Ionic Liquid Phase (SILP) Catalysis

Supported Ionic Liquid Phase (SILP) catalysis describes the heterogenization of a homogeneous catalyst system by confining an ionic liquid solution of catalytically active complexes on a solid support [43, 44]. .

SILP concept is more advantageous than biphasic ionic liquid methodology because of high price of ionic liquid and high viscosity leading to slow mass transfer. In the SILP concept only very thin films of the ionic liquids are used in a highly efficient manner. Since solubility for CO and H_2 in the ionic liquid is low (and diffusion of CO is slow), liquid-liquid biphasic catalysis with a ionic catalyst solution does not take place in the bulk of the ionic liquid, but in the diffusion layer between the gas phase and the ionic liquid. This results in a large amount of the precious catalyst metal not being utilized during the reaction. An ideal system would consist of a bulk phase of the size of the diffusion layer thus allowing for the complete rhodium inventory to be catalytically active; a condition that is fulfilled in an ideal manner by the SILP concept [45].

5.3. Hydroformylation Using IL as Media

The first example of homogeneous transition metal catalysis in an ionic liquid dates back to 1972 when Parshall described the platinum catalysed hydroformylation of ethene in tetraethylammonium trichlorostannate (mp. 78°C) [46]. The reaction conditions applied were harsh, with 90 °C and 400 bar syngas pressure.

This work was followed by the pioneering studies of Knifton who reported the ruthenium- and cobalt catalyzed hydroformylation of internal and terminal alkenes in molten $[Bu_4P]Br$ in 1987 [47]. The ionic medium was found to stabilize the catalyst species, indicated by improved catalyst lifetime at low syngas pressures and higher temperatures.

The first investigations of the biphasic rhodium-catalyzed hydroformylation in room temperature ionic liquids were published by Chauvin et al. in 1995 [48]. The hydroformylation of 1-pentene with the neutral catalysts $[Rh(CO)_2(acac)]$/ triarylphosphine was carried out in a biphasic reaction using $[BMim][PF_6]$ as the ionic liquid with a 99 % yield and a TOF of 333 h^{-1} under relatively mild conditions (80 ºC, 2 MPa of syn gas (H_2/CO, 1:1) pressure, 2 h). However, in this case, a small part of the rhodium catalyst is leached into the organic phase.

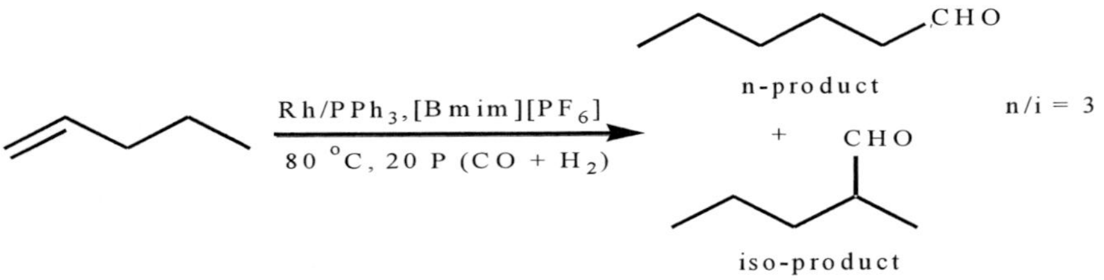

Scheme 9.

Later on, Favre et al. and Kottsieper et al. found that the leaching of the Rh catalyst can be suppressed by the modification of P-ligands by cationic (guanidium or pyridinium) and anionic (sulphonate) groups (Figure 5) [49, 50].

A series of modified imidazolium and pyridinium ionic phosphines have been prepared by the P-nucleophilic addition to 1-vinyl-imidazole [51] or 4-vinylpyridine [52]. The corresponding Rh complexes, immobilized in various imidazolium and pyridinium ionic liquids, have been employed in the hydroformylation of various olefins

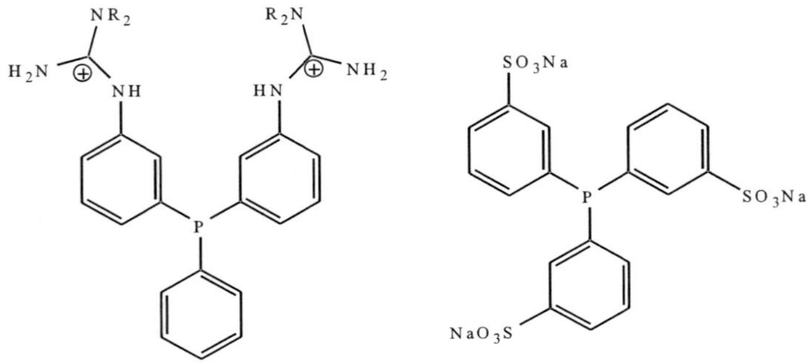

Figure 5.

The ionic liquid $[BMim][PF_6]$ appears to be the of choice ionic liquid in the hydroformylation of olefins in particular for the product isolation and catalyst recycling steps. $[BMim][PF_6]$ stabilizes the Rh catalyst under thermal stress during product distillation. Also, since this ionic liquid is almost immiscible with water as well as nonpolar solvents such as hexanes, extraction methods or supported ionic liquid phase (SILP) systems can be used for the separation and catalyst recycle steps and this circumvents the problems associated with distillation or other traditional methods of catalyst-product separation [53].

5.4. Hydroformylation Using Supported Ionic Liquid-Phase (SILP) Systems

5.4.1. Supported Ionic Liquid Catalysis by Impregnation

Supported ionic liquid catalysts by impregnation were successfully applied to Rh catalyzed hydroformylation reaction using silica [7] as support. By taking the advantages of SILP our group has reported, the hydroformylation of methyl acrylate using Rh/PPh_3-SILP catalyst system in water (Scheme 10) [54, 55].

The generated branched aldehydes can be further converted without purification to 4-methyl-substituted pyrazolin- 5-ones in the presence of an acid catalyst (Path A). Using the same system, the regioselective hydroformylation of allyl alcohol was successfully performed in water, followed by subsequent hydrogenation of the hydroformylation products using a Ru based SILPC system gives 1,4-butanediol in a good yield (Path B).

This technique is advantageous because rhodium leaching into the aqueous phase can be avoided and easy catalyst/product separation is available. The catalyst also does not require modification of PPh_3 ligand to air-sensitive TPPTS ligand. The rates of reaction and the product distributions strongly depend on various reaction variables such as temperature, pressure, catalyst concentration and nature of the ligand.

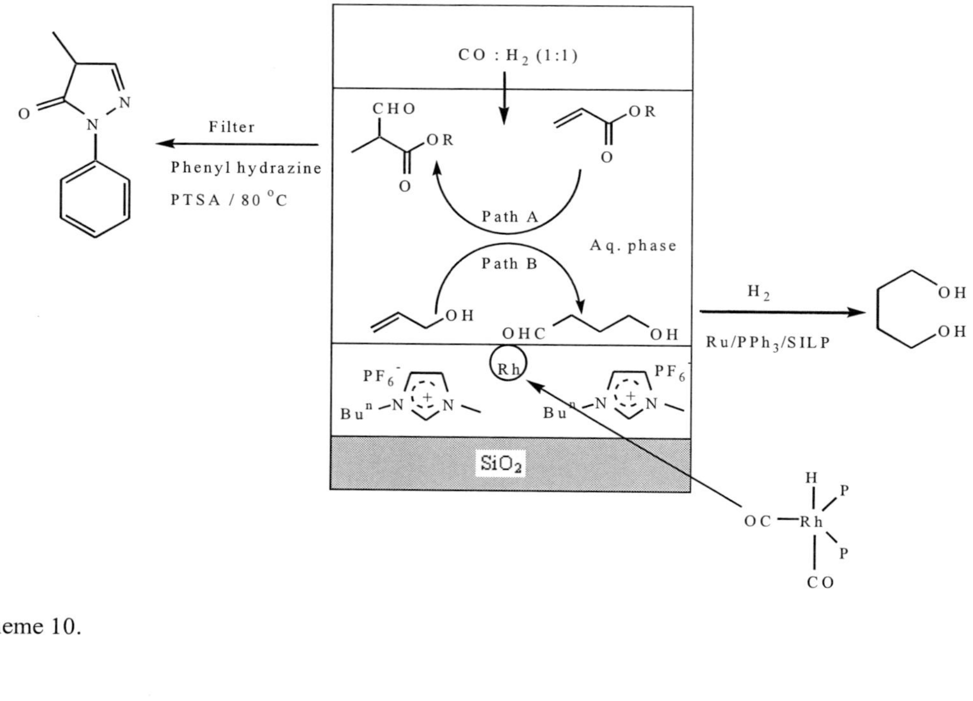

Scheme 10.

5.4.2. Supported Ionic Liquids by Chemical Bonds

In 2002, Mehnert and co-workers were the first to apply SILP-catalysis to Rh catalyzed hydroformylation [56]. Rh-SILP catalysts based on silica gel support modified with a monolayer of covalently anchored ionic liquid fragments of 3-triethoxysilylpropylimidazolium (average of 0.4 ionic liquid fragments per nm^2) were prepared and the results obtained in 1-hexene hydroformylation were compared with ionic liquid-liquid biphasic system (Scheme 11). Mehnert et al. found that the SILP-catalyst exhibited a higher TOF (63 min^{-1}) compared with the biphasic system (23 min^{-1}) due to the higher concentration of Rh at the interface and the generally larger interface of the supported system

.

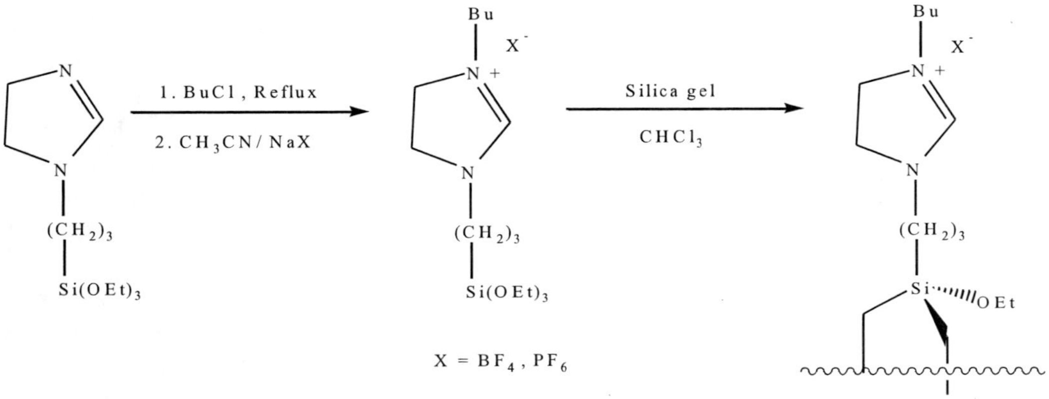

Scheme 11.

5.5. Hydrogenation Using IL as Media

The catalytic hydrogenation of unsaturated bonds is an important transformation in the industrial production of fine and bulk chemicals. However, separation of product from the reactants remains problematic.

The first example of catalytic hydrogenation using ionic liquid as a media was reported by Chauvin et al. in 1995 [57]. A solution of the cationic [Rh(nbd)($Ph_3P)_2$]PF_6 complex [nbd = norbornadiene (bicyclo[2.2.1]hepta-2,5-diene)] in [BMim]PF_6 or [BMim]SbF_6 was shown to be an effective catalyst for the biphasic hydrogenation of pent-1-ene. Reaction rates were up to five times higher than in acetone as solvent, in contrast, poor results was obtained with [BMim][BF_4] which was ascribed to the presence of trace amounts of strongly coordinating chloride ions in their sample of this ionic liquid.

Same group taken advantage of the biphasic system to perform the selective hydrogenation of cyclohexadiene to cyclohexene (Scheme 12). Dupont and coworkers [58] performed the biphasic hydrogenation of cyclohexene with Rh(cod)$_2BF_4$ (cod = cycloocta-1,5-diene) in ionic liquids. They observed roughly equal rates (turnover frequencies of ca. 50 h^{-1}) in [BMim][BF_4] and [BMim][PF_6](presumably their [BMim][BF_4] was chloride-free).

The same group showed that $RuCl_2(Ph_3P)_3$ in [BMim][BF_4] is an effective catalyst for the biphasic hydrogenation of olefins, with turnover frequencies up to 540 h^{-1} [59].

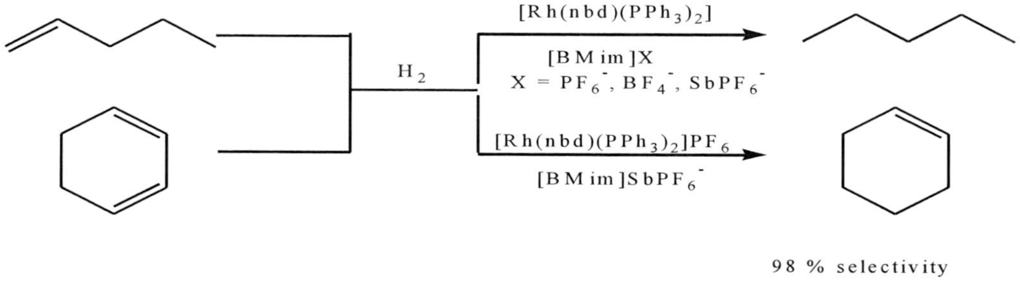

Scheme 12.

Similarly, [(BMim)$_3$]-Co(CN)$_5$ dissolved in [BMim][BF_4] catalyzed the hydrogenation of butadiene to but-1-ene, in 100% selectivity at complete conversion. Further the same group reported the enantioselective hydrogenation of 2-(6-methoxy-2-naphthyl) acrylic acid using chiral [$RuCl_2$(S)-BINAP]$_2$NEt$_3$ complex in [BMim][BF_4]–isopropyl alcohol system [60] to afford anti-inflammatory drug, (S)-naproxen, in 80% ee (Scheme 13).

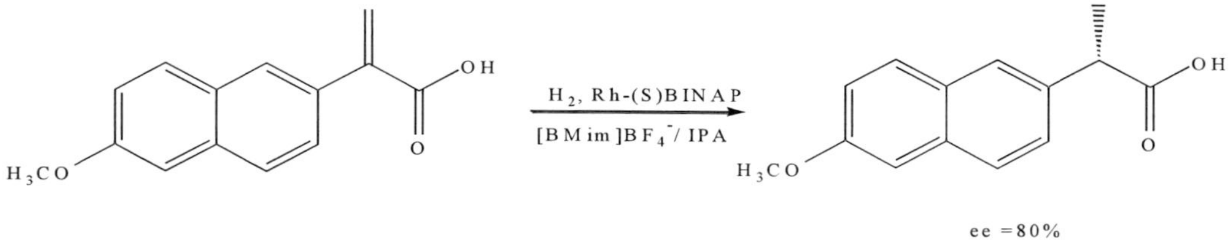

Scheme 13.

Tumas and co-workers investigated, biphasic ionic liquid-supercritical CO_2 system for hydrogenation of olefins [61]. Jessop et al reported, asymmetric hydrogenation of tiglic acid and the precursor of the anti-inflammatory drug ibuprofen by using Ru(OAc)$_2$((R)-tolBINAP) as catalyst [62].

5.6. Hydrogenation Using SILP and Functionalized Ionic Liquids System

Hydrogenation and hydroformyaltion using supported ionic liquid phase (SILP) was first reported and reviewed by Mehnert and co worker [44, 63]. Mehnert et al. reported hydrogenation of 1-hexene, cyclohexene, and 2,3-dimethyl-2-butene using Rh-complex [Rh(norbornadiene)(PPh$_3$)$_2$][PF$_6$] and [BMim][PF$_6$] were impregnated on silica gel (289 m^2g^{-1}). For comparison the Rh-catalyst was also investigated in unsupported ionic liquid medium [63].

In comparison to homogeneous and liquid–liquid biphasic system, the amount of the required ionic liquid phase was drastically reduced and the new catalyst system enabled the usage of the preferred fixed-bed technology.

Later on, Mikkola et al suppoeted Pd(acac), on a structurally active carbon cloth (ACC) in three different ionic liquids [BMim][PF_6], [BMim][BF_4] and [A336][PF_6] respectively. Theses catalyst system were further applied for hydrogenation of citral, the selectivity of citral hydrogenation depend on the type of ionic liquid and amount of transition metal [64].

Further on various group reported diverse heterogeneous support for metal complexes or nanoparticle (Ru and Pd.), Such as molecular sives [65], carbon nanofibers (CNF) [66], alumina [67], multiwalled carbon nanotubes (MWCNTs) [68], these supports were attractive because of inherent advantageous properties such as good mechanical strength, high chemical stability, and large surface area-to-volume ratio.

Recently, Zhang et al reported hydrogenation of carbon dioxide using basic task specific ionic liquid (*N,N*-dimethylaminoethyl)-2,3-dimethylimidazolium trifluoromethanesulfonate ([mammim][TfO] as promoter in heterogeneous catalyst "Si"-$(CH2)_3$-NH($CSCH_3$)-RuCl3-PPh_3 [69]. Hu et al prepared palladium nanoparticles using the functionalized ionic liquid as a ligand.

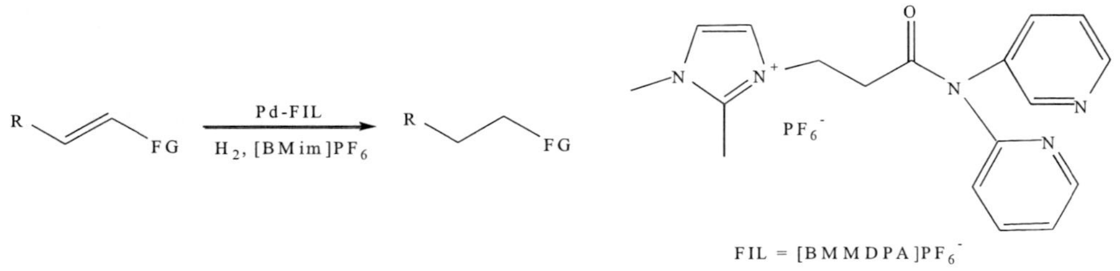

Scheme 14.

The ionic liquid (2,3-dimethyl-1-[3-*N,N*-bis(2-pyridyl)-propylamido] imidazolium hexafluorophosphate, ([BMMDPA][PF_6]) plays a very important role in stabilizing, dispersing, as well as modifying the palladium nanoparticles towards the selective hydrogenation of various functionalized alkenes(Scheme 14), [70].

5.6.1. Asymmetric Hydrogenation

Wasserscheid and co-worker recently reported hydrogenation of a prochiral ionic liquid cation in presence of achiral catalyst and enantiomerically pure anion (Scheme 15).

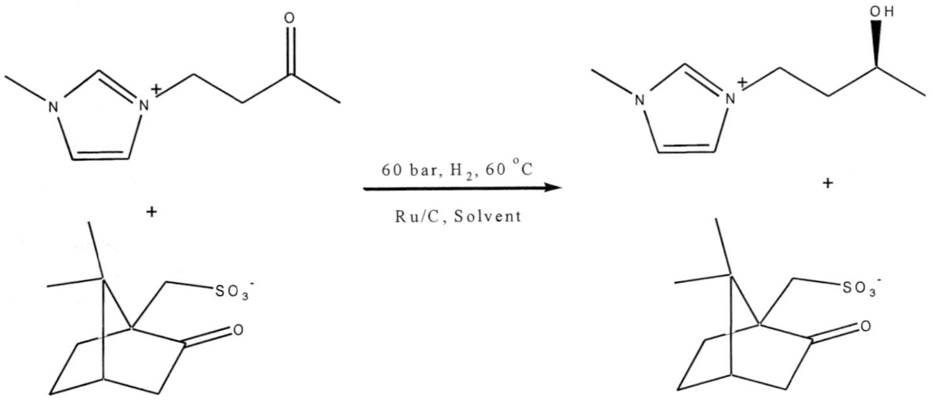

Scheme 15.

Chirality transfer in imidazolium camphorsulfonate ionic liquids through ion pairing effects to give hydrogenated cation with up to 80% ee [71].

5.7. Hydroamination Reaction

Hydroamination of olefins is used to synthesize numerous classes of organo-nitrogen molecules such as alkylated amines, enamines or imines that are used as chemical intermediates in the area of natural products, pharmaceuticals, fine chemicals, dyes, polymers and surfactants.

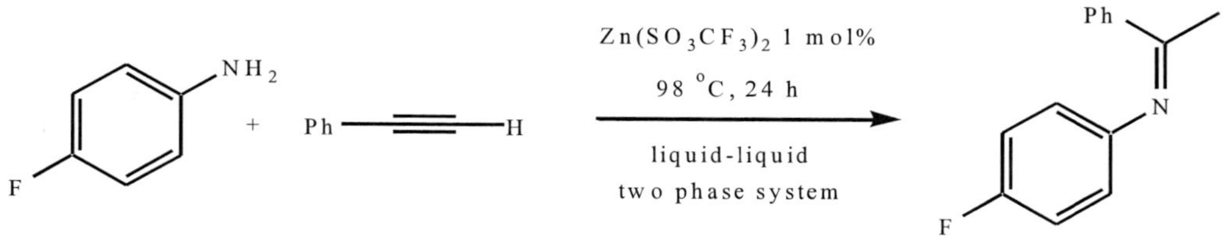

Scheme 16.

Lercher and Co-worker reported, $Zn(OTf)_2$-catalyzed hydroamination reactions of phenylacetylene with alkyl or arylamines; the ionic liquid 1-ethyl-3-methylimidazolium trifluoromethanesulfonate was used as the polar phase that contains the polar catalyst ($Zn(OTf)_2$) while heptane was used as the non-polar solvent for the starting materials and the product (Scheme 16).

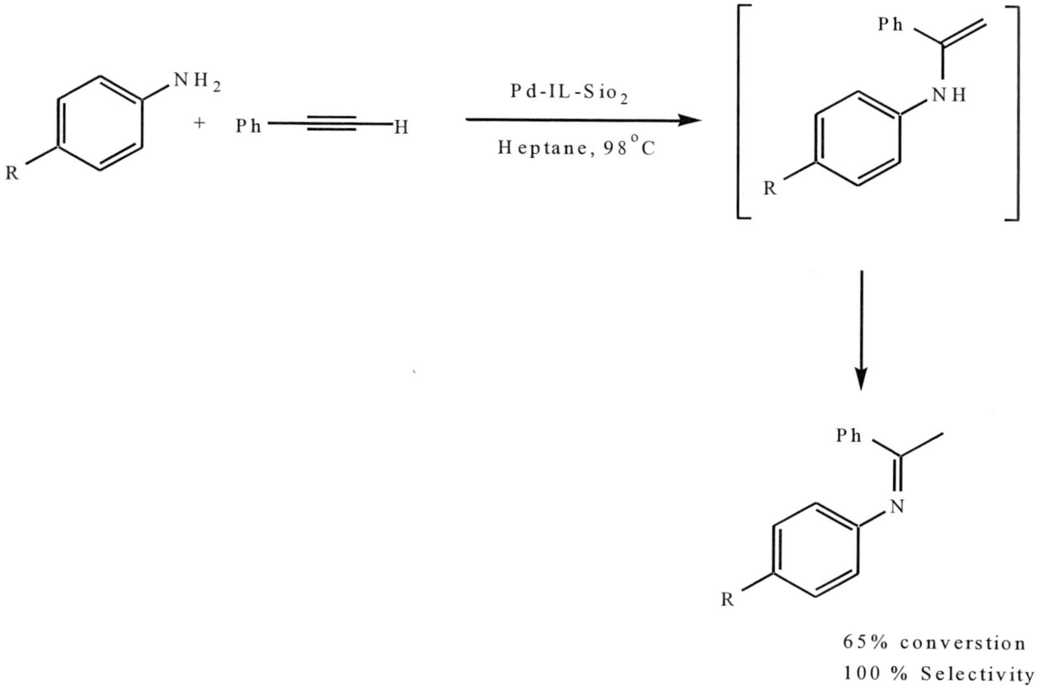

Scheme 17.

After the reaction, the catalyst remained quantitatively in the ionic liquid which was easily separated from the heptane phase. Subsequent concentration of the organic phase under vacuum gave direct access to the pure hydroamination product [72, 73].Later on, Breitenlechner et al explored scope for the immobilization of homogeneous catalysts in supported ionic liquids. In this study, cationic metal complexes of Rh(I), Pd(II) and Zn(II) were supported on silylated diatomic earth (10 m^2 g^{-1}) using 1-ethyl-3-methylimidazolium trifluoromethanesulfonate [EMim][OTf] as immobilizing agent. The resulting catalysts were tested in the direct addition of 4-isopropylaniline to phenylacetylene in n-heptane as the co-solvent [74]. The reaction yields the enamine which in situ isomerizes to the corresponding imine (Scheme 17)

With the same concept films of silica supported ionic liquids Jimenez et al. reported novel bi-functional catalytic system combining soft Lewis acidic and strong Brønsted acidic functions (Scheme 18).

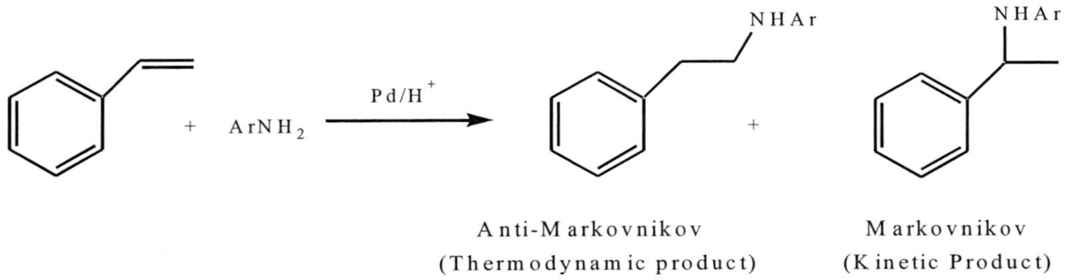

Scheme 18.

The synthesized catalyst showed exceptional catalytic activity for the addition of aniline to styrene, providing the Markownikoff product under kinetically controlled conditions and mainly the anti-Markownikoff product in the thermodynamic regime [75]. Recently Yang et al reported hydroamination of nonactivated alkenes with sulfonamides, carboxamides, p-nitroaniline and carbamates using SO_3H-functionalized ionic liquids as efficient and reusable catalysts. The hydroamination could be performed on a large scale and the acidic ionic liquid catalyst could be reused successfully [76].

5.8. Heck Reaction

Palladium catalyzed coupling of aryl halides with alkenes is known as Heck reaction. The palladium-catalyzed Heck reaction of olefins with aryl and vinyl halides is actually a method of broad scope that has found application throughout organic chemistry. However, the high consumption of the expensive palladium catalyst makes it a relatively impractical process on an industrial scale. Recycling the catalyst is therefore a key objective.

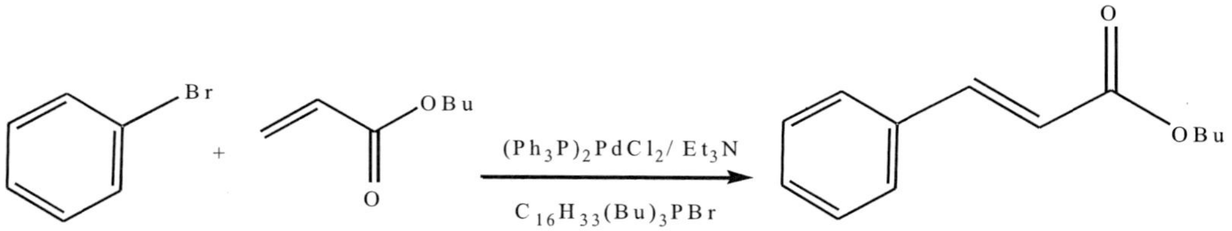

Scheme 19.

Kaufmann et al [69, 77], firstly reported Heck reaction of bromobenzene with butyl acrylate in molten tetraalkylammonium and tetraalkylphosphonium bromide salt to produce butyl trans-cinnamate in high yield (Scheme 19). No formation of elemental palladium metal was observed and the product was isolated by distillation from the ionic liquid.

Later on, Herrmann and Bohm [78-79] reported Heck reaction of chlorobenzene with styrene in molten Bu_4NBr with 0.5 mol% phospha-palladacycle catalysts loading resulting in easy product separation, catalyst recycling and further increases in catalyst productivity. The author reported increase in yield of stilbene from 20 % in DMF to 99% in Bu_4NBr.

Seddon and coworkers performed Heck couplings in two different ionic liquid [BMim][PF_6] or [hexylpyridinium][Cl] and interestingly observed siginificant difference in reactivity. In the case of imidazoium ionic liquids, the 2-H proton is acidic and in the presence of a base can be deprotonated to form a carbene. These imidazolylidine carbenes formed from imidazolium cation may take part in the catalytic cycle [80]. This was later confirmed by Xiao and coworkers. Author observed a significantly enhanced rate of the Heck coupling in [BMim][Br] compared to the same reaction in [BMim][BF_4]. This difference could be explained by the formation of the corresponding palladium–carbene complexes (which were isolated and characterized) in the [BMim][Br] but not in the latter [BMim][BF_4]. The isolated carbene complexes were shown to be active catalysts when redissolved in [BMim][Br]. Presumably, formation of the carbene in [BMim][Br] can be attributed to the stronger basicity of bromide compared to tetrafluoroborate.

Further, Srinivasan et al. studied Heck reaction of iodobenzene with ethyl acrylate in [BBim][BF_4] carried out under ultrasound conditions to in situ transmission electron microscopy.These studies showed the presence of highly stabilized clusters of Pd (0) nanoparticles. Further of this authors also demonstrated the formation of a Pd–carbene complex by subjecting a mixture of $Pd(OAc)_2$ or $PdCl_2$ and NaOAc in the ionic liquids, [BBim][Br] and [BBim][BF_4], to ultrasonication for an hour. The formation of the carbene complex was confirmed by ^1H NMR analysis [81]. So, it was suggested that, the active catalyst in heck reaction is a palladium nanoparticle.

5.8.1. Heck Reaction by SILP

Hagiwara et al reported $Pd(OAc)_2$ supported on amorphous silica using [Bmim][PF_6] as a liquid support was prepared and used as a heterogeneous catalyst [82]. Catalyst was characterized using SEM, EPMA, and AFM technique. The obtained Pd-catalyst was used at least six cycles. Later on various group have reported diverse solid support with the aid of ionic liquids for different pd-precursors such as reversed phase silica gel [83], Ordered mesoporous silica MCM-41 [84], ILmodified mesoporous SBA-15 [85], 1,1,3,3-

tetramethylguanidinium (TMG)-modified molecular sieve SBA-15 (SBA-TMG-Pd) [86], Ionogels [87].

5.8.2. Heck Reaction in Functionalized Ils

Zhao et al, synthesized various nitrile functionalized ionic liquids and shown their application in coupling reaction. The author reported the formation of Pd (0) nanoparticle in imidazolium-based ionic polymer and the nitrile-functionalized ILs which was further isolated and applied for coupling reaction [88-91].

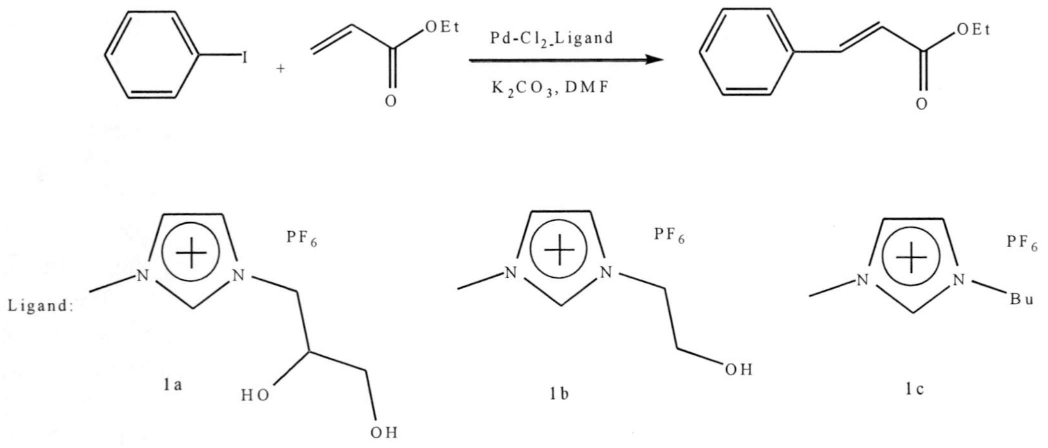

Scheme 20.

Later on, Shreeve and co-worker reported a series of pyrazolyl- and 3,5-dimethylpyrazolyl-functionalized 2-methylimidazolium-based salts and their thermal properties were determined by DSC and TGA. These salts reacted easily with palladium (II) chloride to generate mononuclear palladium ionic liquid complexes. The catalytic activity and recyclability of the palladium complexes in the corresponding ionic liquids were examined for Heck, Suzuki and Sonogashira cross-coupling reactions in the absence of phosphine ligands [92, 94].

Recently, wang et al reported diverse type of functionalized ionic liquid such as porphyrin functionalized pyridinium-based IL [95], hybrid P,N-ligand functionalized imidazolium-based IL [96] and diol-functionalized imidazolium based IL (Scheme 20), [97] shown their application in palladium-catalyzed Heck reactions as a ligand or solvent.

Recently, Wang et al designed and synthesized novel task-specific ionic liquid based on ethanolamine-functionalized quaternary ammonium salt, 4-Di(hydroxyethyl)aminobutyl tributylammonium bromide (DHEABTBAB), which acts as a base, ligand and reaction medium, and exhibits a very high activity and recyclability to palladium-catalyzed olefinations of iodoarenes, bromoarenes and chloroarenes with olefins to yield excellent product yields under phosphine-free reaction conditions (Scheme 21).

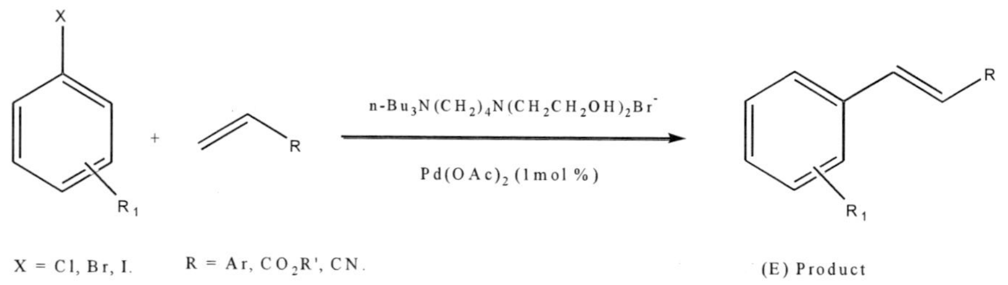

Scheme 21.

It is notable that palladium and DHEABTBAB could be repeatedly recycled and reused for six consecutive trials without significant loss of their activities [98].

5.9. Suzuki Reaction

Palladium catalyzed coupling between aryl halides or aryl trifaltes and aryl boronic acid is known as Suzuki reaction. This is the other coupling reaction mostly studied in ionic liquids. Alonso et al well reviewed the methodologies of Suzuki reaction in non aqueous medium [99]. Welton and co-worker firstly reported palladium catalysed Suzuki cross-coupling reactions in ambient temperature [BMim][BF_4] ionic liquid shown advantages over conventional solvent exhibiting unprecedented reactivity in addition to easy product isolation and catalyst recycling [100]. Author reported increase in reaction rate as compared with the conventional Suzuki conditions. The reaction of bromobenzene with phenylboronic acid under conventional Suzuki conditions result gave an 88% yield in 6 h (TON, 5 h^{-1}), while the equivalent reaction in [BMim][BF_4] gave 93% in 10 min (TON, 455 h^{-1}).

Subsequently, Srinivasan et al. reported reaction of halobenzenes with phenylboronic acid under mild conditions in ionic liquid with methanol as a co-solvent using ultrasound [101]. Calo et al studied the effects exerted by different ionic liquids on catalyst stability, reaction rates and regio- and stereoselectivity based on the Coulombic interaction between the cations and anions in ILs. Eg. the tetraalkylammonium halides and acetate, the bulkiness of tetrahedral tetraalkylammonium cations, which forces the anions away from the cation, renders these anions available for a good activity and stability of the palladium catalysts. On the contrary, the planar structures of imidazolium and pyridinium cations, due to a strong Coulombic interaction that binds the anions tightly, decrease their availability for stabilisation and activity of the catalysts [102].

5.9.1. Suzuki Reaction Using SILP and Functionalized Ionic Liquids System

Hagiwara et al. extensively studied the Suzuki reaction using diverse support for the immobilization of $Pd(OAc)_2$-catalyst such as amorphous silica [103], reversed phase amorphous silica [104], diethylaminopropylated (NDEAP) alumina pores [105], reversed phase alumina [106], nanosilica dendrimers [107]. The author reported that among ionic liquids tested, [BMim][PF_6] was better to hold $Pd(OAc)_2$ than [BMim][Br] [BMim][$(CF_3SO_2)_2N$] or [HMim][PF_6].

Recently, Karimi, et al. novel palladium-supported periodic mesoporous organosilica based on an alkylimidazolium ionic liq. (Pd@PMO-IL) as a efficient and reusable support for Suzuki Reaction of a variety of activated and deactivated haloarenes with arylboronic acids [108].

Dyson and co-worker first time shown the application of nitrile-functionalized pyrrolidinium-based ionic liquids as reaction media for Suzuki and Stille C-C cross-coupling reaction, the results was compared with imidazolium and pyridinium systems (including those with and without nitrile functionalities). Furthermore the nature of the ionic liquid strongly influences the catalyzed reaction and it would appear that, in addition to the nitrile group, the strength of anion-cation pairing in the ionic liquid and the viscosity of the ionic liquid play critical roles. In situ generation of nanoparticle was also reported [109, 110]. Same group reported the hydroxyl-functionalized ILs [111] and ether/polyether functionalized imidazolium/ pyridinium-based ILs [112] (Scheme 22). The reactions proceed more efficiently in these solvents due to better stabilization of the palladium catalyst by the hydroxyl/ether groups through weak coordinative interactions.

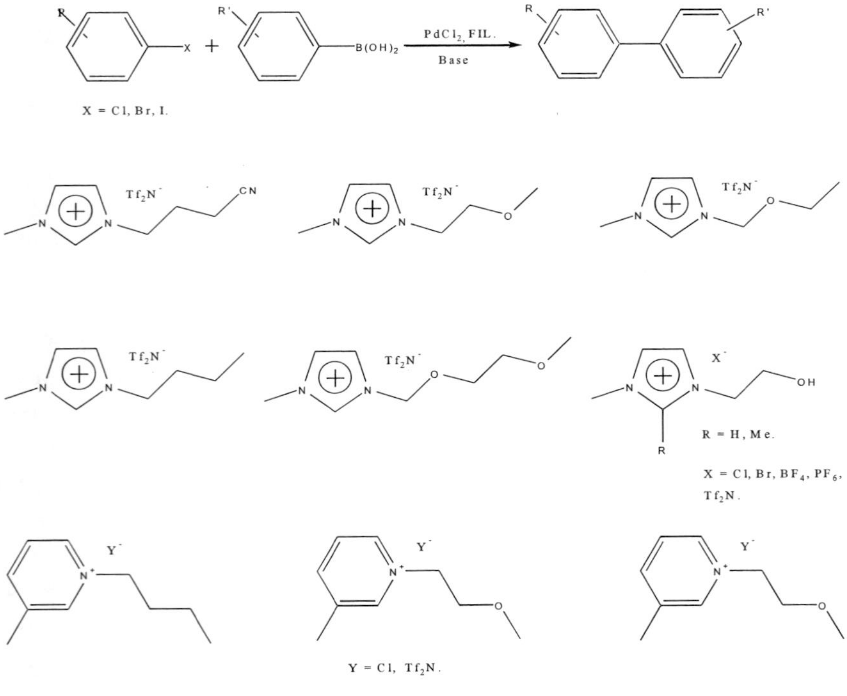

Scheme 22.

The position and the number of oxygen atoms in the ether side chain strongly influence the outcome of the coupling reactions in the imidazolium-based ILs. Although imidazolium and pyridinium-based ILs provide the same activity for the Suzuki reaction, The pyridinium-based ether-functionalized ILs are preferred over the imidazolium counterparts due to their increased stability and inability to form carbene-containing species. Similarly, Shreeve and co-worker reported the pyrazolylfunctionalized ILs with Pd-catalyst were efficient system for Suzuki reaction [113, 114].

5.10. Negishi Reaction

As compared to Suzuki and Heck reaction Negishi coupling is less explored in ionic liquids. Sirieix et al reported, Palladium catalyzed Negishi cross-coupling reactions between aryl- or benzylzinc halides and various aryl iodides in 1-butyl-2,3-dimethylimidazolium tetrafluoroborate [BDMim][BF_4] ionic liquids using a novel phosphine (Scheme 23) prepared by the reaction of PPh_2Cl with [BMim][PF_6][115].

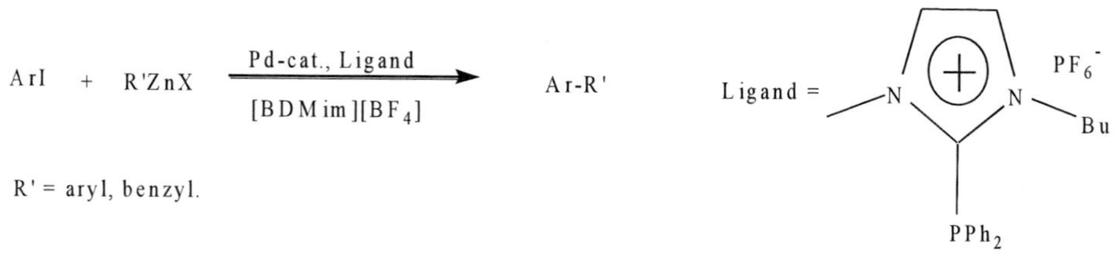

Scheme 23.

This solvent allows a facile work-up and rapid cross-coupling reactions at room temperature

6. Task Specific Ionic Liquids in Organic Synthesis

The term "Task specific ionic liquids was first coined by Davis and successfully applied for the CO_2 capture [20,21]. This novel salt readily and reversibly sequesters CO_2 (Scheme 24).

The incorporation of functional groups in either cation or anion can impart a particular capability to the ionic liquids, enhancing their ability for catalyst like Bronsted acidic ionic liquids and Bronsted basic ionic liquids.

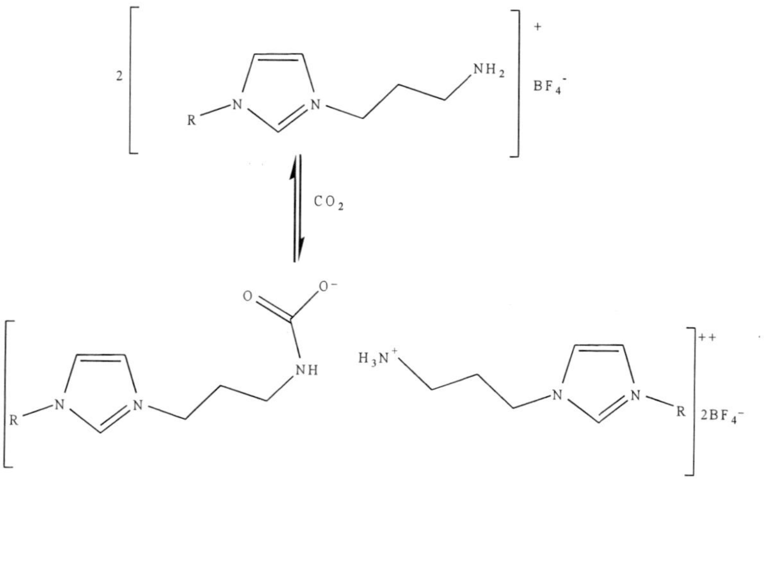

Scheme 24.

6.1. Bronsted Acidic Ionic Liquids (BAILS)

6.1.1. Bronsted Acidity Determination

Organic transformations are sensitive to the presence of protons. Therefore, knowledge of the *p*Ka value is important in order to decide about their possible application as reaction media or catalyst. Gilbert et al. reported acidity scale for Brønsted acids such as $HNTf_2$ and HOTf soluble in non-chloroaluminate ionic liquids [116]. With the same concept Duan et al reported the Bronsted acidity of pure ionic liquids in dichloromethane on the basis of the Hammett acidity function (Hammett method), wherein a basic indicator was used to trap the acidic proton, using UV-Vis spectrophotometer [117].

Hammett method consists of evaluating the protonation extent of uncharged indicator bases (named I) in a solution, in terms of the measurable ratio [I]/[IH+]. Further Fei et al. have synthesized a series of Brønsted-acidic ionic liquids and determined their solid-state structure as well as *p*Ka values. The *p*Ka values of all the salts, was determined by potentiometric titration with KOH [118].

Table 1. *p*Ka values of carboxylic acids in H_2O at 25°C

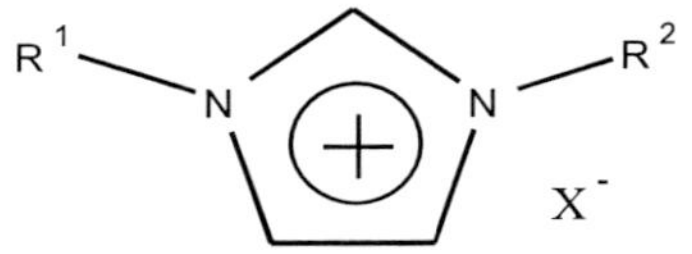

Entry	R^1	R^2	X	*p*Ka
1	CH_3	CH_2COOH	Cl	1.90
2	CH_3	$(CH_2)_3COOH$	Cl	3.83
3	CH_2COOH	CH_2COOH	Cl	1.33
4	$(CH_2)_3COOH$	$(CH_2)_3COOH$	Cl	3.46
5	CH_3	CH_2COOH	BF_4	2.00
6	CH_3	CH_2COOH	CF_3SO_3	2.30
7	CH_2COOH	CH_2COO^-	-	2.92

Increasing the alkyl chain length leads to a decrease in acidity (entry 1, 2). This indicates presence of positive charge on the imidazolium ring has only a minor effect on the acidity. As with aliphatic dicarboxylic acids, the dicarboxylic acid with the shortest chain length (entry 3) was the most acidic in the series. The *p*Ka of the acid with R^2=CH_2COOH (entry 1) is much lower than that of chloroacetic acid, suggesting that the positively charged imidazolium group is a stronger electron withdrawing group than a chloro group. In aliphatic carboxylic acids, the inductive effect of the imidazolium ring drops dramatically with increasing chain length (entry 4). Simultaneously Yang and Kou have determined the Lewis acidity of several chloroaluminate ionic liquids by means of an IR spectroscopic probe method [119]. The first ionic liquid ethyl ammonium nitrate (EAN) reported by Paul Walton consider to be as a Bronsted acidic ionic liquids [17]. However, in the last decade the BAILs have been specifically designed as catalysts, reagents and/or solvents in organic synthesis was explored [120, 121].

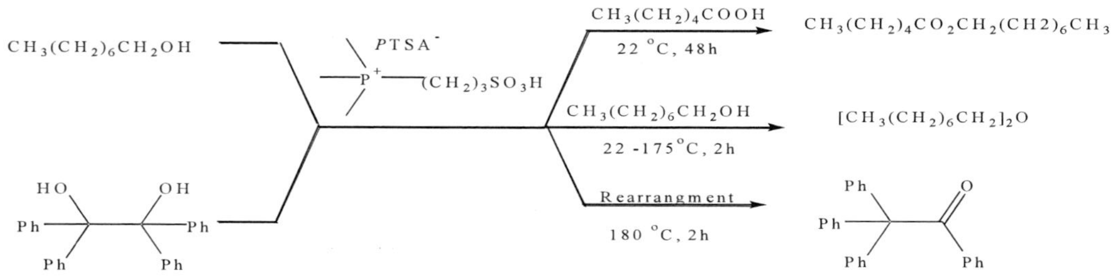

Scheme 25.

Cole et al. designed (BAILs) containing an alkane sulfonic acid group covalently attached to the ionic liquid cation and applied for the transformations such as esterifications, ether formation, and the pinacol–pinacolone rearrangement as a catalyst as well as solvent (Scheme 25) [122].

6.2. Esterification Reaction

Esterification of carboxylic acids using liquid inorganic acid catalysts is well known. However, removal of water /use of excess amount of the reactants is generally needed for satisfactory conversion, also it is very difficult to recycle the liquid inorganic acid catalysts. Zhu et al reported, practical and efficient procedure for esterification of carboxylic acids with alcohols is reported, using a Brønsted acidic ionic liquid, 1-methylimidazolium tetrafluoroborate [HMim][BF_4] as recyclable catalyst and solvent (Scheme 26).

$$RCOOH \quad + \quad R'OH \xrightarrow[110^{\circ}C]{[HMim][BF_4]} RCOOR'$$

Scheme 26.

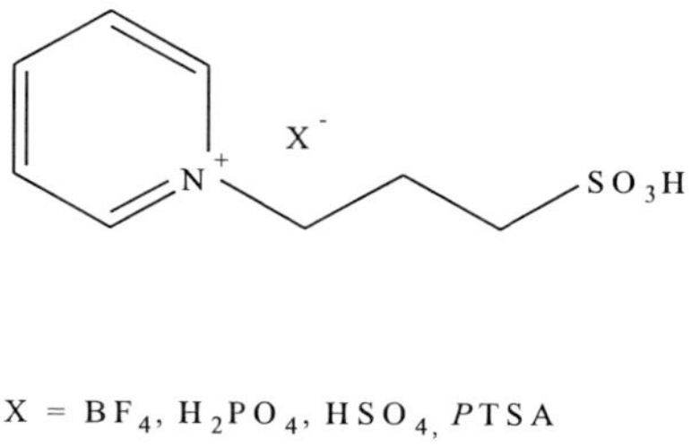

X = BF_4, H_2PO_4, HSO_4, *PTSA*

Figure 6.

The product ester was insoluble and separates from the ionic liquid as the reaction proceeds, presumably helping to drive the equilibrium towards products. High yields of products were achieved with a simple separation, and the ionic liquid reused several times [123]. Subsequently, various group reported the esterification reaction in ionic liquids Xing et al reported several water-stable, SO_3H-functional BAILs that bear an alkane sulfonic acid group in a pyridinium cation (Figure 6).

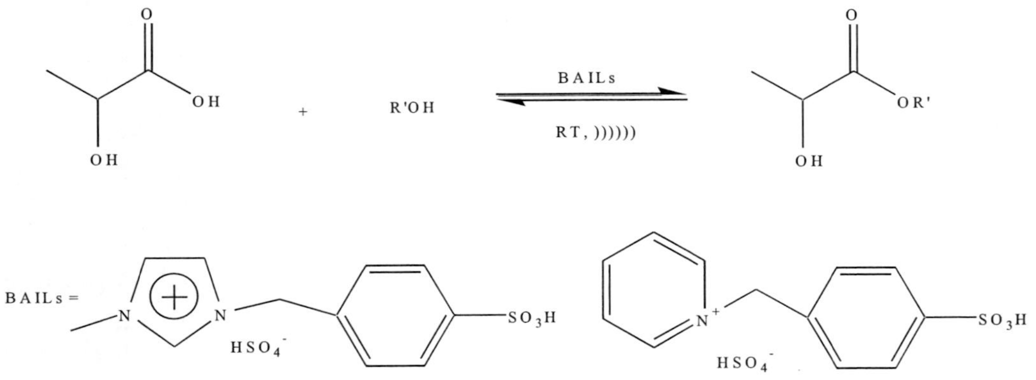

Scheme 27.

The catalytic activity of these BAILs depends on its anions increase of the anions Brønsted acidity improves its catalytic activity; it was proven with experiment [124].

Recently, Li et al. reported esterification reactions of lactic acid with a variety of straight chain aliphatic alcohols, cyclohexanol and benzyl alcohol using two novel BAILs that bear an aromatic sulfonic acid group on the imidazolium or pyridinium cation under ultrasound irradiation (Scheme 27).

The product esters were separated by easy decantation and BAILs could be reused for five times after simple treatments [125].

6.3. Transesterification Reaction

Transesterification of methyl acetoacetate (MAA) with alcohols was first reported by Ming et al. using ionic liquid-regulated sulfamic acid [126]. Bronsted acidic ionic liquid NH_2SO_3H $[C_3MIm]Cl$ shows satisfactory conversion rate and selectivity for transesterification, however it suffers through the disadvantages like IL is used as solvent and it is required in a large amount (i.e. 10 g of ionic liquid and 1 g of sulfamic acid), moreover the halogen content makes it less attractive from greener prescriptive.

Recently, we have demonstrated transesterification of β-ketoesters with variety of alcohols using halogen free Bronsted acidic ionic liquid $[NMP][HSO_4]$ as a novel catalyst Scheme 28 [127].

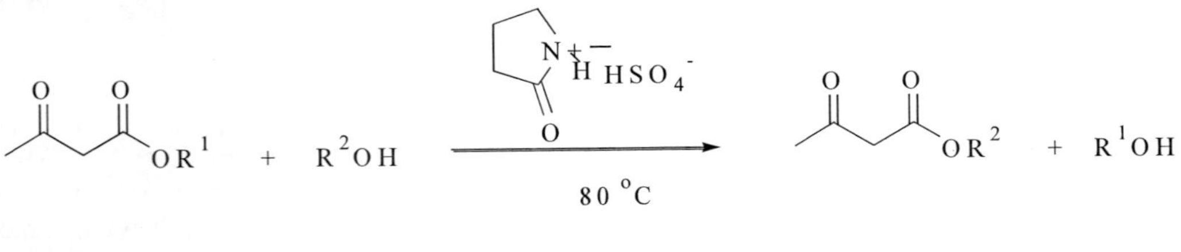

Scheme 28.

Later on we reported, transesterification reaction of dimethyl carbonate with phenol to methylphenyl carbonate and diphenyl carbonate using dibutyltin oxide catalyst in conjunction with Brønsted and Lewis acidic ionic liquids (Scheme 29).

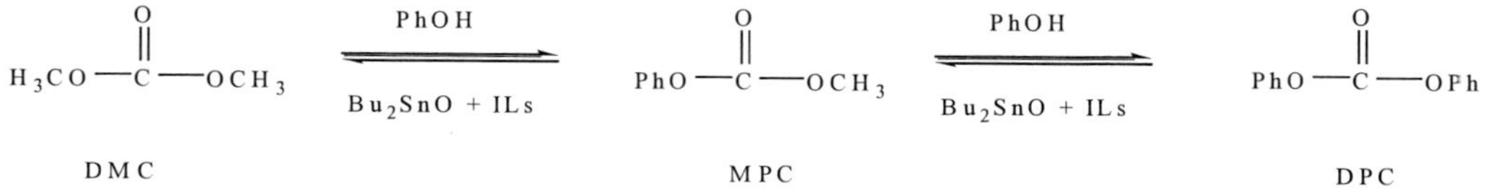

Scheme 29.

It was observed that the use of BAILs significantly enhances the yield of diphenyl carbonate. The ionic liquid having *p*-toluenesulfonate as anion and metal halide (e.g. $ZnCl_2$) as Lewis acid precursor exhibited higher activity and selectivity for diphenyl carbonate formation. Furthermore, the Brønsted and Lewis acidity of ionic liquids was measured by IR spectroscopy using pyridine as a probe and their Lewis acidity order was also determined [128]. Synthesis of biodiesel were reported in various functionalized ionic liquids such as Ether-functionalized ionic liquids [129], alkane sulfonic acid group [130,131].

6.4. Protection of Carbonyl and Amine

Protection of carbonyl and amines plays an important role in organic synthesis.

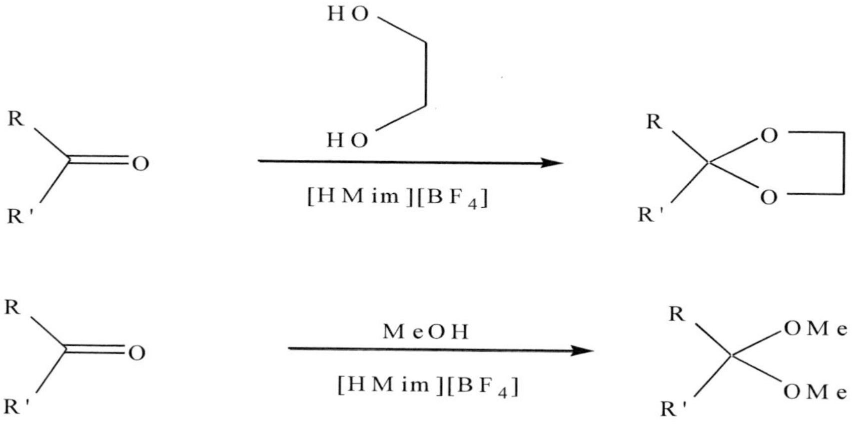

Scheme 30.

He and co-worker investigated protection of carbonyls as acetals or ketals using Brønsted acidic ionic liquid [Hmim][BF_4] as catalyst as well as solvent (Scheme 30). The product was separated conveniently from the reaction system, and the ionic liquid reused after removal of water [132]. Subsequently, Du end Duan et al reported protection of carbonyl using BAILs [133] and Lewis acidic ionic liquid Choline chloride · $xZnCl_2$ (x = 1-3) or benzyltrimethylammonium chloride · $2ZnCl_2$ [134].Later on Sadula et al reported first

Brønsted acidic ionic liquid-catalyzed *N*-tert-butoxy carbonylation of amines.The reported protocol is efficient, inexpensive, chemoselective and the IL is reusable [135].

6.5. Reductive Amination of Carbonyl Compounds

Reductive amination of carbonyl compounds using sodium borohydride was conducted in Bronsted acidic ionic liquid, 1-methyl imidazolium tetrafluoroborate [HMim][BF_4] [136].

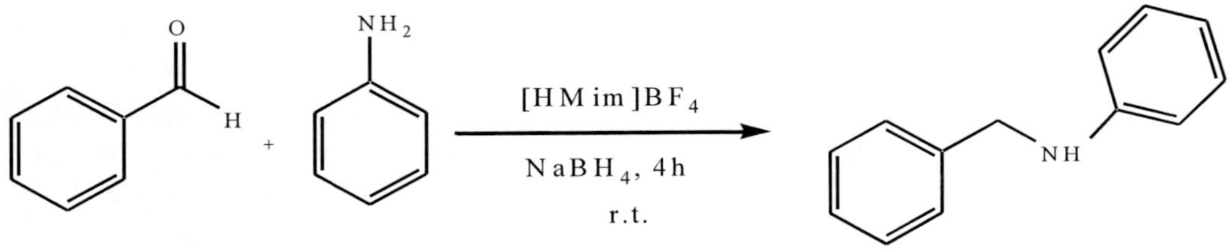

Scheme 31.

The ionic liquid played the dual role of solvent as well as catalyst for efficient conversion of aldehydes and ketones to amines in excellent yields without the formation of side products. The reported protocol was mild, efficient and user and environmentally friendly (Scheme 31).

6.6. Friedlander Reaction

The reaction of o-aminoarylaldehyde with active methylene compounds for the synthesis of quinoline is known as Friedlander reaction or annulations. Palimkar et al explored the use of BAILs as green reaction media as well as promoters for synthesis of biologically active quinolines and related polyheterocycles by using Friedlander heteroannulation protocol in the absence of any added catalyst (Scheme 32).

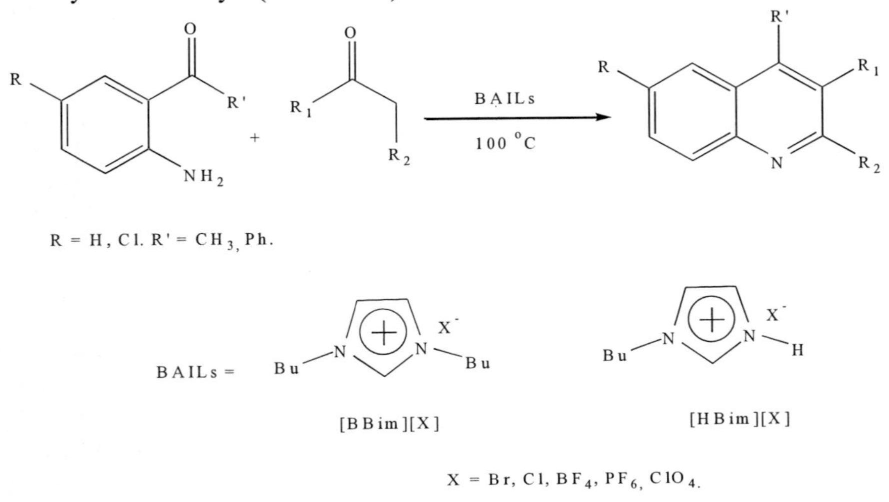

Scheme 32.

The efficacy of the BAILs for the heterocyclization reaction has been correlated to the acidity of the BAILs in terms of basicity of the anions and ^{1}H NMR chemical shifts [137].Later on Karthikeyan et al reported Lewis acidic ionic liquid [138] and Akbari et al investigated sulfonic acid functionalized ILs [139] as a catalyst for Friedlander reaction.

6.7. Mannich Reaction

This is a classical method for the synthesis of β-amino carbonyl compounds.In mannich reaction, an amine, two carbonyl compounds and acid (or base) catalysts are used to produce β-amino carbonyl compounds.

Zhao et al synthesized several BAILs were and successfully used as solvents and catalysts Mannich reactions at 25 °C (Scheme 33). The used Brønsted acidic ionic liquids includes [BMim][HSO_4], [BMim][H_2PO_4], [HMim][*P*TSA] and [HMim][Tfa]. Out of these BAILs screened higher yields were obtained in the presence of [HMim][Tfa]. Further [HMim][Tfa] was reused four times without considerable loss of activity [140]

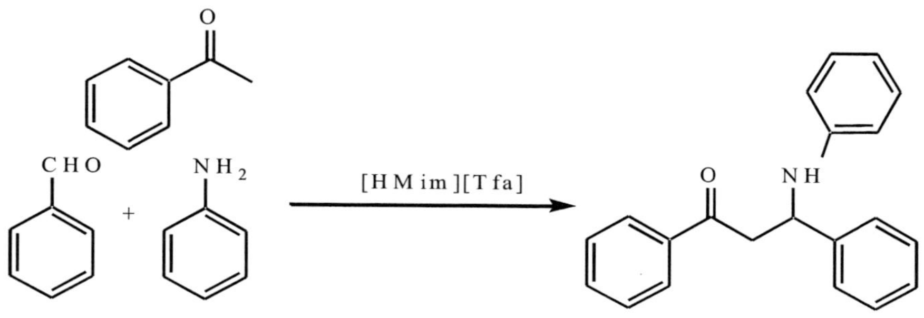

Scheme 33.

Sahoo and others reported, imidazolium, tetraalkylammonium and triphenylphosphine bearing sulfonic group based ILs and investigated for the mannich reaction [141-142].

More recently Zheng et al. investigated asymmetric Mannich reaction using [EMIm][Pro] as a catalyst. A variety of optically active β-amino carbonyl compounds were synthesized in up to 99% yield with up to 99 dr and 99% ee [143].

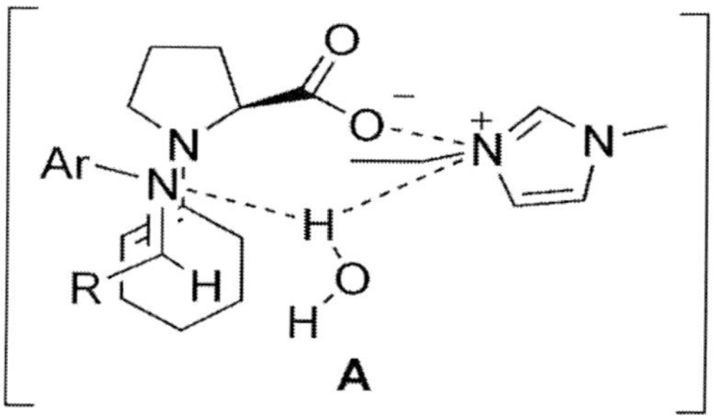

Figure 7.

Author proposed transition state on the basis of the stereochemistry of the corresponding Mannich products (Figure 7).

6.8. Prins Reaction

The Prins reaction involves formation of a C-C bond, is a notable method for the formation of tetrahydropyran derivatives. The condensation of olefins with aldehydes under strongly acidic conditions and high reaction temperatures, which limits its potential as an effective synthetic methodology.

Wang et al first time reported, six water stable Brønsted acidic task-specific ionic liquids ([HMim][BF_4],[$(CH_2)_4SO_3$HMim][HSO_4], [$(Ac)_2$Bim][Br], [NMP][HSO_4], [BMim][HSO_4] and [BMim][H_2PO_4] were synthesized and used as environmentally benign catalysts for Prins reaction under mild reaction conditions [144].

Recently, Fuchigami and co-worker [145] investigated Prins cyclization of homoallylic alcohols, thiols, and amines with various aldehydes in an ionic liquid HF salt (Et_4NF.5HF) afforded the corresponding 4-fluorinated heterocycles in excellent yields (Scheme 34).

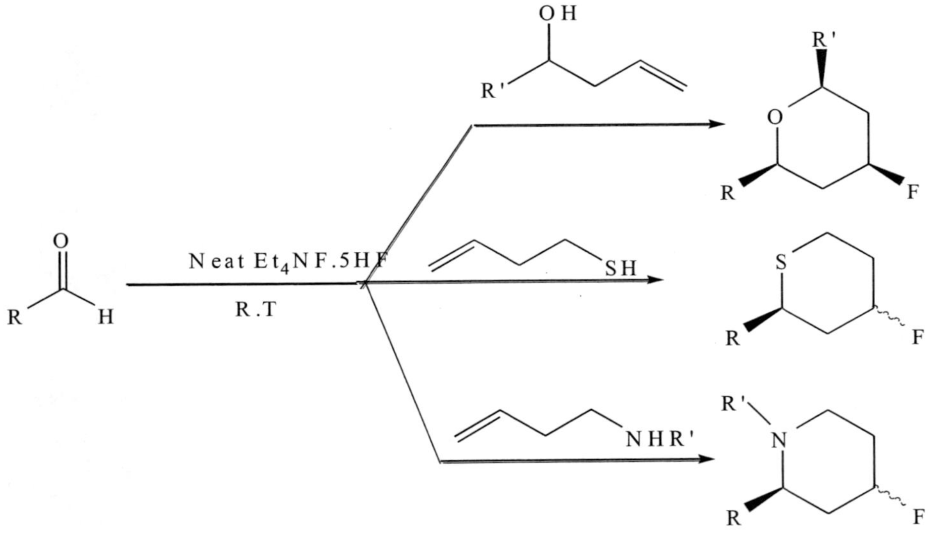

Scheme 34.

6.9. Kabachnik-Fields Reaction

In which the amine, aldehyde and di or trialkylphosphite are reacted in a single-pot in the presence of catalyst to afford the corresponding products.Shingare and co-worker demonstrated a one-pot three component coupling of an aldehydes, an amines, and triethyl phosphite using 1-benzyl-3-methyl imidazolium hydrogen sulphate [BnMim][HSO_4] as an efficient catalyst (Scheme 35).

The reactions proceed under solvent-free conditions at room temperature. This methodology afforded the corresponding α-aminophosphonates in shorter reaction times with excellent yields [146].

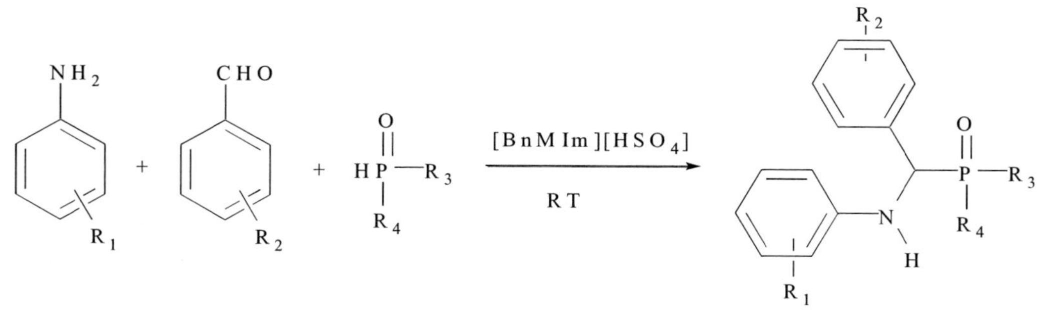

Scheme 35.

6.10. Dakin-West Type Reaction

Conventional Dakin–West reaction is a condensation of an α-amino acid with acetic anhydride in the presence of a base provides the α-acetamido ketones through an intermediate azalactone. Iqbal et al. developed a new route using aromatic aldehydes, enolizable ketones or β-keto esters and acetonitrile in the presence of acetyl chloride and a catalytic amount of a Lewis acid catalyst such as $CoCl_2$ to provide β-amido carbonyl compounds [147].

We have developed Bronsted acidic ionic liquid [HMim][HSO_4] catalyzed protocol for synthesis of various β-amido carbonyl compounds (Scheme 36).

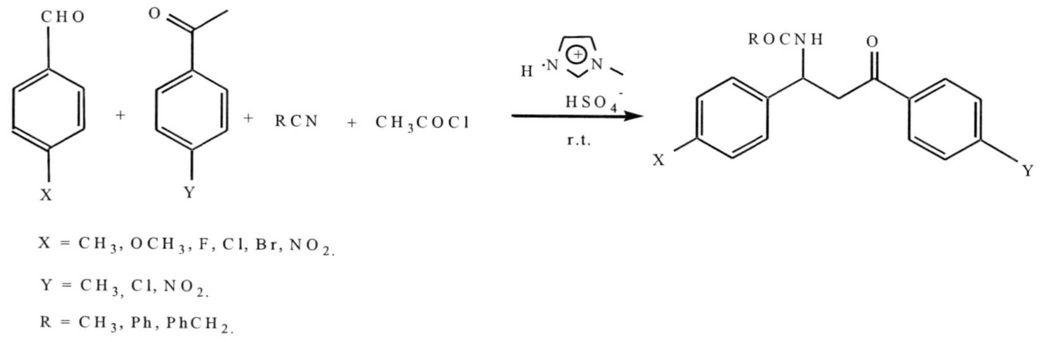

Scheme 36.

The system tolerates a wide range of functional groups affording the desired products in good to excellent yields. The BAILs were reused without any significant loss in product yields [148].

7. BIGINELLI REACTION

Biginelli reaction is an important method for the synthesis of dihydropyrimidinones, which involves a one pot reaction between aldehyde, 1,3-dicarbonyl compound and urea/thiourea in the presence of an acid catalyst. Dong et al. demonstrated a green protocol for Biginelli reaction catalyzed by a novel task-specific room temperature ionic liquid [149],

Cheap and reusable task-specific ionic liquids bearing an alkane sulfonic acid group in an acyclic trialkylammonium cation was found to be effective catalyst for synthesizing 3,4-dihydropyrimidine-2(1H)-ones (Scheme 37).

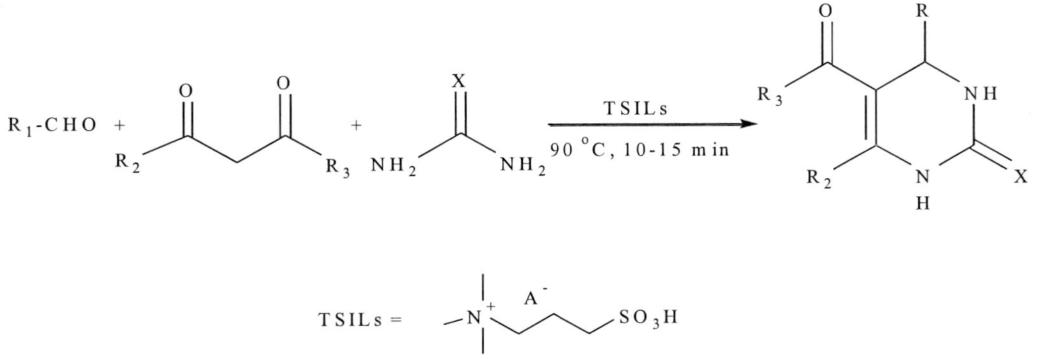

Scheme 37.

The protocol has several advantages: good yields, short reaction time and simplicity in the experimental procedure. The catalyst could be recycled and reused six times without noticeable decrease in the catalytic activity. Recently, Yadav et al investigated chiral ionic liquid-catalyzed, efficient and unprecedented version of the Biginelli reaction using new variants of its active methylene component, viz. 2-phenyl-1,3-oxazol-5-one/2-methyl-2-phenyl-1,3-oxathiolan-5-one, with aromatic aldehydes and urea/thiourea enantio- and diastereoselectively, yields 5-amino-/mercaptoperhydropyrimidines (Scheme 38).

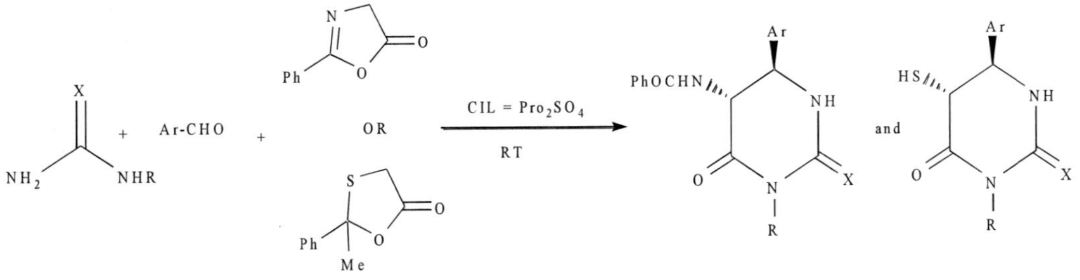

Scheme 38.

This three-component domino cyclocondensation reaction is effected via ring transformation of an isolable intermediate in a one-pot procedure [150].

7.1. Bronsted Basic Ionic Liquids

Ranu and co-worker introduced a task-specific Bronsted basic ionic liquids 1-butyl-3-methylimidazolium hydroxide ([BMim][OH]), as a catalyst and reaction medium for Michael addition of 1,3-dicarbonyl compounds, cyano esters, and nitro alkanes to a variety of conjugated ketones, carboxylic esters, and nitriles (Scheme 39).

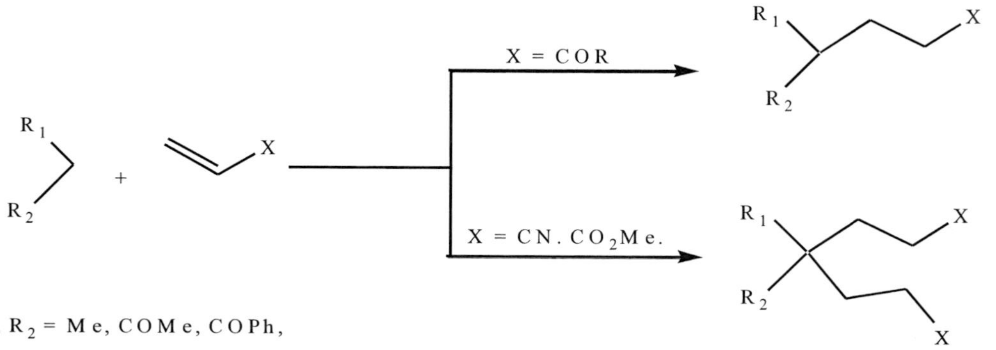

R_1, R_2 = Me, COMe, COPh, CO_2Et, Co_2Me, NO_2

Scheme 39.

Interestingly, the addition to α, β-unsaturated ketones proceeds in the usual way, giving the monoaddition products, but reaction of open-chain 1,3-dicarbonyl compounds with α,β-unsaturated esters and nitriles toward bis-addition to produce exclusively bis-adducts in one stroke [151].

7.2. Knoevenagel Reaction

Knoevenagel condensation is widely employed to synthesize intermediates of fine chemicals. Ranu and co-worker explored the same basic ionic liquids for the Knoevenagel reaction. A wide range of aliphatic and aromatic aldehydes and ketones easily undergo condensations with diethyl malonate, malononitrile, ethyl cyanoacetate, malonic acid and ethyl acetoacetate (Scheme 40).

The reactions proceed at room temperature and are very fast. The significant advantages offered by this methodology were: (a) general applicability to a large number of substrates with very facile reaction of aliphatic aldehydes with diethyl malonate, which was difficult to achieve by other methods, (b) mild reaction conditions (room temperature), (c) clean and fast (7-40 min) reaction, (d) high isolated yields of products and (e) reusability of catalyst and cost-effectiveness [152].

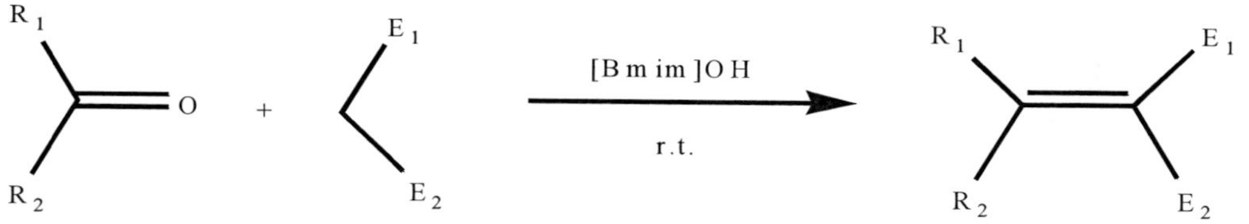

R1, R_2, = alkyl, aryl, H

E_1, E_2 = CN, COMe, COOMe, COOEt, COOH

Scheme 40.

More recently Sharma et al developed, Cost-effective and carbon dioxide absorbing ionic liquid, tri-(2-hydroxyethyl) ammonium acetate, was shown to perform multiple roles in Knoevenagel condensation [153]. It acted solvent, as well as catalyst for the less reactive cyanoacetic acid and also as a risk reduction medium for the unevenly generated large amount of CO_2 gas for large scale reactions. The reaction was scaled up for multi-gram synthesis of commercially important alpha cyanoacrylic acids.

7.3. Henry Reaction

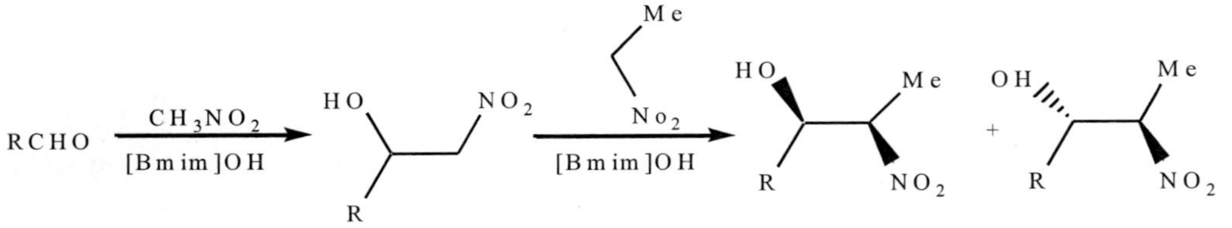

Scheme 41.

The Henry reaction is one of the most useful carbon-carbon bond forming reactions and has wide synthetic applications in organic synthesis by which 2-nitroalcohols are formed on treatment of nitroalkanes and carbonyl derivatives with a basic catalyst. Wu et al. developed an efficient protocol for Henry reaction using basic ionic liquid [BMim][OH] as catalyst and reaction medium (Scheme 41) [154].

7.4. Markovnikov Addition

Xu et al. developed ([BMim][OH] catalyzed efficient and convenient Markovnikov addition of *N*-heterocycles to vinyl esters without the requirement of any other catalyst and organic solvent (Scheme 42).

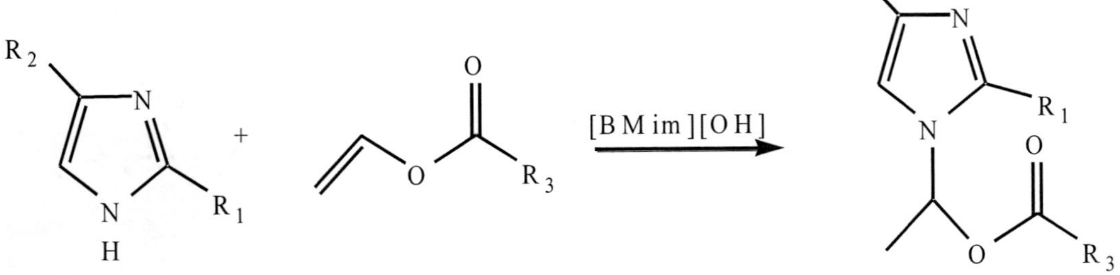

Scheme 42.

This strategy was quite general and it worked with a broad range of *N*-heterocycles as addition substrates, including five-membered *N*-heterocycles, pyrimidines and purines [155].

7.5. Mannich-Type Reaction

Liu et al. developed basic ionic liquid [Bmim][OH] catalyzed one-pot Mannich-type reaction for three-component synthesis of β- amino carbonyl compounds (Scheme 43).

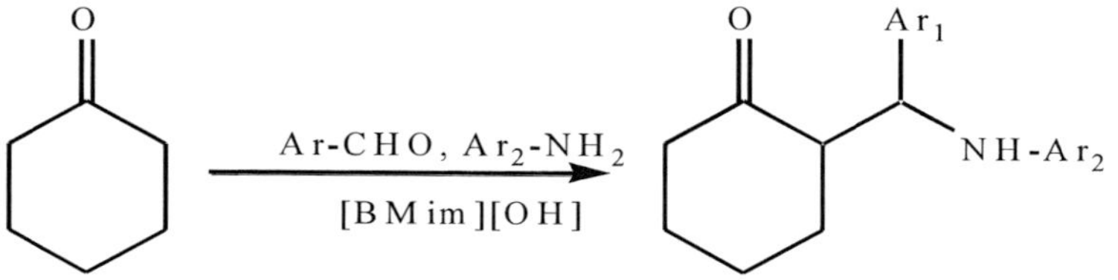

Scheme 43.

The ionic liquid, which was environmentally friendly and recycled at least 5 times without significant loss of activity [156].

7.6 Feist-Benary Reaction

Ranu et al. investigated conversion of hydroxydihydrofurans to furans (Feist-Benary products) using the ionic liquid, 1-methyl-3- pentylimidazolium bromide at 70-75 °C (Scheme 44).

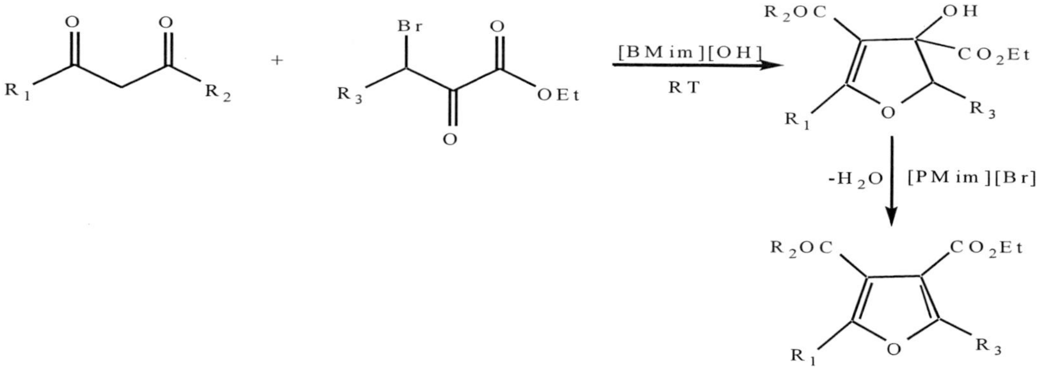

Scheme 44.

The significant advantages offered by this procedure were room temperature operation, considerably fast reaction, high yields and excellent cis stereo selectivity for the IFB products [157].

7.7. Synthesis of Quinazoline-2,4(1*H*,3*H*)-diones

We developed the [BMim][OH] catalyzed the synthesis of quinazoline-2,4(1*H*,3*H*)-diones from CO_2 and 2-aminobenzonitriles (Scheme 45).This method offers marked improvements with regard to operational simplicity, high isolated yields of products, greenness of the procedure, avoiding hazardous organic solvents and toxic catalyst [158]. Considering the economical value of the quinazoline-2,4(1H,3H)-diones derivatives we developed a new methodology which minimizes the number of unit operations and waste streams.

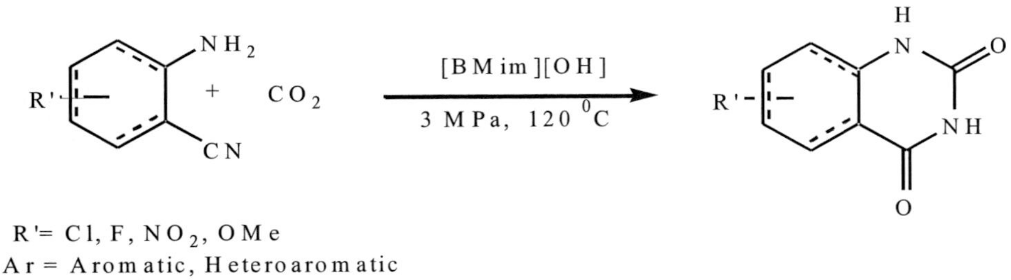

Scheme 45.

7.8. Other Application of Ionic Liquids

7.8.1. Ionic Liquid as an Electrolyte

The whole electroplating sector is based on the use of conventional aqueous solutions as electrolytes. Which has limitation such as liberation of hydrogen molecule during electrolysis, narrow electrochemical windows, low thermal stability, and evaporation; these prevent aqueous solutions being applied to the deposition of several technically important materials [159].

RTILs have several attractive properties such as insignificant vapor pressure, solubility of a wide range of organic and inorganic compounds, wide electrochemical window and tunability of properties through the change in the combination of cation and anion [160]. Consequently, there has been a significant increase in the studies on the use of RTILs for various separation processes. The use of RTILs as a substitute to molecular diluents in solvent extraction processes has been shown to result in unusual trends in extraction and enhanced selectivity. In addition, the wide electrochemical window makes it possible to use them as the electrolytic medium for electro winning of several metals at near ambient conditions.

7.8.2. Chiral Pool for Asymmetric Synthesis

Over the last decades, ionic liquid were boomed as a reaction media and catalyst for the various organic transformation. Research involving chiral ionic liquids (CILs) has been much more limited and only recently has come to the forefront. Chiral ionic liquids have their potential application to asymmetric synthesis and optical resolution. Recently, Armstrong and co-worker well reviewed the literature on the synthesis and application of CILs [161]. Today, this is an area of research that is poised for rapid development and expansion.

7.8.3. Media for Nanoparticle Synthesis

Ionic liquid play a synergetic role in the nanoparticle synthesis. Various functionalized and non-functionalized imidazolium cationic moiety of the ionic liquids may act as a stabilizer as well as capping agent or covalent support for the nanoparticle. Recently, Neouze, et al. reviewed diverse aspect of the ionic liquids in nanoparticle synthesis based on (i) ILs as solvents for the stabilization of nanoparticles, (ii)ILs as solvents in ''non-conventional'' activated syntheses of nanoparticles, (iii) Imidazolium interaction directly with nanoparticle surfaces, (iv) Imidazolium as supports for functional groups interacting with the nanoparticles, (v) Imidazolium poisoning of nanoparticle surfaces [162].

CONCLUSION

The present review summarized the recent developments in the area of ionic liquids as catalyst, support or reaction media in organic synthesis. Various aspects of the ionic liquids from their synthesis to application in diverse field have considered. The possibility to adjust the properties of ILs such as the hydrophobicity, viscosity, thermal stability, density, polarity and solubility to suit to the particular process is one of their key advantages and thus they can be truly described as designer solvents. The current scenario of ionic liquids as electrolyte, reaction media for nanoparticle, use of chiral ionic liquids in asymmetric synthesis put the ionic liquid field at a new horizon.

REFERENCES

[1] *Ionic Liquids in Synthesis;* Wasserscheid, P., Welton, T., Wiley-VCH:Weinheim, 2003.

[2] *Green Solvents for Organic Synthesis;* Ahluwalia V.K.; Varma, R.S. Narosa Pub. House 2009.

[3] Welton, T. *Chem. Rev.* 1999, 99, 2071-2083.

[4] Gujar, A.C.; White. M.G. *Catalysis*, 2009, 21, 154–190.

[5] Ionic Liquids; Stark, A.; Seddon, K.R. *Kirk-Othmer Encyclopedia of Chemical Technology,* John Wiley & Sons,

[6] Jain, N.; Kumar, A.; Chauhan, S.; Chauhan, S.M S. *Tetrahedron* 2005, 61, 1015– 1060.

[7] Martins, M.A. P.; Frizzo, Clarissa P.; Moreira, D.N., Zanatta, N.; Bonacorso, H.G. *Chem. Rev.* 2008, 108, 2015–2050.

[8] Rantwijk, F. V.; Sheldon, R.A. *Chem. Rev.* 2007, 107, 2757-2785.

[9] Parvulescu, V.I.; Hardacre, C. *Chem. Rev.* 2007, 107, 2615-2665.

[10] Zhao, D.; Wu, M.; Kou, Y.; Min, E. Catal. Today 2002, 74, 157–189.

[11] Olivier-Bourbigou, H.; Magna, L.; Morvan, D. *Appl. Catal. A: Gen.* 2010, 373, 1–56.

[12] Song, C. E. *Chem. Commun.* 2004, 1033–1043.

[13] Wasserscheid, P., Keim, W. *Angew. Chem. Int. Ed.* 2000, 39, 3773-3789.

[14] Liu, Y., Wang, S.-S., Liu, W., Wan, Q.-X., Wu, H.-H., Gao, G.-H. *Curr,Org. Chem.* 2009, 13, 1322-1346

[15] Lee, J.W.; Shin, J.Y.; Chun, Y.S.; Jang, H.B.; Song, C.E.; Lee, S.-G. *Acc. Chem. Res.* 2010, 43, 985-994.

[16] Gu, Y.; Li, G. *Adv.Syn. Catal.* 2009, 351, 817-847.

[17] Walden, P. *Bull. Acad. Imper. Sci.* 1914, 1800-1802.

[18] Chiappe, C.; Pieraccini, D. *J. Phys. Org. Chem.* 2005, 18, 275–297.

[19] Zhao, H. *Phy. Chem. Liq* .2003, 41, 545–557.

[20] Bates, E.D.; Mayton, R. D.; Ntai, Ioanna.; Davis, Jr J. H. J. Am. Chem. Soc. 2002, 124, 926-927.

[21] Davis, J. H. *Chem. Lett.* 2004, 33, 1072-1077.

[22] Wilkes, J. S.; Levisky, J. A.; Wilson, R. A.; Hussey C. L. *Inorg. Chem.* 1982, 21, 1263-1264.

[23] Bonhote, P. Dias, A.P.; Papageorgiou, N.; Kalyanasundaram, K.; Grätzel, M. *Inorg. Chem.,* 1996, 35, 1168-1178.

[24] Earle, M.J.; Katdare, S.P.; Seddon, K.R. *Org. Lett.* 2004, 6, 707-710.

[25] Olofson, R.A.; Thompson, W. R.; Michelman, J.S. *J. Am.Chem. Soc.* 1964, 86, 1865-1866.

[26] Arduengo, A.J.; Harlow, R.L.; Kline, M.*J. Am. Chem. Soc.* 1991, 113, 361-363.

[27] Arduengo, A. J. *Acc. Chem. Res.* 1999, 32, 913-921.

[28] Boon, J.A.; Levisky, J.A.; Wilkes, J.S. *J. Org. Chem.* 1986, 51, 480-483.

[29] Adams, C.J.; Earle, M.J.; Roberts, G.; Seddon, K.R. *Chem. Commun.* 1998, 19, 2097-2098.

[30] Surette, J.K.D.; Green, L.; Singer, R.D. *Chem. Commun.*1996, 24, 2753-2754.

[31] Joni, J.; Haumann, M.; Wasserscheid, P. *Adv. Syn. Catal.* 2009, 351, 423-431.

[32] Nara, S.J.; Harjani, J.R..; Salunkhe, M.M. *J. Org. Chem.* 2001, 66, 8616-8620.

[33] Naik, P.U.; Harjani, J.R..; Nara, S.J.; Salunkhe, M.M. *Tetrahedron Lett.* 2004, 45, 1933–1936.

[34] Lee, C.W. *Tetrahedron Lett.* 1999, 40, 2461-2464.

[35] Kumar, A.; Pawar S. S. *J. Org. Chem.* 2004, 69, 1419-1420.

[36] Abbott, A.P.; Capper, G.; Davies, D.L.; Rasheed, R.K.; Tambyrajah, V. *Green Chem.* 2002, 4, 24-26.

[37] Kumar, A.; Pawar S.S. *J. Mol. Catal. A: Chem.* 2005, 235, 244–248.

[38] Rebeiro, G.L., Khadilkar, B.M. *Synthesis* 2001, 3, 370-372.

[39] Morales, R.C.; Tambyrajah, V.; Jenkins, P.R.; Davies, D.L.; Abbott, A.P. *Chem. Commun.* 2004, 10. 158-159.

[40] Sheldon, R. *Chem. Commun.* 2001, 2399–2407

[41] Dysan, P.J.; Geldbach, T.J. (Eds.), Metal Catalyzed Reaction in Ionic Liquids; Catalysis by Metal Complexes, Vol. 29, Springer, Dordrecht, 2005.

[42] Zhao, D.; Wu, M.; Kou, Y. Min, E. *Catal. Today* 2002, 74, 157-162.

[43] Doorslaer, C.V.; Wahlen, J.; Mertens, P.; Binnemans K.; Vos, D.D. *Dalton Trans. 2010, 39, 8377–8390*

[44] Mehnert, C.P. Chem. Eur. J. 2004, 11, 50-56.

[45] Haumann, M. Riisager A. *Chem. Rev.* 2008, 108, 1474-1497.

[46] Parshall, G.W. *J. Am. Chem. Soc.,* 1972, 94, 8716-8719.

[47] Knifton, J.F. *J. Mol. Catal.* 1987, 43, 65-78.

[48] Chauvin, Y.; Mussmann, L.; Olivier, H. *Angew. Chem. Int. Ed. Engl.* 1995, 34, 2698-2700.

[49] Favre, F.; Olivier-Bourbigou, H.; Commereuc, D.; Saussine, L. *Chem. Comm.* 2001, 15, 1360-1361.

[50] Kottsieper, K.W.; Stelzer, O.; Wasserscheid P. *J. Mol. Catal. A: Chem.* 2001, 175, 285-288.

[51] Brauer, D.J.; Kottsieper, K.W.; Liek, C.; Stelzer, O.; Waffenschmidt, H.; Wasserscheid, P. *J. Organomet. Chem.* 2001, 630, 177-184.

[52] Dupont, J.; de Souza, R. F.; Suarez, P. A. Z. *Chem. Rev.* 2002, 102, 3667-3692.
[53] Mehnert, C.P.; Cook, R.A.; Dispenziere, N.C.; Afeworki, M.J. *J. Am. Chem. Soc.* 2002, 124, 12932-12933.
[54] Panda, A.G.; Bhor, M.D. Jagtap, S.R.; Bhanage, B.M. *Appl. Catal. A: Gen.,* 2008, 347, 142 – 147;
[55] Panda, A.G.; Jagtap, S.R, Nandurkar, N.S.; Bhanage, B.M. *Ind. Eng. Chem. Res.* 2008, 47, 969 – 972.
[56] Yang, Y.; Deng, C.; Yuan, Y.; *J. Catal.* 2005, 232, 108-116.
[57] Chauvin, Y.; Mussman, L.; Olivier, H. *Angew. Chem., Int. Ed. Engl.* 1995, 34, 2698-2700.
[58] Suarez, P.A.Z.; Dullins, J.E.L.; Einloft, S.; de Souza, R.F.; Dupont, J. *Polyhedron* 1996, 15, 1217-1219.
[59] Suarez, P. A. Z.; Dullins, J. E. L.; Einloft, S.; de Souza R. F.; Dupont, J. *Inorg. Chim. Acta* 1997, 255, 207-209.
[60] Monteiro, A.L.; Zinn, F.K.; De Souza, R.F.; Dupont, J. *Tetrahedron Asy.* 1997, 8,177-179.
[61] Liu, F.; Abrams, M. B.; Baker, R. T.; Tumas,W. *Chem. Commun.* 2001, 433-434.
[62] Brown, R. A.; Pollet, P.; McKoon, E.; Eckert, C. A.; Liotta, C. L.; Jessop, P. G. *J. Am. Chem. Soc.* 2001, 123, 1254-1255.
[63] Mehnert, C.P.; Mozeleski, E.J.; Cook, R.A. *Chem. Commun* 2002. 8, 3010-3011.
[64] Mikkola, J.P.; Virtanen, P.; Karhu, H.; Salmi, T.; Murzin, D.Y. *Green Chem.* 2006, 8, 197-205.
[65] Huang, J.; Jiang, T.; Gao, H.; Han,Buxing.; Liu, Z.; Wu, W.; Chang, Y.; Zhao G. *Angew. Chem., Int. Ed. Engl.* 2004, 43, 1397-1399.
[66] Ruta, M.; Laurenczy, G.; Dyson, P.J.; Kiwi-Minsker, L. *J. Phy. Chem. C* 2008, 112, 17814-17819.
[67] Arras, J.; Ruppert, D.; Claus, P. *Appl. Catal. A: Gen.* 2009, 371, 73-77.
[68] Rodrıguez-Perez, L.; Teuma, E.; Falqui, A.; Gomez, M.; Serp,P. *Chem. Commun.,* 2008, 4201-4203.
[69] Zhang, Z.; Xie, Y.; Li, W.; Hu, S.; Song, J.; Jiang, T.; Han, B. *Angew. Chem., Int. Ed. Engl.* 2008, 47, 1127-1129.
[70] Hua, Y.; Yang, H.; Zhang, Y.; Hou, Zhenshan.; Wang , X.; Qiao, Y.; Li, Huan .; Feng, B.; Huang Q. *Catal. Commun.* 2009, 10, 1903-1907.
[71] Schneiders,K.; Bosmann, A.; Schulz,P.S.; Wasserscheida, P. *Adv. Synth. Catal.* 2009, 351, 432 – 440
[72] Neff, V.; Müller, T.E.; Lercher, J.A. *Chem. Commun.* 2002, 8, 906-907.
[73] Bodis, J.; Müller, T.E.; Lercher, J.A. *Green Chem.* 2003 5, 227-231.
[74] Breitenlechner, S.; Fleck, M.; Müller, T.E.; Suppan, A. *J. Mol. Catal. A: Chem.* 2004 214, 1,175-179
[75] Jimenez, O.; Müller, T.E.; Sievers, C.; Spirkl, A.; Lercher, J.A. *Chem.Commun.* 2006 28, 2974-2976.
[76] Yang, L.; Xu, L.-W.; Xia, C.-G. *Synthesis* 2009, 12, 1969-1974.
[77] Kaufmann, D. E.; Nouroozian, M.; Henze, H. *Synlett* 1996, 1091-1092.
[78] Herrmann, W.A.; Böhm, V.P.W. *J. Organomet. Chem.* 1999, 572, 141-145.
[79] Böhm, V.P.W.; Herrmann, W.A. *Chem. Eur. J.* 2000, 6, 1017-1025.

[80] Carmichael, A.J.; Earle, M.J.; Holbrey, J.D.; McCormac, P.B.; Seddon, K.R. *Org. Lett.* 1999, 1, 997-1000.
[81] Deshmukh, R. R.; Rajagopal, R.; Srinivasan, K. V. *Chem. Commun.* 2001, 1544-1545.
[82] Hagiwara, H.; Sugawara, Y.; Isobe, K; Hoshi, T; Suzuki, T. *Org. Lett.* 2004, 6, 2325-2328.
[83] Hagiwara, H.; Sugawara, Y.; Hoshi, T.; Suzuki, T. *Chem. Commun.* 2005, 2942-2944.
[84] Sarkar M. S.; Qiu H.; Jin M.-J. *J.Nanosci. Nanotechnol.* 2007, 7, 3880-3883.
[85] Jung, J.-Y.; Taher, A.; Kim, H.-J.; Ahn, W.-S.; Jin, M.-J. *Synlett* 2009, 39-42.
[86] Ma, X.; Zhou, Y.; Zhang, J.; Zhu, A.; Jiang, T.; Han, B: *Green Chem.* 2008, 10, 59-66.
[87] Volland, S.; Gruit, M.; Régnier, T.; Viau, L.; Lavastre, O.; Vioux, A. *New J. Chem.* 2009 33, 2015-2021
[88] Dongbin , Z.. Design, synthesis and applications of functionalized ionic liquids. Theses no 3531 2006.
[89] Zhao, D.; Fei, Z.; Geldbach, T.J.; Scopelliti, R.; Dyson, P.J. *J. Am. Chem. Soc.* 2004,126, 15876-15882.
[90] Zhao, D.; Fei, Z.; Ohlin, C.A.; Laurenczy, G.; Dyson, P.J. *Chem. Commun.* 2004, 10, 2500-2501.
[91] Dubbaka, S.R.; Zhao, D.; Fei, Z.; Volla, C.M.R.; Dyson, P.J.; Vogel, P. *Synlett* 2006, 18, 3155-3157.
[92] Wang, R.; Piekarski, M.M.; Shreeve, J.M. *Org. Bio. Chem.* 2006, 4, 1878-1886.
[93] Wang, R.; Xiao, J.-C.; Twamley, B.; Shreeve, J.M. *Org. Bio. Chem.* 2007, 5, 671-678
[94] Ye, C.; Xiao, J.-C.; Twamley, B.; LaLonde, A.D.; Norton, M.G.; Shreeve, J.M. Eur. J.Org. Chem. 2007, 30, 5095-5100.
[95] Wan, Q.-X.; Liu, Y. *Catal. Lett.* 2009, 128, 487-492.
[96] Wan, Q.-X.; Liu, Y.; Cai, Y.-Q. *Catal. Lett.* 2009, 127, 386-391.
[97] Wan, Q.-X.; Liu, Y.; Lu, Y.; Li, M.; Wu, H.-H. *Catal. Lett.* 2008, 121, 331-336.
[98] Wang, L.; Li, H.; Li, P. *Tetrahedron,* 2009, 65, 364-368.
[99] Alonso, F.; Beletskay, I.P.; Yus , M. *Tetrahedron* 2008, 64 , 3047-3101.
[100] Mathew, C. J.; Smith, P. J.; Welton, T. *Chem. Commun.* 2000, 1249-1250.
[101] Rajagopal, R.; Jarikote, D.V.; Srinivasan, K.V.; *Chem. Commun.* 2002, 6, 616-617.
[102] Calo, V.; Nacci, A.; Monopoli, A. *Eur. J. Org. Chem.* 2006, 17, 3791-3802.
[103] Hagiwara, H.; Sugawara, Y.; Isobe, K., Hoshi, T.; Suzuki, T. *Org. Lett.* 2004, 6, 2325-2328.
[104] Hagiwara, H.; Sugawara, Y.; Hoshi, T.; Suzuki, T. *Chem. Commun.* 2005, 23, 2942-2944.
[105] Hagiwara, H.; Ko, K.H.; Hoshi, T.; Suzuki, T. Chem. Commun. 2007, 27, 2838-2840.
[106] Hagiwara, H.; Keon, H.K.; Hoshi, T.; Suzuki, T. *Synlett* 2008, 4, 611-613.
[107] Hagiwara, H.; Sasaki, H.; Tsubokawa, N.; Hoshi, T.; Suzuki, T.; Tsuda, T.; Kuwabata, S. *Synlett* 2010, 13, 1990-1996.
[108] Karimi, B.; Elhamifar, D.; Clark, J.H.; Hunt, A.J. *Chem. Eur. J.* 2010, 16, 8047-8053.
[109] Zhao, D.; Fei, Z.; Geldbach, T.J.; Scopelliti, R.; Dyson, P.J. *J. Am. Chem. Soc.* 2004, 126, 15876-15882
[110] Cui, Y.; Biondi, I.; Chaubey, M.; Yang, X.; Fei, Z.; Scopelliti, R.; Hartinger, C.G.,Dyson, P.J. *Phys.Chem. Chem. Phys.* 2010, 12, 1834-1841.
[111] Yan, N.; Yang, X.; Fei, Z.; Li, Y.; Kou, Y.; Dyson, P. J. *Organometallics* 2009, 28, 937-939.

[112] Yang, X.; Fei, Z.; Geldbach, T. J.; Phillips, A. D.; Hartinger, C. G.; Li, Y.; Dyson, P. J. *Organometallics* 2008, 27, 3971-3977.

[113] Wang, R.; Twamley, B.; Shreeve, J. M. *J. Org. Chem.* 2006, 71, 426-429.

[114] Wang, R.; Piekarski, M. M.; Shreeve, J. M. *Org. Biomol. Chem.* 2006, 4, 1878-1886.

[115] Sirieix, J.; Ossberger, M.; Betzemeier, B.; Knochel, P. *Synlett* 2000, 11, 1613-1615.

[116] Thomazeau, C.; Bourbigou, H. O.; Magna, L.; Luts, S.; Gilbert B. *J. Am. Chem. Soc.*2003, 125, 5264-5265.

[117] Duan, Z.; Gu, Y.; Zhang, J.; Zhu, L.; Deng, Y. *J. Mol. Catal. A: Chem.* 2006,250, 163–168.

[118] Fei, Z.; Zhao, D.; Geldbach, T. J.; Scopelliti, R.; Dyson, P. J. *Chem.—Eur. J.* 2004, 10, 4886-4893.

[119] Yang, Y.; Kou, Y. *Chem. Commun.* 2004, 226-227.

[120] Hajipour, A. R.; Rafiee, F.*Org. Prep. Proc. Int.* 2010, 42, 285–362.

[121] Johnson, K.E.; Pagni, R. M.; Bartmess, J. *Mon. fur Chem.* 2007,138, 1077–1101.

[122] Cole, A. C.; Jensen, J. L.; Ntai, I.; Tran, K. L. T.; Weaver,K. J.; Forbes, D. C.; Davis, J. H., Jr. *J. Am. Chem. Soc.*2002, 124, 5962-5963.

[123] Zhu, H.-P.; Yang, F.; Tang, J.; He, M.-Y. *Green Chem.*2003, 5, 38–39.

[124] Xing, H.; Wang, T.; Zhou, Z.; Dai, Y. *Ind. Eng. Chem. Res.* 2005, 44, 4147-4150.

[125] Li, X.; Ma, Q.L.L. *Ultrason. Sonochem.*2010,17, 752–755.

[126] Bo, W.; Ming, Y.L.; Shaun, S.J. *Tetrahedron Lett.* 2003, 44 5037-5039.

[127] Qureshi, Z.S.; Deshmukh, K.M.; Bhor, M.D.; Bhanage, B.M. *Catal. Commun.* 2009, 10, 833–837;

[128] Deshmukh, K.M., Qureshi, Z.S., Dhake, K.P., Bhanage, B.M. *Catal. Commun.* 2010, 12, 207–211.

[129] Zhao, H.; Song, Z.; Olubajo, O.; Cowins J. V. *Appl Biochem Biotechnol.*2010, 162,13–23.

[130] Wu,Q.; Chen,H.; Han, M.; Wang, D.; Wang, J. *Ind. Eng. Chem. Res.* 2007, 46, 7955-7960.

[131] Liang, x.; Yang, J. *Green Chem.* 2010, 12, 201–204.

[132] Wu, H.-H.; Yang, F.; Cui, P.; Tang, J.; He, M.-Y. *Tetrahedron Lett.* 2004, 45, 4963-4965.

[133] Du, Y.; Tian, F. *Syn. Commun.* 2005, 35, 2703-2708.

[134] Duan, Z.; Gu, Y.; Deng, Y. *Catal. Commun.* 2006, 7, 651-656.

[135] Sunitha, S.; Kanjilal, S.; Reddy, P.S.; Prasad, R.B.N.*Tetrahedron Lett.* 2008, 49, 2527-2532.

[136] Reddy, P.S.; Kanjilal, S.; Sunitha, S.; Prasad, R.B.N. *Tetrahedron Lett.* 2007, 48, 8807-8810

[137] Palimkar, S.S.; Siddiqui, S.A.; Daniel, T.; Lahoti, R.J.; Srinivasan, K.V. J. Org. Chem. 2003, 68, 9371-9378.

[138] Karthikeyan, G.; Perumal,P.T. *J. Heterocyclic Chem.* 2004, 41, 1039-1041.

[139] Akbari, J.; Heydari, A.; Kalhor, H.R.; Kohan, S.A. *J. Comb. Chem.* 2010, 12, 137–140.

[140] Zhao, G.; Jiang, T.; Gao, H.; Han, B.; Huang, J.; Sun, D. *Green Chem.* 2004, 6, 75-77.

[141] Sahoo, S.; Joseph, T.; Halligudi S.B. *J. Mol. Catal. A: Chem.* 2006, 244, 179–182.

[142] Dong, F.; Zhenghao, F.; Zuliang, L. Catal. Commun. 2009, 10, 1267–1270.

[143] Zheng, Xin.; Qian, Y.-B.; Wang Y. *Eur. J. Org. Chem.* 2010, 515–522

[144] Wang, W.; Shao, L.; Cheng, W.; Yang, J. He, M. *Catal. Commun.* 2008, 9, 337–341.

[145] Kishi, Y.; Nagura, H.; Inagi, S.; Fuchigami, T. *Chem. Commun.* 2008, 3876–3878.
[146] Sadaphal, S.A.; Sonar, S.S.; Kategaonkar, A.H.; Shingare, M.S. *Bull. Korean Chem.Soc.*2009 30, 1054-1056.
[147] Rao, I.N.; Prabhakaran, E.N.; Das, S.K.; Iqbal, J. *J.Org. Chem.* 2003, 68, 4079–4082,
[148] Deshmukh, K.M., Qureshi, Z.S., Nandurkar, N.S; Bhanage, B.M. Can. J. Chem. 2009, 81, 401–405.
[149] Dong, F.; Jun, L.; Xinli, Z.; Zhiwen, Y.; Zuliang, L. *J. Mol. Catal. A: Chem.* 2007,274, 208-211.
[150] Yadav, L.D.S.; Rai; A.; Rai, V.K.; Awasthi, C. Tetrahedron 2008, 64, 1420-1429.
[151] Ranu, B. C.; Banerjee, S. Org. Lett. 2005, 7, 3049-3052.
[152] Ranu, B.C.; Jana, R.; *Eur. J. Org. Chem.* 2006, 3767-3770.
[153] Sharma, Y.O.; Degani, M.S. Green Chem., 2009, 11, 526–530.
[154] Wu, H.; Zhang, F.R.; Wan, Y.; Ye, L. *Lett. Org. Chem.* 2008, 5, 209-211.
[155] Xu, J.M.; Liu, B.K.; Wu, W.B.; Qian, C.; Wu, Q.; Lin, X.F.; *J. Org. Chem.* 2006, 71, 3991-3993.
[156]Gong, K.; Fang, D.; Wang, H.L.; Liu, Z.L.; *Monatsh. Chem.* 2007, 138, 1195-1998.
[157] Ranu, B.C.; Adak, L.; Banerjee, S.; *Tetrahedron Lett.* 2008, 49, 4613-4617.
[158] Patil, Y. P.; Tambade, P. J.; Deshmukh, K. M.; Bhanage B. M. Catal.Today 2009,148, 355–360.
[159] Electrodepostion from ionic liquids , Endres, F.; MacFarlane D.; Abbott, A. *Wiley-VCH* Verlag Gmbh, Weinheim 2008, ISBN 978-3-527-31565-9.
[160] Simka, W.; Puszczyk, D.; Nawrat G.; *Electrochimica Acta* 2009, 5307–5319.
[161] Ding, J.; Armstrong, D. W. Chirality 2005,17, 281–292 .
[162] Neouze, M.-A. *J. Mater. Chem.,* 2010, 20, 9593-9607.

In: Focus on Catalysis Research: New Developments
Editors: Minjae Ghang and Bjørn Ramel
ISBN: 978-1-62100-45?-4

Chapter 5

THE ENZYMATIC CATALYSIS OF NEURAMINIDASE AND DE NOVO DESIGNS OF NOVEL INHIBITORS

Zhiwei Yang, Gang Yang*, Yuangang Zu, Yujie Fu
Key Laboratory of Forest Plant Ecology, Ministry of Education,
Northeast Forestry University, Harbin 150040, P. R. China

ABSTRACT

The influenza infection pandemic has spread on a world scale and become a major threat to human health. Neuraminidase (NA), a major surface glycoprotein of influenza virus with well-conserved active sites, offers an ideal target for the development of antiviral drugs. In this chapter, we will review the recent results on the NA active sites, NA-inhibitor interactions, proton transfers of NA inhibitors as well as respective roles of the core templates, functional groups and synergistic effects of NA inhibitors through the fragment approach. On such basis, the *de novo* designs were carried out in order to discover novel NA inhibitors as potential antiviral drugs. The calcium ion rather than conserved water molecules was found to be crucial to stabilize the NA active site; nonetheless, the conserved water molecules can greatly alter the NA-inhibitor interactions. The interaction mechanisms of NA were studied with two current antiviral drugs oseltamivir and peramivir as well as two potent lead compounds 4-(N-acetylamino)-5-guanidino-3-(3-pentyloxy) benzoic acid (BA) and 5-[(1R,2S)-1-(acetylamino)-2-methoxy-2-methylpentyl]-4-[(1Z)-1-propen-yl)-(4S,5R]-D-proline (BL). It indicated that for some NA inhibitors, for the proton transfers to form the zwitterions are not facile in aqueous solutions, which explains the low oral bioavailability of current antiviral drugs. With the fragment approach, it was revealed that the core templates rather than functional groups play a larger role during the inhibitor interactions with the NA proteins; moreover, the binding qualities are largely determined by the synergistic effects

---

* Corresponding author: Email: theobiochem@gmail.com. Foundation item: Foundation for the Talented Funds of Northeast Forestry University (No. 220-602042), Special Fund for Basic Scientific Research of Central Colleges (No. DL09EA04-2) and the Cultivated Funds of Excellent Dissertation of Doctoral Degree Northeast Forestry University (grap09).

of the core templates and functional groups. Based on the understanding of NA's catalytic mechanisms and structure-activity relationships, seven tripeptides were successively designed as NA inhibitors. It was found that some of them show much higher inhibiting activities than the lead compound BA. As the tripeptides have the advantages such as low toxicities and side effects, we believe that this work will guide synthetic and medicinal chemists to the tripeptide-based antiviral drugs. It may throw light on the treatment and prevention of influenza infections, since the resistances have ever been reported for the current commercial drugs.

I. INTRODUCTION

The influenza virus, one of the main causes of acute respiratory infections to humans, may result in annual epidemics and infrequent pandemics. In the past century, three influenza pandemics occurred and millions of people were killed. Each outbreak was brought by a new viral strain in humans [1, 2], which was often introduced by an existing flu virus transmitted from other animal species, or an existing human strain picking up new genes, usually from the infected birds or pigs [3, 4]. The latest 2009 flu pandemic is due to the emergence of a novel strain that combines the genes from humans, pig and bird flu [5]. As the viruses tend to mutate in order to escape the immune system, vaccinations can provide only a limited control. Antiviral drugs are probably the most effective method to prevent and treat influenza infections, where the neuraminidase (NA) inhibitors have found wide applications [1, 6-8].

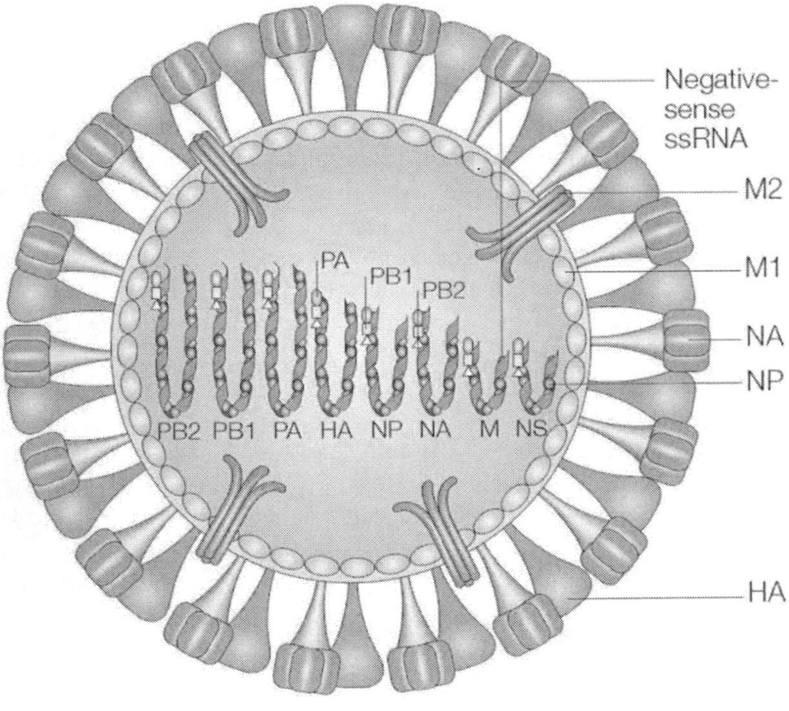

Figure 1. Schematic structure of influenza A virion [21].

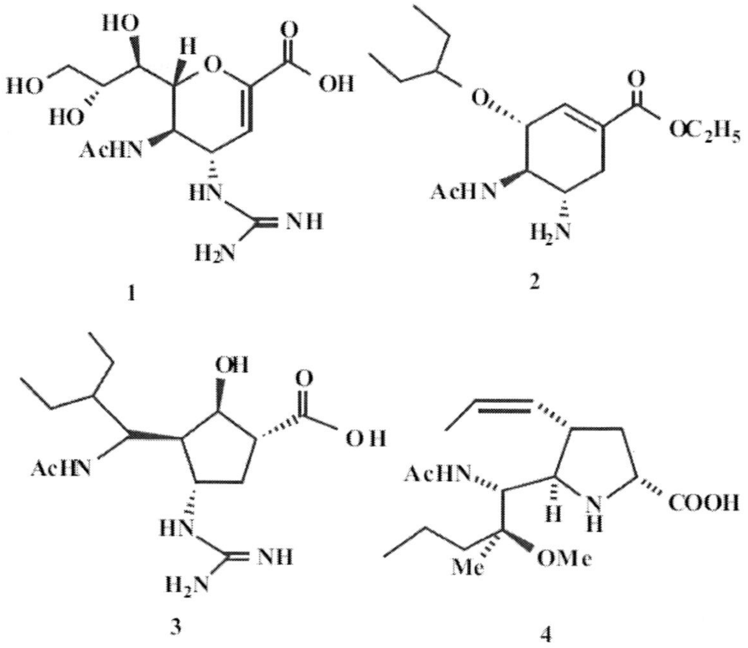

Scheme 1. Chemical structures of some NA inhibitors [49].

Influenza viruses are negatively stranded RNA viruses with a segmented genome, belonging to the orthomyxoviridae family. They can be divided into three serologically distinct types: influenza viruses A and B are concerned with human pathogens, especially influenza A being responsible for regular outbreaks, such as the 1918's and 2009's pandemics [9, 10], while influenza virus C seldom causes disease symptoms. Influenza virus A can be further classified on basis of the different antigenic properties of the surface glycoproteins (Figure 1) [11, 12]: haemagglutinin (HA) recognizes the sialic-acid containing receptors in the host cells and leads to the fusion of the endosomal and viral membranes [9, 13]. Instead, NA plays an essential role during the transport of virus particles through the upper respiratory tract and during the release of virion progeny from infected cells [14-17]. Both of the two surface glycoproteins (HA and NA) are vital to the viral infections and thus provide exciting opportunities for the development of anti-influenza agents with broad spectrum [4, 7, 18-20].

Influenza virus neuraminidase (NA, EC 3.2.1.18) is a major surface glycoprotein of influenza viruses A and B (about 50 copies per virion). It is a tetramer consisting of four identical disulfide linked subunits (Mr 60 kDa). Each monomer contains one calcium co-factor and induces the catalysis independently [22]. X-ray crystallography reveals that the monomer of NA is a β-sheet propeller structure consisting of six blades, and its catalytic site is in the loop region on the top of the propeller, composed of 14 conserved residues [23-26]. NA specifically identifies the sialic acid (SA) species and then cleaves the α-glycosidic bond between SA and glycoconjugate, which, in turn, destroys the binding between the virion and host-cell receptor [27]. This motion is particularly essential during the releasing of virions from the infected cells [16], assisting the movement of viruses through the mucus in the respiratory tract, as well as reducing the aggregation of virus particles [19, 28-30]. Owing to the essential roles during the influenza virus replications and high conservations of various

NA active sites, wide interest has arisen as to the targeting of potent inhibitors for NA proteins [31-43].

Currently, zanamivir (Relenza by Glaxo Wellcome/Biota) and oseltamivir (Tamiflu by Hoffman-La Roche/Gilead) are recommended for the prevention and treatment of influenza in human clinical trials by the World Health Organization (WHO) (Compounds 1 and 2 in Scheme 1) [1, 44-46]. They are designed as NA inhibitors by structure-based drug design (SBDD) methods and with the available information of NA active sites [7]. Besides, peramivir [40, 47] and A-315675 (BL) (Compounds 3 and 4 in Scheme 1) [48] were also designed in this way, which are under the phase III trials in North America and Europe. In this chapter, we will summarize the recent achievements of ours on the understanding towards NA-inhibitor interactions, mainly focusing on

(a) The NA active site;
(b) Proton transfer of NA inhibitors;
(c) Synergistic effects during NA-inhibitor interactions;
(d) De novo designs of novel NA inhibitors.

II. The NA Active Site

With the aid of structure-based rational drug design methods and abundant structural information of neuraminidases (NA), varieties of inhibitors have been designed recently [40, 50-52]. Docking ranks as one of the most popular structure-based rational drug design methods [53-61]. Generally, the calcium ion and conserved water molecules are retained during the docking procedures, for they have been recognized to affect the interaction strengths and rates between substrates and receptors [59-63]. The X-ray experiments [59] indicated that the absence of the calcium ion in the N9 sub-type NA protein causes conformational disruption of several nearby loops. The conserved water molecules have also been observed to play an important role by helping to build up H-bonded networks [60, 64-67].

The key residues of various NA proteins are nearly the same, consisting of ten polar residues (Arg118, Glu119, Asp151, Arg152, Arg224, Glu227, Glu276, Glu277, Arg292 and Arg371) and four hydrophobic residues (Trp178, Ile222, Ala246, and Tyr406) [19, 68]. Note that the residue numbering of N9 sub-type NA protein is used throughout. As Figure 2 shows, the NA active site can be divided into four sub-regions with different properties [69]. Sub-region 1 (S1) consists of a triad of arginine residues (Arg118, Arg292 and Arg371), which is positively charged and forms an H-bonded surface; Sub-region 2 (S2) contains negatively charged residues Asp151, Glu119 and Glu227; Sub-region 3 (S3) is constituted by residues Trp178 and Ile222 and forms a hydrophobic cave with the nearby residue Arg152; Sub-region 4 (S4) is a mixed polarity pocket containing residues Glu276 and Glu277. S1 and S3 are responsible for the locations of substrates whereas S2 and S4 tune the orientations of substrates [32, 34, 37, 40, 69-77]. There is another hydrophobic pocket in the NA active site formed by the sidechains of residues Ile222 and Ala246 and the hydrophobic side face of residue Arg224 [78]. However, this hydrophobic pocket has almost no contacts with substrates and is seldom considered [32, 34, 37, 40, 69-77].

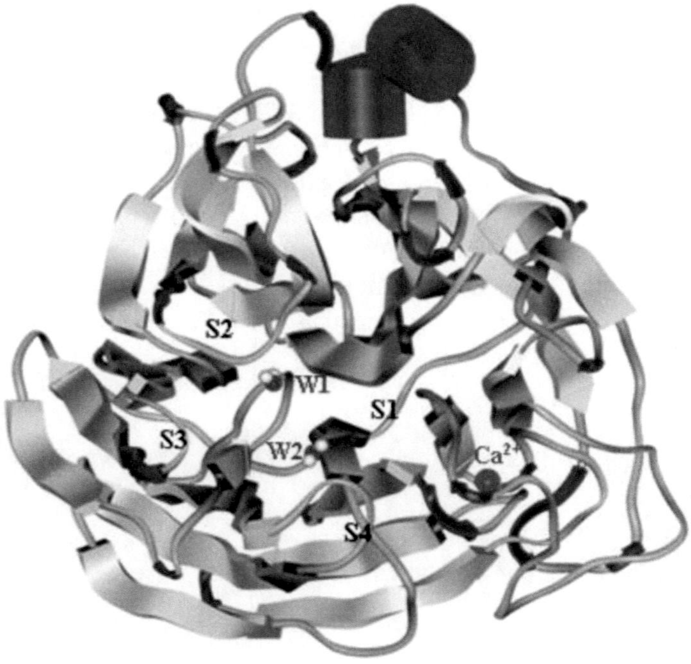

Figure 2. Ribbon diagram of N9-subtype neuraminidase structure with the calcium ion and conserved water molecules [75].

The four sub-regions in the active site were marked as S1–S4. The calcium ion (Ca^{2+}) and two conserved water molecules (W1 and W2) were drawn in ball and stick models.

The calcium ion is crucial to the maintenance of three-dimensional structure of NA protein whereas the conserved water molecules exert slight influences [75, 79]. Compared with the crystal structure, the backbone-atom root-mean-square deviations (bRMSD) are calculated at 2.1 Å for NA1, 0.9 Å for NA2 and 1.1 Å for NA3, respectively. Where NA1 is the structure with neither of the calcium ion and conserved water molecules; NA2 is the structure with only the calcium ion, and NA3 is the structure with both of the calcium ion and conserved water molecules. Moreover, the calcium ion and conserved water molecules can affect the shapes of the NA active sites [75]. As Figure 3 shows, the active sites of NA2 and NA3 match finely with each other but differ obviously from that of NA1.

The secondary structures of the NA active-site residues are altered by the presence of the calcium ion or/and conserved water molecules [75]. For example, the key residues Asn146, Gly147 and Tyr148 are of random coils in NA1. With the calcium ion (NA2), the three residues all change to H-bonded turns and with the addition of conserved water molecules (NA3), residues Asn146 and Gly147 change to random coils. The electrostatic potential surfaces (EPS), solvent accessible surfaces (SAS) areas of the NA active sites are also altered greatly [75]. It has been clear that the NA active site is stabilized by the calcium ion which is necessary for the bioactivities, through strong electrostatic effects with negatively charged residues Glu276 and Glu277 [59, 75, 80]. In comparison with NA2, the positively charged S1 in NA1 is sharply reduced and covered up by the residues near the active site (Figure 3). The positively charged S4 is also reduced somewhat and meanwhile, extruded towards the active site. The polar and non-polar as well as the total SAS areas are greatly reduced in the case of

NA1: the polar and non-polar areas of the residues in the active sites are summed to 49.47 and 42.43 Å^2 in NA1 and 259.52 and 118.14 Å^2 in NA2, respectively. As shown in Figure 3, the conserved water molecules show electrostatic interactions towards the key residues and result in the reduction of positive charges in S1, S2 and S4 [75]. In contrast to the positive-charge dominated active site in NA2, more negatively charged regions are present in the NA3 active site. Synchronously, the conserved water molecules tune the hydrophilic and hydrophobic properties of the active site [75]. With the conserved water molecules, the polar and nonpolar areas of the NA3 active site amount to 250.79 and 108.76 Å^2, respectively.

The active sites were defined through the Binding Site Analysis module. The NA structures were colored by electrostatic potentials. The BA substrate was shown in stick model.

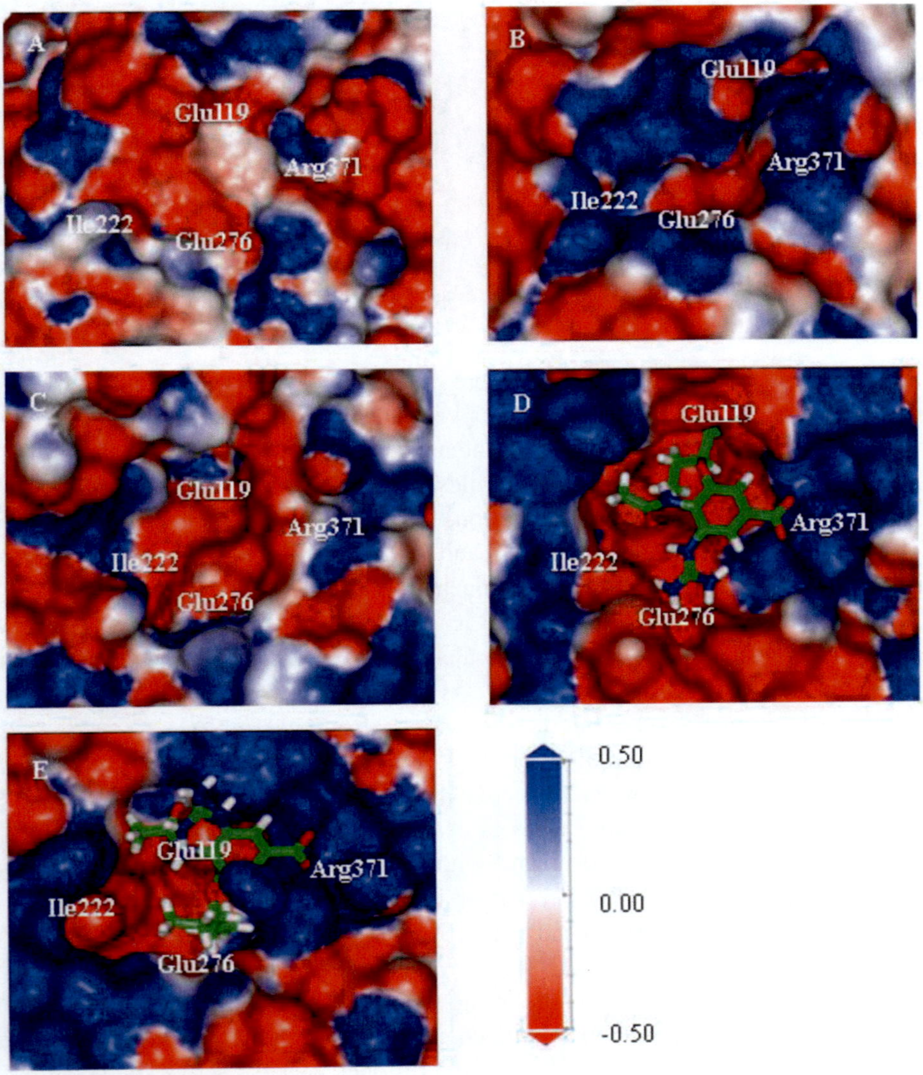

Figure 3. Electrostatic potential surfaces of the active sites of (a) NA1, (b) NA2, (c) NA3, (d) NA2^{BA} and (e) NA3^{BA} [75].

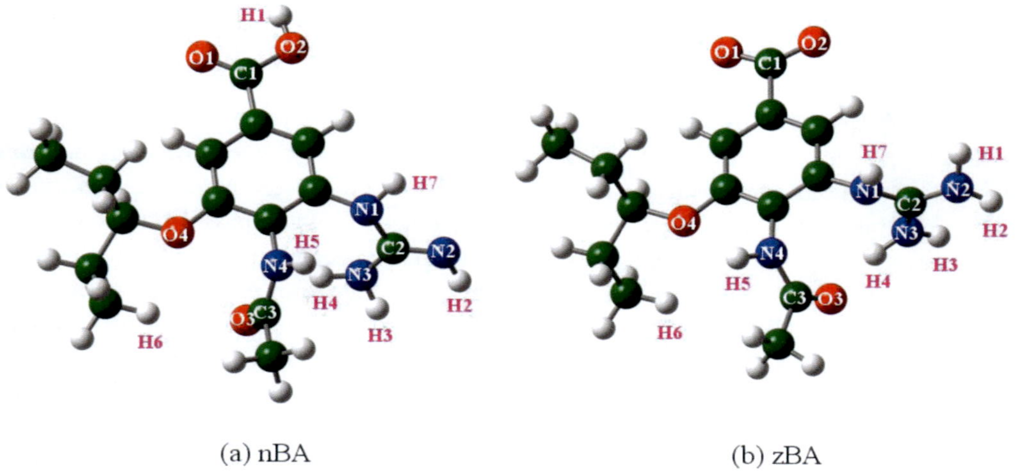

Figure 4. The neutral and zwitterionic isomers of BA [76].

The respective roles of the calcium ion and conserved water molecules on the NA active sites are further clarified by the binding differences of 4-(N-acetylamino)-5- guanidino-3-(3-pentyloxy) benzoic acid (BA) [75-77, 81]. BA can not be docked into the NA active site in absence of the calcium ion (NA1). But it is well contained in the NA2 active site whereas some portions protrude outside the NA3 active site (Figure 3). The total interaction energies are equal to -217.78 and -117.76 kcal·mol^{-1} for NA2 and NA3, respectively [75]. The interacting modes and strengths between BA and NA2, NA3 have remarkable differences; e.g., the guanidino group of BA is bound with S4 through electrostatic interactions with residues Glu276 and Glu277 in NA2; however, its functional group faces opposite to the NA3 active site (Figure 3). In addition, with the inclusion of conserved water molecules, the docking simulations produce satisfactory agreement with the experimental results [60, 81-84].

III. Proton Transfer of NA Inhibitors

Quantitative structure-activity relationship (QSAR), docking and molecular dynamics (MD) studies are widely used to investigate the inhibiting activities of the conceivable NA inhibitors [53, 54, 58, 60, 83-94]. The potent NA inhibitors, such as zanamivir, oseltamivir carboxylate (the active form of oseltamivir) and BA, can usually exist in two distinct isomers; i.e., neutral and zwitterionic ones. Take BA as an example. In the zwitterionic form, the carboxyl group is deprotonated and the guanidino group is protonated, see Figure 4b. Density functional calculations revealed quite different geometries for the neutral and zwitterionic isomers [76, 95], thus being expected to show distinct bioactivities. The neutral form has higher relative stability, while the zwitterion is generally considered to be bioactive.

The largest structural difference between the neutral (nBA) and zwitterionic (zBA) isomers of BA is the position of the H1 atom, which bonds the carboxyl O2 atom in the former whereas the guanidino N2 atom in the latter (Figure 4) [76]. The C1-O2 and C1-O1 distances in nBA are 1.215 and 1.357 Å, respectively [96]. When the carboxyl group is deprotonated (zBA), the two C-O distances are degenerated at ca. 1.250 Å. As to the guanidino sites, the N2, C2, N3 and H2 atoms in nBA fall nearly within one plane, with the ψ(N1C2N2H2) dihedral of -179.48°. However, the protonation of the guanidino group (zBA) distorts this plane, where the ψ(N1C2N2H2) dihedral was optimized at 170.59°. Besides, the other parts of nBA and zBA also have geometric differences: e.g., the N-acetylamino group is neutralized between the guanidino and 3-pentyloxy groups in nBA whereas moves greatly towards the guanidino group in zBA (Figure 4). It has been proven that the proton transfer process that transforms the nBA to the zBA structure needs a water passage that contains at least four water molecules [76, 97, 98]. When four water molecules are added, the neutral BA geometry of the neutral isomer (nBA4) is ready to transform to the zwitterion. The ψ(N1C2N2H2) dihedral in nBA4 was calculated at 173.10°, close to the value 170.59° in zBA rather than -179.48° in nBA. Meanwhile, the N-acetylamino groups are oriented different in nBA and nBA4, but similarly in zBA and nBA4. The zwitterionic geometry (zBA) is perfectly conserved in zBA4 [76].

As Table 1 shows, the neutral isomer is much more stable than the zwitterion in the gas phase, and the energy difference (E_{rel}) equals 24.76 kcal mol^{-1} at B3LYP/6-311++G(d,p) level of theory. The population analysis reveals that the neutral form predominates [76]. The relative stability of the zwitterion is greatly enhanced by the addition of water molecules. The value (E_{rel}) of zBA4 and nBA4 amounts to 2.54 kcal mol^{-1}, much less than that in the absence of water. When the effects of water solvents are further considered within the implicit solvation models, the E_{rel} values were calculated to be -7.53 kcal mol^{-1} for zBA(s) and nBA(s) and -6.84 kcal mol^{-1} for zBA4(s) and nBA4(s), respectively (Table 1). It means the zwitterion is more stable than the neutral isomer in aqueous solutions, and the population analysis indicated that the zwitterionic isomer should predominate in aqueous solutions [76].

The reaction energy diagram of the proton transfer process is shown in Figure 5. IN4 is the intermediate of the proton transfer process nBA4 → zBA4. The geometry difference between IN4 and nBA4 is mainly the opposite orientation of the carboxyl H atoms [76]. The transition states of Step 1 (nBA4 → IN4) and Step 2 (IN4 → zBA4) are characterized by the $H_7O_3^+$ and $H_5O_2^+$ ion species [96, 99-102], and their energy barriers are equal to 14.85 and 9.13 kcal mol^{-1}, respectively. With the implicit solvation models, the values change to 12.34 and 5.28 kcal mol^{-1}, respectively. It means Step 1 rather than Step 2 plays a more decisive role, but Step 2 is more facilitated to take place in aqueous solutions. Step 1 (rotation of the carboxyl O-H group) can proceed in the absence of water [76]. As indicated in Table 1, the energy differences of IN4 and nBA4, IN(s) and nBA(s), IN4(s) and nBA4(s) are close to each other. It reveals that more water molecules than four may have no obvious stabilizing effects on IN. In addition, the energy barriers of nBA → IN and nBA(s) → IN(s) are respectively calculated at 12.50 and 10.25 kcal mol^{-1}, which are both less than the values 14.85 kcal mol^{-1} for nBA4 → IN4 and 12.34 kcal mol^{-1} for nBA4(s) → IN4(s), see Figure 5. Hence, Step 1 should be regarded as water-insensitive, even if not water-suppressed. The proton transfer process to form the zwitterion is not facilitated in aqueous solutions, albeit it is more stable than the neutral isomer.

Among the known NA inhibitors, the proline-based drug A-315675 (BL, Compound 4 in Scheme 1) is special in that the direct proton transfers from the carboxyl group to the amine group, which may facilitate the zwitterion formation [48, 103, 104]. However, neither of the BL and proline zwitterions exists independently in the absence of water and will spontaneously convert to the canonical isomers [95, 105-108]. In gas phase, three isomers of BL (BLa, BLc and BLd) exist on the potential energy surface (PES), with BLa that can convert to the zwitterion (BLb) of the lowest energy (Figure 6). The zwitterions of BL and proline were estimated 13.25 and 15.01 kcal mol^{-1} higher in energy than their corresponding canonical isomers [95]. It means the zwitterion stabilities are somewhat enhanced by the introduction of functional groups. With the addition of two and three water molecules, the energy differences between the zwitterions and corresponding canonical isomers change to 3.13 and -1.54 kcal mol^{-1}, respectively. That is, three water molecules make the zwitterion stability superior to the canonical isomer, probably due to the stabilization of H-bonds with the water molecules [95]. With two and three water molecules, the energy barriers for the conversion from the canonical isomers to the zwitterions were calculated to be 4.96 and 3.13 kcal mol^{-1}, respectively. Hence, the zwitterion formation of BL is facile to take place with the addition of two molecules and further facilitated by more water molecules.

Table 1. The relative energies (in kcal mol^{-1}) of the zwitterionic and neutral BA isomers under various conditions

Basis sets	6-31G(d)	6-31G(d)$^{ZPE\ a}$	6-311++G(d,p)//6-31G(d)
zBA vs. nBA	28.61 (24.85) ^{b}	28.32	24.76
zBA(s) vs. nBA(s)	-2.89		-7.53
zBA4 vs. nBA4	2.19	2.63	2.54
zBA4(s) vs. nBA4(s)	-5.23		-6.84

^{a} Zero point energy corrections (ZPE) are made at B3LYP/6-31(d) level of theory.
^{b} The value was obtained at B3LYP/6-311++G(d,p) level of theory.

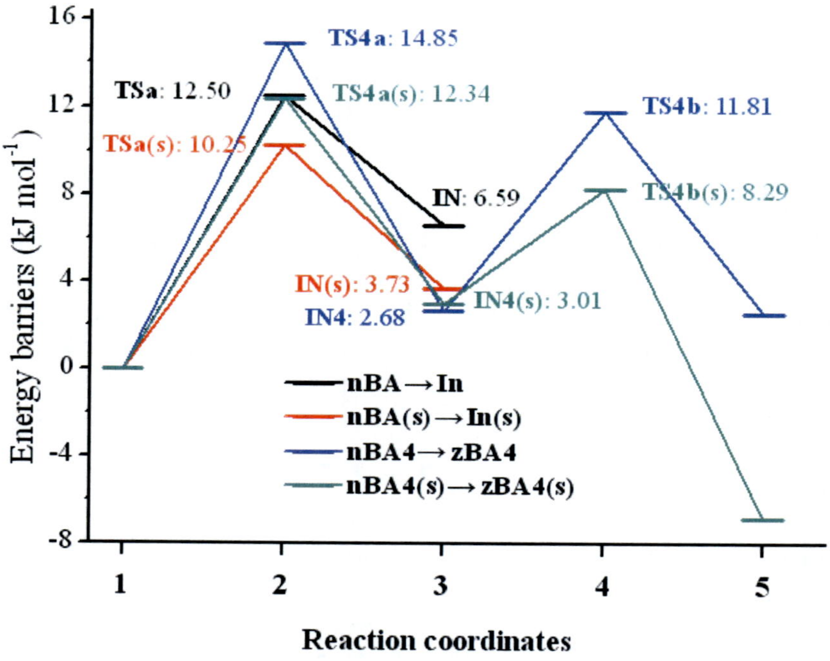

Figure 5. The reaction energy profiles for the proton transfer nBA4 → zBA4 obtained at B3LYP/6-311++G(d,p)//6-31G(d) level of theory [76]. Step 1 in the absence of water (nBA → IN) is also shown for comparisons.

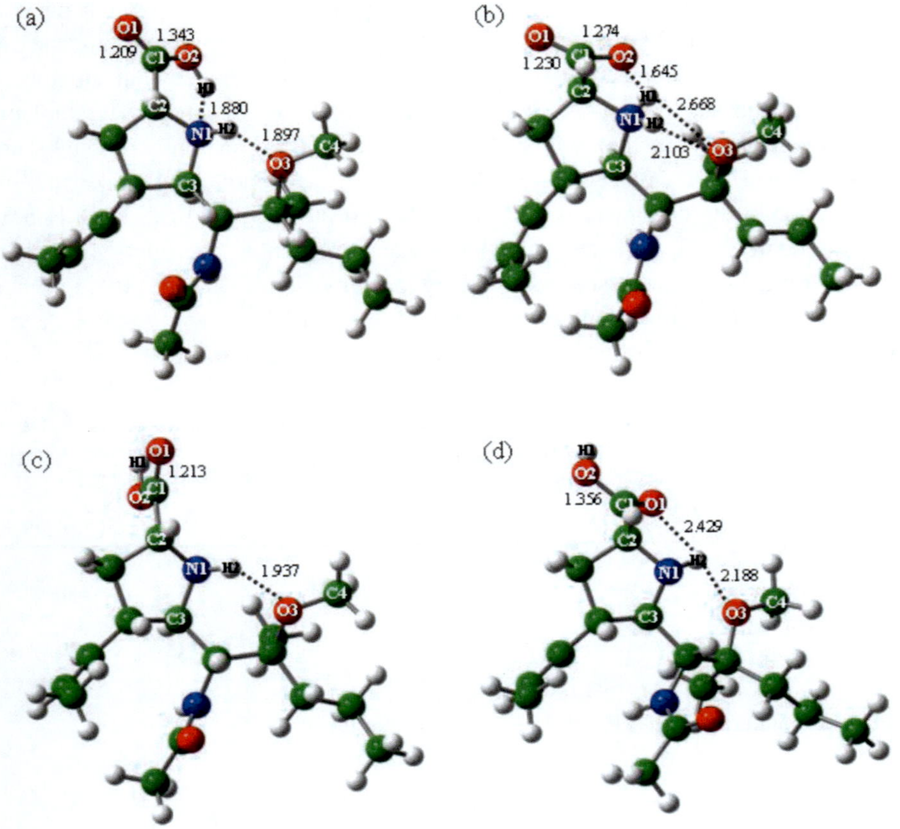

Figure 6. Structures of BL isomers in the absence of water: (a) BLa, (b) BLb, (c) BLc and (d) BLd [95]. The N1-H1 distance in the zwitterionic isomer (BLb) was fixed at 1.030 Å.

IV. Synergistic Effects during NA-inhibitor Interactions

With the in-depth understanding of the NA active sites and geometric aspects of inhibitors, a series of novel NA inhibitors have been designed in order to overcome the low bioavailability, rapid metabolizability and emerging resistances of zanamivir and oseltamivir [7, 32, 38]. Among them, 4-(N-acetylamino)-5-amido-3-(3-pentyloxy) benzoic acid (*2A* in Scheme 2) achieves a micromolar *in vitro* bioactivity against the type-A influenza virus. *2A* has identical functional groups as *1A* (oseltamivir carboxylate) [6, 8, 45, 46, 109, 110]; however, *1A* contains a different core template, as cyclohexene. Peramivir (*3C* in Scheme 2) is an injectable drug with the cyclopentane core template [40, 47]. Hence, the benzene, cyclohexene and cyclopentane core templates show their respective merits in the rational drug designs. Besides, some functional groups are frequently used, such as carboxyl, guanidino and 3-pentyloxyl, see Scheme 2 [32-43]. How the core templates and functional groups be assembled into a novel potential NA inhibitor is currently an essential and focusing issue in synthetic and medicinal chemistry.

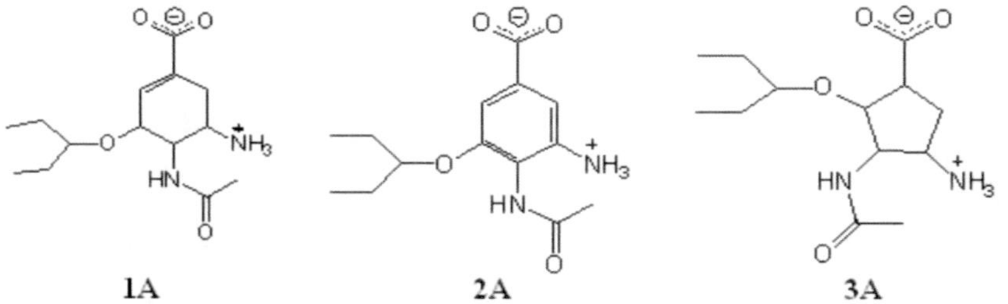

1A
oseltamivir carboxylate
IC_{50}, 1 nM

2A
IC_{50}, 15000 nM

3A

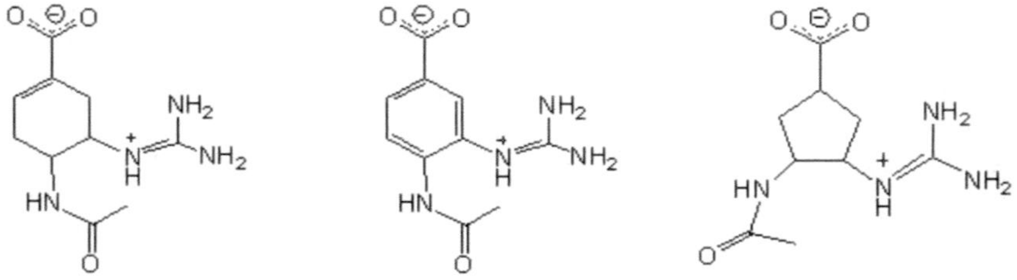

1B

2B
IC_{50}, 2500 nM

3B
IC_{50}, 115000 nM

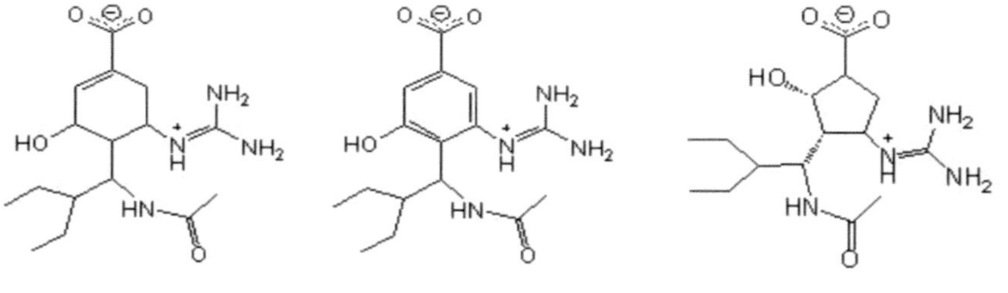

1C

2C

3C
peramivir
IC_{50}, 0.75 nM

Scheme 2. Chemical structures and bioactivities of some potential NA inhibitors [111].

The nine NA inhibitors listed in Scheme 2 have been proven to occupy the identical binding pocket [111], as observed in the previous simulations [1, 7, 19, 42, 51, 52, 75, 87, 112-114]. However, their binding poses at the NA active site differ somewhat from each other. As shown in Figure 7a, *1A* is characterized by strong electrostatic effects of its amido and N-acetylamino groups with residues Glu119, Asp151, Glu227 and Glu277, associated with numerous H-bonds [1, 7, 51, 113, 114]. The *1A* carboxyl group forms three H-bonds with residue Arg292 and one H-bond with residue Tyr406, respectively. The 3-pentyloxy group is adjusted to fit the hydrophobic cave containing residues Glu276 and Glu277. However, the *1B* guanidino group forms ionic interactions with the negatively charged carboxyl groups of residues Glu119, Asp151 and Glu227, with the two H-bonds formed with residue Asp151 (Figure 7b). Owing to the addition of the large 1-acetamido-2-ethylbutyl group, *1C* moves somewhat out of the NA active-site pocket and its guanidino group is nearly opposite to the case of *1B*, see Figure 7c.

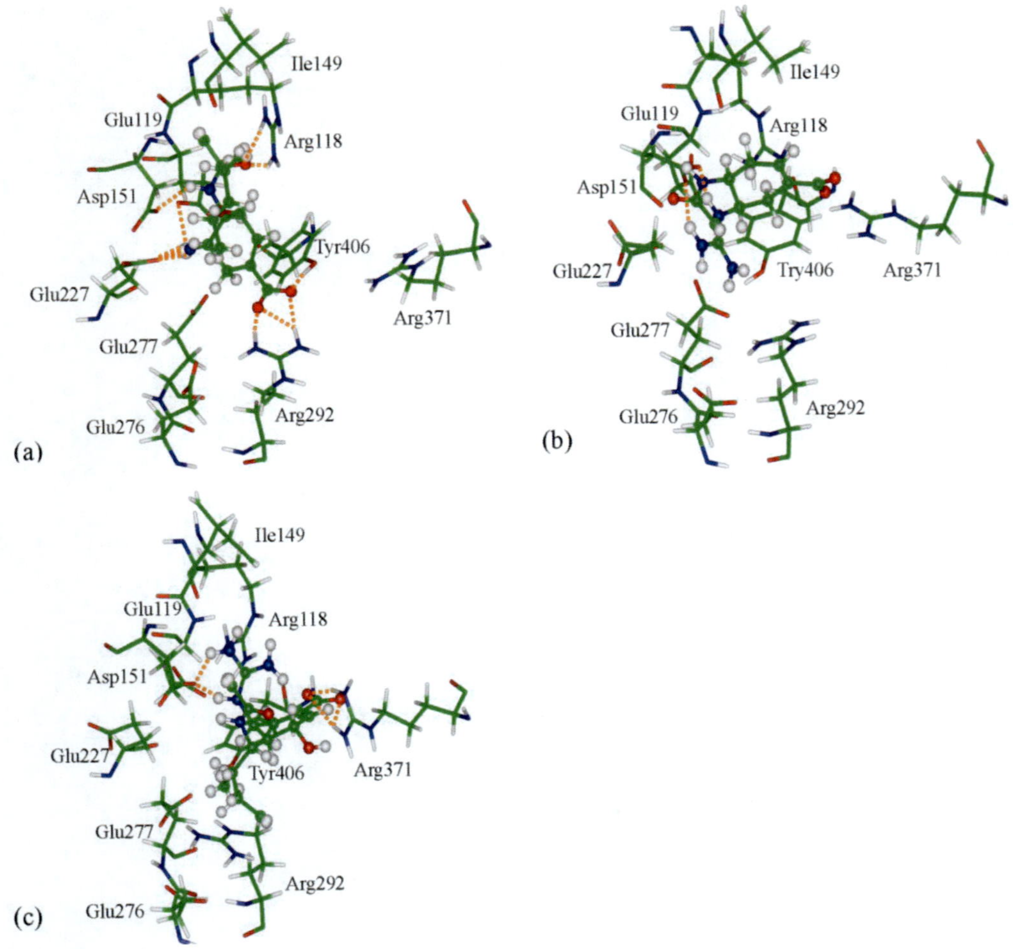

Figure 7. Views of the binding modes of the NA active-site residues with the inhibitors (a) *1A*, (b) *1B* and (c) *1C* [111]. Key residues are represented by stick models. Inhibitors are represented by ball and stick models. The important H-bonds are labeled in the dashed orange lines.

As to *2A*, its carboxyl group is docked towards the arginine triad (Arg118, Arg292 and Arg371), with four H-bonds formed, see Figure 8a. Similar to the case of *1A* (oseltamivir carboxylate), the 3-pentyloxy group of *2A* orients towards residues Glu276 and Glu277 and strong ionic effects are observed involving the amido and N-acetylamino groups with residues Asp151 and Glu227. As Figure 8b shows, the carboxyl group of *2B* is bound in a similar way as that of *2A*. But two additional H-bonds are formed between the *2B* N-acetylamino group and residue Asp 151. Consistent with *1C*, *2C* is somewhat out of the NA active-site pocket, with the large 1-acetamido-2-ethylbutyl group (Figure 8c).

Similar in *1A*-NA (Figure 7a), there are numerous charge-transfer interactions between *3A* and residues Glu119, Asp151, Glu227 and Glu277 (Figure 9a). The carboxyl group of *3A* is stabilized by residues Arg118 and Tyr406, with one H-bond for each residue (Figure 9a). However, the carboxyl group of *3B* is bound towards residues Arg292 and Tyr406, with the formation of two H-bonds with each residue (Figure 9b). In light of the current pandemic threat [7, 8, 40, 47], *3C* (peramivir) has received fast-track designation by FDA. The binding pose of *1C* is close to that of *1A* and agrees with the previous results [7, 40, 52]. However, three and one H-bonds are formed between residues Arg371, Tyr406 and the carboxyl group of *3C*, respectively. The guanidino group of *3C* forms many H-bonds with residues Glu119, Asp151, Glu227 and Glu277 (Figure 9c).

Through the fragment approach [88, 115-120], the energy contributions of the respective core templates (E_{core}), functional groups ($E_{functional}$) and synergistic effects ($E_{synergistic}$) during the interaction processes with the NA proteins were evaluated by the following equation [111]:

$$E_{inter} = E_{core} + \sum E_{functional} + E_{synergistic} \quad (1)$$

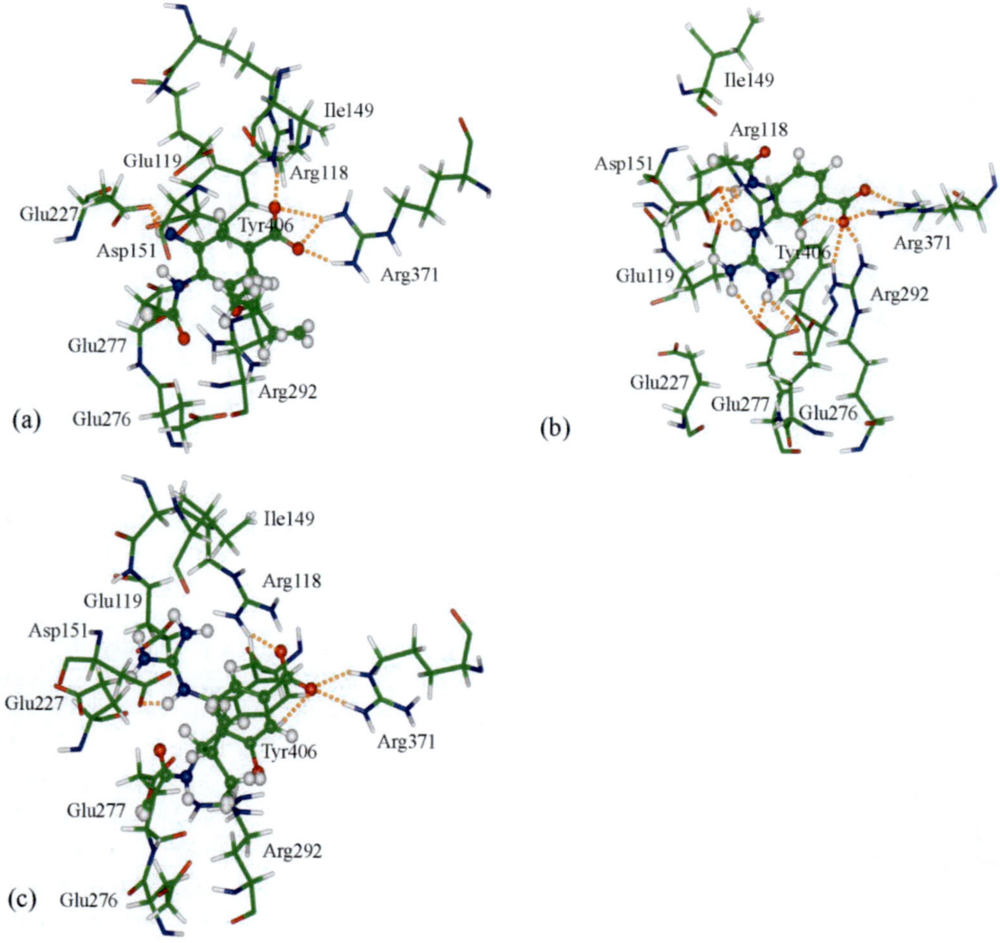

Figure 8. Views of the binding modes of the NA active-site residues with the inhibitors (a) *2A*, (b) *2B* and (c) *2C* [111]. Key residues are represented by stick models. Inhibitors are represented by ball and stick models. The important H-bonds are labeled in the dashed orange lines.

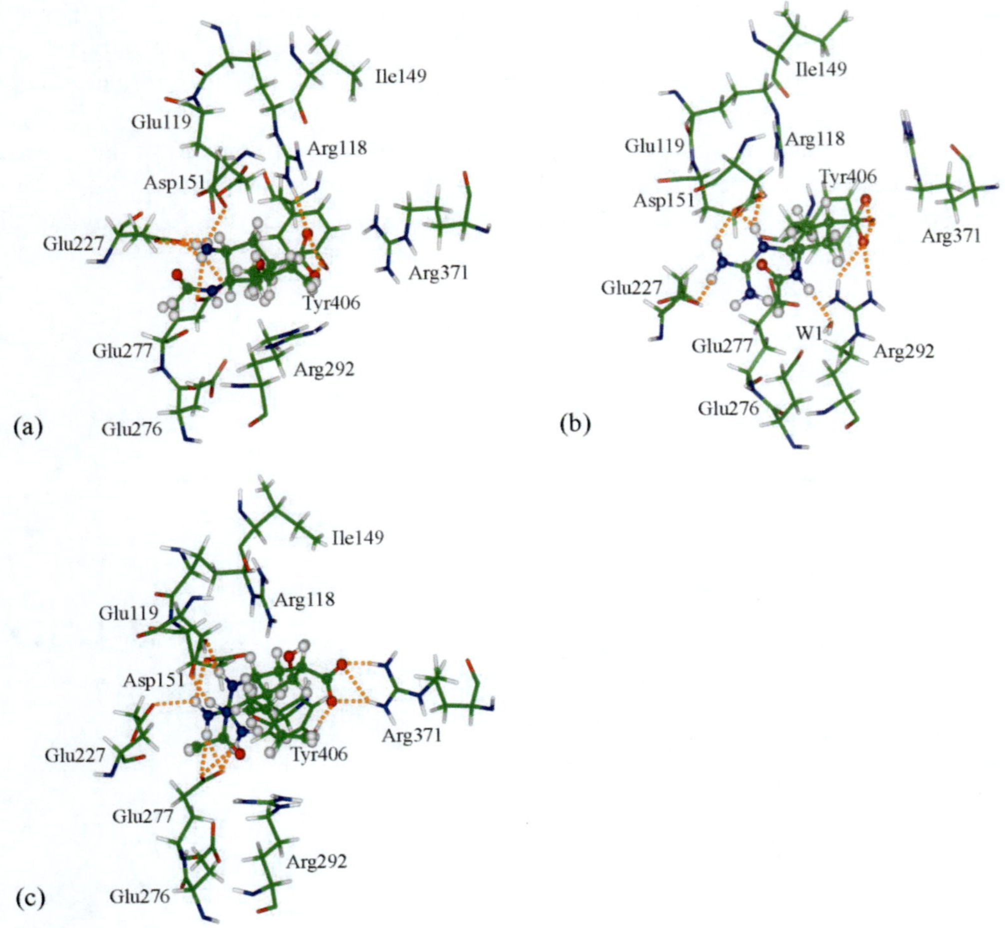

Figure 9. Views of the binding modes of the NA active-site residues with the inhibitors (a) *3A,* (b) *3B* and (c) *3C* [111]. Key residues are represented by stick models. Inhibitors are represented by ball and stick models. The important H-bonds are labeled in the dashed orange lines.

The energy contributions of carboxyl ($E_{carboxyl}$), guanidino ($E_{guanidino}$), N-acetylamino ($E_{N\text{-}acetylamino}$), 1-acetamido-2-ethylbutyl ($E_{1\text{-}acetamido\text{-}2\text{-}ethylbutyl}$), hydroxyl ($E_{hydroxyl}$), 3-pentyloxyl ($E_{3\text{-}pentyloxyl}$) and amido ($E_{amido}$) were calculated to be 61.76, 60.70, 140.64, 136.77, -7.79, 137.95 and 12.07 kcal mol^{-1}, respectively [111]. Obviously, a single functional group does not have inhibiting activity. The inhibitors with the same functional groups but different core templates may show obvious binding differences. For example, *1C*, *2C* and *3C* all contain the four carboxyl, guanidino, 1-acetamido-2-ethylbutyl and hydroxyl functional groups, but with different core templates. Their interaction energy differences with NA can be larger than 100 kcal mol^{-1} [111]. Besides, the core templates of the three inhibitors all play an important role during the binding processes, with the energy contributions of the cyclohexene, benzene and cyclopentane core templates accounting for 77.74%, 58.64% and 44.38%, respectively. Hence, the core templates are crucial to the bioactivities [111].

We found that *1A, 2A* and *3A* with the carboxyl, N-acetylamino, 3-pentyloxyl and amido groups have exceptionally good synergistic effects [111]. Especially *1A*, which has the largest synergistic energy ($E_{synergistic}$=-863.39 kcal mol^{-1}) among the nine inhibitors. The N-acetylamino groups of the three inhibitors finely fit the hydrophobic cave consisting of residues Trp178 and Ile222; in addition, their 3-pentyloxyl groups improve the synergistic effects with NA through more contacts with key active-site residues. Accordingly, the synergistic effects are probably responsible for the different binding qualities of the NA inhibitors. It thus affords valuable insights into the interactions between NA and its inhibitors.

V. De Novo Designs of Novel NA Inhibitor

Recently, the peptides against influenza viruses have shown potential as therapeutic agents [121-124]. The 12-mer peptides (54-N1 and 69-N2) display the broad-spectrum inhibitory activities to the influenza strains, through the interactions with the NA proteins [122]. However, these peptides are only partially docked into the NA active sites and will not form compact binding structures. Accordingly, it is of high urgency to discover novel peptides of low molecular weights, in order to efficiently control and treat the influenza infections. Tripeptides play an important role in biological processes and drug designs [125, 126] and there are many reports that tripeptides contribute to the clinical researches, such as thrombin [127], HIV protease [128], HCV protease [129] and immune systems [130]. On the basis of NA's catalytic mechanisms and structure–activity relationships [7, 42, 112, 122, 131], seven tripeptides were successively explored as the NA inhibitors.

The interaction energy (E_{inter}) of FRG and NA was calculated to be -249.83 kcal mol^{-1}, with electrostatic effects dominating the binding process. Whereas the lack of sidechains in Gly (H atoms) makes it unfit for the hydrophobic portion of NA active site (Figure 10a). Accordingly, the C-terminus Gly of FRG was improved by the G→V mutation. Compared with Gly, the sidechain of Val is more suitable with the NA active site (Figure 10b). The interaction energy (E_{inter}) between FRV and NA equals -289.88 kcal mol^{-1}, about 40 kcal mol^{-1} higher than that of FRG. Similar to FRG, the electrostatic rather than vdW interactions dominate during the binding process. The binding pose of FRV within the NA active site is close to those of current NA drugs (oseltamivir, zanamivir and peramivir) [1, 7, 51, 112-114]. Accordingly, FRV rather than FRG fits better with the NA active site.

In order to clarify the roles of the Arg and Phe portions, tripeptides FHV and YRV were designed as the NA inhibitors (Figs. 10 c and d). The interaction energies (E_{inter}) of FHV and YRV with NA were calculated to be -254.00 and -224.51 kcal mol^{-1}, respectively. Obviously, the Arg and Phe portions of the tripeptides, especially the latter, are crucial to the binding processes and should be reserved in the designs of tripeptide inhibitors. Accordingly, the improvement of tripeptide-based NA inhibitors was centered on the mutations of the C-terminus amino acid Val. In this way, the tripeptides FRT, FRS and FRI were designed as NA inhibitors. Owing to the addition of hydrophilic property (V→T mutation) (Figure 10e), the interaction energy (E_{inter}) of FRT with NA dropped to -255.45 kcal mol^{-1}. The V→S mutation to form FRS increases the hydrophilicity and its spatial size is less than FRT with the V→T mutation (Figure 10f). The interaction energy (E_{inter}) of FRS with NA was calculated at -250.04 kcal mol^{-1}, somewhat less than those of FRV and FRT. Accordingly, FRV rather than FRT and FRS suits best with the NA active site. Owing to the increases of spatial size and hydrophobic property by the V→I mutation, FRI moves closer to the NA active-site pocket

(Figure 11a). The interaction energy (E_{inter}) of FRI equals -291.56 kcal mol^{-1} and is larger than any of the above six tripeptides. Moreover, FRI has complementary properties against the geometrical and biophysical environment of the NA active site, which can also be observed in the cases of current potent NA drugs; e.g., oseltamivir, zanamivir and peramivir [1, 7, 51, 112-114].

For all the seven tripeptides, the electrostatic effects play a dominant role during the binding processes, which is consistent with current NA drugs [1, 7, 51, 112-114]. The carboxyl groups of the seven tripeptides except FRG and YRV have strong electrostatic interactions with the Arg triad (Arg118, Arg292 and Arg371). Residues Asp151, Glu119, Glu227 and Glu277 contribute a lot in all the seven cases. These four residues have already received enough attention from rational drug designs [7, 112, 113]. The catalytic residue Asp151 is crucial to NA functions and the three glutamic acid residues (Glu119, Glu227 and Glu277) are important to stabilize the NA active sites [131, 132]. With the low toxicity and viral resistance, tripeptides throw new light on the rational designs of novel anti-influenza drugs [122]. Especially, the interaction energy (E_{inter}) of FRI with NA amounts to -291.56 kcal mol^{-1} and ranks the largest among the seven tripeptides, also much larger than that of BA (-160.64 kcal mol^{-1}) [76]. Accordingly, it shows great potential as an ideal lead compound.

We then studied the chemical states of the FRI N-terminus during the binding processes. Besides the normal NH_3^+ form around the physiological pH values, the deprotonated (NH_2) and acetylated ($NHCOCH_3$) forms were also considered and designated as FRI_{dep} and FRI_{Ac}, respectively. The deprotonation causes the benzene group to move out of the NA active-site pocket (Figure 11b), with the affinity being obviously reduced to -138.22 kcal mol^{-1}. Accordingly, the depronation is not favored for the inhibiting activity. When the N-terminus was protected by acetylization, the interaction energy (E_{inter}) with the NA protein was reduced to -267.32 kcal mol^{-1}, indicating the decrease of inhibiting activity by acetylization. Accordingly, FRI rather than FRI_{dep} and FRI_{Ac} matches satisfactorily with the NA active site. In addition, the decrease of interaction energy by acetylization is in agreement with experiments that long peptides may not form compact binding complexes with the NA receptors [121, 122, 124]. Nonetheless, FRI_{Ac} still displays reasonable space orientation at the NA active site (Figure 11c) and medium inhibiting activity. The simulation results revealed the possibility of discovering tripeptides as potent antiviral drugs.

VII. CONCLUSION

The influenza infection pandemic has spread on a world scale and become a major threat to human health. Neuraminidase (NA, EC 3.2.1.18) offers a potential target for the development of antiviral drugs. There are continuing efforts in the design and synthesis of novel and potent NA inhibitors.

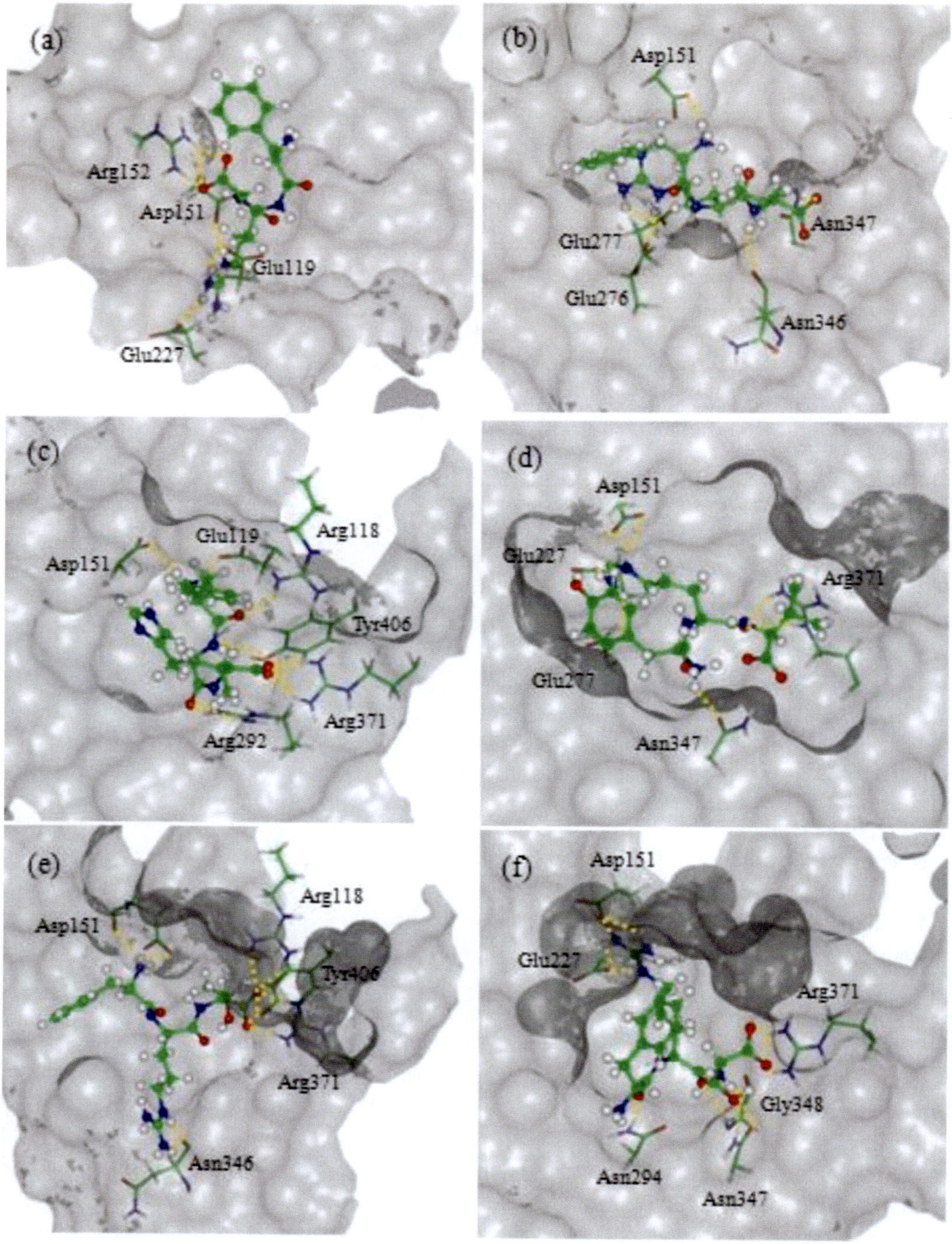

Figure 10. Views of the binding modes of the tripeptides with the NA active-site residues: (a) FRG, (b) FRV, (c) FHV, (d) YRV, (e) FRT and (f) FRS. Key residues are represented by stick models. Tripeptides are represented by ball and stick models. The Connolly surfaces of the NA active-site (in grey) are created using the InsightII 2005 scripts. The important H-bonds are labeled in the dashed gold lines.

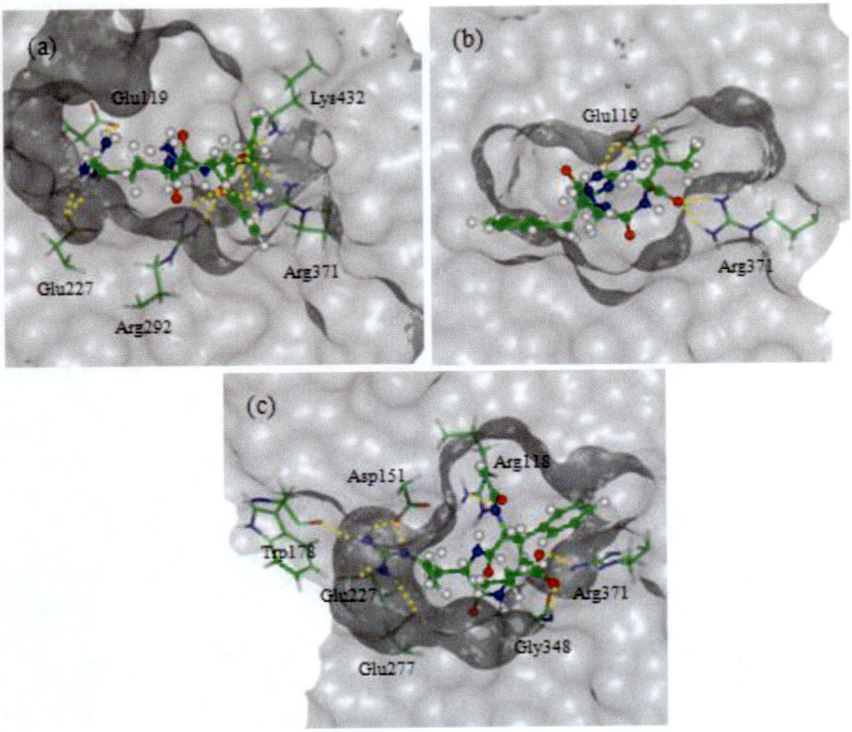

Figure 11. Views of the binding modes of FRI, FRI_{dep} and FRI_{Ac} with the NA active-site residues: (a) FRI, (b) FRI_{dep} and (c) FRI_{Ac}. Key residues are represented by stick models. Tripeptides are represented by ball and stick models. The Connolly surfaces of the NA active-site (in grey) are created using the InsightII 2005 scripts. The important H-bonds are labeled in the dashed gold lines.

The understanding of NA active sites helps to guide the studies of NA-inhibitor interactions and structure-based rational drug designs. The active sites of various NA proteins are structurally alike and can be divided into four sub-regions, involving ten polar residues and four hydrophobic residues. The calcium ion rather than conserved water molecules is crucial for stabilizing the NA active site. Without the calcium ion, the NA active site shrinks remarkably so that the substrates can not be docked into the "pocket". The binding modes can be well-predicted through structural and property analyses: sub-regions 1 and 3 determine the locations of substrates and sub-regions 2 and 4 tune the space orientations of substrates.

The zwitterions of the NA inhibitors are responsible for the inhibiting activities albeit the neutral isomers predominate in the absence of water solvents. With the addition of four water molecules, the BA geometry of the neutral isomer (nBA4) is induced to resemble the zwitterions and remains rather stable throughout the proton transfer process; in addition, the relative stability of the zwitterion vs. neutral isomer will be enhanced and reversed in aqueous solutions. Owing to the larger energy barrier, Step 1 rather than Step 2 of the proton transfer plays a decisive role but it is not facilitated by water solvents. Similar to BA, the zwitterion of BL does not represent a local minimum on the potential energy surface, but with the addition of three water molecules, it is more stable than the neutral isomer. In addition, the zwitterion is facile to form when the water molecules exceed one, which makes it an ideal lead compound for antiviral drugs.

Compared with oseltamivir carboxylate, the changes of the core template or/and functional groups in the other inhibitors cause the reduction of interaction energies and numbers of H-bonds with the NA proteins. The contributions of each core template and functional group, evaluated by the fragment approach, indicate that the core templates rather than functional groups play a larger role during the binding processes In addition, the binding qualities are determined by the synergistic effects of the core templates and functional groups. Among the nine ligands in Scheme 2, oseltamivir carboxylate has the largest synergistic energy with its functional groups fitting perfectly with the NA active site. It is in good agreement with the largest interaction energy, numerous H-bonds with the NA active-site residues as well as the lowest IC_{50} value.

Based on the in-depth understanding of NA's catalytic mechanisms and structure-activity relationships, seven tripeptides were successively designed as NA inhibitors. Their interaction energies decrease in the order of FRI > FRV > FRT > FHV > FRS > FRG > YRV. The Arg and Phe portions of these tripeptides play important roles during the binding processes. Among the seven tripeptides, FRI has the largest interaction energy, much superior to the potential drug BA. Besides, the protonation of N-terminus (NH_3^+) is crucial for the inhibitory activities and longer peptides may not be suitable as NA inhibitors. The studies of NA-inhibitor interactions and designs of NA inhibitors will provide new insights into the rational design of anti-influenza drugs.

REFERENCES

[1] Moscona A. Neuraminidase inhibitors for influenza. *N Engl J Med* 2005, *353,* 1363-1373.

[2] Collier R., *The Plague of the Spanish Lady. The Influenza Pandemic of 1918-1919.* Atheneum, New York.: 1974.

[3] Skehel J. An overview of influenza haemagglutinin and neuraminidase. *Biologicals* 2009, *37,* 177-178.

[4] Das K.; Aramini J. M.; Ma L. C.; Krug R. M.; Arnold E. Structures of influenza A proteins and insights into antiviral drug targets. *Nat Struct Mol Biol* 2010, *17,* 530-538.

[5] Girard M. P.; Tam J. S.; Assossou O. M.; Kieny M. P. The 2009 A (H1N1) influenza virus pandemic: A review. *Vaccine* 2010, *28,* 4895-4902.

[6] Gubareva L. V.; Kaiser L.; Hayden F. G. Influenza virus neuraminidase inhibitors. *Lancet* 2000, *355,* 827-835.

[7] von Itzstein M. The war against influenza: discovery and development of sialidase inhibitors. *Nat. Rev. Drug. Discov.* 2007, *6,* 967-974.

[8] Beigel J.; Bray M. Current and future antiviral therapy of severe seasonal and avian influenza. *Antivir. Res.* 2008, *78,* 91-102.

[9] Wagner R.; Matrosovich M.; Klenk H. D. Functional balance between haemagglutinin and neuraminidase in influenza virus infections. *Reviews in Medical Virology* 2002, *12,* 159-166.

[10] Garten R. J.; Davis C. T.; Russell C. A.; Shu B.; Lindstrom S.; Balish A.; Sessions W. M.; Xu X. Y.; Skepner E.; Deyde V.; Okomo-Adhiambo M.; Gubareva L.; Barnes J.; Smith C. B.; Emery S. L.; Hillman M. J.; Rivailler P.; Smagala J.; de Graaf M.; Burke

D. F.; Fouchier R. A. M.; Pappas C.; Alpuche-Aranda C. M.; Lopez-Gatell H.; Olivera H.; Lopez I.; Myers C. A.; Faix D.; Blair P. J.; Yu C.; Keene K. M.; Dotson P. D.; Boxrud D.; Sambol A. R.; Abid S. H.; George K. S.; Bannerman T.; Moore A. L.; Stringer D. J.; Blevins P.; Demmler-Harrison G. J.; Ginsberg M.; Kriner P.; Waterman S.; Smole S.; Guevara H. F.; Belongia E. A.; Clark P. A.; Beatrice S. T.; Donis R.; Katz J.; Finelli L.; Bridges C. B.; Shaw M.; Jernigan D. B.; Uyeki T. M.; Smith D. J.; Klimov A. I.; Cox N. J. Antigenic and Genetic Characteristics of Swine-Origin 2009 A(H1N1) Influenza Viruses Circulating in Humans. *Science* 2009, *325,* 197-201.

[11] Palese P.; Shaw M. L., *Orthomyxoviridae: the viruses and their replication.* Lippincott Williams and Wilkins Publishers: Philadelphia, 2007; Vol. 47, p 1647-1689.

[12] Roberts N. A.; Govorkova E. A., *The activity of neuraminidase inhibitor oseltamivir against all subtypes of influenza viruses.* Nova Science Publishers: New York, 2009; p 93-118.

[13] Skehel J. J.; Wiley D. C. Receptor binding and membrane fusion in virus entry: The influenza hemagglutinin. *Annual Review of Biochemistry* 2000, *69,* 531-569.

[14] De Clercq E.; Neyts J. Avian influenza A (H5N1) infection: targets and strategies for chemotherapeutic intervention. *Trends Pharmacol Sci* 2007, *28,* 280-285.

[15] Palese P.; Tobita K.; Ueda M.; Compans R. W. Characterization of temperature-sensitive influenza virus mutants defective in neuraminidase. *Virology* 1974, *61,* 397-410.

[16] Liu C. G.; Eichelberger M. C.; Compans R. W.; Air G. M. Influenza type-a virus neuraminidase does not play a role in viral entry, replication, assembly, or budding. *J. Virol.* 1995, *69,* 1099-1106.

[17] Ohuchi M.; Asaoka N.; Sakai T.; Ohuchi R. Roles of neuraminidase in the initial stage of influenza virus infection. *Microbes. Infect.* 2006, *8,* 1287-1293.

[18] Colman P. M. Neuraminidase inhibitors as antivirals. *Vaccine* 2002, *20,* S55-S58.

[19] Garman E.; Laver G. Controlling influenza by inhibiting the virus's neuraminidase. *Curr. Drug. Targets* 2004, *5,* 119-136.

[20] Moscona A. Medical management of influenza infection. *Annu. Rev. Med.* 2008, *59,* 397-413.

[21] Horimoto T.; Kawaoka Y. Influenza: lessons from past pandemics, warnings from current incidents. *Nat Rev Micro* 2005, *3,* 591-600.

[22] Laver W. G.; Valentine R. C. Morphology of the isolated hemagglutinin and neuraminidase subunits of influenza virus. *Virology* 1969, *38,* 105-119.

[23] Colman P. M.; Laver W. G.; Varghese J. N.; Baker A. T.; Tulloch P. A.; Air G. M.; Webster R. G. Three-dimensional structure of a complex of antibody with influenza virus neuraminidase. *Nature* 1987, *326,* 358-363.

[24] Varghese J. N.; Colman P. M. Three-dimensional structure of the neuraminidase of influenza virus A/Tokyo/3/67 at 2.2 A resolution. *J Mol Biol* 1991, *221,* 473-486.

[25] Varghese J. N.; Epa V. C.; Colman P. M. Three-dimensional structure of the complex of 4-guanidino-Neu5Ac2en and influenza virus neuraminidase. *Protein Sci* 1995, *4,* 1081-1087.

[26] vonItzstein M. Design and synthesis of influenza virus sialidase inhibitors: Novel anti-influenza drugs. *Options for the Control of Influenza Iii* 1996, *1123,* 697-701-860.

[27] Chong A. K. J.; Pegg M. S.; Taylor N. R.; Vonitzstein M. Evidence for a sialosyl cation transition-state complex in the reaction of sialidase from influenza-virus. *Eur. J. Biochem.* 1992, *207,* 335-343.

[28] Air G. M.; Laver W. G. The neuraminidase of influenza virus. *Proteins* 1989, *6,* 341-356.

[29] Colman P. M.; Tulip W. R.; Varghese J. N.; Tulloch P. A.; Baker A. T.; Laver W. G.; Air G. M.; Webster R. G. Three-dimensional structures of influenza virus neuraminidase-antibody complexes. *Philos Trans R Soc Lond B Biol Sci.* 1989, 511-518.

[30] Oxford J. S.; Novelli P.; Sefton A.; Lambkin R. New millennium antivirals against pandemic and epidemic influenza: the neuraminidase inhibitors. *Antivir. Chem. Chemother.* 2002, *13,* 205-217.

[31] Colman P. M.; Hoyne P. A.; Lawrence M. C. Sequence and structure alignment of paramyxovirus hemagglutinin-neuraminidase with influenza virus neuraminidase. *J Virol* 1993, *67,* 2972-2980.

[32] Chand P.; Babu Y. S.; Bantia S.; Chu N.; Cole L. B.; Kotian P. L.; Laver W. G.; Montgomery J. A.; Pathak V. P.; Petty S. L.; Shrout D. P.; Walsh D. A.; Walsh G. M. Design and Synthesis of Benzoic Acid Derivatives as Influenza Neuraminidase Inhibitors Using Structure-Based Drug Design1. *J. Med. Chem.* 1997, *40,* 4030-4052.

[33] Kim C. U.; Lew W.; Williams M. A.; Liu H.; Zhang L.; Swaminathan S.; Bischofberger N.; Chen M. S.; Mendel D. B.; Tai C. Y. Influenza neuraminidase inhibitors possessing a novel hydrophobic interaction in the enzyme active site: design, synthesis, and structural analysis of carbocyclic sialic acid analogues with potent anti-influenza activity. *J. Am. Chem. Soc.* 1997, *119,* 681-690.

[34] Williams M. A.; Lew W.; Mendel D. B.; Tai C. Y.; Escarpe P. A.; Laver W. G.; Stevens R. C.; Kim C. U. Structure-activity relationships of carbocyclic influenza neuraminidase inhibitors. *Bioorg. Med. Chem. Lett.* 1997, *7,* 1837-1842.

[35] Smith B. J. A conformational study of 2-oxanol: insight into the role of ring distortion on enzyme-catalyzed glycosidic bond cleavage. *J. Am. Chem. Soc.* 1997, *119,* 2699-2706.

[36] Smith P. W.; Sollis S. L.; Howes P. D.; Cherry P. C.; Starkey I. D.; Cobley K. N.; Weston H.; Scicinski J.; Merritt A.; Whittington A. Dihydropyrancarboxamides Related to Zanamivir: A New Series of Inhibitors of Influenza Virus Sialidases. 1. Discovery, Synthesis, Biological Activity, and Structure? Activity Relationships of 4-Guanidino-and 4-Amino-4H-pyran-6-carboxamides. *J. Med. Chem.* 1998, *41,* 787-797.

[37] Lew W.; Wu H.; Mendel D. B.; Escarpe P. A.; Chen X.; Laver W. G.; Graves B. J.; Kim C. U. A new series of C3-aza carbocyclic influenza neuraminidase inhibitors: synthesis and inhibitory activity. *Bioorg. Med. Chem. Lett.* 1998, *8,* 3321-3324.

[38] Atigadda V. R.; Brouillette W. J.; Duarte F.; Babu Y. S.; Bantia S.; Chand P.; Chu N.; Montgomery J. A.; Walsh D. A.; Sudbeck E. Hydrophobic benzoic acids as inhibitors of influenza neuraminidase. *Bioorg. Med. Chem.* 1999, *7,* 2487-2497.

[39] Sears P.; Wong C. H. Carbohydrate mimetics: a new strategy for tackling the problem of carbohydrate-mediated biological recognition. *Angew. Chem. Int. Ed.* 1999, *38,* 2300–2324.

[40] Babu Y. S.; Chand P.; Bantia S.; Kotian P.; Dehghani A.; El-Kattan Y.; Lin T. H.; Hutchison T. L.; Elliott A. J.; Parker C. D. BCX-1812 (RWJ-270201): discovery of a

novel, highly potent, orally active, and selective influenza neuraminidase inhibitor through structure-based drug design. *J. Med. Chem.* 2000, *43,* 3482-3486.

[41] Steindl T.; Langer T. Influenza virus neuraminidase inhibitors: generation and comparison of structure-based and common feature pharmacophore hypotheses and their application in virtual screening. *J. Chem. Inf. Comp. Sci.* 2004, *44,* 1849-1856.

[42] Russell R. J.; Haire L. F.; Stevens D. J.; Collins P. J.; Lin Y. P.; Blackburn G. M.; Hay A. J.; Gamblin S. J.; Skehel J. J. The structure of H5N1 avian influenza neuraminidase suggests new opportunities for drug design. *Nature* 2006, *443,* 45-49.

[43] Platis D.; Smith B. J.; Huyton T.; Labrou N. E. Structure-guided design of a novel class of benzyl-sulfonate inhibitors for influenza virus neuraminidase. *Biochem. J.* 2006, *399,* 215-223.

[44] von Itzstein M.; Wu W. Y.; Kok G. B.; Pegg M. S.; Dyason J. C.; Jin B.; Van Phan T.; Smythe M. L.; White H. F.; Oliver S. W. Rational design of potent sialidase-based inhibitors of influenza virus replication. *Nature* 1993, *363,* 418-423.

[45] Carr J.; Ives J.; Kelly L.; Lambkin R.; Oxford J.; Mendel D.; Tai L.; Roberts N. Influenza virus carrying neuraminidase with reduced sensitivity to oseltamivir carboxylate has altered properties in vitro and is compromised for infectivity and replicative ability in vivo. *Antivir. Res.* 2002, *54,* 79-88.

[46] Ward P.; Small I.; Smith J.; Suter P.; Dutkowski R. Oseltamivir (Tamiflu) and its potential for use in the event of an influenza pandemic. *J. Antimicrob. Chemother.* 2005, *55 (Suppl. S1),* i5-i21.

[47] Young D.; Fowler C.; Bush K. RWJ-270201 (BCX-1812): a novel neuraminidase inhibitor for influenza. *Philos Trans R Soc Lond B Biol Sci* 2001, *356,* 1905-1913.

[48] Kati W. M.; Montgomery D.; Carrick R.; Gubareva L.; Maring C.; McDaniel K.; Steffy K.; Molla A.; Hayden F.; Kempf D.; Kohlbrenner W. In vitro characterization of A-315675, a highly potent inhibitor of A and B strain influenza virus neuraminidases and influenza virus replication. *Antimicrob Agents Chemother* 2002, *46,* 1014-1021.

[49] Zhang H.; Xu W. F. Recent advances in anti-influenza agents with neuraminidase as target. *Mini Rev Med Chem.* 2006, *6,* 429-448.

[50] Taylor N. R.; Cleasby A.; Singh O.; Skarzynski T.; Wonacott A. J.; Smith P. W.; Sollis S. L.; Howes P. D.; Cherry P. C.; Bethell R.; Colman P.; Varghese J. Dihydropyrancarboxamides related to zanamivir: a new series of inhibitors of influenza virus sialidases. 2. Crystallographic and molecular modeling study of complexes of 4-amino-4H-pyran-6-carboxamides and sialidase from influenza virus types A and B. *J. Med. Chem.* 1998, *41,* 798-807.

[51] Varghese J. N.; Smith P. W.; Sollis S. L.; Blick T. J.; Sahasrabudhe A.; McKimm-Breschkin J. L.; Colman P. M. Drug design against a shifting target: a structural basis for resistance to inhibitors in a variant of influenza virus neuraminidase. *Structure* 1998, *6,* 735-746.

[52] Smith B. J.; McKimm-Breshkin J. L.; McDonald M.; Fernley R. T.; Varghese J. N.; Colman P. M. Structural Studies of the Resistance of Influenza Virus Neuramindase to Inhibitors. *J. Med. Chem.* 2002, *45,* 2207-2212.

[53] Murray C. W.; Baxter C. A.; Frenkel A. D. The sensitivity of the results of molecular docking to induced fit effects: Application to thrombin, thermolysin and neuraminidase. *J. Comput. Aided. Mol. Des.* 1999, *13,* 547-562.

[54] Yi X.; Guo Z. R.; Chu F. M. Study on molecular mechanism and 3D-QSAR of influenza neuraminidase inhibitors. *Bioorg. Med. Chem.* 2003, *11,* 1465-1474.

[55] Mann M. C.; Islam T.; Dyason J. C.; Florio P.; Trower C. J.; Thomson R. J.; von Itzstein M. Unsaturated N-acetyl-D-glucosaminuronic acid glycosides as inhibitors of influenza virus sialidase. *Glycoconj J.* 2006, *23,* 127-133.

[56] Platis D.; Smith B. J.; Huyton T.; Labrou N. E. Structure-guided design of a novel class of benzyl-sulfonate inhibitors for influenza virus neuraminidase. *Biochem. J.* 2006, *399,* 215-223.

[57] Sun C.; Zhang X.; Huang H.; Zhou P. Synthesis and evaluation of a new series of substituted acyl(thio)urea and thiadiazolo [2,3-a] pyrimidine derivatives as potent inhibitors of influenza virus neuraminidase. *Bioorg Med Chem* 2006, *14,* 8574-8581.

[58] Zheng M.; Yu K.; Liu H.; Luo X.; Chen K.; Zhu W.; Jiang H. QSAR analyses on avian influenza virus neuraminidase inhibitors using CoMFA, CoMSIA, and HQSAR. *J. Comput. Aided. Mol. Des.* 2006, *20,* 549-566.

[59] Smith B. J.; Huyton T.; Joosten R. P.; McKimm-Breschkin J. L.; Zhang J. G.; Luo C. S.; Lou M. Z.; Labrou N. E.; Garrett T. P. Structure of a calcium-deficient form of influenza virus neuraminidase: implications for substrate binding. *Acta. Crystallogr. D.* 2006, *62,* 947-952.

[60] Barillari C.; Taylor J.; Viner R.; Essex J. W. Classification of water molecules in protein binding sites. *J. Am. Chem. Soc.* 2007, *129,* 2577-2587.

[61] Rungrotmongkol T.; Udommaneethanakit T.; Frecer V.; Miertus S. Combinatorial design of avian influenza neuraminidase inhibitors containing pyrrolidine core with a reduced susceptibility to viral drug resistance. *Comb Chem High Throughput Screen* 2010, *13,* 268-277.

[62] Chong A. K.; Pegg M. S.; von Itzstein M. Influenza virus sialidase: effect of calcium on steady-state kinetic parameters. *Biochim Biophys Acta* 1991, *1077,* 65-71.

[63] Cheng L. S.; Amaro R. E.; Xu D.; Li W. W.; Arzberger P. W.; McCammon J. A. Ensemble-Based Virtual Screening Reveals Potential Novel Antiviral Compounds for Avian Influenza Neuraminidase. *J. Med. Chem.* 2008, *51,* 3878-3894.

[64] Dunitz J. D. The entropic cost of bound water in crystals and biomolecules. *Science* 1994, *264,* 670-670.

[65] de Graaf C.; Pospisil P.; Pos W.; Folkers G.; Vermeulen N. P. E. Binding mode prediction of cytochrome P450 and thymidine kinase protein-ligand complexes by consideration of water and rescoring in automated docking. *J. Med. Chem.* 2005, *48,* 2308-2318.

[66] Verdonk M. L.; Chessari G.; Cole J. C.; Hartshorn M. J.; Murray C. W.; Nissink J. W.; Taylor R. D.; Taylor R. Modeling water molecules in protein-ligand docking using GOLD. *J Med Chem* 2005, *48,* 6504-6515.

[67] Garcia-Sosa A. T.; Mancera R. L. The effect of a tightly bound water molecule on scaffold diversity in the computer-aided de novo ligand design of CDK2 inhibitors. *J. Mol. Model.* 2006, *12,* 422-431.

[68] Wade R. C. 'Flu' and structure-based drug design. *Structure* 1997, *5,* 1139-1145.

[69] Wang G. T.; Chen Y. W.; Wang S.; Gentles R.; Sowin T.; Kati W.; Muchmore S.; Giranda V.; Stewart K.; Sham H.; Kempf D.; Laver W. G. Design, synthesis, and structural analysis of influenza neuraminidase inhibitors containing pyrrolidine cores. *J. Med. Chem.* 2001, *44,* 1192-1201.

[70] Kim C. U.; Lew W.; Williams M. A.; Wu H.; Zhang L.; Chen X.; Escarpe P. A.; Mendel D. B.; Laver W. G.; Stevens R. C. Structure-activity Relationship Studies of Novel Carbocyclic Influenza Neuraminidase Inhibitors. *J. Med. Chem.* 1998, *41,* 2451-2460.

[71] Atigadda V. R.; Brouillette W. J.; Duarte F.; Ali S. M.; Babu Y. S.; Bantia S.; Chand P.; Chu N.; Montgomery J. A.; Walsh D. A.; Sudbeck E. A.; Finley J.; Luo M.; Air G. M.; Laver G. W. Potent inhibition of influenza sialidase by a benzoic acid containing a 2-pyrrolidinone substituent. *J Med Chem* 1999, *42,* 2332-2343.

[72] Chand P.; Babu Y. S.; Bantia S.; Rowland S.; Dehghani A.; Kotian P. L.; Hutchison T. L.; Ali S.; Brouillette W.; El-Kattan Y.; Lin T.-H. Syntheses and Neuraminidase Inhibitory Activity of Multisubstituted Cyclopentane Amide Derivatives. *J Med Chem* 2004, *47,* 1919-1929.

[73] Maring C. J.; Stoll V. S.; Zhao C.; Sun M.; Krueger A. C.; Stewart K. D.; Madigan D. L.; Kati W. M.; Xu Y.; Carrick R. J.; Montgomery D. A.; Kempf-Grote A.; Marsh K. C.; Molla A.; Steffy K. R.; Sham H. L.; Laver W. G.; Gu Y. G.; Kempf D. J.; Kohlbrenner W. E. Structure-based characterization and optimization of novel hydrophobic binding interactions in a series of pyrrolidine influenza neuraminidase inhibitors. *J. Med. Chem.* 2005, *48,* 3980-3990.

[74] Yang G.; Yang Z. W.; Wu X. M.; Zu Y. G. A novel anti-influenza drug: molecular docking of trihydroxymethoxyflavone. *Chin. Comput. Appl. Chem.* 2008, *25,* 409-414.

[75] Yang G.; Yang Z. W.; Zu Y. G.; Wu X. M.; Fu Y. J. The calcium ion and conserved water molecules in neuraminidases: roles and implications for substrate binding. *Internet. Electron. J. Mol. Des.* 2008, *7,* 97-113.

[76] Yang Z. W.; Yang G.; Zu Y. G.; Fu Y. J.; Zhou L. J. The conformational analysis and proton transfer of the neuraminidase inhibitors: a theoretical study. *Phys. Chem. Chem. Phys.* 2009, *11,* 10035-10041.

[77] Yang Z. W.; Zu Y. G.; Wu X. M.; Liu C. B.; Yang G. A computational investigation on the interaction mechanisms of neuraminidases and 3-(3-pentyloxy)benzoic acid. *Acta. Chimica. Sinica.* 2010, *14,* 1370-1378.

[78] Stoll V.; Stewart K. D.; Maring C. J.; Muchmore S.; Giranda V.; Gu Y. G. Y.; Wang G.; Chen Y. W.; Sun M. H.; Zhao C.; Kennedy A. L.; Madigan D. L.; Xu Y. B.; Saldivar A.; Kati W.; Laver G.; Sowin T.; Sham H. L.; Greer J.; Kempf D. Influenza neuraminidase inhibitors: Structure-based design of a novel inhibitor series. *Biochemistry* 2003, *42,* 718-727.

[79] Lawrenz M.; Wereszczynski J.; Amaro R.; Walker R.; Roitberg A.; McCammon J. A. Impact of calcium on N1 influenza neuraminidase dynamics and binding free energy. *Proteins.* 2010, *78,* 2523-2532.

[80] Takahashi T.; Suzuki T.; Hidari K. I.; Miyamoto D.; Suzuki Y. A molecular mechanism for the low-pH stability of sialidase activity of influenza A virus N2 neuraminidases. *FEBS Lett* 2003, *543,* 71-75.

[81] Atigadda V. R.; Brouillette W. J.; Duarte F.; Babu Y. S.; Bantia S.; Chand P.; Chu N.; Montgomery J. A.; Walsh D. A.; Sudbeck E.; Finley J.; Air G. M.; Luo M.; Laver G. W. Hydrophobic benzoic acids as inhibitors of influenza neuraminidase. *Bioorg Med Chem* 1999, *7,* 2487-2497.

[82] Mancera R. L. De novo ligand design with explicit water molecules: an application to bacterial neuraminidase. *J. Comput. Aided Mol. Des.* 2002, *16,* 479-499.

[83] Masukawa K. M.; Kollman P. A.; Kuntz I. D. Investigation of Neuraminidase-Substrate Recognition Using Molecular Dynamics and Free Energy Calculations. *J. Med. Chem.* 2003, *46,* 5628-5637.

[84] Bonnet. P.; Bryce. R. A. Molecular dynamics and free energy analysis of neuraminidase-ligand interactions. *Protein Sci.* 2004, *13,* 946-957.

[85] Oakley A. J.; Barrett S.; Peat T. S.; Newman J.; Streltsov V. A.; Waddington L.; Saito T.; Tashiro M.; McKimm-Breschkin J. L. Structural and functional basis of resistance to neuraminidase inhibitors of influenza B viruses. *J Med Chem* 2010, *53,* 6421-6431.

[86] Thomas A.; Jourand D.; Bret C.; Amara P.; Field M. J. Is There a Covalent Intermediate in the Viral Neuraminidase Reaction? A Hybrid Potential Free-Energy Study. *J. Am. Chem. Soc.* 1999, *121,* 9693-9702.

[87] Wall I. D.; Leach A. R.; Salt D. W.; Ford M. G.; Essex J. W. Binding constants of neuraminidase inhibitors: An investigation of the linear interaction energy method. *J. Med. Chem.* 1999, *42,* 5142-5152.

[88] Wang T.; Wade R. C. Comparative binding energy (COMBINE) analysis of influenza neuraminidase-inhibitor complexes. *J. Med. Chem.* 2001, *44,* 961-971.

[89] Birch L.; Murray C. W.; Hartshorn M. J.; Tickle I. J.; Verdonk M. L. Sensitivity of molecular docking to induced fit effects in influenza virus neuraminidase. *J Comput Aided Mol Des* 2002, *16,* 855-869.

[90] Fornabaio M.; Cozzini P.; Mozzarelli A.; Abraham D. J.; Kellogg G. E. Simple, intuitive calculations of free energy of binding for protein-ligand complexes. 2. Computational titration and pH effects in molecular models of neuraminidase-inhibitor complexes. *J. Med. Chem.* 2003, *46,* 4487-4500.

[91] Zhang J.; Yu K.; Zhu W.; Jiang H. Neuraminidase pharmacophore model derived from diverse classes of inhibitors. *Bioorg Med Chem Lett* 2006, *16,* 3009-3014.

[92] Verma R. P.; Hansch C. A QSAR study on influenza neuraminidase inhibitors. *Bioorg Med Chem* 2006, *14,* 982-996.

[93] Wei D. Q.; Du Q. S.; Sun H.; Chou K. C. Insights from modeling the 3D structure of H5N1 influenza virus neuraminidase and its binding interactions with ligands. *Biochem Biophys Res Commun* 2006, *344,* 1048-1055.

[94] Malaisree M.; Rungrotmongkol T.; Decha P.; Intharathep P.; Aruksakunwong O.; Hannongbua S. Understanding of known drug-target interactions in the catalytic pocket of neuraminidase subtype N1. *Proteins* 2008, *71,* 1908-1918.

[95] Yang Z. W.; Wu X. M.; Zhou L. J.; Yang G. A proline-based neuraminidase inhibitor: DFT studies on the zwitterion conformation, stability and formation. *Int J Mol Sci* 2009, *10,* 3918-3930.

[96] Yang G.; Wu X.; Zu Y.; Liu C.; Fu Y.; Zhou L. Proton transfer at the carboxylic sites of amino acids: A single water molecule catalyzed process. *Int. J. Quantum Chem.* 2009, *109,* 320-327.

[97] Balta B.; Aviyente V. Solvent effects on glycine II. Water-assisted tautomerization. *J. Comput. Chem.* 2004, *25,* 690-703.

[98] Yang G.; Zu Y. G.; Liu C. B.; Fu Y. J.; Zhou L. J. Stabilization of amino acid zwitterions with varieties of anionic species: The intrinsic mechanism. *J Phys Chem B* 2008, *112,* 7104-7110.

[99] Jorgensen W. L.; Chandrasekhar J.; Madura J. D.; Impey R. W.; Klein M. L. Comparison of simple potential functions for simulating liquid water. *J. Chem. Phys.* 1983, *79,* 926-935.

[100] Okumura M.; Yeh L. I.; Myers J. D.; Lee Y. T. Infrared-Spectra of the Cluster Ions H7o3+ .H-2 and H9o4+ .H-2. *J Chem Phys* 1986, *85,* 2328-2329.

[101] Headrick J. M.; Diken E. G.; Walters R. S.; Hammer N. I.; Christie R. A.; Cui J.; Myshakin E. M.; Duncan M. A.; Johnson M. A.; Jordan K. D. Spectral signatures of hydrated proton vibrations in water clusters. *Science* 2005, *308,* 1765-1769.

[102] Markovitch O.; Agmon N. Structure and energetics of the hydronium hydration shells. *J Phys Chem A* 2007, *111,* 2253-2256.

[103] Hanessian S.; Bayrakdarian M.; Luo X. Total synthesis of A-315675: a potent inhibitor of influenza neuraminidase. *J Am Chem Soc* 2002, *124,* 4716-4721.

[104] Abed Y.; Nehme B.; Baz M.; Boivin G. Activity of the neuraminidase inhibitor A-315675 against oseltamivir-resistant influenza neuraminidases of N1 and N2 subtypes. *Antivir. Res.* 2008, *77,* 163-166.

[105] Sapse A. M.; Mallah-Levy L.; Daniels S. B.; Erickson B. W. The gamma. turn: beginning calculations on proline and N-acetylproline amide. *J. Am. Chem. Soc.* 1987, *109,* 3526-3529.

[106] Stepanian S. G.; Reva I. D.; Radchenko E. D.; Adamowicz L. Conformers of non-ionized proline. Matrix-isolation infrared and post-Hartree-Fock beginning study. *J. Phys. Chem. A* 2001, *105,* 10664-10672.

[107] Czinki E.; Csaszar A. G. Conformers of gaseous proline. *Chem. Eur. J.* 2003, *9,* 1008-1019.

[108] Allen W. D.; Czinki E.; Csaszar A. G. Molecular structure of proline. *Chem. Eur. J.* 2004, *10,* 4512-4517.

[109] Yeung Y. Y.; Hong S.; Corey E. J. A Short Enantioselective Pathway for the Synthesis of the Anti-Influenza Neuramidase Inhibitor Oseltamivir from 1,3-Butadiene and Acrylic Acid. *J. Am. Chem. Soc.* 2006, *128,* 6310-6311.

[110] Nie L. D.; Shi X. X.; Ko K. H.; Lu W. D. A short and practical synthesis of oseltamivir phosphate (Tamiflu) from (-)-shikimic acid. *J Org Chem* 2009, *74,* 3970-3973.

[111] Yang Z.; Nie Y.; Yang G.; Zu Y.; Fu Y.; Zhou L. Synergistic effects in the designs of neuraminidase ligands: Analysis from docking and molecular dynamics studies. *J. Theor. Biol.* 2010, *267,* 363-374.

[112] Stoll V.; Stewart K. D.; Maring C. J.; Muchmore S.; Giranda V.; Gu Y.-g. Y.; Wang G.; Chen Y.; Sun M.; Zhao C.; Kennedy A. L.; Madigan D. L.; Xu Y.; Saldivar A.; Kati W.; Laver G.; Sowin T.; Sham H. L.; Greer J.; Kempf D. Influenza neuraminidase inhibitors: structure-based design of a novel inhibitor series. *Biochemistry* 2003, *42,* 718-727.

[113] Aruksakunwong O.; Malaisree M.; Decha P.; Sompornpisut P.; Parasuk V.; Pianwanit S.; Hannongbua S. On the lower susceptibility of oseltamivir to influenza neuraminidase subtype N1 than those in N2 and N9. *Biophys J* 2007, *92,* 798-807.

[114] Amaro R. E.; Cheng X.; Ivanov I.; Xu D.; McCammon J. A. Characterizing Loop Dynamics and Ligand Recognition in Human- and Avian-Type Influenza Neuraminidases via Generalized Born Molecular Dynamics and End-Point Free Energy Calculations. *J. Am. Chem. Soc.* 2009, *131,* 4702–4709.

[115] Verlinde C. L.; Rudenko G.; Hol W. G. In search of new lead compounds for trypanosomiasis drug design: a protein structure-based linked-fragment approach. *J Comput Aided Mol Des* 1992, *6,* 131-147.

[116] Rees D. C.; Congreve M.; Murray C. W.; Carr R. Fragment-based lead discovery. *Nat. Rev. Drug. Discov.* 2004, *3,* 660-672.

[117] Zartler E. R.; Shapiro M. J. Fragonomics: fragment-based drug discovery. *Curr. Opin.Chem. Biol.* 2005, *9,* 366-370.

[118] Du Q. S.; Huang R. B.; Wei Y. T.; Pang Z. W.; Du L. Q.; Chou K. C. Fragment-based quantitative structure-activity relationship (FB-QSAR) for fragment-based drug design. *J. Comput. Chem.* 2009, *30,* 295-304.

[119] Erlanson D. A.; McDowell R. S.; T. O. B. Fragment-based drug discovery. *J. Med. Chem.* 2004, *47,* 3463-3482.

[120] Leach A. R.; Hann M. M.; Burrows J. N.; Griffen E. J. Fragment screening: an introduction. *Mol Biosyst* 2006, *2,* 430-446.

[121] Jones J. C.; Turpin E. A.; Bultmann H.; Brandt C. R.; Schultz-Cherry S. Inhibition of influenza virus infection by a novel antiviral peptide that targets viral attachment to cells. *J Virol* 2006, *80,* 11960-11967.

[122] Kuang Z. Z.; Zheng L. S.; Li S.; Duan Z. J.; Qi Z. Y.; Qu X. W.; Liu W. P.; Zhang W. J.; Li D. D.; Gao H. C.; Hou Y. D. Screening of peptides as broad-spectrum neuraminidase inhibitors against influenza viruses. *Chin. J. of Virol.* 2007, *23,* 165-171.

[123] Matsubara T.; Sumi M.; Kubota H.; Taki T.; Okahata Y.; Sato T. Inhibition of influenza virus infections by sialylgalactose-binding peptides selected from a phage library. *J Med Chem* 2009, *52,* 4247-4256.

[124] Rajik M.; Jahanshiri F.; Omar A. R.; Ideris A.; Hassan S. S.; Yusoff K. Identification and characterization of a novel anti-viral peptide against avian influenza virus H9N2. *Virol J* 2009, *6,* doi: 10.1186/1743-1422X-1186-1174.

[125] Bender F. C.; Whitbeck J. C.; Lou H.; Cohen G. H.; Eisenberg R. J. Herpes simplex virus glycoprotein B binds to cell surfaces independently of heparan sulfate and blocks virus entry. *J Virol* 2005, *79,* 11588-11597.

[126] Townsend D. M.; Tew K. D.; Tapiero H. The importance of glutathione in human disease. *Biomedecine & Pharmacotherapy* 2003, *57,* 145-155.

[127] Shen G. X. Development and current applications of thrombin-specific inhibitors. *Curr Drug Targets Cardiovasc Haematol Disord* 2001, *1,* 41-49.

[128] Kiso Y. Design and synthesis of substrate-based peptidomimetic human immunodeficiency virus protease inhibitors containing the hydroxymethylcarbonyl isostere. *Biopolymers* 1996, *40,* 235-244.

[129] Njoroge F. G.; Chen K. X.; Shih N. Y.; Piwinski J. J. Challenges in modern drug discovery: a case study of boceprevir, an HCV protease inhibitor for the treatment of hepatitis C virus infection. *Acc Chem Res* 2008, *41,* 50-59.

[130] Luger T. A.; Scholzen T. E.; Brzoska T.; Bohm M. New insights into the functions of alpha-MSH and related peptides in the immune system. *Ann N Y Acad Sci* 2003, *994,* 133-140.

[131] Colman P. M.; Varghese J. N.; Laver W. G. Structure of the catalytic and antigenic sites in influenza virus neuraminidase. *Nature* 1983, *303,* 41-44.

[132] Colman P. M. Influenza virus neuraminidase: structure, antibodies, and inhibitors. *Protein Sci* 1994, *3,* 1687-1696.

In: Focus on Catalysis Research: New Developments
Editors: Minjae Ghang and Bjørn Ramel
ISBN: 978-1-62100-455-4

Chapter 6

HYBRID ORGANIC-INORGANIC MATERIALS: APPLICATION IN OXIDATIVE CATALYSIS

Graça M. S. R. O. Rocha*
Department of Chemistry, Campus de Santiago, University of Aveiro, 3810-193, Aveiro, Portugal

ABSTRACT

Oxidative catalysis has been and will be playing an important role in the production of large quantities of intermediates. This technology has a great potential for improvement and this has led to a series of better processes, including the development of a great diversity of suitable catalysts. Within these compounds, metal phosphates and phosphonates possessing layered and pillared structures were found to constitute a very good alternative as catalysts to be used in oxidative catalysis.

Metal phosphates and phosphonates can be considered as strong inorganic solid acids and much of their catalytic activity has been attributed to the Brønsted acidity of the interlayered hydroxyl groups and to the Lewis acidity of the metal center. Metal phosphates and phosphonates can be obtained at low temperatures, in aqueous media, using soft chemical routes and their preparation is quite accessible if the correct phosphonic acids and phophates are available. These heterogeneous materials are very interesting from the economical as well as from the environmental points of view because of the high yields and short reaction times, easy recovery from the reaction mixtures and the possibility to perform reactions in solvent-free conditions.

Due to their particular physical and chemical properties and high versatility, in the last forty years, this class of organic-inorganic materials has been captivating the attention of many researchers involved in heterogeneous catalysis. The main research effort in the metal phosphate and phosphonate field was initially directed towards compounds with tetravalent metal cations, but a wide variety of divalent and trivalent metals have also been reported. Nowadays, a large number of metal phosphates and phosphonates of the α- and γ-type are known.

* E-mail addresses: grrocha@ua.pt; Tel.: +351 234 401 542; Fax: +351 234 370 084

The general interest in the chemistry of metal phosphates and phosphonates is mainly due to their unusual compositional and structural diversity varying from one-dimensional arrangements to three-dimensional microporous frameworks, passing by the most common layered networks.

Good results have been obtained with a large diversity of metal phosphates and phosphonates in a variety of organic reactions, and particularly in oxidative catalysis. The importance of these systems has also been recognized in research areas such as electrochemistry, microelectronics, biological membranes and photochemical mechanisms.

1. Introduction

Metal phosphates, $M(HZO_4)_2.nH_2O$, were initially obtained in their amorphous form by the reaction between aqueous solutions of a metal cation and a polybasic acid. Metals (M) include Zr [1-3], Sn [4, 5], Ti [6-8], Mo [9, 10], Th [11], Ce [12, 13], V [14], Pb [15, 16], Hf [17] and Ge [16] and polybasic acids (Z) include phosphoric and arsenic acid [17]. Since 1964, metal phosphates have also been synthesized in its crystalline form by reflux of the respective amorphous phosphates in concentrated solutions of the appropriate acid.

The preparation of metal phosphates may undergo several variations. The basic structure of a phosphate can be obtained in the crystalline, amorphous and pellicular form; protons in the phosphate groups may be substituted by other metals, $nH^+ \leftrightarrow M^{n+}$ (Cu, V, W, Mo); simple phosphates can be replaced by phosphite or organic phosphonic acids and phosphates; mixed polymers of general formula $M(RZO_3)_{2-x}(R'ZO_3)_x$ can be synthesized with phosphate / phosphonate groups and it is possible to leave "spaces" in the structures with the aim of increasing the selectivity of the reactions; the properties of the phosphonates can be varied from neutral (e. g. $P–CH_3$) or weakly acidic (e. g. $P–CH_2COOH$) to strong acidic (e. g. $P–C_6H_4SO_3H$) or, even to basic compounds (e. g. $P–C_2H_4NH_2$). Metal phosphates can also be intercalated with polar organic molecules, such as alcohols, glycols and amines. This huge possibility of variations makes practicable the synthesis of a wide variety of metal phosphates and phosphonates that, due to their characteristics (acidic nature, high capacity for ion-exchange, high resistance to temperature, radiation, oxidizing solutions and insolubility), can be used successfully in a vast range of research areas. The preparation of mixed metal phosphates or arsenates, as Zr-Ti phosphate [18], Zr-Ge phosphate [19], Zr-Sn phosphate [20] and arsenophosphates of Zr [21] has also received special attention.

The vast majority of research work has been developed with zirconium *bis*(monohydrogen orthophosphate), $[Zr(HPO_4)_2.H_2O]$ because it is one of the best characterized phosphates and under thermal features, is the most stable member of the tetravalent metal phosphates.

In the following sections, the structures, preparation methods and applications of metal phosphates and phosphonates are presented with more detail.

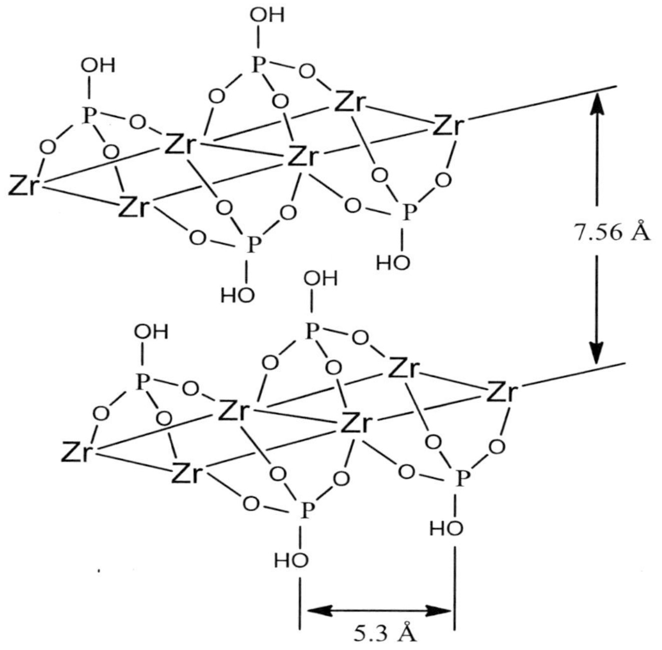

Figure 1. Schematic representation of the α-ZrP structure. The zeolitic cavities and the water molecules were omitted.

2. Structures of Phosphates and Phosphonates

The polyhydrated zirconium *bis*(monohydrogen orthophosphate), [$Zr(HPO_4)_2.nH_2O$], usually designated by zirconium phosphate (ZrP), may occur in the dried, β, and two hydrated forms, the monohydrate, α, and the dihydrated, γ. The crystal structure of the zirconium phosphate was initially determined by X-ray diffraction analysis [22, 23]. The α-zirconium phosphate in the crystalline form has an ordered structure in layers, as represented in Figure 1. The interlayer distance is 7.56 Å and the thickness of the layer, calculated as the shortest distance between the centers of the oxygen atoms of the P—OH groups present on opposite sides of one layer, is 6.3 Å. The distance between adjacent P—OH groups on one side of the layer is 5.3 Å then the "free area" around each phosphate group is 24 Å. The arrangement of the pendent phosphate groups creates six-sided cavities, each containing one water molecule in the interlayer region. These water molecules form a hydrogen bonding network with the P—OH groups belonging to one of the sides of the layer. There are no hydrogen bonds between the layers. Only Van der Waals and electrostatic forces are responsible for keeping the layers together [23].

The β and γ phases of the crystalline zirconium phosphate are prepared by refluxing a Zr^{4+} salt in a mixture of sodium dihydrogenphosphate and hydrochloric acid [2]. The β-$Zr(HPO_4)_2$ and γ-$Zr(HPO_4)_2.2H_2O$ phases differ from the α-ZrP on the packing and structure of the layers [6, 24, 25]. The distance between the layers of the β and γ phases is 9.4Å and 12.3Å, respectively. The proposed structures for the β- and γ-ZrP are more open than for the α-ZrP due to the greater separation between the layers [26]. Due to the lack of "water crystallization" in the β-ZrP, it was proposed that the neighbor mono-hydrogenphosphate groups of adjacent layers must be aligned in opposite positions to each other to allow the formation of hydrogen bonding between the layers. The structure of γ-ZrP is achieved by the insertion of water molecules between the layers of the β-ZrP. It is possible that the water molecules are located near the mono-hydrogenphosphate groups promoting the formation of hydrogen bonds between these groups with the help of the water molecules. The solid state ^{31}P magic angle (MAS) NMR spectroscopy study of the γ-$Zr(HPO_4)_2.2H_2O$ showed that half of the phosphate groups are dihydrogenphosphates $H_2PO_4^-$, and the other half are orthophosphates, PO_4^{3-}, γ-$Zr(PO_4)(H_2PO_4).2H_2O$.

When the zirconium phosphate is subjected to an elevation of temperature, different phases are formed [27, 28]. The P—OH groups present in the interlayers condense into pyrophosphates at a temperature ranging from 200 to 450° C (depending on the degree of crystallinity of the sample) while the condensation of the P—OH groups on the surface occurs at much higher temperatures [29]. Between 150 and 235°C, the ζ phase is formed without loss of water (this phase is a stable form of α-ZrP). For this phase, the distance between layers is 7.41Å. A second change occurs between 235 and 400° C, resulting in the η phase by the loss of one mole of water of the ζ phase. For this phase, the distance between layers is 7.17Å [27, 30]. Above 450° C, zirconium pyrophosphate is formed due to the release of water and condensation of adjacent phosphate groups [31, 32]. Other forms of zirconium phosphate were also referred [2, 30], which differ in the mode of linking of the layers and in the influence of the interlayer water of hydration. The θ-ZrP phase arises from the acidification of α-ZrP that has been half-exchanged with sodium ions.

The structures of α- and γ-TiP are identical to the α- and γ-ZrP with a layered arrangement. The interlayer spacing is 7.6Å for α-TiP and 11.6Å for γ-TiP [33]. However, like in the case of α- and γ-ZrP, some differences between the α- and γ-titanium phosphate must be taken into account. In the α-TiP, each phosphorus atom has an acid group. The γ-$Ti(H_2PO_4)(PO_4).2H_2O$ has two different types of phosphorus (one orthophosphate and one dihydrogenphosphate). Only the dihydrogenphosphate groups have acid groups, subsequently each phosphorus atom of this type has two active sites to participate in specific reactions.

There are other series of complex metal oxides in layers, having the general formula $MOZO_4$ (M = V, Nb, Ta and Mo; Z = P, As and S). The α form of $VOPO_4$ is a member of this series and crystallizes in a tetragonal layer structure consisting of distorted VO_6 octahedra and PO_4 tetrahedra linked by corner-shared oxygen atoms. The oxygen atoms in each VO_6 unit form a near regular octahedron (Figure 2) [34]. The vanadium coordination is similar to that observed for V_2O_5. However, the interactions between the layers in the α-$VOPO_4$ are much weaker than in the V_2O_5 because the number of V=O···V interactions is reduced due to the presence of the PO_4 groups.

Metal phosphonates (or organophosphates), $M(O_3ZR)_2.nS$; R = inorganic (H) or organic radical (—CH_3, —C_6H_5, —$O(CH_2)_n$—CH_3, etc.), Z = (P and As) and S = a polar solvent intercalated in the interlayer region, can be considered as organic derivatives of metal phosphates. The concatenation of ZrO_6 octahedra and O_3P—R tetrahedra are similar to that occurring in the metal phosphates justifying the layer structure of these compounds. Due to the short lateral distance between adjacent O_3P—R groups on each side of the α-layer, interpenetrating of the R groups belonging to adjacent layers cannot occur for steric reasons and a double film of R groups is expected for all members of this class [35].

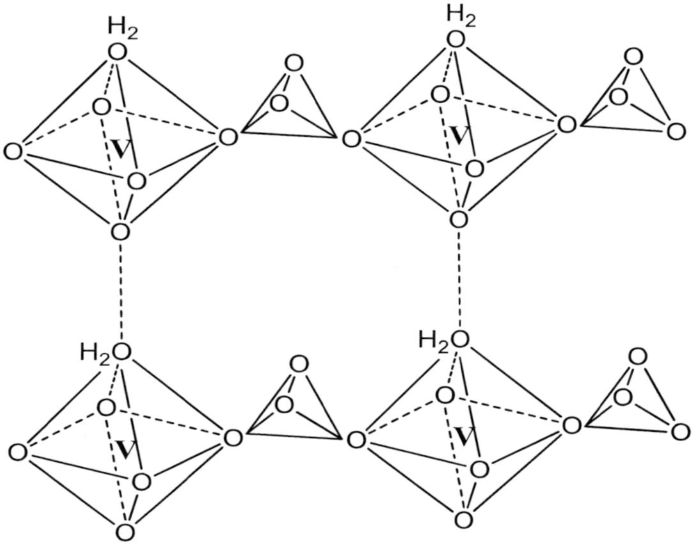

Figure 2. Schematic representation of the α-$VOPO_4.H_2O$.

Several types of reactions can be carried out to obtain a variety of organically pillared metal phosphonates with a range of properties and chemical behaviors. The general procedure is to prepare the diphosphonic acid of the pillar adequate to the objectives of the phosphonate to be synthesized. Some of these compounds include pillared zirconium aryldiphosphonates and alkyldiphosphonates, pillared structures with divalent elements, combined aryl-alkyl pillars and γ-ZrP layers and functionalized pillared hybrids: polyethers and polyamines. These materials are produced with controlled pore sizes, varying degrees of hydrophilic and hydrophobic character and chemical behavior as a consequence of the incorporation of varied functional groups as part of the pillars [36].

New zirconium phosphates and phosphonates with different structures from the traditional α- and γ-type zirconium phosphates, recently appear in the literature [37]. The structure and properties of compounds with different dimensionality, ranging from single or composite linear chains, to layered compounds, and open-framework structures have been presented. Low-dimensional 1D and 2D structures are easily accessible and able to be functionalized for many applications and the 3D open-framework compounds are attractive due to their modular structure suitable to host different classes of guest molecules. In many cases, the presence of groups different from phosphate or phosphonate tetrahedra, coordinated to zirconium, such as fluoride, chloride, or hydroxyl groups, and neutral ligands such as dimethylsulfoxide, were found. Some of these new compounds showed unusual characteristics and reactivity offering new possibilities for the preparation of tailor-made compounds, with structure and reactivity that can be tuned for specific purposes [37].

Two types of Sn(IV) phosphonates have been prepared recently; monophosphonates that form porous spherical aggregates and diphosphonates that form layered three-dimensional structures. The use of spacer groups such as methyl, phenyl and phosphite were found to change the pore structure and/or increase the surface area. The structure of $Sn(O_3PCH_3)_2$ was determined from its powder pattern obtained from a sample treated hydrothermally at 220°C for 30 days. The structure is layered with pendant methyl groups forming a bilayer similar to the structure of zirconium phenylphosphonate [38].

Two new zirconium aminophosphonates have been obtained by the reaction of Zr(IV) with piperazine-*N,N'*-*bis*-(methylenephosphonate) building blocks. Although prepared in similar conditions, their composition and crystal structure is markedly different. The compound of formula $Zr_2H_4[(O_3PCH_2)_2N_2C_4H_8]_3.9H_2O$, has a three dimensional structure but the compound of formula $ZrF_2(O_3PCH_2)_2.(NH)_2C_4H_8$, has a pillared layered structure. The effect of the various synthesis parameters was discussed. A probable structure-directing parameter seems to be the pH value of the starting precipitation solution that can influence the protonation of N atoms of piperazine moiety [39].

3. Preparation of Phosphates and Phosphonates

Metal phosphates can be prepared in the amorphous, crystalline and pellicular forms [22]. The mixture of aqueous solutions of a metal cation and a polybasic acid result in the precipitation of amorphous gels of variable composition (Equation 1).

$$M + H_3ZO_4 \xrightarrow{H_2O} M(HZO_4)_2.nH_2O \qquad \text{Equation 1}$$

The crystalline phosphates are obtained by reflux of the amorphous phosphates in phosphoric acid [3]. However, if the reflux is carried out in a mixture of phosphoric acid and sodium hydrogenphosphate, the γ and β phases are obtained [2]. This process is extremely slow, requiring 14 days for complete crystallization in an aqueous solution of phosphoric acid 12M molar [40]. The reaction may be sped up either by the addition of hydrofluoric acid (HF) [8, 41] to the solution or by raising the temperature under high pressure [42]. The resulting crystalline polymers adopt a layered structure, as shown in Figure 1.

The α-ZrP and γ-ZrP that suffered an half exchange with sodium ions (θ-ZrP) have a greater interlayer distance than the original form and for this reason they are easily intercalated with alcohols [43], polar solvents or other potential substitutes. Some organic substituents intercalated by direct or indirect methods were urea, amides, aminoacids and heterocyclic bases [43, 44], as well as inorganic substituents [45, 46].

When γ-ZrP is immersed in an aqueous solution of ethylene oxide, this epoxide is intercalated in its interlayers. The epoxy groups react with the hydrogenphosphate groups in the surface of the interlayers suffering ring opening. As a result, a layered organic derivative of the phosphate is formed, having phosphoric ester bonds, $Zr(HOC_2H_4OPO_3)_2.H_2O$ (Figure 3) [47].

These phosphoric ester groups can be exchanged with phosphate ions from NaH_2PO_4 solutions, so that the ester ions are released into the solution. The initial γ-ZrP is regenerated after being treated with a hydrochloric acid solution [48]. All of this sequence is represented by Equations 2, 3 and 4.

$$\gamma\text{-}Zr(O_3POH)_2.2H_2O + H_2C\underset{O}{\text{—}}CH_2 \longrightarrow Zr(O_3POCH_2CH_2OH)_2.H_2O$$

Equation 2

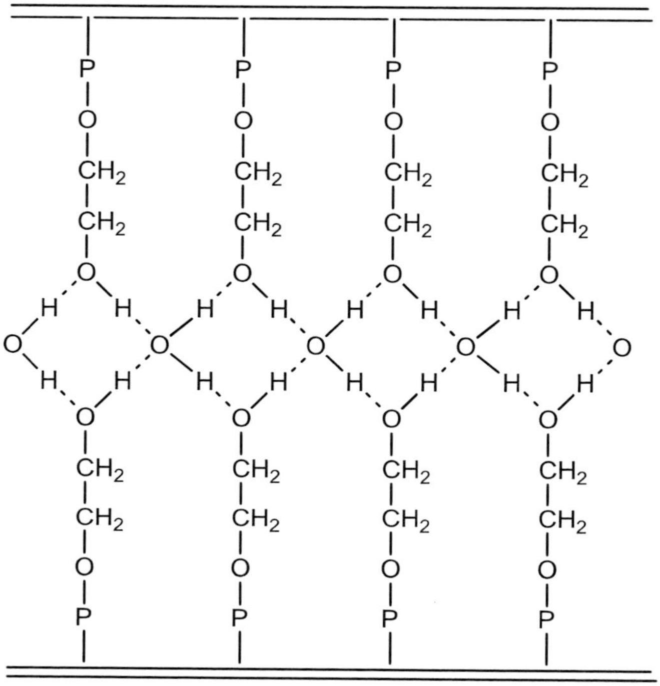

Figure 3. Schematic representation of $Zr(HOC_2H_4OPO_3)_2.H_2O$.

$$Zr(O_3POCH_2CH_2OH)_2.H_2O + NaH_2PO_4 \longrightarrow Zr(O_3PONa)(O_3POH) + 2(OH)_2POO(CH_2)_2OH$$

Equation 3

$$Zr(O_3PONa)(O_3POH) + HCl \longrightarrow \gamma\text{-}Zr(O_3POH)_2 + NaCl$$

Equation 4

The original structure of the zirconium phosphate can also be modified by reaction with *n*-alkylamines. The *n*-alkylamines are protonated by the acidic groups of the ZrP and the intercalate is formed by a bilayer of *n*-alkylammonium ions packed between the negatively charged layers of the ZrP [1, 49, 50]. It is possible to intercalate a maximum of two moles of *n*-alkylamine per mole of phosphate [51]. The process of intercalation of an *n*-alkylamine by ZrP occurs step by step with the formation of different immiscible phases, characterized by a well-defined composition and interlayer distance [52].

Some organic derivatives of the zirconium phosphate, such as the zirconium *bis*(hydroximethanophosphonate), form intercalation compounds with *n*-propylamine, *n*-butylamine and benzylamine. The study of the interlayer distance of these compounds also suggests that two organic molecules are intercalated per mole of ZrP, creating biomolecular layers [53]. Due to the length of the pendant organic groups in the interlayers, some mixed polymers with pending hydroxyl groups are very selective relatively to the absorption of amines.

The bonds between the layers of the α-$VOPO_4.2H_2O$ are weak and this phosphate is easily intercalated by *n*-alkylamines [54, 55]. These species are intercalated as neutral molecules without any change in the charge of the layers involved in the reaction. A second type of intercalation reaction involves the reduction of vanadium(V) to vanadium(IV) with the concomitant intercalation of cations to balance the negative charge of the layers. This process of "redox intercalation" can be conveniently done with the help of iodides. This method was used to intercalate *n*-alkylammonium ions by the α-$VOPO_4.2H_2O$ [56]. The reaction of either the $VOPO_4$ or the $VOPO_4.2H_2O$ with pyridine leads to the formation of $VOPO_4$.py, in which all the vanadium atoms in the metal oxide layer are coordinated with the pyridine [57].

As already mentioned, there is a structural difference between α- and γ-titanium phosphate. For this reason, during the intercalation of *n*-alkylamines by the α-$Ti(HPO_4)_2.H_2O$ [58] it was found that the acid-base reaction between acid groups of the interlayer and the intercalated basic molecules, influence all the acid groups of the phosphate. However, the reaction between *n*-alkylamines and γ-TiP shows that only 66% of the P-OH groups can interact with the amino groups of the molecules to be intercalated due to the structural characteristics of the γ-TiP [59].

The basic structural unit of organo substituted metal phosphonates is represented by $M(O_3ZR)_n$, where M is a metal such as Zr, Sn or Ti and Z is a pentavalent atom such as phosphorus or arsenic. R is an organic group which may be saturated or unsaturated, substituted or unsubstituted and includes aryl, haloalkyl, alkylaryl, aminoalkyl, carboxialkyl, or fenoxyalkyl or cianoalkyl groups. The phenyl derivative, zirconium *bis*(benzenophosphonate), $[Zr(O_3PPh)_2]$, (Figure 4) [53] was the first phosphonate to be prepared but at present several derivatives are known.

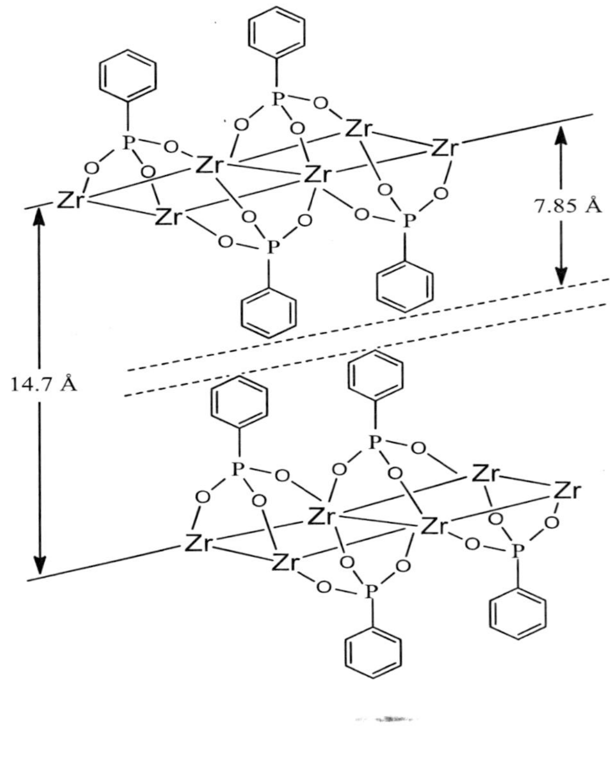

Figure 4. Schematic representation of the zirconium *bis*(benzenophosphonate).

The metal organo substituted phosphates are identical to the phosphonates, but the organic group (R) is connected to Z through an oxygen, having the following structure, $M(O_3ZOR)_n$. From the chemical point of view, there is an important difference between the phosphonates and the phosphates. In the first case, the C—P bond does not suffer hydrolysis in acidic media, but in the second case, the C—O—P bond undergoes hydrolysis with the consequent formation of C—OH and HO—P. As expected, organic phosphates are thermally unstable due to the organic groups present in the interlayers, so its use can only be done at low temperatures. However, some organic phosphates with aromatic groups in the interlayers are extremely stable until 400 or 500° C [60].

The preparation of "pillared" and "no pillared" organic phosphates and phosphonates is done by the same method indicated for the preparation of simple phosphates, with the difference that the phosphoric acid is replaced by an organic *bis*-phosphonic acid ($H_2O_3P—R—PO_3H_2$), organic *bis*-phosphate ($H_2O_3P—O—R—O—PO_3H_2$), organic phosphonic acid (RPO_3H_2) or an organic phosphate ($ROPO_3H_2$). The resulting polymers have a layered structure similar to those of the simple phosphates but the —P—OH group in the interlayer is replaced by —P—R—P—, —P—O—R—O—P—, —P—R and —P—OR groups, respectively (R = organic radical) [53] (Figure 5).

Mixed pillared organophosphonates and organophosphates result from the mixture of different types of organic *bis*-phosphonic acids or organic *bis*-phosphates. Mixed organophosphonates, "pillared" and "no pillared", can be prepared by the use of a solution of a metal with a mixed solution of an organic *bis*-phosphonic acid and organic phosphonic acid [61, 62], or a mixed solution of an organic *bis*-phosphonic acid and organic *bis*-phosphate followed by hydrolysis. Figure 6 represents schematically these two types of mixed phosphonates / phosphates.

As can be assumed from Figure 6, these types of solids possess organic groups with different lengths, R and R', in the interlayers, corresponding to the organic *bis*-phosphonic acids or *bis*-phosphates and organic phosphonic acids or phosphates used. Thus, it is possible to prepare porous crystalline compounds in which the size of the cavities can be systematically changed. The pore size depends on the distance X between two neighboring pillars and the difference in lengths between pillars (size of the R group) and the pendant groups (size of the R' group) represented by Y (Figure 6).

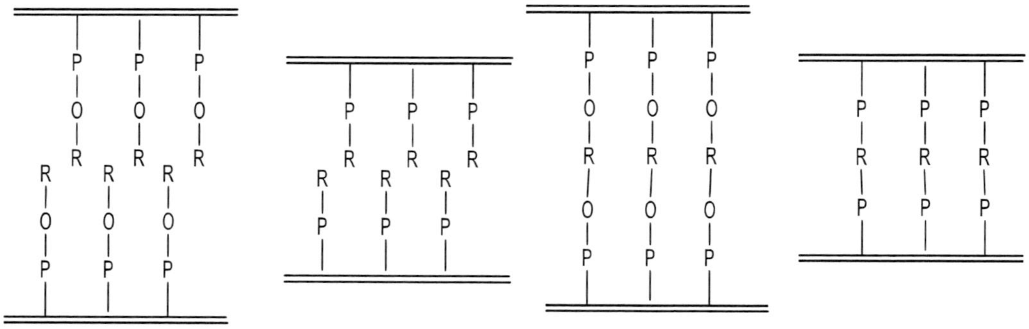

Figure 5. "No pillared" and "pillared" zirconium organophosphates and phosphonates.

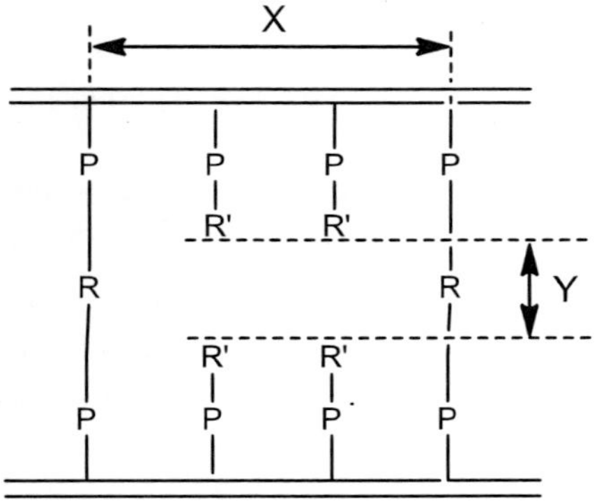

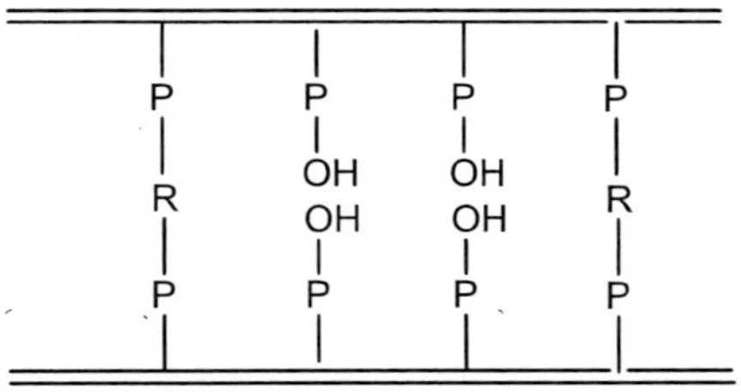

Mixed zirconium organophosphonate

Mixed zirconium organophosphonate/phosphate

X= distance between pillars
Y= distance between pendent groups

Figure 6. Idealized structures of mixed zirconium phosphates / phosphonates with different sizes of cavities.

In recent years, there has been a significant progress in the preparation and characterization of molecular thin films on surfaces. Layered metal phosphonates are particularly suitable compounds for the layer-by-layer growth of very thin films, which are of great interest due to its structural analogy with type-Y Langmuir-Blodgett films [63].

Some crown ethers have been attached as mono or bilayers to the layers of γ-ZrP and these compounds may also be of interest for supramolecular chemistry [63].

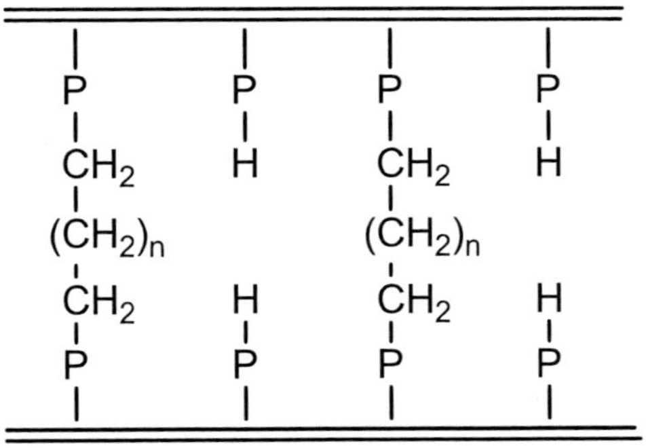

Figure 7. Cavities in a pillared ZrP with 50% of the pillars replaced by phosphite groups.

4. Applications of the Metal Phosphates and Phosphonates

4.1. Use of the Phosphates and Phosphonates in Catalysis

An active catalyst results not only from the design of the polymer and its active sites but, most importantly, if the synthesized polymer has a large percentage of its catalytic groups "available" on the surface instead of being inaccessible in a impenetrable net of hydrocarbon chains.

As referred in previous sections, the interlayer region in both metal phosphates and its derivatives is almost completely filled by counter-ions, solvated molecules or by pillars. Therefore, the available surface area is limited to the exterior of the materials and represents the main disadvantage for their wider application in catalysis. This drawback can be solved if the porosity of the catalysts is increased and consequently a larger number of active groups are accessible to the reagents [64]. In an attempt to achieve materials with higher porosity to be used as catalysts and/or molecular sieves, several methods have been attempted [65-67]. One method consists in the replacement of the organic radicals in some "pillars" by smaller groups such as ≡P—OH or ≡PH with the formation of structures like the one represented in Figure 7.

The length of the organic radical determines the "height" of the pores (Figure 8), and the ratio between the number of "pillars" and the ≡PH groups determines its "width" and "depth" (Figure 7).

It has been demonstrated by various chemical and physico-chemical techniques [68-70] that the surface of the metal phosphates have acidic sites that are responsible for their participation in catalytic processes and also that the acid catalytic activity is related to the acid strength of these sites [70]. Several reviews have been published about the determination of the acidity of solid catalysts, where a detailed discussion of the most used methods is described [71]. There are two types of acidic sites in the α-ZrP with very different acid strengths [69]. Similarly, it was found that the acidity depends upon the degree of crystallinity of the catalyst [72] and the thermal treatment to which it was submitted. The acidity increases with thermal treatment up to 400° C, beyond which it decreases suddenly.

Due to its characteristics, the α-ZrP is a good solid for acid catalysis at medium temperatures and some reactions involving dehydration [72-74], isomerization [75], dehydrogenation [73] and oxidation [76, 77] have been investigated. Some of such applications are gas phase catalysis processes, thus only few examples of the applications of this catalyst in liquid phase organic synthesis have been reported.

Crystalline zirconium phosphate can catalyze the epoxide ring opening by attack of a nucleophile. For example, in nucleophilic solvents or in neat on refluxing, the reaction of cyclohexane oxide in methanol or in acetic acid was complete at room temperature in two days. This catalyst was also used for the synthesis of cyclic ethers from suitable unsaturated alcohols or diols. The intramolecular dehydration of 1,n diols to the corresponding cyclic ethers was efficiently catalyzed by amorphous zirconium phosphate between 175° and 210°C in neat [35].

Zirconium phosphate, in the α and γ phase, was also employed to selectively transform carbohydrates, such as fructose and inulin, into 5-hydroxymethyl-2-furaldehyde (HMF), an important intermediate for fine chemicals and furanic polymers. The activity of the amorphous and crystalline zirconium phosphates as Friedel-Crafts catalysts have been investigated in the alkylation of anisole with alcohols in both liquid and gas phase, and compared with the activities shown by the superacidic Nafion-H. It was found that the amorphous form was better in the liquid phase and at a lower temperature than in the gas phase, but with lower selectivity [35].

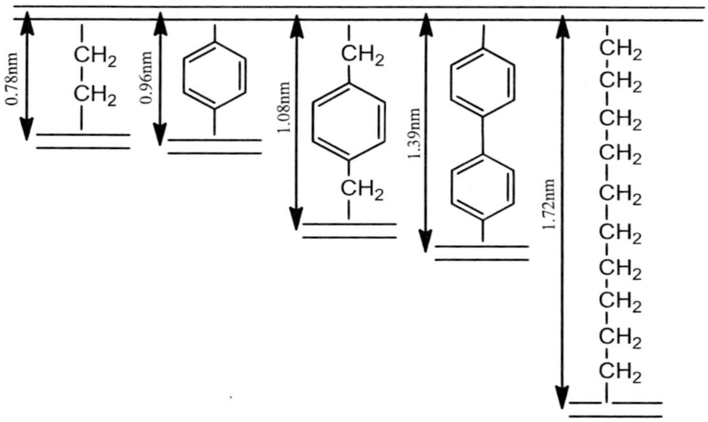

Figure 8. Interlayer distances of some pillared ZrP, showing the control of the height of the pores.

Zirconium phosphonates also exhibit good catalytic activity, particularly if they have some organic groups (carboxylic or sulphonic acid) linked to the inorganic structure [73]. In compounds with pillars and cavities, metal species can be linked to the organic groups in the interlayers, if these groups are able to be complexed by the metal species. This method has been applied successfully to bind Pd(II) to α-ZrP derivatives with pillars formed by pyridyl and allyl groups [73].

Nitrogen-containing derivatives such as oximes, semicarbazones and tosylhydrazones are important compounds in organic synthesis and are used in the protection and characterization of carbonyl compounds. Deprotection of these derivatives has been carried out using various catalysts by acidic or oxidative processes [35], in homogeneous or heterogeneous systems. Zirconium sulfophenylphosphonate $[Zr(O_3PCH_3)_{1.2}(O_3PC_6H_4SO_3H)_{0.8}]$ has been used for this purpose in acetone/water at reflux. Similarly, dithioacetals such as 1,3-dithiolanes and 1,3-dithianes were converted into the corresponding carbonyl compounds at 60°C in the presence of glyoxylic acid as exchange reagent. Zirconium sulfophenylphosphonate was also used in the preparation of cyclic ketals from carbonyl compounds and appropriate diols (dithiols) in dichloromethane at reflux [35].

The protection of hydroxyl groups of alcohols is of great importance in synthetic organic chemistry. Three procedures for the selective protection of alcohols and phenols as tetrahydropyranil, acetyl and tritmethylsilyl derivatives have been reported using zirconium sulfophenylphosphonate as a catalyst. The reported methods occur at room temperature using 3,4-dihydro-*2H*-pyran, acetic anhydride or hexamethyldisilazane as reactants, respectively [35]. In contrast with the data reported for other heterogeneous catalysts such as montmorillonite Clay and zeolite HSZ-360, the acetylation of tertiary alcohols occurs without the formation of dehydration products [35].

Recently, a simple method for the protection of aldehydes as 1,1-diacetates derivatives and their subsequent deprotection mediated by zirconium sulfophenylphosphonate using different reaction conditions has been described. When aldehydes were treated with acetic anhydride at room temperature in solvent-free conditions, 1,1-diacetates were obtained in good to excellent yield. Ketones were not acylated under these conditions meaning that this is a selective method for the preparation of 1,1-diacetates from aldehydes in the presence of ketones [35].

The ring opening of epoxides by amines is an important route for the preparation of β-aminoalcohols, and a number of examples of such ring opening have been reported, including the use of montmorillonite K-10 under microwave irradiation. When zirconium sulfophenylphosphonate was used in solvent-free conditions in the presence of amines, the corresponding β-aminoalcohols were obtained under mild conditions [35].

The preparation of pyrroles by the condensation of 1,4-diones with primary amines (Paal-Knorr condensation) was tested using zirconium sulfophenylphosphonate and the results were compared with the ones obtained with potassium exchanged zirconium phosphate. Zirconium sulfophenylphosphonate and potassium exchanged zirconium phosphate do not present any differences in the reaction time and yield when aliphatic amines are used. Otherwise, zirconium sulfophenylphosphonate is more efficient than potassium-exchanged zirconium phosphate in the reaction of 2,5-hexanedione with less basic aromatic amines, due to the activation of the dione by the acid catalyst, via oxygen protonation [35].

Aluminium, niobium, vanadium, titanium, tin, cerium and potassium phosphates, pyrophosphates and zirconium phosphates oxynitrides have also been referred by its catalytic ability [78]. For example, jasminaldehyde was prepared by the condensation between benzaldehyde and heptanal using amorphous aluminophosphate as a catalyst. Aluminium phosphate also shows to be useful as support for enzyme immobilization and was used in the enzymatic hydrolysis of 1-naphthylphosphate at pH 5.6 to afford 1-naphtol. Vanadyl phosphate was used as acid catalyst in the dehydration of fructose aqueous solutions to afford 5-hydroxymethyl-2-furaldehyde (HMF). The selective oxidation of HMF to furan-2,5-dicarboxaldehyde has been also studied with the vanadyl phosphate alone and modified with trivalent metals [78].

4.2. Phosphates and Phosphonates as Inorganic Ion Exchangers

Several ion exchanged phosphates have been obtained with different acid/base properties and these new forms have been used to catalyze a number of oxidation reactions [73]. Protons can be exchanged at different rates depending on the size and hydration degree of the cations. The exchange of large and highly hydrated cations has been achieved by using precursors with higher interlayer distance [79, 80].

Alkali metal exchanged forms of zirconium phosphate, that is compounds of formula $Zr(YPO_4)_2$; (Y=Li, Na, K, Cs), should act as a good basic catalyst, capable to pull off protons and generating carboanions that may attack an electron acceptor center forming a new carbon-carbon bond. Potassium-exchanged zirconium phosphate was tested in the Michael and Henry reaction with good yields and in shorter reaction times than when alumina and alumina-supported potassium fluoride were used. This catalyst was also found to be useful in the Knoevenagel condensation, desilylation of phenolsilylethers, cyanosilylation of carbonyl compounds, synthesis of 2-nitroalkanols and for the preparation of pyrroles starting from 1,4-hexanedione and aliphatic or aromatic amines (Paal-Knorr condensation) with better results comparatively with other basic solid catalysts (e. g. Al_2O_3, zeolites, montmorillonites). Substituted 4H-chromene derivatives, were synthesized with better results than those from the experiments carried out with Al_2O_3 or molecular sieves [35].

Recent publications refer the use of Pd, Pt and Ni ion-exchanged Zr, Ti and Sn and their application to the selective heterogeneous catalytic hydrogenation of alkenes [81] and the use of potassium and sodium exchanged zirconium phosphate amorphous as heterogeneous catalysts for the oxidation of (+)-3-carene [82] with good results. Ru(III) has been anchored onto ZrP, TiP, SnP, ZrW, TiW and SnW by an ion exchange method to give Ru(III)ZrP, Ru(III)TiP, Ru(III)SnP, Ru(III)ZrW, Ru(III)TiW, and Ru(III)SnW, respectively. The materials have been tested, with good results, for the catalytic oxidation of benzyl alcohol and styrene. In addition, Ru(III) has been reduced to Ru(0) and its catalytic activity was investigated for hydrogenation of 1-octene, nitrobenzene, and cyclohexanone [83].

The ion exchange properties vary with the degree of crystallinity of the phosphates [84]. The exchange capacity of the amorphous zirconium phosphate is identical to that of the crystalline phosphate. However, the crystalline phosphate does not exchange with cesium ion while the amorphous phosphate has a high exchange capacity with this ion [3]. The γ-ZrP allows the exchange with univalent and multivalent cations without the addition of a base [42, 85] due to the greater distance between the layers. Many of the properties of other phosphates, such as titanium [22, 86], hafnium [87], germanium [16], tin [4, 5] and lead [16] are identical to those referred for zirconium phosphate.

One of the first studies on the catalytic capacity of polymers that experienced ion exchange was carried out by the exchange of Cu^{2+} with γ-$Zr(O_3POH)_2.2H_2O$. The interlayer distance of the resulting polymer is 9.51Å compared with the 12.3Å of the starting γ-ZrP, fully hydrated. This ion exchange polymer efficiently catalyzes the oxidation of substrates such as carbon monoxide, methanol and sulphur dioxide and was more active than the species that have suffered a similar exchange with Co^{2+} [88]. However, the species of Co^{2+} are more selective regarding the oxidation of *n*-butanol. The catalytic oxidation of propene by α-ZrP that underwent ion exchange with divalent metals was analyzed and good results have been obtained in comparison with zirconium phosphate that did not suffered ionic exchange [89] and the oxidative dehydrogenation of benzene to cyclohexene catalyzed by $ZrCu(PO_4)_4$ have also been referred with success [90].

The affinities of ion exchange of other phosphates and arsenates, have also shown good properties as ion sieves relative to alkali metals Li^+ and Cs^+ if the interlayer spacing is less than 7.8Å. For example, tin phosphate [4] and crystalline arsenate of zirconium [91, 92], have no properties as ionic sieves relative to the Li^+, Na^+, K^+ and Cs^+ ions, because the interlayer distance of the tin phosphate (7.9Å) is identical to that of zirconium arsenate (7.82Å). But, titanium arsenate (7.77Å) [8] and α-zirconium phosphate (7.6Å) [92] exchange with Li^+, Na^+ and K^+ while titanium phosphate (7.56Å) [86] exchange with Li^+ and Na^+ and thorium arsenate (7.05Å) [93] is specific to Li^+.

The cation exchange capacity and the stability of zirconium phosphate and its derivatives, when subjected to exchange with radioactive ions have also been the target of several studies [94].

The study of the system cerium(IV)-phosphoric acid is very complex because several crystalline precipitates can be obtained [13, 95], including the fibrous precipitate of general formula $Ce(HPO_4)_2.3H_2O$. The properties of ion exchange of fibrous phosphates relative to transition metal ions, alkali ions and alkaline earth ions [96] showed a high selectivity for some cations. The fibrous thorium phosphate [11] and the cerium(IV) phosphate have shown similar properties relative to ion exchange.

Comparatively, with organic ion exchange resins, the phosphates and arsenates mentioned above have the advantage of being more resistant to temperature, radiation and oxidizing solutions. Comparatively, with zeolites, they are much more resistant to acidic medium. For these reasons, amorphous zirconium phosphate has been used in dialysis machines to remove ammonium ions from blood [97] as well as support in gas-solid chromatography [97-99]. The great selectivity of the fibrous cerium phosphate in relation to Pb^{2+} and other divalent cations is a great advantage for the practical application of this phosphate [86]. The thorium arsenate is specific to the exchange with Li^+, so this material has been used for the separation of Li^+ from other cations [93]. The titanium phosphate has been used for the separation of radioactive Cs from strongly acidic nuclear fuel recycling solutions [31] and in the separation between sodium and potassium through thin layer chromatography [100, 101]. The recovery of Ca^{2+} from the waste water from the industry of detergents [102] has been performed by the γ-titanium phosphate because this phosphate is able to hold calcium ions from acidic solutions at room temperature [33].

The organophosphonates with terminal sulphonic groups between the layers are potential acidic ion exchangers [53, 103]. When this type of polymer is placed in contact with an aqueous solution of copper(II) sulphate [104], its color changed from white to blue. The behavior of ion exchange of organophosphonates with terminal carboxylic groups in the interlayers was also studied with some metals including Cu, Ni and Co [105, 106]. The phosphates with cyano groups in the interlayers are useful as complexing agents for the immobilization of catalytically active metals, such as Pd^{2+}, Ir^{1+}, Rh^{1+}, Ru^{2+}, Os^{2+}, Cu^{2+} and Ag^{2+} [105, 107]. Polymers with nitrile and mercapto groups in the interlayers can complex with copper and silver ions at room temperature with the formation of compounds that show good catalytic activity [108, 109].

4.3. Intercalates of Phosphates and Phosphonates

The majority of investigations on the intercalation chemistry of layered acid salts have been performed with α–zirconium phosphate. This phase exhibits a rich and varied intercalation chemistry [42, 43, 73]. Its reactivity arises from the high acidity in the protonic form and the weak forces between the layers. Its ion-exchange capacity, 6 meq/g, represents a density of one proton per 24Å [110] on either side of each layer.

Recent work has focused on the intercalation of organic amines, with the intention of using the size of the organic group to control the size of the interlayer region. The motivation is to produce host materials for catalytic reactions by propping the layers open with large organic cations [111].

n-Alkylamines are readily intercalated into α–ZrP from aqueous solutions, gas phase or organic solvents. A maximum of 2 moles of amine per formula weight is intercalated, forming a bilayer of guest molecules in the Van der Waal's gap [51]. The experimentally observed increase of 2.21Å in the interlamellar separation for each additional carbon added to the alkyl chain of the *n*-alkylamine suggest that the molecules are not perpendicular to the layers but are inclined by about 56°, as illustrated schematically in Figure 9.

α–ZrP does not intercalate alcohols directly and the crystals do not swell, either in water or alcohol. However, when the half-sodium exchanged phase, $Zr(NaPO_4)(HPO_4).5H_2O$, which has an interlayer spacing of 11.8 Å, is treated with methanol in an acid solution, the methanol is intercalated and the Na^+ cations are exchanged with the protons resulting in the formation of $Zr(HPO_4)_2.MeOH$. The methanol in this phase is easily displaced by other alcohols [43] or polar solvents. Other organic guests which have been intercalated by both direct and indirect methods are urea, amides, biopharmaceuticals such as aminoacids, and heterocyclic bases such as imidazole [43, 112]. Organometallic guests such as cobaltocene [45], and $FcCH_2CH_2NH_2$ (Fc=Fe[Cp][η-C_5H_4]) have also been intercalated [46].

Although only two crystalline forms of layered titanium phosphates are known, α-$Ti(HPO_4)_2 \cdot H_2O$ and γ-$Ti(H_2PO_4)(PO_4) \cdot 2H_2O$, many derivatives have been reported, including partially and completely substituted ion-exchanged forms and intercalation compounds. Monoalkylamines are usually employed as templates for the synthesis of titanium phosphate metastable phases, but it was not until 2005 that the preparation of a γ-titanium phosphate alkylamine-intercalated compound was reported, $(C_6H_{13}NH_3)[Ti(HPO_4)(PO_4)].H_2O$ [113]. Other three organically templated titanium phosphates with the formulas $Ti_2(HPO_4)_2(PO_4)_2.C_2N_2H_{10}$, $Ti_3(H_2PO_4)(HPO_4)_{3.5}(PO_4)_2.C_2N_2H_{10}$, and $Ti_7(HPO_4)_6(PO_4)_6.C_3N_2H_{12}$ have been prepared hydrothermally from titanium powder, phosphoric acid and structure-directing organic amines. The first compound is an ethylenediamine intercalated layered titanium phosphate in which the layers are similar to that of γ-type titanium phosphate. The other two compounds are of open-framework type, and the diprotonated organic amine molecules are entrapped in 12- and 8-membered-ring channels [114].

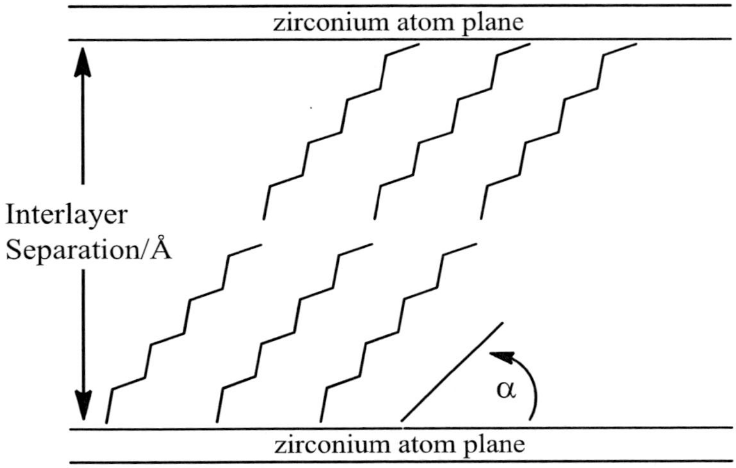

Figure 9. Schematic diagram of the structure of linear chain amines inside the lamellae of α–ZrP [43]. α represents the tilt angle of the chains.

Vanadium phosphate shows a wide range of intercalation reactions which parallel, in many aspects, the chemistry of the covalent oxide systems [115]. $VOPO_4$ readily forms reversible hydrated phases; the dihydrate is an intercalate where one water molecule is coordinated to the vanadium and the second is located in the interlayer space (Figure 2). Similar intercalation reactions have been described for larger donor ligands such as pyridine and bipyridines [57]. $VOPO_4.2H_2O$ readily undergoes redox intercalation reactions with IA and IIA metal cations group in the presence of a reducing agent [116, 117]. The oxidizing power of V^{5+} is such that only mild reducing agents such as metal iodides are required [118]. The iodide is oxidized to iodine, V(V) is reduced to V(IV) and the metal cations are inserted between the layers. An important difference between $VOPO_4.2H_2O$ and other oxide lattices is that redox intercalation of the type discussed above gives materials with low electrical conductivity [115].

Recently, several metalloporphyrins have been intercalated into α-ZrP [119-121]. The metalloporphyrins included the metal complexes of TMPyP [5, 10, 15, 20-tetrakis (1-methylpyridinium-4yl-)-porphyrin], TMAPP [5, 10, 15, 20-tetrakis (N, N, N, trimethylanilinium-4yl-)-porphyrin], TAPP [5, 10, 15, 20-tetrakis (anilinium-4yl-)-porphyrin] and TPyP [5, 10, 15, 20-tetrakis (pyridinium-4yl-)-porphyrin]. The intercalation of the porphyrins is difficult and the use of a spacer to expand the zirconium phosphate interlayer region is but one practical approach. Some spacers like the *p*-methoxyaniline, *n*-butylamine or *n*-propylamine have been used to exchange with porphyrins. Experimental results confirmed that the catalytic activity of the intercalated compounds is superior to that of the free porphyrins in solution.

The intercalation of a series of ionic liquids into α- and γ-zirconium phosphate was referred for the first time in 2007 [122, 123]. The ionic liquids chosen were the 1-alkyl-3-methylimidazolium chloride, [$C_n$mim]Cl (n = 2,4,6,8). α- and γ -Zirconium phosphate were preintercalated with *n*-butyl and *n*-octyl amines. Based on XRD data, the interlayer distances of the resulting [$C_n$mim]Cl (n = 2,4,6,8) intercalation compounds were similar, suggesting that the ionic liquids were arranged in an approximately planar manner, as confirmed by molecular modeling.

5. Conclusion

A good part of this review has been dedicated to the description of the structures, preparation and applications of metal phosphates and phosphonates, mainly Zr(IV). The aim was to enlighten the versatility of this class of compounds and the many possibilities offered to researchers involved in heterogeneous catalysis to prepare materials tailored for specific reactions.

In the first two sections, a description of the structures and preparation of phosphates and phosphonates is done considering the most recent results published in the literature.

In the section of the applications of the phosphates and phosphonates, a survey of their most recent applications as catalysts, inorganic ion exchangers and hosts has been considered.

This class of materials are a friendly alternative to clays and modified clays, zeolites and zeolite-like systems such as Ti-silicalite. Although, in the last years, the application of these compounds as catalysts in liquid phase synthesis have shown an increase it is predictable that its application will grow and will produce new synthetic procedures to be used successfully in academic and industrial laboratories.

References

[1] Alberti, G. (1978). Syntheses, crystalline-structure, and ion-exchange properties of insoluble acid salts of tetravalent metals and their salt forms. *Accounts of Chemical Research,* 11, 4, 163-170.

[2] Clearfield, A., Blessing, R.H. & Stynes, J.A. (1968). New crystalline phases of zirconium phosphate possessing ion-exchange properties. *Journal of Inorganic and Nuclear Chemistry,* 30, 8, 2249-2258.

[3] Clearfield, A. & Stynes, J.A. (1964). The preparation of crystalline zirconium phosphate and some observations on its ion exchange behavior. *Journal of Inorganic and Nuclear Chemistry,* 26, 1, 117-129.

[4] Fuller, M.J. (1971). Ion exchange properties of tin(IV) materials. 4. Crystalline tin(IV) phosphate. *Journal of Inorganic and Nuclear Chemistry,* 33, 2, 559-566.

[5] Costantino U. & Gasperoni A. (1970) Crystalline insoluble acid salts of tetravalent metals. 11. Synthesis and ion-exchange properties of tin(IV) phosphate and tin(IV) arsenate. *Journal of Chromatography, 51*, 2, 289-296.

[6] Allulli S., Ferragina C., La Ginestra A., Massucci M.A. & Tomassini N. (1977) Preparation and ion-exchange properties of a new phase of crystalline titanium phosphate, $Ti(HPO_4)_2.2H_2O$. *Journal of Inorganic and Nuclear Chemistry,* 39, 6, 1043-1048.

[7] Alberti G., Costantino U. & Giovagnotti M.L.L. (1979) Crystalline insoluble acid salts of tetravalent metals. 28. Synthesis of crystalline $Ti(HPO_4)_2.2H_2O$ by the HF procedure and some comments on its formation and structure. *Journal of Inorganic and Nuclear Chemistry*, 41, 5, 643-647.

[8] Alberti G. & Torracca E. (1968) Crystalline insoluble acid salts of polivalent metals and polybasic acids. 6. Preparation and ion-exchange properties of crystalline titanium arsenate. *Journal of Inorganic and Nuclear Chemistry,* 30, 11, 3075-3080.

[9] Haushalter R.C., Mundi L.A. & Strohmaier K.G. (1991) A new microporous (ca. 25 vol. permill. void space) molybdenum phosphate based on the octahedral tetrahedral $Mo_2O_2(PO_4)_2(H_2PO_4)$ - framework. *Inorganic Chemistry,* 30, 2, 153-154.

[10] Mundi L.A., Strohmaier K.G., Goshorn D.P. & Haushalter R.C. (1990) Vacant coordination sites within the tunnels of a microporpous, neutral framework molybdenum phosphate with 35 vol-percent void space-structure of $Mo_8O_{12}(PO_4)_4(HPO_4)_2.13H_2O$. *Journal of the American Chemical Society,* 112, 22, 8182-8183.

[11] Alberti G. & Costantino U. (1970) Crystalline insoluble acid salts of tetravalent metals. 10. Fibrous thorium phosphate, a new inorganic ion-exchange material suitable for making (support-free) inorganic sheets. *Journal of Chromatography,* 50, 3, 482-486.

[12] Alberti G., Costantino U. & Zsinka L. (1972) Crystalline insoluble acid salts of tetravalent metals. 12. Synthesis and ion-exchange properties of microcrystalline cerium(IV) phosphate. *Journal of Inorganic and Nuclear Chemistry,* 34, 11, 3549-3560.

[13] Alberti G., Costantino U., Di Gregorio F., Galli P. & Torracca E. (1968) Crystalline insoluble acid salts of polybasic metals. 3. Preparation and ion-exchange properties of cerium(IV) phosphate of various crystallinities. *Journal of Inorganic and Nuclear Chemistry,* 30, 1, 295-304.

[14] Morris M., Adams J.M. & Dyer A. (1991) Mechanism of *n*-alkylammonium ion intercalation into the layered host α-$VOPO_4.2H_2O$. *Journal of Material Chemistry,* 1, 1, 43-49.

[15] Torracca E., Alberti G., Allulli S., Costantino U. & Massucci M.A. (1969) *Ion Exch. Ind. Pap Conf.,* 318.

[16] Winkler A. & Thilo E. (1966) Uber eine reihe saurer verbindungen $HXVP_2O_8$ und $H_2X_4P_2O_8$ mit schichtstruktur XV=As und SB X_4=Si, Ge, Sn, Pb, Ti und Zr. Zeitschrift fur anorganische und allgemeine chemie, 346, 1-2, 92.

[17] Qureshi M. & Nabi S.A. (1970) Preparation and properties of titanium arsenates. *Journal of Inorganic and Nuclear Chemistry,* 32, 6, 2059-2068.

[18] Frianeza T.N. & Clearfield A. (1984) Catalytic studies on the dehydration of cyclohexanol by crystalline titanium phosphate and mixed titanium-zirconium phosphates. 1. Effect of thermal phase-changes on catalytic activity. *Journal of Catalysis,* 85, 2, 398-404.

[19] Galli P., La Ginestra A., Berardelli M.L., Massucci M.A. & Patrono P. (1985) Ge-Zr phosphate system - the advantages of thermal methods in the investigation of phase mixtures or solid-solutions. *Thermochimica Acta,* 92 615-618.

[20] Bagnasco G., Ciambelli P., Turco M., La Ginestra A. & Patrono P. (1991) Layered zirconium-tin phosphates. 1. Chemical and physical characterization. *Applied Catalysis,* 68, 1-2, 55-68.

[21] Berardelli M.L., Galli P., La Ginestra A., Massucci M.A. & Varshney K.G. (1985) Crystalline zirconium(IV) hydrogenarsenate hydrogenphosphate monohydrate - synthesis, ion-exchange properties, and thermal-behavior. *Journal of the Chemical Society.,* Dalton Transactions, 5, 9, 1737-1742.

[22] Smith G. D. & Clearfield A. (1969) Crystallography and structure of α-zirconium *bis*(monohydrogen orthophosphate) monohydrate. *Inorganic Chemistry,* 8, 3, 431-436.

[23] Clearfield A. & Troup J.M. (1977) Mechanism of ion-exchange in zirconium-phosphates. 20. Refinement of crystal-structure of α-zirconium phosphate. *Inorganic Chemistry* 16, 12, 3311-3314.

[24] Clearfield A. & Djuric Z. (1979) Mechanism of ion-exchange in zirconium-phosphates. 25. Exchange of surface protons with ammonium ion. *Journal of Inorganic and Nuclear Chemistry*, 41, 6, 903-904.

[25] Clearfield A. & Berman J. (1981) On the mechanism of ion-exchange in zirconium-phosphates. 34. Determination of the surface-areas of α-$Zr(HPO_4)_2.H_2O$ by surface exchange. *Journal of Inorganic and Nuclear Chemistry*, 43, 9, 2141-2142.

[26] Yamanaka S. & Tanaka M. (1979) Formation region and structural model of γ-zirconium phosphate. *Journal of Inorganic and Nuclear Chemistry,* 41, 1, 45-48.

[27] Clearfield A. & Pack S.P. (1975) Factors determining ion-exchange selectivity. 1. High-temperature phases formed by α-zirconium phosphate and its sodium and potassium exchanged forms. *Journal of Inorganic and Nuclear Chemistry,* 37, 5, 1283-1290.

[28] Albertsson J., Ahrland S., Johansson L., Nihlgard B. & Nilsson L. (1964) Inorganic ion exchangers. 2. Sorption rate + dehydration studies on zirconium phosphate + tungstate gels. *Acta Chemica Scandinavica,* 18, 6, 1357.

[29] Torracca E., Alberti G. & Conte A. (1966) Stoicheiometry of ion exchange materials containing zirconium and phosphate. *Journal of Inorganic and Nuclear Chemistry,* 28, 2, 607-613.

[30] Clearfield A., Landis A.L., Medina A.S. & Troup J.M. (1973) More on crystalline zirconium phosphates. Journal of Inorganic and Nuclear Chemistry, 35, 4, 1099-1108.

[31] Perárek V. & Vesely V. (1972) Synthetic inorganic ion-exchangers. 1. *Hydrous oxides and acidic salts of multivalent metals. Talanta,* 19, 3, 219-262.

[32] Catalytica Associates Inc. (1983) *New Catalytic Materials-Layered Structures as Novel Catalyst,* Vol.7, Multiclient Study Nº 4183 LS.

[33] Llavona R., Alvarez C., Garcia J.R., Suarez M. & Rodriguez J. (1987) Lamellar inorganic ion exchangers. H^+/Ca^{2+} ion-exchange in γ-titanium phosphate. *Inorganic Chemistry,* 26, 7, 1045-1049.

[34] Calvo C. & Jordan B. (1973) Crystal-structure of α-VPO_5. *Canadian Journal of Chemistry,* 51, 16, 2621-2625.

[35] Curini M., Rosati O. & Costantino U. (2004) Heterogeneous catalysis in liquid phase organic synthesis, promoted by layered zirconium phosphates and phosphonates. *Current Organic Chemistry,* 8, 7, 591-606.

[36] Clearfield, A. & Wang, Z. (2002) Organically pillared microporous zirconium phosphates. *Journal of the Chemical Society, Dalton Transactions,* 15, 2937-2947.

[37] Vivani R., Alberti G., Costantino, F. & Nocchetti M. (2008) New advances in zirconium phosphate and phosphonate chemistry: Structural archetypes. *Microporous and Mesoporous Materials,* 107, 1-2, 58-70.

[38] Kirumakki S., Huang J., Subbiah A., Yao J., Rowland A., Smith B., Mukherjee A., Samarajeewa S. & Clearfield A. (2009) Tin(IV) phosphonates: porous nanoparticles and pillared materials. *Journal of Materials Chemistry,* 19, 17, 2593-2603.

[39] Taddei M., Costantino F. & Vivani R. (2010) Synthesis and crystal structure from X-ray powder diffraction data of two zirconium diphosphonates containing piperazine groups. *Inorganic Chemistry*, 49, 20, 9664-9670.

[40] Kullberg L., Clearfield A. & Oskarsson A. (1974). Mechanism of ion-exchange in crystalline zirconium-phosphates. 11. Variation in unit-cell dimensions and sodium ion-hydrogen ion-exchange behavior in highly crystalline α-zirconium phosphates. *Journal Physical Chemistry,* 78, 12, 1150-1153.

[41] Costantino U., Alberti G. & Giulietti R. (1980). Preparation of large crystals of α-$Zr(HPO_4)_2.H_2O$. *Journal of Inorganic and Nuclear Chemistry*, 42, 7, 1062-1063.

[42] Clearfield A. (1990) *Comments in Inorganic Chemistry,* 10, 89-128.

[43] Costantino U. (1979) Intercalation of alkanols and glycols into zirconium(IV) hydrogen-phosphate monohydrate. *Journal of the Chemical Society-Dalton Transactions,* 2, 402-405.

[44] Beneke K., Behrendt D. & Lagaly G. (1976) Intercalation compounds of zirconium-phosphate. *Angewandte chemie-international edition in English,* 15, 9, 544-545.

[45] Johnson J.W. (1980) Cobaltocene intercalates of zirconium hydrogen phosphate. *Journal of the Chemical Society-Chemical Communications,* 6, 263-265.

[46] Kurmoo M., O'Hare D., Formstone C., FitzGerald E. & Cox P.A. (1991) Single-crystal conductivity study of the tin dichalcogenides $SnS_{2-x}Se_x$ intercalated with cobaltocene. *Journal of Materials Chemistry,* 1, 1, 51-57.

[47] Yamanaka S. (1976) Synthesis and characterization of organic derivatives of zirconium-phosphate. *Inorganic Chemistry,* 15, 11, 2811-2817.

[48] Yamanaka S., Maeda H. & Tanaka M. (1979) Exchange-reaction of phosphoric ester ions with phosphate ions in a heterogeneous system containing the organic derivative

of γ-zirconium phosphate. *Journal of Inorganic and Nuclear Chemistry*, 41, 8, 1187-1191.

[49] Costantino U. (1981) Intercalation of alkanols and alkylamines in insoluble acid salts of tetravalent metals having a layered structure of γ-type. *Journal of Inorganic and Nuclear Chemistry, 43*, 8, 1895-1902.

[50] Michel E. & Weiss A. (1965) Kristallines zirkonphosphat ein kationenaustauscher mit schichtstruktur und innerkristallinem quellungsvermogen. *Zeitschrift fur naturforschung, Teil* B; 20, 12, 1307.

[51] Tindwa R.M. & Clearfield A. (1979) Mechanism of ion-exchange in zirconium-phosphates. 21. Intercalation of amines by α-zirconium phosphate. *Journal of Inorganic and Nuclear Chemistry,* 41, 6, 871-878.

[52] Alberti G., Casciola M. & Costantino U. (1985) Inorganic ion-exchange pellicles obtained by delamination of α-zirconium phosphate crystals. *Journal of Colloid and Interface Science,* 107, 1, 256-263.

[53] Costantino U., Alberti G., Allulli S. & Tomassini N. (1978) Crystalline $Zr(R\text{-}PO_3)_2$ and $Zr(R\text{-}OPO_3)_2$ compounds (R=organic radical) - new class of materials having layered structure of zirconium-phosphate type. *Journal of Inorganic and Nuclear Chemistry,* 40, 6, 1113-1117.

[54] Ladwig G. (1965) Uber die konstitution des $VPO_5(nH_2O)$. *Zeitschrift fur anorganische und allgemeine chemie,* 338, 5-6, 266.

[55] Beneke K. & Lagaly G. (1983) Intercalation into $NbOPO_4.3H_2O$ and comparison with $VOPO_4.2H_2O$. *Inorganic Chemistry,* 22, 10, 1503-1507.

[56] Martinez-Lara M., Jimenez-Lopez A., Moreno-Real L., Brusque S. & Ruiz-Hitzky E. (1985) Redox intercalation of alkylammonium ions into $VOAO_4.nH_2O$ (A=P, As). *Materials Research Bulletin*, 20, 5, 549-555.

[57] Jacobson A.J., Johnson J.W., Brody J.F. & Rich S.M. (1982) Coordination intercalation reactions of the layered compounds $VOPO_4$ and $VOAsO_4$ with pyridine. *Inorganic Chemistry,* 21, 10, 3820-3825.

[58] Menendez F., Espina A., Trabajo C. & Rodriguez J. (1990) Intercalation of normal-alkylamines by lamellar materials of the α-zirconium phosphate type. *Materials Research Bulletin,* 25, 12, 1531-1539.

[59] Menéndez A., Bárcena M., Jaimez E., García J.R. & Rodríguez J. (1993) Intercalation of *n*-alkylamines by γ-titanium phosphate - synthesis of new materials by thermal-treatment of the intercalation compounds. *Chemistry of materials*, 5, 8, 1078-1084.

[60] Avila, C, Y, O. & Clearfield A. (1989) Zirconium-phosphate ester interchange reactions. Journal of the Chemical Society, *Dalton Transactions,* 8, 1617-1623.

[61] Clearfield, A. & Thakur, D. S. (1986) Zirconium and titanium phosphates as catalysts - a review, *Applied Catalysis*, 26, 1-2, 1-26.

[62] Dines, M. B. & Griffith, P. C. (1983) *Synthesis and characterization of layered tetravalent metal terphenyl mono-phosphonate and bis-phosphonate, Polyhedron,* 2, 7, 607-611.

[63] Alberti G., Casciola M., Costantino U. & Vivani R. (1996) Layered and pillared metal(IV) phosphates and phosphonates. *Advanced Materials,* 8, 4, 291-303.

[64] Gates B.C. (1992) *Catalytic Chemistry*, New York, John Wiley & Sons, Inc., 220.

[65] Phadtare S., Megati S. & Zemlicka J. (1992). Unsaturated phosphonates as acyclic nucleotide analogs - anomalous Michaelis-Arbuzov and Michaelis-Becker reactions with multiple bond systems. *Journal of Organic Chemistry*, 57, 8, 2320-2327.

[66] Glaser H. & Scholder R. (1964) Uber lithiumuranate(v) und natriumuranate(v) und uber strukturelle beziehungen zwischen den verbindungstypen Li_7AO_6 und Li_8AO_6. *Zeitschrift fur anorganische und allgemeine chemie,* 327, 1-2, 15-27.

[67] Cooksey R.E., Dines M.B., Griffith P.C. & Lane R.H. (1983) Mixed-component layered tetravalent metal phosphonates phosphates as precursors for microporous materials. *Inorganic Chemistry,* 22, 6, 1003-1004.

[68] Chao E. & Moffat J.B. (1977) Oxidation of iodide to iodine on boron phosphate. *Journal of Catalysis,* 46, 2, 151-159.

[69] Ishiguro A., Hattori T. & Murakami Y. (1978) Acidity of crystalline zirconium-phosphate. *Journal of Inorganic and Nuclear Chemistry*, 40, 6, 1107-1111.

[70] Moffat J.B. & Jewur S.S. (1979) Role of surface-acidity of boron phosphate in the activity and selectivity of the dehydration of alcohols. *Journal of Catalysis,* 57, 1, 167-176.

[71] Winquist B.H.C. & Benesi H.A. (1979) Surface acidity of solid catalysts. *Advances in Catalysis,* 27, 97- 182.

[72] Clearfield A. (1980) The acidity of zirconium-phosphates in relation to their activity in the dehydration of cyclohexanol. *Journal of Catalysis,* 65, 1, 185-194.

[73] Costantino U. & Alberti G. (1984) Recent progress in the intercalation chemistry of layered α-zirconium phosphate and its derivatives, and future perspectives for their use in catalysis. *Journal of Molecular Catalysis,* 27, 1-2, 235-250.

[74] Clearfield A. & Thakur D. (1981) Cyclohexanol dehydration over zirconium-phosphates of varying crystallinity. *Journal of Catalysis,* 69, 1, 230-233.

[75] Kurusu Y., Segawa K., Nakajima Y. & Kinoshita M. (1985) Characterization of crystalline zirconium-phosphates and their isomerization activities. *Journal of Catalysis,* 94, 2, 491-500.

[76] Rocha, G. M. S. R. O., Johnstone, R. A. W. & Neves, M. G. P. M. S. (2002) Catalytic effects of metal(IV) phosphates on the oxidation of phenol and 2-naphthol, *Journal of Molecular Catalysis A: Chemical,* 187, 1 , 95-104.

[77] Rocha, G. O., Rocha, J. & Lin, Z. (2003) Study of catalyst selectivity in the oxidation of phenol, *Catalysis Letters,* 89, 1-2, 69-74.

[78] Sebti, S., Zahouily M., Lazrek H., Mayoral, J. A. & Macquarrie D. J. (2008) Phosphates: New generation of liquid-phase heterogeneous catalysts in organic chemistry, Cu*rrent Organic Chemistry,* 12, 3, 203-232.

[79] Alberti G., Bertrami R., Casciola M., Costantino U. & Gupta J. P. (1976) Crystalline insoluble acid salts of tetravalent metals. 21. Ion-exchange mechanism of alkaline-earth metal-ions on crystalline $ZrHNa(PO_4)_2.5H_2O$. *Journal of Inorganic and Nuclear Chemistry,* 38, 4, 843-848.

[80] Alberti G., Costantino U. & Gill J. S. (1976) Crystalline insoluble acid salts of tetravalent metals. 23. Preparation and main ion-exchange properties of highly hydrated zirconium *bis* monohydrogen orthophosphates. *Journal of Inorganic and Nuclear Chemistry*, 38, 9, 1733-1738.

[81] Álvaro V. F. D. & Johnstone R. A. W. (2008) High surface area Pd, Pt and Ni ion-exchanged Zr, Ti and Sn(IV) phosphates and their application to selective heterogeneous catalytic hydrogenation of alkenes. *Journal of Molecular Catalysis A: Chemical,* 280, 1-2, 131-141.

[82] Rocha G. M. S. R. O., Domingues R. M. A., Simões M. M. Q. & Silva A. M. S. (2009) Catalytic activity of tetravalent metal phosphates and phosphonates on the oxidation of (+)-3-carene. *Applied Catalysis A: General,* 353, 2, 236-242.

[83] Joshi R. & Chudasama U. (2010) Hydrogenation and oxidation reactions involving ruthenium supported catalysts. *Industrial Engineering Chemical Research,* 49, 6, 2543-2547.

[84] Kumar V. & Qureshi M. (1970) Synthesis and ion-exchange characteristics of titanium antimonates. *Journal of Chemical Society* (A), 9, 1488-1491.

[85] Kalnins J.M. & Clearfield A. (1978) Mechanism of ion-exchange in zirconium-phosphates .23. Exchange of 1st row divalent transition-elements on γ-zirconium phosphate. *Journal of Inorganic and Nuclear Chemistry*, 40. 11, 1933-1936.

[86] Alberti G., Cardini-Galli P., Costantino U. & Torracca E. (1967) Crystalline insoluble salts of polybasic metals .1. Ion-exchange properties of crystalline titanium phosphate. *Journal of Inorganic and Nuclear Chemistry,* 29, 2, 571-578.

[87] Clearfield A. & Thomas J.R. (1969) Solubility of α-zirconium phosphate and hafnium phosphate in strong phosphoric acid. *Inorganic Nuclear Chemistry Letters,* 5, 9, 775-779.

[88] Kalman T.J., Dudukovic M. & Clearfield A. (1974) Copper-substituted zirconium-phosphate - new oxidation catalyst. *Advances in Chemistry Series,* 133, 645-668.

[89] Iwamoto M., Nomura Y. & Kagawa S. (1981) Catalytic-oxidation of propene over zirconium-phosphates. *Journal of Catalysis*, 69, 1, 234-237.

[90] Clearfield A. (1984) Group-IV phosphates as catalysts and catalyst supports. *Journal of Molecular Catalysis,* 27, 1-2, 251-262.

[91] Torracca E., Costantino U. & Massucci M.A. (1967) Crystalline insoluble salts of polybasic metals. V. Ion-exchange properties of crystalline and amorphous zirconium arsenate. *Journal of Chromatography,* 30, 2, 584-592.

[92] Smith G.D., Clearfield A. & Hammond B. (1968) Zirconium arsenates and their ion exchange behavior. *Journal of Inorganic and Nuclear Chemistry,* 30, 1, 277-285.

[93] Massucci M.A. & Alberti G. (1970) Crystalline insoluble acid salts of tetravalent metals. 9. Thorium arsenate, a new inorganic ion exchanger specific for lithium. *Journal of Inorganic and Nuclear Chemistry,* 32, 5, 1719-1727.

[94] Komarneni S. & Roy R. (1982) Use of γ-zirconium phosphate for Cs removal from radioactive-waste. *Nature,* 299, 5885, 707-708.

[95] Herman R.G. & Clearfield A. (1975) Crystalline cerium(IV) phosphates—I Preparation and characterization of crystalline compounds. *Journal of Inorganic and Nuclear Chemistry,* 37, 7-8, 1697-1704.

[96] Alberti G., Casciola M., Costantino U. & Luciani M.L. (1976) Crystalline insoluble acid salts of tetravalent metals. 24. Ion-exchange behavior of fibrous cerium(IV) phosphate. *Journal of Chromatography,* 128, 2, 289-299.

[97] Gordon M., Popvtzer M., Greenbaum M., McArthur M., DePalma J.R. & Maxwell M.H. (1968) *Proceedings European Dialysis Transplant Association,* 5, 86.

[98] Dyer A., Leigh D. & Sharples W.E. (1976) Gas adsorption studies on ion-exchanged forms of crystalline zirconium-phosphate. *Journal of Chromatography,* 118, 3, 319-329.

[99] Allulli S., Tomassini N., Bertoni G. & Bruner F. (1976) Synthetic inorganic-ion exchangers as adsorbents for gas-chromatography. *Analytical Chemistry,* 48, 8, 1259-1261.

[100] Alberti G., Bertrami R. & Costantino U. (1976) Crystalline insoluble acid salts of tetravalent metals. 22. Effect of small amounts of Na^+ on ion-exchange of alkaline-earth metal-ions on crystalline $Zr(HPO_4)_2H_2O$. *Journal of Inorganic and Nuclear Chemistry,* 38, 9, 1729-1732.

[101] Giammari G., Alberti G. & Grassini-Strazza G. (1967) Chromatographic behavior of inorganic ions on crystalline titanium phosphate or zirconium phosphate thin layers. *Journal of Chromatography,* 28, 1, 118-123.

[102] Clearfield A. (1982) *Inorganic Ion Exchange Materials,* Boca Raton, CRC Press, Inc., 304.

[103] Digiacomo P.M. & Dines M.B. (1982) Lamellar zirconium phosphonates containing pendant sulfonic-acid groups. P*olyhedron*, 1, 1, 61-68.

[104] Digiacomo P.M. & Dines M.B. (1980) *U.S. Patent,* 4, 235, 991, Nov.25

[105] Digiacomo P.M. & Dines M.B. (1984) *U.S. Patent,* 4, 436, 899, Mar.13

[106] Digiacomo P.M. & Dines M.B. (1980) *U.S. Patent,* 4, 235, 990, Nov.25

[107] Digiacomo P.M. & Dines M.B. (1981) *U.S. Patent,* 4, 276, 409, Jun.30

[108] Digiacomo P.M. & Dines M.B. (1981) *U.S. Patent,* 4, 299, 943, Nov.10

[109] Digiacomo P.M. & Dines M.B. (1981) *U.S. Patent,* 4, 276, 410, Jun.30

[110] Leigh D. & Dyer A. (1972) Structure of zirconium phosphate. *Journal of Inorganic and Nuclear Chemistry,* 34, 1, 369-372.

[111] Roberts B. D. & Clearfield A. (1988) Pillaring of layered zirconium and titanium phosphates. *Inorganic Chemistry,* 27, 18, 3237-3240.

[112] Beneke K., Behrendt D. & Lagaly G. (1976) Intercalation compounds of zirconium-phosphate. *Angewandte Chemie-International Edition in English*, 15, 9, 544-545.

[113] Mafra L., Rocha J. & Garcia J. R. (2005) Structural characterization of layered γ-titanium phosphate $(C_6H_{13}NH_3)[Ti(HPO_4)(PO_4)].H_2O$. *Chemistry of Materials,* 17, 25, 6287-6294.

[114] Liu Y., Shi Z., Fu Y. & Pang W. (2002) Hydrothermal Synthesis and Structural Characterization of Three Organically Templated Titanium Phosphates: $Ti_2(HPO_4)_2(PO_4)_2.C_2N_2H_{10}$, $Ti_3(H_2PO_4)(HPO_4)_{3.5}(PO_4)_2.C_2N_2H_{10}$, and $Ti_7(HPO_4)_6(PO_4)_6.C_3N_2H_{12}$. *Chemistry of Materials,* 14, 4, 1555-1563.

[115] O'Hare D. & Bruce D. W. (1992) *Inorganic Materials,* Wiley, Chichester, 210-235.

[116] Adams J., Morris M. & Dyer A. (1991) Mechanism of *n*-alkylammonium ion intercalation into the layered host α-$VOPO_4.2H_2O$. *Journal of Materials Chemistry,* 1, 1, 43-49.

[117] Lopez A. J., Lara M. M., Real L. M., Bruque S., Casal B. & R-Hitzky E. (1985) Redox intercalation of alkylammonium ions into $VOPO_4.nH_2O$, $VOAsO_4.nH_2O$. *Materials Research Bulletin,* 20, 5, 549-555.

[118] Jonhson J. W., Jacobson A. J., Brody J. F., Scanlon J. C. & Lewandowski J. T. (1985) Redox intercalation reactions of $VOPO_4.2H_2O$ with mono-valent and divalent-cations. *Inorganic Chemistry,* 24, 12, 1782-1787.

[119] Wang H. Y., Ji W. D. & Han D. X. (2008) Layered zirconium phosphate-supported metalloporphyrin: Synthesis and catalytic application. *Chinese Chemical Letters,* 19, 11, 1330-1332.

[120] Kim R. M., Pillion J. E., Burwell D. A., Groves J. T. & Thompson M. E. (1993) Intercalation of aminophenyl-substituted and pyridinium-substituted porphyrins into zirconium hydrogen phosphate - evidence for substituent-derived orientational selectivity. *Inorganic Chemistry,* 32, 21, 4509-4516.

[121] Wang H., Han D., Li N. & Li K. (2005) Molecular modeling of the intercalation of porphyrins into α-zirconium phosphate. *Journal of Molecular Modeling,* 12, 1, 9-15.

[122] Wang H., Zou M. & Li N. (2007) Preparation and characterization of ionic liquid intercalation compounds into layered zirconium phosphates. *Journal of Materials Science,* 42, 18, 7738-7744.

[123] Wang H. Y. & Han D. X. (2007) A new method of immobilizing ionic liquids into layered zirconium phosphates. *Chinese Chemical Letters,* 18, 6, 764-767.

In: Focus on Catalysis Research: New Developments
Editors: Minjae Ghang and Bjørn Ramel

ISBN: 978-1-62100-455-4

Chapter 7

CATALYTIC ROLE OF BIMETALLIC CORE TOWARDS OLEFIN POLYMERIZATIONS

Srinivasa Budagumpi and Il Kim *

The WCU Center for Synthetic Polymer Bioconjugate Hybrid Materials, Department of Polymer Science and Engineering, Pusan National University, Busan 609-735, Republic of Korea

ABSTRACT

Binuclear bridged and non-bridged transition metal catalysts with definite electronic and steric modulations currently attract significant attention in the catalysis community, mainly because cooperative effects of adjacent metal ions are expected to lead to unique activation modes towards olefin polymerizations and to novel reactivity patterns. Cooperative effects between adjacent catalytic centers in binucleating ligands with aliphatic/aromatic spacers were shown to induce significant rate and selectivity enhancements in ethylene and α-olefin oligo/polymerizations. The binuclear metallocene catalysts are fascinating because of their beneficial effects in the formation of polymers for various explicit applications. In this chapter, we essentially made an attempt to focus on the catalytic role of bimetallic core towards olefin polymerizations. A comprehensive overview of this topic, which the present review is aimed at giving, seems appropriate and timely. In the main stream of this chapter, we tried to cover the recent applications of atom/bond bridged bimetallic coordination and organometallic catalysts with well defined architectures over their monometallic counter parts and an account on the usage and precise applications of supported bimetallic metallocene catalysts.

* Corresponding author. E-mail: ilkim@pusan.ac.kr, Tel.: +82-51-510-2466; Fax: +82-51-513-7720

1. INTRODUCTION

Catalysts containing two or more active centers are employed with the hope of synergistic effects of two active metal centers in the catalysts and of possible occurrence of tandem reactions promoted by two active centers [1]. Activity of these compounds depends on the nature of ligand, metal, chelating atoms, and metal-metal distance which internally affect the metal-metal interactions. Compounds encompassing bimetallic core may demonstrate explicit reactivity patterns for the conversion of olefin monomers into polymers are directly interrelated to a subtle interplay between the metals and the Lewis bases called 'ligands'.

Both bridged and non-bridged bimetallic late first row transition metal coordination catalysts are the current interests for the polymerization of olefins with definite electronic and steric modulations. Homo and hetero bimetallic non-metallocene organometallic compounds have found to be more proficient catalysts for ethylene polymerization using methyalumoxane (MAO) as cocatalyst over a series of monometallic catalysts. However, hetero and homonuclear bimetallic catalysts show an increased activity up to three and four fold with compared to the monometallic species respectively [2,3]. On the other hand, the melting temperature of the formed polymer is also increased by 3-6 °C. Perhaps, the rise in activity and melting temperature of the polymer are mainly because of presence of most active bimetallic core.

The binuclear metallocene catalyst/MAO systems especially, zirconium catalysts, are most effective in ethylene and propylene polymerizations known to chemists long back. Since then, a large number of contributions on the synthesis, reactivity and applications of new members of this imperative kin of aromatic nitrogen containing metallocenes have appeared. As a result, the synthesis and reactivity towards olefins in order to produce desired polymer are well known in catalysis, and publications account for this spectacular development. On the other hand, π-complexation of aromatics to electron-withdrawing transition-metal moieties such as tricarbonylchromium, tricarbonylmanganese, cyclopen-tadienyliron/ruthenium and pentamethylcyclopentadienyliron/ruthenium subunits, alters their chemical reactivity and provides a variety of synthetically, industrially useful compounds when reacted for olefin polymerization/oligomerization reactions. The combination of outstanding cocatalytic activity of MAO towards metallocenes especially with group IV, VI and VIII, with the high electronic and steric variability of the latter, led to a novel class of olefin oligo/polymerization catalysts which is currently revolutionizing the polymer industry. The polymers containing about 20 mol% propylene have broader polydispersity than polyethylene homopolymer. As a result of propylene content versus molecular weight measurement, it was found that the propylene content in the polymer was higher at low molecular weight region and decreased with increasing the molecular weight. In contrast, by using mononuclear metallocene catalysts, the propylene content of the polymer was almost invariable from low molecular weight region to high molecular weight region due to the homogeneity of active species [4-6].

Various $MgCl_2$, silica or carbon nanotube supported catalysts containing different amounts of titanium-hafnium or titanium-zirconium were found to be active compounds for the polymerization of ethylene and α-olefins as well as for their copolymerizations. This results in the formation of polymers with much broader molecular weight distributions as compared with similar systems containing monometallic species [7]. It was found that the activity of the unsupported catalysts was much greater than that of the supported catalysts. However, the catalytic activity of the supported catalysts formed by the remote donor or spacer ligands was much higher than that of the monometallic supported catalysts.

In this chapter we address the recent advances made in the chemistry and catalytic role of bimetallic complexes, homogeneous and supported, towards olefin polymerizations. The cooperative effects of adjacent metal ions leading to unique activation modes towards olefin polymerizations and to novel reactivity patterns are main concerns of this chapter.

2. Synthesis of Bimetallic Core towards Olefin Polymerizations

2.1. Ligand Design

Within the past decade, studies on bridged and non-bridged bimetallic coordination/ organometallic compounds with side-off, end-off and compartmental/remote donor multidentate ligands have attained the status of a major scientific discipline. The investigation and application of these compounds has led to the development of safer and more benign diagnostic agents, catalysts for various chemical processes, many new technologies being developed each year. In today's world, the synthetic chemists in research and development field are highly challenged to consider more environmentally and economically benign methods for the generation of desired target molecules for the polymer preparations with defined compositions. On the other hand, many industrial chemical processes for the preparation of polymers has a major intension are to get more yield with the less or same ingredients in fewer steps with high safety profiles. Although there are still considerable debate and speculations on the nature and existence of available single metal-based catalysts in the industry that could provide a validation for the often observed significant rate, yield enhancement and decreased time and man power consumption, there is little doubt that bimetallic catalysts will became standard tools in order to decrease the excess consumption of time, man power and mono-metallic catalysts.

The bimetallic catalysts suitable for the olefin polymerizations can be prepared by keeping a central group viz., pyrazolyl, triazolyl, phthalazyl, phenoxo, thiophenoxo, etc., as bridging module between two similar/dissimilar metal ions. These bridging components provide bond/atom bridging elements for two metal ions. By selecting an appropriate bridging component between the metal ions; it is trouble-free to tune the metal-metal distances and so produced reactivities, which intern helpful to select the appropriate activity poses by the entire host-guest inorganic assembly. The ligands contain defined separate binding subunits, each being able to coordinate one metal ion and the bridging units like pyrazolyl, phenolate, and other functional groups; in their deprotonated forms, they have a strong tendency to bridge the two metal ions. For this reason, the two metal ions are forced to remain close to each other, and the bridging groups play a key role in determining the molecular geometry of the binuclear species. Moreover, because of the number of binding sites, the ligand does not completely saturate the coordination sites of the metal ions and the complexes formed can be used to assemble at least one secondary labile ligand which is associated with the metal salt.

Another suitable care should be taken while designing the ligands, that the presence of alkyl groups like, methyl, ethyl, isopropyl or tertiary butyl in the ligand system, in such a way, that the said groups should hinder the formed polymer to continue the polymerization by the metal ion. If the alkyl groups are failed to perform their activity, than the chain termination occurred to stop the further polymerization by the metal ion, which ultimately results in the formation of oligomers. Therefore, to get an apt product from the polymerizations by the bimetallic core, it's necessary to choose the presence and positions of the alkyl groups in the ligand entities.

On the other hand, the ligands which have an ability to form binuclear complexes suitable for olefin polymerization without any metal-metal interactions are also of great importance as two individual metal ions do act as two individual same type complexes in the same catalyst. The aromatic or aliphatic spacer is expected to enhance the rigidity of the ligand backbone and to enable fine-tuning of the electronic character of the *N* or/and *O*-donors by ring substitutions, thus broadening the possible polymerization/ oligomerization activities. In particular, Salophan ligands are close analogues of the Salans in which the aliphatic spacer has been replaced by an aromatic one. The catalysts derived from these kinds of ligands do retain their dianionic character in different environments and are amenable to fine-tuning of the electronic character of the *N*-donors and finally, have led to promising catalysts for polymerization of α-olefins. Apart from Salans, to date, most of the binucleating ligands having aralkyl bridging components tend to be positioned at the extremities of the ligand manifold due to their nature of the design, and thus would be expected to limit close cooperative metal-to-metal interactions. With the intent of forcing the metal centers into a closer proximity, spatially confined M_2 centers (M = Fe, Co, Ni, Zn) on bulky aryl-bridged pyridyl-imine compartmental ligands have been studied for ethylene oligomerization, yielding low molecular weight materials with moderate activity in combination with methylaluminoxane.

2.2. Metal Ion Selectivity

Metal ion selectivity for the olefin polymerization is very important and decides the nature and extent of product formed. The discovery of homogeneous catalysts with transition metal ions for olefin polymerization has brought a revolution in polymer synthesis since the homogeneous catalysts with transition metals-based polymers can possess excellent physical properties or stereo-regularities that are difficult or impossible to be achieved by other known polymerization methods. In order to get the polymer with excellent physical properties or stereo-regularities from the metal based catalysts, it is important that to consider the nature of metal ion viz., size, charge, coordination number and oxidation state, and the ligand field viz., denticity, donor atoms and the chelate ring size over it.

In a usual manner, non-metallocene octahedral zirconium(IV) and titanium(IV) catalysts with N_2O_2 coordination sites are forming polyethylene with compatible yields. Similarly, bi- and mono- cyclopentadienyle zirconium(IV) and titanium(IV) catalysts with constrained tetrahedral geometry do form the stereo-regular polyethylene from ethane, however, with the same or similar ligands, iron (II), cobalt(II) and nickel(II) catalysts produces oligomers. Diimine coordination catalysts derived from the condensation of 2,6-biacetyl pyridine with 2,4-disubstitued anilines with MAO as co-catalyst display high activities in ethylene oligomerizations with good selectivities to linear α-olefins. The catalytic activities of catalysts and distribution of resultant α-olefins closely depend on substituent of aryl ring and Al/Fe ratio. Whereas, with the same or similar ligands with cobalt(II) and nickel(II) catalysts, produces stereo specific polymers in high yields along with higher oligomers. However, this property is almost opposite incase of butadiene polymerizations with iron(II), cobalt(II) and nickel(II) coordination catalysts.

Although olefin polymerizations have been the domain of early and late transition metal catalysts, there has been growing interest recently for the late transition metal catalysts with the solid silica or carbon nanotube support. These supported catalysts are found to be efficient heterogeneous catalysts towards olefin polymerizations. Heterogeneous catalysts are the workhorse of many industrial processes, since they have many processing and economic advantages over their soluble counterparts. For the preparation of supported catalysts, the catalyst precursor molecule should have at least one primary amine or hydroxyl group which could be interacted with 3-(triethoxysilyl)propylisocyanate in THF, further with tetraethyl orthosilicate in NH_4OH yield the silica supported catalysts for the olefin polymerizations. Similar preparation methods can be followed for the preparation of carbon nanotube supported transition metal catalysts. Metal –based pre-catalyst bearing free hydroxyl or primary amine groups were made to interact with functionalized multi-walled carbon nano-tubes dispersed in ethanol at 70-80 ^{o}C to get the desired supported compounds. After completion of the reaction, the supported catalyst was purified by repeated washing with water followed by acetone to remove the unreacted or excess carbon nanotube fractions. Magnesium chloride supported catalysts can also be prepared using quite methods expressed for silica supported catalysts. However, the studies related this field is restricted only to titanium and zirconium metals and very rarely to some other metals at special conditions. The catalysts were prepared by reaction of a spherical adduct $MgCl_2$.nEtOH with $TiCl_4$ in the presence of diisobutyl phthalate (DIBP) at ambient temperatures. The invention discloses a dicaryon nickel series compound and a preparation method thereof. The dicaryon nickel catalyst, after activation with methylaluminoxane, has higher catalytic activity while catalyzing ethylene polymerization, thus obtaining branching polyethylene with high-molecular weight. The dicaryon catalyst contains two metallic centers and a special synergistic effect exists between two centers and ensures the dicaryon catalyst to have different catalyst prosperities compared with mononuclear catalyst; moreover improves catalyst activation and changes molecular weight of polymer, etc.

3. Reactivity of Bimetallic Core towards Olefin Polymerizations

Olefin polymerizations and co-polymerizations using late transition metal catalysts have blossomed in the last several years. While the commercial impact remains to be seen, the investment of research effort by both academic and industrial laboratories is rapidly increasing. Late transition metal catalysts that possess high electrophilicity and/or steric hindrance remain the focus for producing high molecular weight polymer. The detailed mechanistic picture developed around the cationic α-diimine nickel and palladium catalysts provides a mechanistic justification for the said physical qualities of the produced polymers.

3.1. Binuclear Coordination Core

Catalysts with one active transition metal centre have their advantage in the synthesis of polymers with regulated molecular weights and structures. In synthetic organic reactions, the catalysts containing two active centers are employed with the expectation of synergistic effects of two close active metal centers in the catalysts and of possible occurrence of tandem reactions promoted by two active centers. Homo- and hetero binuclear bridged transition metal catalysts have their own reactivity mode towards α-olefins and produce different kind of oligo/polymers depending upon their synergistic effects, and produces different kind of tandem reactions. In tandem catalysis, one catalyst produces α-olefin oligomers which are incorporated into high-molecular weight polyethylene by a second catalyst in the reaction mixture, utilizing the *same* ethylene feed. Since this type of polymerization requires intermolecular processes at low catalyst concentrations. The tandem reaction by two catalysts is probable also in olefin polymerization because many metal-catalyzed olefin polymerization reactions involve chain transfer *via* β-hydrogen elimination of polymer (or oligomer) with a terminal vinyl group, and re-insertion of the unsaturated molecule into a new metal–carbon bond occurs commonly in many metal-catalyzed olefin polymerizations.

3.1.1. Atom Bridged Binuclear Core

Two similar transition metal ions can bridge between a negatively charged bridging ligands viz., chloride, bromide, phenolate, pyrazolate, triazolate, etc., produces different sort of metal-metal interactions which in turn depends on the metal-metal distances (Figure 1). Recently Kunrath F.A. *et al.* [8] reported a new class of oligomerization catalysts based on chloro bridged nickel(II) complexes with distorted trigonal bipyramidal geometry containing sterically hindered tris(pyrazolyl)borate ligands which produce selectively 1-butene upon activation with alkylaluminum cocatalysts such as MAO and TMA. Both turnover frequency and selectivity are dependent on the nature of the alkylaluminum cocatalyst employed in the oligomerization reactions, suggesting that these activators act as ligands attached to the nickel center. Further investigations into the reactivity of these nickel catalysts with propylene will be the subject of a future report. These ligands have been used in the preparation of highly active Ti(IV) and Zr(IV) catalysts for ethylene polymerization [9]. Moreover, these highly sterically demanding hydrotris-(pyrazolyl)borate ligands can control coordination environments of the metal centers, preventing the formation of coordinatively saturated, less reactive complexes and favoring the synthesis of tetrahedral complexes. The active species formed by the action of alkylaluminum salts on the chloro-bridged binuclear nickel catalysts are dependents on the nature of alkylaluminum salts used. It is worth noting that the lifetime of the nickel active species obtained with TMA is much lower than that for species issued from MAO. At 1.1 atm the system binuclear nickel catalyst/TMA is active until ca. 30 min and the system binuclear nickel catalyst/MAO is still active during the 60 min with no indication of deactivation. These results strongly suggest, as mentioned before, that the active species maintains the alkylaluminum species bonded to the nickel center, as observed in recent studies using XAS spectroscopy [10].

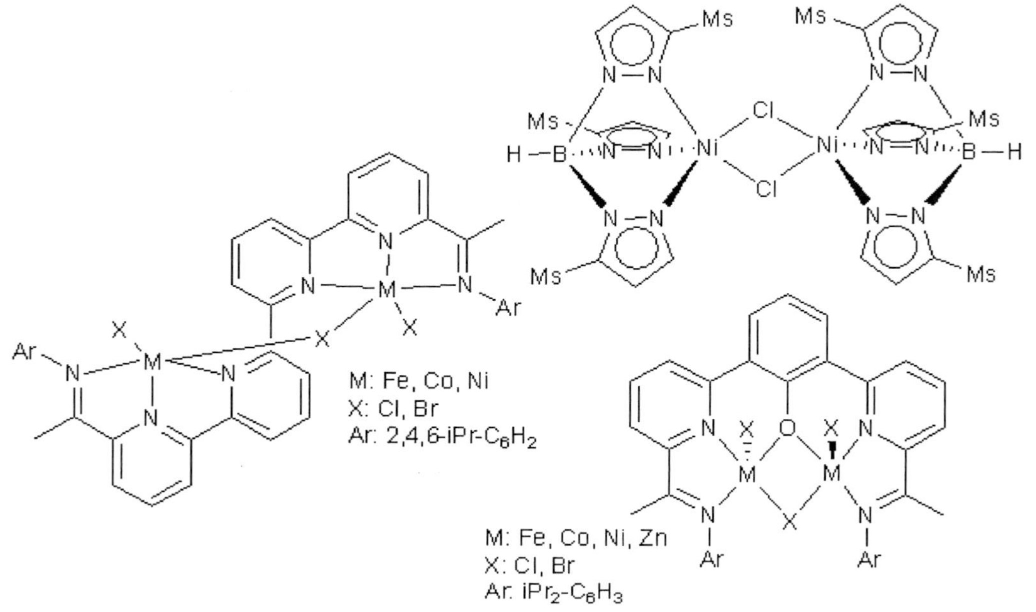

Figure 1. Atom bridged bimetallic coordination compounds.

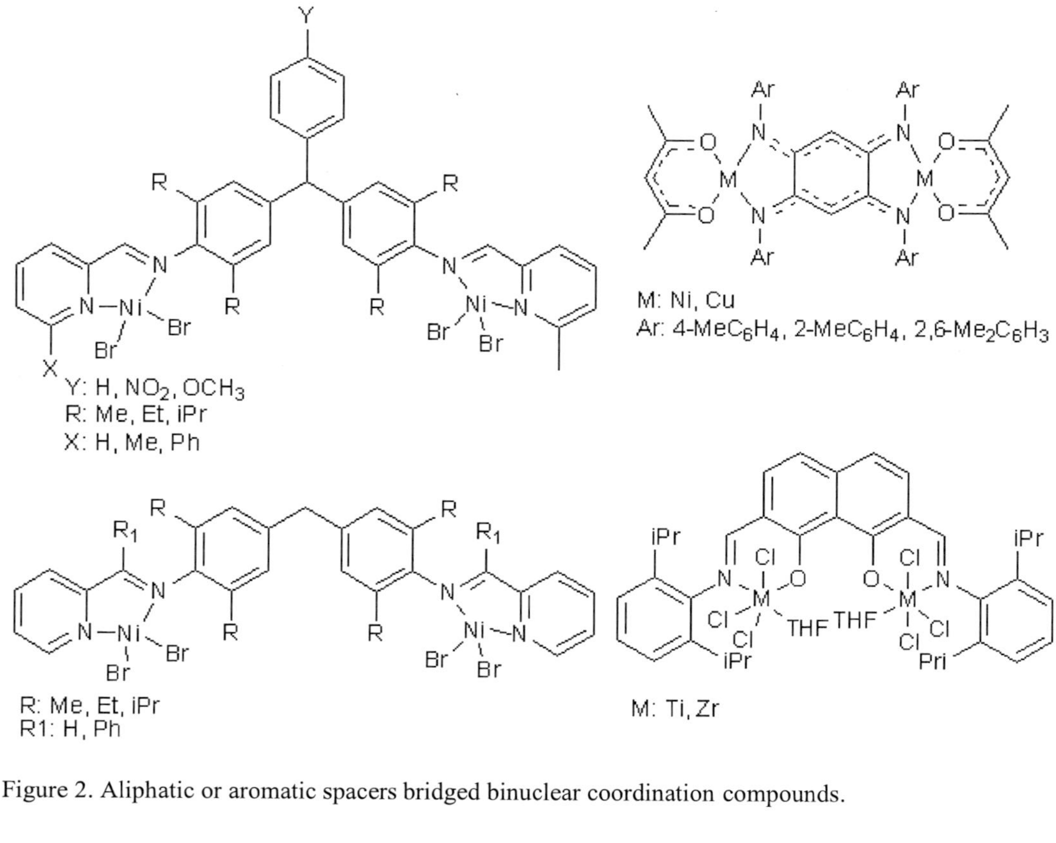

Figure 2. Aliphatic or aromatic spacers bridged binuclear coordination compounds.

Recently, Champouret Y. D. M. *et al.* [11] reported two new bulky aryl and phenoxo-bridged pyridyl-imine compartmental ligands which have been prepared in moderate to good overall yields via a Stille-type cross-coupling approach. The molecular structure of phenoxo ligand reveals a transoid configuration within the pyridyl-imine units with a hydrogen-bonding interaction maintaining the phenol which provides endogenous bridging between metal ions is being coplanar with one of the adjacent pyridine rings. Both the ligands vigorously reacts with Fe(II), Co(II), Ni(II) and Zn(II) salts to produce bridged binuclear catalysts for the effective ethylene oligomerizations. In all the cases one of the halide ion acts as the exogenous bridging module between the phenoxo-bridged metal centres along with/without a solvent (acetonitrile) molecule. The ligands can be readily accessed in which sterically bulky groups can be introduced at the termini of the ligand manifolds for the hindrance purpose. Significantly, all the dicobalt and dinickel systems display appreciable activity for alkene oligomerization upon activation with MAO with the cobalt systems forming mainly a mixture of linear α-olefins and internal olefins while the Ni systems additionally promote methyl branched structures. For both the binuclear nickel and cobalt systems, a coordination-insertion mechanism seems likely, followed by *β*-hydride elimination (chain transfer). Furthermore, an isomerization process (chain walking) appears operational for both binuclear cobalt and nickel systems. However, only in the case of the nickel systems is the chain walking competitive with the chain propagation with the effect that further insertions of ethylene into a secondary metal-alkyl bond can occur to give the distinctive branched microstructures is observed [12].

Khamker Q. *et al.* [13] have reported a novel bis(imino)quaterpyridine ligand can be formed in good yield by the condensation reaction of 6,6'''-bis(acetyl)quaterpyridine with 2,4,6-triisopropylaniline; the 2,6-diisopropylphenyl derivative. Interaction of these ligands with late first row transition metal salts viz., $FeCl_2$, $CoCl_2$, $CoBr_2$ and $(DME)_2NiCl_2$ yield halo-bridged binuclear catalysts for the ethylene polymerizations. When these catalysts activated with MAO for ethylene polymerizations yield the broad product distribution with multiple peaks found in addition to those corresponding to the linear a-olefins (in the range C6-C26). A chain-walking mechanism has been previously used to account for the observed isomerisation/branching in related nickel-based catalysts [14,15] and the low catalytic activity is comparable to that observed using the mononuclear counterpart [14]. Quarter-pyridine binucleating ligand is found as an effective scaffold for two oligomerisation-active metal(II) centres [metal = cobalt, nickel, iron]. Despite the metals centres occupying similar tridentate cavities within the hexadentate ligand framework, the metal centres can react independently with donor ligands (*e.g.*, MeCN, H_2O). Upon activation with MAO, the dicobalt species have proved to be the most active systems for ethylene oligomerisation affording highly linear a-olefins while the nickel-based binuclear nickel catalyst/MAO is less active and less selective affording a broad distribution of olefinic products with the ^1H NMR spectrum revealing a mixture of alkene-containing compounds composed of mostly internal olefins (52%) along with lower levels of a-olefins (22%), tri-substituted (21%) and vinylidenes (5%). In addition, the ^{13}C NMRspectrum of these oligomeric products indicates the presence of mainly methyl branches, along with very low levels of longer chain branches (*e.g.*, ethyl and propyl) [16]. However, dinickel systems generate branched materials with a range of different vinylic chain ends.

Alkoxo-bridged binuclear octahedral titanium(IV) complexes with monoclinic system space group $P2_1/n$ have been reported by Rhodes B. *et al.* [17], demonstrating that the ligands are the derivatives of amino alcohols with different substitutions. Alcoholic hydroxyl group provides bridging between the two titanium(IV) centres upon deprotonation and the remaining coordinations are from the different donors present in the ligand to satisfy octahedral field around metal ions. These two Ti(IV) complexes were studied as Ziegler–Natta type polymerization catalysts in the presence of an excess of methylaluminoxane (MAO) for the polymerizations of ethylene. It was found that ethylene polymerization activity for both catalysts increased by increasing the polymerizations temperature from 25 to 70 °C. If the catalysts were treated with trimethylaluminum (TMA), the activities increased at 25 °C, but slightly decreased at 70 °C. However, the complexes are found to be totally ineffective for the oligo/polymerizations of propylene. It is proposed that methylation of the catalyst precursors is difficult at room temperature by MAO, or the TMA in it, but is spontaneous at 70°C. If the catalysts were preactivated (methylated) with TMA prior to the initiation of the polymerization, the activities were increased at room temperature compared to when the catalysts were not treated with TMA. The activities at higher temperatures were somewhat decreased, perhaps the greater instability of the methylated titanium precursor at this temperature.

Gibson V.C. *et al.* [18] described a series of coordinatively unsaturated dichloro bridged chromium(III) ethylene polymerization precatalysts bearing either β-diketimate ligands. A general feature of these ligands is the presence of bulky aryl substituents which offer protection to the active centres. Solid polyethylene is obtained with samples displaying high molecular weights and virtually no branching of the polymer was observed. The bridged chromium catalyst displayed highest activity being 75 g $mmol^{-1}$ h^{-1} bar^{-1} using diethylaluminium chloride (Et_2AlCl) as activator. The new catalyst types described herein represent a notable addition to the limited list of non-cyclopentadienyl chromium ethene polymerization catalysts [19] and highlight the importance of the choice of co-catalyst for optimal catalyst performance. These catalysts show good activity towards ethylene polymerizations compared to their mononuclear counter parts.

3.1.2. Aliphatic or Aromatic Spacers Bridged Binuclear Core (Remote Donor Ligands)

Cooperativity effects between adjacent catalytic centers in binucleating ligands with aliphatic/aromatic spacers were shown to induce significant rate and selectivity enhancements in ethylene and α-olefin oligo/polymerizations (Figure 2). In the aliphatic/aromatic spacers bridged binuclear catalysts area, new families of transition metal catalysts formed by the remote donor ligands like monophenoxyiminato, methylene-bis-aniline derivatives, etc., have attracted attention due to ease of preparation, ability to support living polymerizations, and utility in producing new polyolefin architectures.

Salata M.R. and Marks T. J. [20] reported the marked ethylene polymerization selectivity characteristics of binuclear phenoxyiminato zirconium catalysts formed by the remote donor diimine-1,8-dihydroxynaphthalene ligand. Ethylene homopolymerizations using said dizirconium catalysts activated with MAO afford high molecular weight linear polyethylenes with activities which are approximately 8 times that of mononuclear zirconium catalysts at the same conditions [21,22]. Further it also investigated that the ethylene and 1-hexene copolymerizations with dizirconium catalyst achieved efficient 1-hexene co-enchainment under mild reaction conditions. The comonomer incorporation is found to increase incrementally over a 5.5-fold 1-hexene concentration range. In marked contrast to these results, mononuclear control catalyst of zirconium yields only traces of copolymer under identical reaction conditions. This is mainly due to the synergetic effect present in the dizirconium active species hosted by the remote donor ligand. Further the same authors [23] carried out the similar studies on the same remote donor ligand with titanium(IV) metal to compare ethylene polymerizations activity among group-4 metal ions. The homo-polymerization studies of ethylene for phenoxyiminato dititanium catalysts are modest and are slightly lower than those of their corresponding dizirconium counterparts. However, at 40 °C, dititanium catalyst produces polyethylene with an activity approximately two times greater than that of mononuclear phenoxyiminato titanium(IV) counter parts. As reported for other mononuclear phenoxyiminato catalysts, [24,25] mononuclear titanium and zirconium produce linear, high *M*w polyethylenes, however the substantial polydispersities indicate that multiple catalytic sites or conformations may be involved. The binuclear complexes also produce very high molecular weight linear polyethylenes at 24 °C, as indicated by the insolubility of the polymeric products.

Not only group-4 bimetal hubs are active against olefin polymerizations, late first row transition bimetal foundations are also found to be active catalysts in the literature. only a few binuclear Ni(II) and Pd(II) complexes with bridging ligands have been used in norbornene polymerization [26,27]. Binuclear, divalent nickel(II) and copper(II) acetylacetonato complexes with bulky sterically hindered N-substituent $6\pi + 6\pi$ amine-imine ligands have been synthesized for catalytic ethylene polymerization studies by Haung Y.-B. *et al.* [28]. The authors explains the synthesis of a series of binuclear, divalent nickel and copper acetylacetonato complexes of the type [M(acac)-{μ-C_6H_2(=NAr)$_4$}M(acac)] (M: Ni, Cu) by reaction of the corresponding M(acac)$_2$ precursor with various bulky steric hindrace π-acceptor N-substituted 2,5-diamino-1,4-benzoquinonediimines, which are metalated and become remote donor bridging ligands. The coordination geometry around the metal ions of the Ni and Cu complexes is square-planar, and a complete electronic delocalization of the quinonoid π-system occurs between the metal centers. When metal based catalysts were activated with MAO as cocatalyst, all the binuclear Ni(II) complexes exhibited high activities both for addition polymerization of norbornene and methyl methacrylate (MMA) polymerization, which produce syndiotactic-rich poly(methyl methacrylate) (PMMA) with broad molecular weight distribution. This activity is mainly because of the two Ni(II) centres susceptible of showing cooperative effects through the spacer. However, the binuclear Cu(II) complexes of the similar ligands show moderate activities for norbornene polymerization and are inactive for MMA polymerization. Binuclear Ni(II) catalyzed polymerization of acrylates is not a coordination-insertion mechanism, but rather follows a free radical polymerization mechanism. Lee S. S. *et al.* [29] reported that bimetallic salicylaldimine-nickel(II) catalysts show higher activities and higher polar monomer incorporations in the ethylene/2-(methoxycarbonyl)norbornene and ethylene/2-(acetoxymethyl)norbornene copolymerizations than the mononuclear complex. There is no binuclear Cu(II) complex that has been used in the addition polymerization of norbornene.

Recently Bahuleyan B. K. et al. [30] reported a series of sterically and electronically modulated iminopyridyl Ni(II) bimetallic catalysts for the catalytic ethylene and 1-hexene oligo/polymerizations. When catalysts are activated with methylaluminoxane or common alkyl aluminums such as ethyl aluminum sesquichloride (EASC), all catalysts oligomerize/ polymerize ethylene with activities exceeding 10^6 g-product/(mol-Ni h bar) at 30 °C to yield C4 as a major product. Among all the cocatalysts, EASC records the best activity and the effectiveness is in the order of EASC >> MAO > methyaluminumdichloride (MADC). The effect of electronic modification of the complexes on catalytic activity was also explored by the group by using UV–vis spectroscopy and cyclic voltammetry. In order to compare the polymeric effects of the bimetallic system, authors also prepared and studied the catalytic activities of monometallic catalysts. Compared to the monometallic system, the bimetallic catalysts produced polymers with high MW (Mw = 8800 s. 2000) and broad MWD of 5.3–8.3 vs. 1.2. The cooperative effect due to two metal centers of bimetallic complexes was evidenced by comparing the molecular weight of polymer and the ^{13}C NMR spectra of poly(1-hexene) obtained by its monometallic counterpart. It is evident from the NMR spectrum that the polyhexene obtained by the bimetallic catalyst can be considered as an all-in-one combination of polymeric structures with regions containing methyl, butyl and higher branches, which are assumed to be due to the presence of ethylene-propylene copolymer, meso and racemic configurations of regio-irregular head-to-head methyl branches, poly(1-hexene), poly(α-olefin)s with longer branches and PE. Poly(1-hexene) synthesized using bimetallic complex showed very complicated microstructure comparing to that of the monometallic catalyst, due to the electronic/spatial communication of adjacent metal centers.

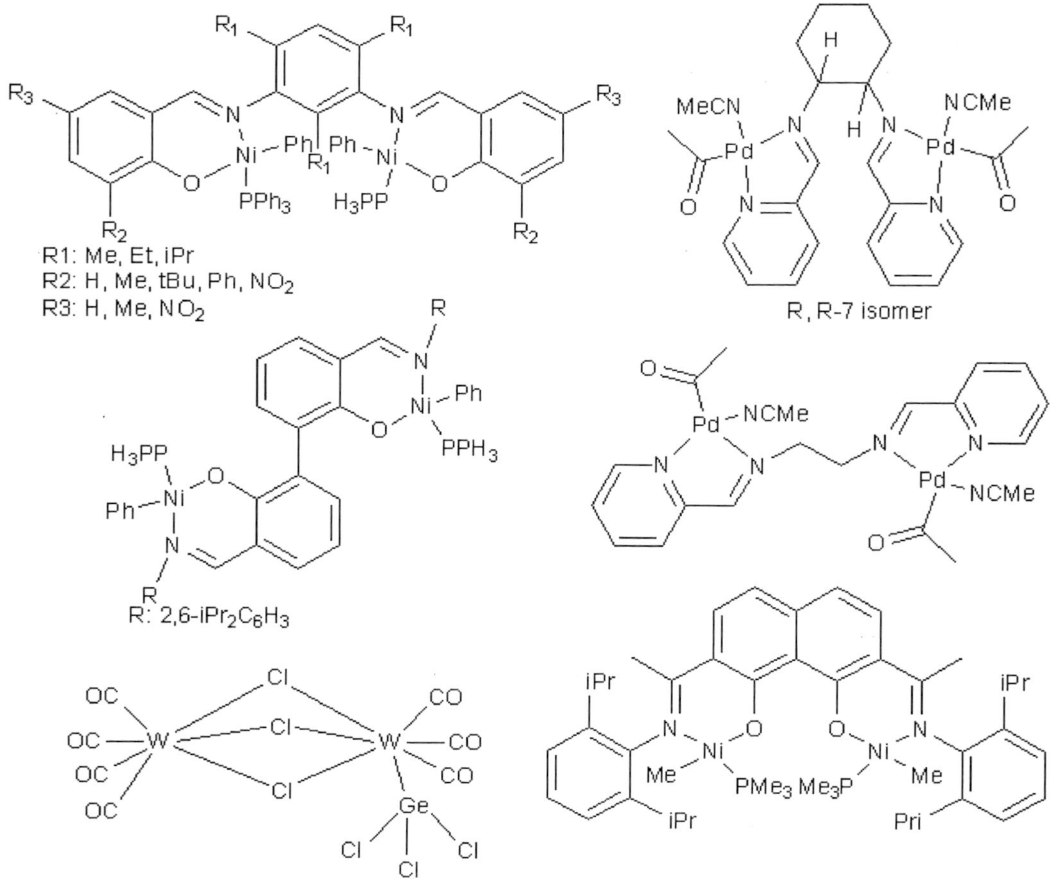

Figure 3. Binuclear non-metallocene organometallic compounds.

More or less similar work has been reported by Jie S. *et al.* [31]. The authors group reported the efficient synthesis of a series of dinickel(II) complexes bridged by bis(pyridinylimino) ligands. The ligands were prepared through the condensation reaction of 4,4'-methylenebis(2,6-disubstituted anilines) with 2-pyridinecarboxaldehyde or 2-benzoylpyridine. The catalytic systems were typically investigated through varying reaction conditions, such as the molar ratio of cocatalyst (MAO) to nickel complex (Al/Ni), reaction temperature, reaction period and the pressure of ethylene, for the optimum. When these dinickel(II) catalyst activated with MAO, showed considerably good activities for ethylene oligomerization and polymerization. Their catalytic activities and the properties of PEs obtained were depended on the arched environment of ligand and reaction conditions. Along with elevating reaction temperature in the range of -10 ^{o}C to 40 ^{o}C, the polymerization activity by the dinickel catalysts is significantly declined, while the amount of oligomers were increased. Up to 60 ^{o}C, only trace of polymer was obtained, in addition, the oligomerization activity was also lowered to 42.8 kg mol^{-}1 (Ni) h^{-1}. At the Al/Ni molar ratio of 100:1, only oligomer C4 was formed with lower activity. When the Al/Ni molar ratio was further enhanced, the polymerization activity efficiently increased. At the Al/Ni molar ratio of 2000, the catalytic system displayed the higher polymerization activity of 712 kg mol^{-1} (Ni) h^{-1}.

3.2. Binuclear Organometallic Core

Binuclear organometallic transition metal compounds are found to be effective and efficient catalysts for the olefin polymerizations compared with their mononuclear counterparts. There are very less efforts have been made for the atom or bond bridged binuclear organometallic catalysts mainly because of the presence of very short distance between metal canters. Among all, aliphatic or aromatic spacers bridged bimetallic metallocene compounds found to be the good catalysts for the ethylene polymerization activity due to the existence of high cooperative effect induced by the spacer and organic moiety around the metal ions.

3.2.1. Non-metallocene Binuclear Organometallic Core

Electrophilic non-metallocene complexes of the late transition metals are of interest because of their catalytic potential in processes such as C-H bond activation and olefin polymerization. Non-metallocene binuclear organometallic core suitable for olefin polymerizations can be constructed in many ways. The alternating copolymerization of olefins with carbon monoxide to give polyketones has been catalyzed by many different electrophilic palladium(II) complexes. Of particular interest is the copolymerization of vinyl monomers with carbon monoxide by binuclear catalysts since this process gives rise to true stereo-centers within the polymer backbone. Representative binuclear compounds are shown in Figure 3.

Bimetallic, neutrally charged dinickel 2,7-diimino-1,8-dioxynaphthalene polymerization catalysts having methyl or 1-naphthyl and a triphenyl phosphine groups attached to the two nickel ions have been found to be the effective ethylene polymerizations catalysts. Along with that ethylene-co-norbornene copolymerizations by the same binuclear non- metallocene organometallic catalysts display increased catalytic activity by methyl branch formation, and comonomer enchainment selectivity versus the monometallic analogues. Furthermore, these systems turn over in the absence of cocatalyst under mild conditions [32]. Room-temperature ethylene homopolymerizations using the binuclear catalysts in presence of the phosphine scavenger/cocatalyst $Ni(cod)_2$ under conditions minimizing mass transport and exotherm effects. Bimetallic organometallic core afford polyethylenes with molecular weights comparable to those produced by the mononuclear analogues and with polydispersities consistent with single-site processes. However, the bimetallic catalysts exhibit a 2-fold greater polymerization activity along with increased methyl branching. The branch density by ^1H NMR is approximately two times than that achieved by the mononuclear catalysts under identical reaction conditions and is confirmed by depressed DSC determined melting points. In the absence of a cocatalyst, the mononuclear systems do not produce polyethylene. In contrast, the bimetallic methyl or 1-naphthyl and a triphenyl phosphine substituted catalysts produce polyethylenes with increased branching densities and concurrently depressed melting points, albeit at somewhat reduced polymerization rates versus the cocatalyzed polymerizations. A detailed cause for the branch formation pathways in the binickel catalysts mediated ethylene homopolymerizations is shown in Figure 4. Ultimately for single-site d^8 Ni(II) aryloxyiminato ethylene polymerization catalysts, a proximate catalytically active Ni site substantially increases activity, degree of and selectivity for methyl group branching, and comonomer incorporation selectivity versus the mononuclear analogues.

The reaction of the bis(bidentate) ligand *R,R/S,S-trans*-iminopyridine ligand formed by the condensation of 1,2-diaminocyclohexane with pyridine-2-carboxaldehyde in 1:2 molar ratio [33]. Bis-imine ligand yields binuclear organometallic Pd(II) catalyst when reacted with $[PdMe_2(\mu\text{- pyridazine})]_2$ metal precursor in stoichometric ratio. So formed bimetallic Pd(II) catalyst is having two methyl groups and NN donor sites around each Pd(II) ion. The cyclohexyl group acts as the cyclo-aliphatic spacer between the two Pd(II) ions and provides the suitable cooperative effects between the metal centres. Both the metal centers can be close together or widely separated depending on the conformation about the N-C bonds. Cooperative effects between the two palladium centers might be expected if the metals are close, but they are widely separated by the cyclo-hexyl spacer. Under these circumstances, the metal centers can act independently. This type of Pd(II) catalysts in the presence of carbon monoxide were found to be active catalyst precursors for the copolymerization of styrene or 4-methylstyrene to a greater extent.

Yet another interesting target is to understand the catalytic mechanism behind the olefin polymerizations from cationic d^8 bimetal systems attached with an organic moiety directly to the active metal centre. Due to their functional group tolerance, ethylene and 1- olefins can be copolymerized with polar monomers such as acrylates. These studies prompted renewed interest in neutral binickel ethylene polymerization catalysts. By comparison to their cationic Ni(II) counterparts, they are more tolerant toward polar reagents. Thus, polymerization can be carried out in aqueous emulsion to afford polyolefin dispersions. Ethylene polymerization with binuclear phosphinoenolate complexes as catalyst precursors was studied extensively [34] due to the higher yield and high molecular weight polymer. Found activities are higher than that of their analogous mononuclear complexes. As with mononuclear Ni(II) phosphinoenolate complexes, number average molecular weights are rather low and activation by a phosphine scavenger is required for these polymerizations. Like their mononuclear analogues, highest average activities are observed at reaction temperatures of 50 to 60 °C. The maximum activities observed are significantly higher for the binuclear complexes. Ni(II) complexes derived from 2,5-disubstituted amino-*p*-benzoqinones, polymerize ethylene with higher catalyst activities and afford polyethylene with a molecular weight of up to $M_n = 1.3 \times 10^5$ g mol^{-1}. Salicylaldiminato binickel complexes, with a bridging moiety derived from 2,4,6-trialkyl- *m*-phenylene diamines, were reported prior to this [35]. By comparison to an analogous mononuclear complex investigated, a slightly higher molecular weight and broader molecular weight distribution were found for these binickel catalysts.

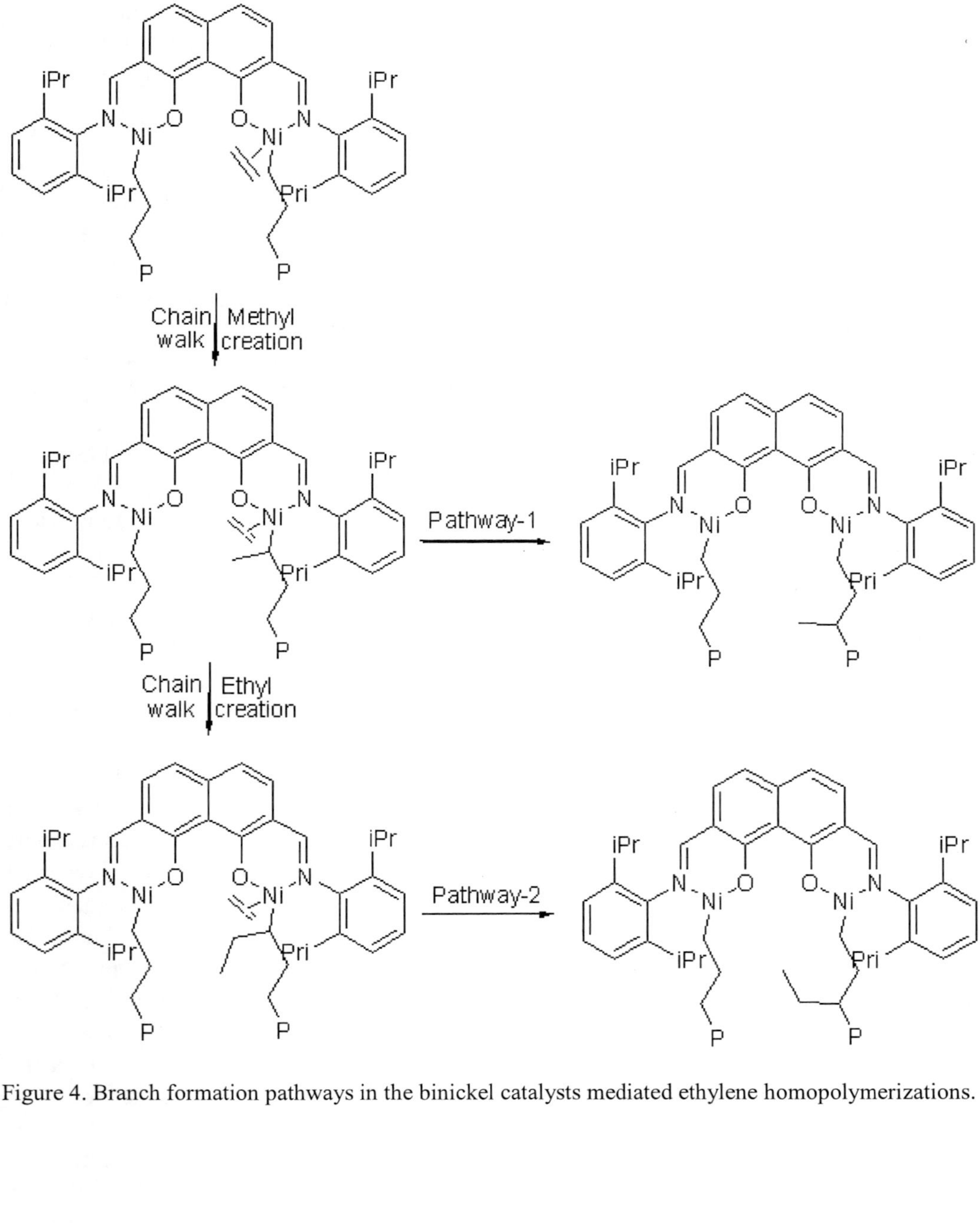

Figure 4. Branch formation pathways in the binickel catalysts mediated ethylene homopolymerizations.

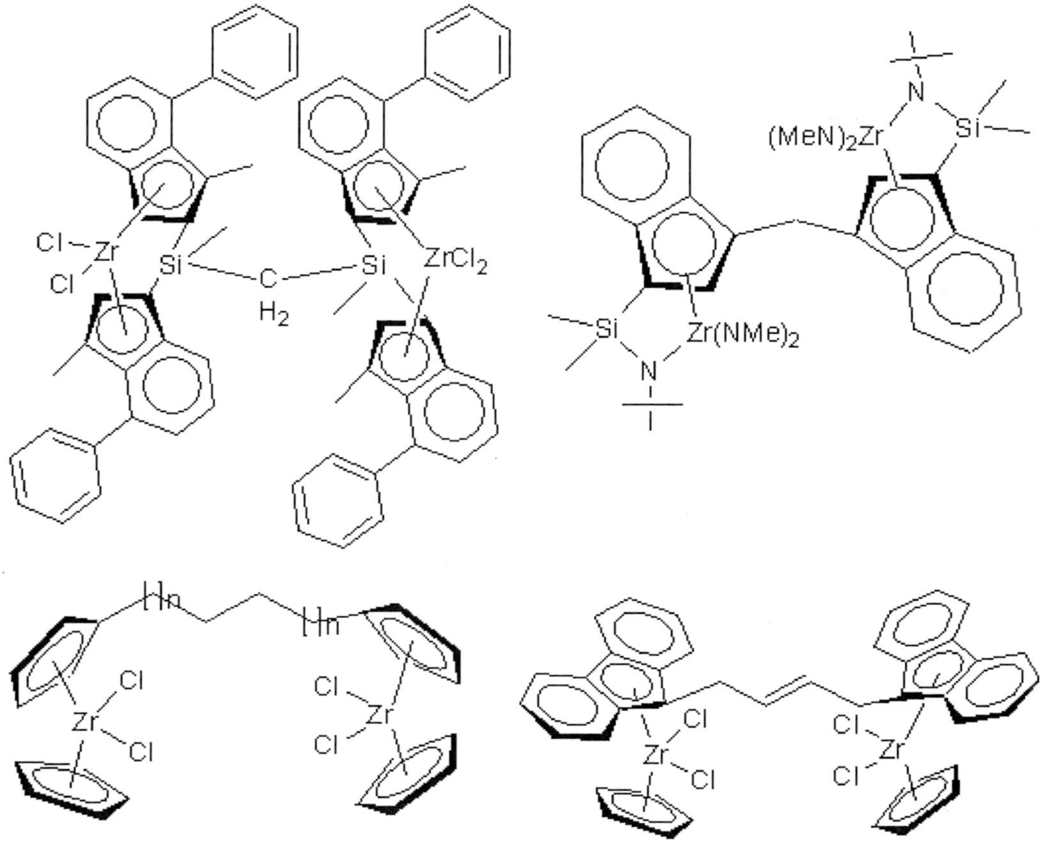

Figure 5. Binuclear metallocene compounds.

On the other hand, polymer molecular weights are influenced by the reaction temperature; with increasing polymerization temperature molecular weights decrease. The molecular weights of the polymers formed at 50 °C with the binuclear complexes are significantly higher than of the polymers formed with the mononuclear analogues. The observation of higher molecular weights and higher polymerization activities at the same time is an indication that the higher activities by comparison to the mononuclear complexes is rather due to an intrinsically higher rate of chain growth, than to a more efficient activation of the catalyst precursors. This concept affects the polymer microstructures, the degree of branching in the formed polymers increases with the polymerization temperature and depends to some extent also on the natures of catalyst precursors.

Very few examples are available in the literature which explore atom or bond bridged binuclear organometallic compounds active for olefin polymerizations. A novel class of dichloro bridged *β*-diketiminato chromium complexes was prepared by MacAdams L. A. *et al.* to serve as homogeneous model system for the heterogeneous Phillips ethylene polymerization catalyst [36]. Dichloro bridged alkyl bichromium(III) catalyst is formed by the addition of dimethyl zinc to [(Ph)nacnacCrCl$_2$(THF)$_2$] (where nacnac is acetylacetonate) in THF. The catalyst lies on a crystallographic inversion center and adopts the geometry of an edge sharing bioctahedron, emphasizing the preference of Cr(III) for octahedral coordination.

When activated with MAO, the dinuclear alkyl catalyzed the polymerization of ethylene. The rate of polymerizations and microstructure of the produced polymer is somewhat same as bichromium coordination catalyst. It did not react with propylene or 1-hexene, even when activated with MAO, nor did it catalyze the copolymerization of 1-hexene or propylene with ethylene.

Trichloro and germyl bridged two binuclear tungsten carbonyl catalysts have been reported by Gorski M. *et al.* [37] which serves as initiators for ring-opening metathesis polymerization of norbornene. Trichloro substituted ditungsten compound reacts rapidly at room temperature with phenylacetylene to give red poly(phenylacetylene) in high yield. In reaction of NBE in benzene solution at 70 °C in the presence of ditungsten catalyst, the yield of ROMP polymer (poly-1,3-cyclopentylenevinylene, poly-NBE) reached *ca.* 50% (by mass), but the rest of NBE transforms to 2,2'- binorbornylidene (bi-(NBE)), and to the hydroarylation product (2-phenylnorbornane) in a 1:6 molar ratio, respectively. The simultaneous formation of four stereo-isomers (anti–cis, syn–cis, anti–trans, and syn–trans) of bi-(NBE) is indicated by the ^{1}H NMR spectroscopic technique. More or less similar results were observed when germyl bridged catalyst is subjected for the norbornene polymerization under the identical reaction conditions.

3.2.2. Metallocene Based Binuclear Core

Metallocene compounds are found to be the most active compounds for the olefin polymerizations with one and two metal ions. Since catalytic activity of the dinuclear complexes (Figure 5) varies depending on the structure and length of the bridging ligand, systematic studies of the polymerization using the catalysts would provide useful information on the cooperative effect of the two metal centers in the catalytic reaction. Mononuclear Zr complexes $CpZrCl_2\{\eta^5\text{-}C_5H_4(CH_2)_nCH{=}CH_2\}$ (n = 1, 2, 3) undergo intermolecular metathesis of the vinyl group catalyzed by a Ru complex to produce dinuclear metallocenyl complexes with allyl spacer bridging modules. Further, hydrogenation of the products catalyzed by Pd/C affords complexes with a flexible polymethylene chain that bridges two Cp_2ZrCl_2 groups suitable for olefin polymerizations [38]. The dinuclear complexes catalyze polymerization of ethylene and propylene in the presence of MAO. The activity of the Zr/Zr dinuclear complexes for ethylene polymerization is higher than that of the mononuclear precursors. The length and flexibility of the bridging spacer group of the bis-cyclopentadienyl ligand also influence the catalytic activity. Marks *et al.* designed the ligand for the new dinuclear Ti and Zr complexes, which has two active centers with a structure similar to that of the constrained geometry catalysts (CGC) [39]. They discussed unique catalytic properties of the dinuclear complexes in the presence of monovalent and divalent borates as the cocatalyst. Polyethylene obtained from the reaction catalyzed by the dinuclear CGC-Zr complex and MAO has a molecular weight that is 600 times higher than that produced by the mononuclear catalyst.

Recently, cross metathesis reactions of terminal olefins and α-carbonyl olefins were achieved by using the Ru carbene complexes as the catalyst [40]. This reaction is applied to the preparation of a Zr/Fe heterodinuclear metallocene complex whose metal centers are bridged by a biscyclopentadienyl ligand which acts as aromatic spacer between two different metals. The dinuclear Zr/Zr and Zr/Fe complexes catalyze the polymerization of ethylene in the presence of MAO. Dinuclear complexes exhibit higher catalytic activity than that of their individual mononuclear matching part. However, the catalytic activity of the homo- or hetero dinuclear metallocene complexes decreases with increasing length of the spacer bridging chain present between the metal ions. On the other hand, The Zr/Zr dinuclear complexes with bridged biscyclopentadienyl ligands were reported to show low catalytic activity for ethylene polymerization, which was ascribed to steric hindrance around the metal center caused by close contact of the two metallocene groups.

Substituents on the Cp ligand of the metallocenes lead to diverse effects on the activity of ethylene polymerization. Mononuclear metallocenes with bulky substituents show low catalytic activity for the reaction. This is ascribed to steric hindrance, which suppresses smooth coordination and insertion of the monomer. The Zr/Zr dinuclear complexes having a bulky ligand group also showed low catalytic activity due to the same reason. A positive cooperative effect of the two Zr centers probably serves to enhance the ethylene polymerization. Major active species of the polymerization catalyzed by the dinuclear complexes should be the dicationic complexes. Since the two bulky anions formed from MAO have steric or electrostatic repulsion with each other, one of the two anions is separated effectively from the cationic dizirconium complex. Chen E. Y-. X *et al.* [41] reported an interaction of the Zr-containing cation and the anion formed from boron compounds or MAO and the relationship between the interaction and catalytic behavior in olefin polymerization. Some of the molecular weight, molecular weight distribution, and activity data appear to be inconsistent with the general relationship among molecular weight, polymerization activity, and rate of chain transfer. This may be ascribed to different numbers of active species depending on the kind of Zr complexes.

With MAO as the cocatalyst, ethylene bridged indinyl Zr catalysts enchains 3.5 times more, and methylene bridged catalysts are 4.2 times more, 1-hexene than mononuclear Zr catalyst does. When chlorobenzene is used as the polymerization medium, thereby weakening the catalyst-cocatalyst ion pairing, substantial alterations in catalyst response and polymer product properties are observed. Both homopolymerization and copolymerization results argue that achievable Zr---Zr spatial proximity, as modulated by the ion pairing, significantly influences chain transfer rates and selectivity for co-monomer enchainment and that such proximity effects are highly cocatalyst and solvent sensitive.

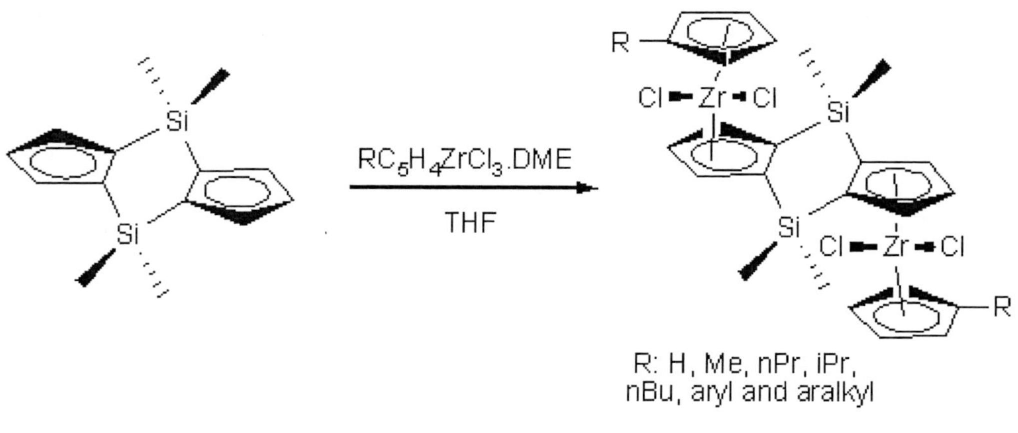

Figure 6. The synthetic route for the preparation of double silylene-bridged binuclear Zr(IV) complexes.

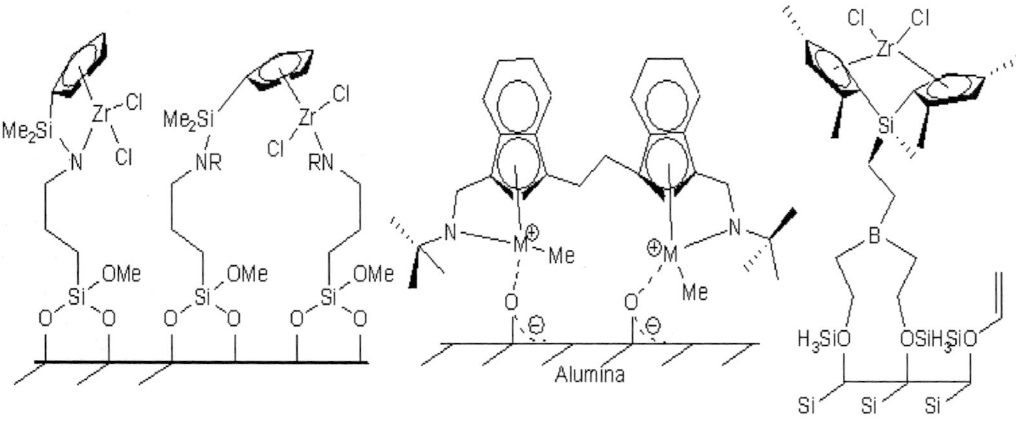

Figure 7. Supported binuclear compounds.

On the other hand, new double silylene-bridged binuclear zirconium complexes for the olefin polyrerizations and copolymerization with 1-hexene activities in the presence of MAO have been reported by Xu S. *et al.* [42] very recently. It is found that the substituent of different nature on the Cp ring plays an important role in the catalytic performance of the series of complexes. Frame of the double Me2Si-bridged cyclopentadienyle ligands improves the stability of the binuclear zirconium complexes to a certain degree, which may be the factor leading to the high catalytic activities of complexes even at very low molar ratio of Al/Zr. Preparation of these sort of compounds is depicted in Figure 6.

3.3. Supported Binuclear Core

Very few reports are available for the supported mononuclear catalyst meant for olefin polymerizations in the literature. However, countable number of reports is on binuclear supported catalysts (Figure 7); hence this field of research needs some more insight by the budding researchers for promising activities which can be helpful in heterogeneous catalysis. Supported catalysts give rise to heterogeneous catalysts which are quite helpful and efficient in the bulk preparations of polymers in industry. However, the catalysts developed by the researchers are often converted into heterogeneous catalysts by drawing suitable changes in the catalysts for the industrial productions of the polymers. Whereas, these supported catalyst directly offer the heterogeneous catalysts, which comprises the further conversion steps.

In recent times, a new generation of magnesium chloride-supported catalysts for propene polymerization has been developed in which a diether especially, 2,2-dialkyl- 1,3-dirnethoxypropane rather than an ester is used as the internal donor in the solid catalyst. Such catalysts are more active than the previously-developed catalysts and are stereospecific even in the absence of an external donor. This has been ascribed to the fact that such diethers, in contrast to the esters, are not extracted from the catalyst by the trialkylaluminium cocatalyst. Addition of Lewis bases (LB) during the preparation (internal LB) or the activation with aluminium alkyls (external LB) of supported, high activity Ziegler-Natta catalysts has been shown to have a great effect on polymerization performance [43]. In the case of 1-alkenes, propylene in particular, LB has been widely used to increase the isotactic stereospecificity of the catalyst, with excellent results. Long back Baraxxoni L. *et al.* [44] have reported a series of bimetallic catalysts obtained by co-supporting hafnium and titanium on $MgCl_2$, were used in ethylene homopolymerization after activation with trialkylaluminum to which variable amounts of different Lewis bases (external bases) were added. The bases used contain oxygen [dibutyl ether (DBE), butyl acetate (BAc) and dibutyl phthalate (DBP)], nitrogen [tetramethylpiperidine (TMP)] and silicon plus oxygen [tetramethoxysilane (TES) and phenyltriethoxysilane (PES)] as heteroatoms with the aim of gaining information on active site type and distribution, as well as on the possibility of modifying the properties of the polyethylene formed. Variable amounts of LB were used with a fixed amount of transition metals (Ti + Hf) and the effect observed at each concentration depends strikingly on the chemical structure of the LB. In all cases GPC curves show a narrowing of the MWD and the disappearance of the higher MW fraction, which is clearly indicated by the lack of the high MW side that is present in the profiles of polyethylene obtained in the absence of LB. Lewis bases seem to be capable of selectively poisoning this last type of site, thus hindering the formation of the highest molecular weight fractions and therefore producing a narrowing of MWD by an opposite mechanism.

Suzuki N. *et al.* [45] have reported the preparation of isospecific metallocene catalysts for olefin polymerization that are covalently tethered on solid surface. They have postulated a novel methodology for preparation of covalently immobilized isospecific metallocene catalysts. Metallocene complexes were tethered on the support surface by siloxane formation. The catalyst performance was improved by masking excess hydroxy groups on each solid surface. Tethering complexes by hydroboration gave the most active catalysts and they produced highly isotactic polypropene. The excellent catalytic performance of the hydroboranated compound could be ascribed to the masking of hydroxyl groups and a longer spacer moiety present between silicon and zirconium.

Williams and Marks reported [46] the utilization of ^{13}C CPMAS NMR spectroscopy for mono- and binuclear CGC-type organo-zirconium complexes that electrophilic, cation-like organo-group 4 species are formed by two chemisorptive pathways on highly acidic sulfated alumina. Polymerization-active cationic species are formed in both the pathways. Along with this they have presented the first example of binuclear CGC-type Ti and Zr complexes supported on a metal oxide. Polymerization studies of these CGC-type group 4 complexes supported on alumina support show moderate activities for ethylene homopolymerization and ethylene/1-hexene copolymerizations, with negligible catalyst nuclearity effects. However, supported Ti catalyst displayed an order of magnitude higher polymerization activity than any of the Zr catalysts.

4. Concluding Remarks

A number of novel binucleating catalysts for olefin polymerization in particular, ethylene polymerization have been disclosed. The research for new catalysis involves the design of new binucleating ligands, nature of metals that has not been used before to produce a combination of two active catalysts for the polymerizations. The present chapter puts focus on the important achievements in controlling the structure of the polymers have been made by the binucleating metal core, such as living stereospecific polymerization of ethylene, propylene and a-olefins. A forth-coming important target would be the preparation of binuclear heterogeneous catalyst on silica or alumina support in order to increase the rigidity, specificity, selectivity and reactivity.

ACKNOWLEDGMENTS

This work was supported by grants-in-aid for the *World Class University Program* (No. R32-2008-000-10174-0) and the *Fundamental R&D Program for Core Technology of Materials* (Ministry of Knowledge Economy, Republic of Korea).

REFERENCES

[1] Takeuchi, D. *Dalton Trans.,* 2010, 39, 311–328.

[2] Tanabiki, M.; Tsuchiya, K.; Motoyama, Y.; Nagashima, H. *Chem. Commun.,* 2005, 3409–3411.

[3] Zhang, D.; Jin, G. -X. Inorg. *Chem. Commun.* 2006, 9, 1322–1325.

[4] Mitani, M.; Oouchi, K.; Hayakawa, M.; Yamada, T.; Mukaiyama, T. *Polymer Bull.* 1995, 35, 677-662

[5] Arriola, D. J.; Carnahan, E. M.; Hustad, P. D.; Kuhlman, R. L.; Wenzel, T. T. *Science,* 2006, 312, 714-719.

[6] Kuhlman, R. L.; Wenzel, T. T. *Macromolecules,* 2008, 41, 4090-4097.

[7] Masi, F.; Malquori, S.; Barauoni, L.; Ferrero, C.; Moalli, A.; Menconi, F.; Invernizzi, R.; Zandona, N.; Altomare, A.; Ciardelli. F. *Makromol. Chem. Suppl.* 1989, 15, 147-165.

[8] Kunrath, F. A.; De-Souza, R. F.; Casagrande, O. L.; Brooks, N. R.; Young, V. G. *Organometallics* 2003, 22, 4739-4743.

[9] (a) Murtuza, S.; Casagrande, O. L., Jr.; Jordan, R. F. Organometallics 2002, 21, 1882-1890. (b) Gil, M. P.; Dos-Santos, J. H. A.; Casagrande, O. L., Jr. Macromol. Chem. Phys. 2001, 202, 319-324. (c) Furlan, L. G.; Gil, M. P.; Casagrande, O. L., *Jr. Macromol. Rapid Commun.* 2000, 21, 1054-1057.

[10] Corker, J. M.; Evans, J. J. *Chem. Soc., Chem. Commun.* 1991, 1104-1106.

[11] Champouret, Y. D. M.; Fawcett, J.; Nodes, W. J.; Singh, K.; Solan, G. A. *Inorg. Chem.* 2006, 45, 9890-9900.

[12] Ittel, S. D.; Johnson, L. K.; Brookhart, M. *Chem. Rev.* 2000, 100, 1169-1204.

[13] Khamker, Q.; Champouret, Y D. M.; Singh, K.; Solan. G. A. *Dalton Trans.,* 2009, 8935–8944.

[14] Armitage, A. P.; Champouret, Y. D. M.; Grigoli, H.; Pelletier, J. D. A.; Singh K.; Solan, G. A. *Eur. J. Inorg. Chem.,* 2008, 4597-4603.

[15] Mecking, S. *Angew. Chem., Int. Ed.* 2001, 40, 534-540.

[16] De-Pooter, M.; Smith, P. B.; Dohrer, K. K.; Bennett, K. F.; Meadows, M. D.; Smith, C. G.; Schouwenaars, H. P.; Geerards, R. A. *J. Appl. Polym. Sci.,* 1991, 42, 399-404.

[17] Rhodes, B.; Chien, J. C. W.; Wood, J. S.; Chandrasekaran, A.; Rausch, M. D. *J. Organomet. Chem. 2001*, 625, 95–100.

[18] (a) Gibson, V. C.; Maddox, P. J.; Newton, C.; Redshaw, C.; Solan G. A.; Andrew J.; White, P.; Williams, D. J. Chem. Commun., 1998, 1651-1652. (b) Gibson, V. C.; Newton, C.; Redshaw, C.; Solan, G.A.; White, A. J. P.; Williams, D. J. *Eur. J. Inorg. Chem.* 2001 1895-1903.

[19] Coles, M. P.; Dalby, C. I.; Gibson, V. C.; Clegg, W.; Elsegood M. R. J. *J. Chem. Soc., Chem. Commun.* 1995, 1709-1711.

[20] Salata M. R.; Marks T. J. *J. Am. Chem. Soc.* 2008, 130, 12-13.

[21] James, D. E. In *Encyclopedia of Polymer Science and Engineering*; Marks, H. F.; Bikales, N. M.; Overberger, C. G.; Menges, G. Eds.; Wiley-Interscience: New York, 1985; 6, 429-454.

[22] Li, L.; Metz, M. V.; Li, H.; Chen, M. -C.; Marks, T. J.; Liable-Sands, L.; Rheingold, A. L. *J. Am. Chem. Soc.*2002, 124, 12725-12741.

[23] Salata, M. R.; Marks T. J. *Macromolecules* 2009, 42, 1920-1933.

[24] Owiny, D.; Parkin, S.; Ladipo, F. T. *J. Organomet. Chem.* 2003, 678, 134–141.

[25] Goyal, M.; Mishra, S.; Singh, A. *Synth. React. Inorg. Met.-Org. Chem.* 2001, 31, 1705–1715.

[26] Hu, T.; Li, Y.; Li, Y. -X; Hu, N. -H. *J. Mol. Catal. A: Chem.* 2006, 253, 155–164.

[27] Lozan, V.; Lassahn, P. -G.; Zhang, C.; Wu, B.; Janiak, C.; Rheinwald, G.; Lang, H. Z. *Naturforsch.* B 2003, 58, 1152-1157.

[28] Huang, Y. –B.; Tang, G. –R.; Jin, G. –Y.; Jin, G. -X. *Organometallics* 2008, 27, 259–269.

[29] Lee, S. S.; Joe, D. J.; Na, S. J.; Park, Y. W.; Choi, C. H.; Lee, B. Y. *Macromolecules* 2005, 38, 10027–10033.

[30] Bahuleyan, B. K.; Lee, U.; Ha, C. –S.; Kim, I. *Appl. Catal. A: General* 2008, 351, 36–44.

[31] Jie, S.; Zhang, D.; Zhang, T.; Sun, W. –H.; Chen, J.; Ren, Q.; Liu, D.; Zheng, G.; Chen, W. *J. Organomet. Chem.* 2005, 690, 1739–1749.

[32] (a) Rodriguez, B. A.; Delferro, M.; Marks, T. J. Organometallics 2008, 27, 2166–2168. (b) Li, H.; Li, L.; Schwartz, D. J.; Stern, C. L.; Marks, T. J. J. Am. Chem. Soc. 2005, 127, 14756–14768. (c) Guo, N.; Li, L.; Marks, T. J. *J. Am. Chem. Soc.* 2004, 126, 6542–6543.

[33] Baar, C. R.; Jennings, M. C.; Puddephatt, R. *J. Organometallics* 2001, 20, 3459-3465.

[34] Wehrmann, P.; Mecking, S. *Organometallics* 2008, 27, 1399–1408.

[35] (a) Zhang, D.; Jin, G.-X. Inorg. Chem. Commun. 2006, 9, 1322–1325. (b) Chen, Q.; Yu, J.; Huang, J. *Organometallics* 2007, 26, 617–625.

[36] MacAdams, L. A.; Kim, W. -K.; Liable-Sands, L. M.; Guzei, I. A.; Rheingold, A. L.; Theopold, K. H. *Organometallics* 2002, 21, 952-960.

[37] Gorski, M.; Kochel, A.; Szyman´ska-Buzar, T. *J. Organomet. Chem.* 2006, 691, 3708–3714.

[38] Kuwabara, J.; Takeuchi, D.; Osakada, K. *Organometallics* 2005, 24, 2705-2712.

[39] (a) Li, L.; Metz, M. V.; Li, H.; Chen, M.-C.; Marks, T. J.; Liable-Sands, L.; Rheingold, A. L. J. Am. Chem. Soc. 2002, 124, 12725-12741. (b) Li, H.; Li, L.; Marks, T. J.; Liable-Sands, L.; Rheingold, A. L. *J. Am.Chem. Soc.* 2003, 125, 10788-10789.

[40] Guo, N.; Li, L.; Marks, T. J. J. *Am. Chem. Soc.* 2004, 126, 6542-6543.

[41] Chen, E. Y.-X.; Marks, T. *J. Chem. Rev.* 2000, 100, 1391-1434.

[42] Xu, S.; Jia, J.; Huang, J. *J. Polym. Sci. : Part A: Polymer Chem.* 2007, 45, 4901–4913.

[43] German Patent 2,347,577, 1974, Montedison; *Chem. Abstr.*, 1975, 83, 59670a.
[44] Baraxxoni, L.; Menconi, F.; Ferrero, C.; Moalli, A.; Masi, F. *J. Mol. Catal.* 1993, 82, 17-27.
[45] Suzuki, N.; Yu, J.; Shioda, N.; Asami, H.; Nakamura, T.; Huhn, T.; Fukuoka, A.; Ichikawa, M.; Saburi, M.; Wakatsuki, Y. *Appl. Catal. A: General* 2002, 224, 63–75.
[46] Williams, L. A.; Marks, T. J. *Organometallics* 2009, 28, 2053–2061.

In: Focus on Catalysis Research: New Developments
Editors: Minjae Ghang and Bjørn Ramel
ISBN: 978-1-62100-455-4

Chapter 8

PREPARATION AND CHARACTERIZATION OF COATED MICROCHANNELS FOR THE SELECTIVE CATALYTIC REDUCTION OF NOx

José R. Hernández Carucci[1,*], Jhosmar Sosa[1,2], Kalle Arve[1], Hannu Karhu[3], Jyri-Pekka Mikkola[1,4], Kari Eränen[1], Tapio Salmi[1], Dmitry Yu. Murzin[1,*]

[1]Laboratory of Industrial Chemistry, Åbo Akademi University, Biskopsgatan 8, Turku/Åbo, 20500, Finland

[2]Departamento de Termodinámica y Fenómenos de Transferencia, Universidad Simón Bolívar, Sartenejas, Apartado Postal 89000, Caracas 1086-A, Venezuela

[3]Department of Physics and Astronomy, University of Turku, Vesilinnantie 5, FIN-20014, Turku, Finland

[4]Faculty of Science and Technology, Umeå University, 90187, Umeå, Sweden

ABSTRACT

Shallow microchannels (Ø= 460 μm) were successfully coated with different catalytically active phases, e.g., Cu-ZSM-5, Cu/(ZSM-5+Al_2O_3), Au/Al_2O_3, Ag/(Al_2O_3+Ionic liquid) and Ag/Al_2O_3, and tested on the hydrocarbon-assisted selective catalytic reduction of NOx (HC-SCR) with different model bio-derived fuels, i.e., methyl- and ethyl laurate produced by transesterification and hexadecane, a paraffinic component that can be produced by decarboxylation and/or decarbonylation of natural oils and fats. Characterization of the washcoats was done by means of X-ray photoelectron spectroscopy (XPS), scanning electron microscopy with energy dispersive X-ray analysis (SEM-EDXA) and laser ablation inductively coupled plasma mass spectrometry (LA-ICP-MS), showing a dependence of the metal loading with the impregnation time and the precursor concentration. The Ag/Al_2O_3 catalysts exhibited, in general, the highest activities towards the NOx reduction. Optima in impregnation time

* Corresponding authors: johernan@abo.fi, dmurzin@abo.fi

and concentration of $AgNO_3$ solution displaying the highest activity in HC-SCR among the prepared Ag/Al_2O_3 washcoats were established. A combination of Cu-ZSM-5 or Cu/(ZSM-5+Al_2O_3) and the optimum Ag/Al_2O_3 catalyst were tested in order to improve the low temperature reduction in SCR with hexadecane as a reducing agent. The enhancement of the activity at low temperatures (< 350 °C) was up to seven-fold compared to the case when only Ag/alumina was used. The effect of the hydrocarbon concentration (hexadecane) and the presence of water in the feed were also investigated.

The reduction results over silver/alumina with bio-derived fuels were compared to those obtained when using fossil-derived fuels, e.g., octane. Octane offered better NOx reduction efficiency compared to hexadecane at temperatures higher than 350 °C. However, for temperatures lower than 350 °C, in which most of the diesel engines operate, the reduction activity of hexadecane was higher compared to octane. The obtained results in the microreactor were compared to those obtained by a conventional Ag/Al_2O_3 catalyst powder in a minireactor (Ø= 10 mm). Similar NOx reduction results were attained when working under similar conditions in both reactor systems, i.e., the same ratio between the total flow rate and the catalyst mass.

Keywords: Microreactor, coating, HC-SCR, Ag/Al_2O_3.

1. INTRODUCTION

According to the Kyoto protocol and new legislations regarding hazardous emissions from light and heavy-duty vehicles, there is a need for the improvement of existing exhaust treatment technologies as well as the development of the new ones. Exhaust emission limits have been set for carbon monoxide, hydrocarbons, nitrogen oxides and other pollutants. Lean burn engines running with gasoline or diesel fuel, operate in a large excess of oxygen, resulting in a decrease of fuel consumption and therefore, less net-CO_2 emissions to the environment. Nonetheless, the strong oxidizing conditions due to O_2 excess tend to increase the production of nitrogen oxides, rendering the traditional approach of a three-way catalyst (TWC) not efficient anymore. Continuous hydrocarbon selective catalytic reduction (HC-SCR) has been found to be a reliable and effective method for the conversion of nitrogen oxides to molecular nitrogen [[1]-[4]].

On the other hand, one of the major trends in developing new technologies is associated with reducing sizes and optimizing the used space as much as possible. Various microdevices have been developed and tested leading to promising results. The theoretical advantages of microreactors over conventional vessels were confirmed by experimental practice, resulting in vast improvements in energy efficiency, faster reactions, better yields, safety, reliability and a larger degree of process control [[5]-[7]].

Still, one of the biggest challenges for microreactor technology is the introduction of catalytically active phases, which display at least the same or preferably even better performance than conventional catalysts.

Ag/alumina catalyst has been found to be one of the best catalysts for the HC-SCR [[1],[8]]. However, the high activity is mainly connected to elevated temperatures (> 300 °C). The need of high temperature represents a severe drawback for the system to be used in combination with modern diesel engines having exhaust gas temperature of 250 °C or less for most of the time during the NEDC test [[9]]. Copper-exchanged zeolites have also been studied for the HC-SCR, giving promising results at low temperatures [[10],[11]]. Some efforts have even been devoted to gold catalyst [[12],[13]] and bimetallic ones, i.e., CeO_2/Cu-ZSM-5 [[14],[15]]. The coating of zeolites in microchannels [[16]-[18]], as well as the coating with aluminium oxides [[19],[20]] was addressed in the literature.

Nonetheless, the studies remain somehow scarce, and the catalytic activity of the washcoatings for relevant chemical systems is still to be investigated, especially regarding the dependence of the washcoat performance and structure with the preparation parameters.

The present work investigates different preparation methods of several metal supported catalysts that have been previously found feasible for the HC-SCR of NOx. Lately, a lot of attention has been put in supported ionic liquid catalyst (SILCA). We even investigated the possibility of using SILCA for our test reaction. The choice of the ionic liquid depends on its thermal stability, viscosity and capacity for dissolving the silver precursor. In this work, 1-butyl-3-methyl-imidazolium trifluoromethanesulfonate was used.

The present investigation combines the benefits of decreasing the size of the equipment, in order to reduce nitrous oxides emissions using well-studied petrofuels, e.g., octane, iso-octane, and comparing their behaviour with newer and less pollutant bio-derived fuels, e.g., methyl- and ethyl laurate, and hexadecane. The results obtained in the microchannels (Ø= 460 μm) are compared with traditional laboratory vessels (minireactors, Ø= 10 mm) [[1],[2]].

2. Experimental

2.1. Catalyst Preparation

2.1.1. Cleaning of the Plates

Stainless steel plates, each with nine parallel shallow microchannels (Ø = 460 μm length 9.5 mm; and depth = 75 μm) were purchased from the Institut für Mikrotechnik Mainz GmbH (IMM). Before the deposition of the washcoat, the plates were cleaned from impurities by using the method described by Howell and Hatalis [[21]]. The plates were immersed in acetone for 5 minutes and rinsed with deionised water. They were further immersed in a solution of 5:1:1 deionised water, hydrogen peroxide (H_2O_2, 29-32%, Merck), ammonium hydroxide (NH_4OH, 32%, Sigma-Aldrich) for 5 minutes and rinsed with deionised water. Finally, the plates were immersed in a solution of 5:1:1:1 deionised water, hydrogen peroxide, phosphoric acid (H_3PO_4, 85%, FF-Chemicals) and acetic acid (CH_3COOH, 99% glacial, J.T. Baker) in an ultrasonic bath for 5 minutes and further rinsed with deionised water.

2.1.2. Washcoating/Impregnation of the Plates with Ag/Al_2O_3

In order to use microstructures for catalytic testing there is a need to introduce an active catalytic phase. Common coating techniques include anodic oxidation, sol-gel, chemical vapour deposition and washcoating [[19]]. The latter has been used in this contribution. Two main methods of preparation were investigated for coating the plates on the catalytic zone (Figure 1). In the first one, alumina-washcoated plates at IMM (specific surface area 70 m^2/g, thickness 10-20 μm, 0.7 mg of γ-alumina per plate) were further impregnated by silver with silver nitrate solutions (J.T. Baker) at our laboratory studying the effect of the silver nitrate concentration and impregnation time on the metal loading. 0.011, 0.022 and 0.04 M solutions were used and the plates were left to impregnate at 12, 24 and 48 h at room temperature and mild stirring. The impregnation times and solution concentrations were chosen based on our previous experience preparing Ag/Al_2O_3 catalysts for exhaust cleaning [[1]]. For the sake of comparison, a second preparation method was investigated. The stainless steel plates purchased from IMM were washcoated with alumina preparing a suspension as described in reference [[19]] and thereafter impregnated by silver with silver nitrate solutions. In this method the channels are covered with a suspension powder, a binder, acid, and water as the main components. In particular, 20 g of ground γ-alumina (A-201, La Roche) were suspended in 75 g of deionised water, 5 g polyvinyl alcohol and 1 g of acetic acid. The suspension was placed on the microplates. The plates were dried during 2 h at 100 °C and further calcined for 3 h at 550 °C in air. More details of the method are found in [[19]]. The alumina-coated plates were further impregnated by silver with silver nitrate solutions (J.T. Baker) as described above.

After the silver impregnation, all the plates were dried during 2 h at 100 °C and further calcined for 3 h at 550 °C in air. The effect of the impregnation time and precursor concentration on the catalytic activity was studied.

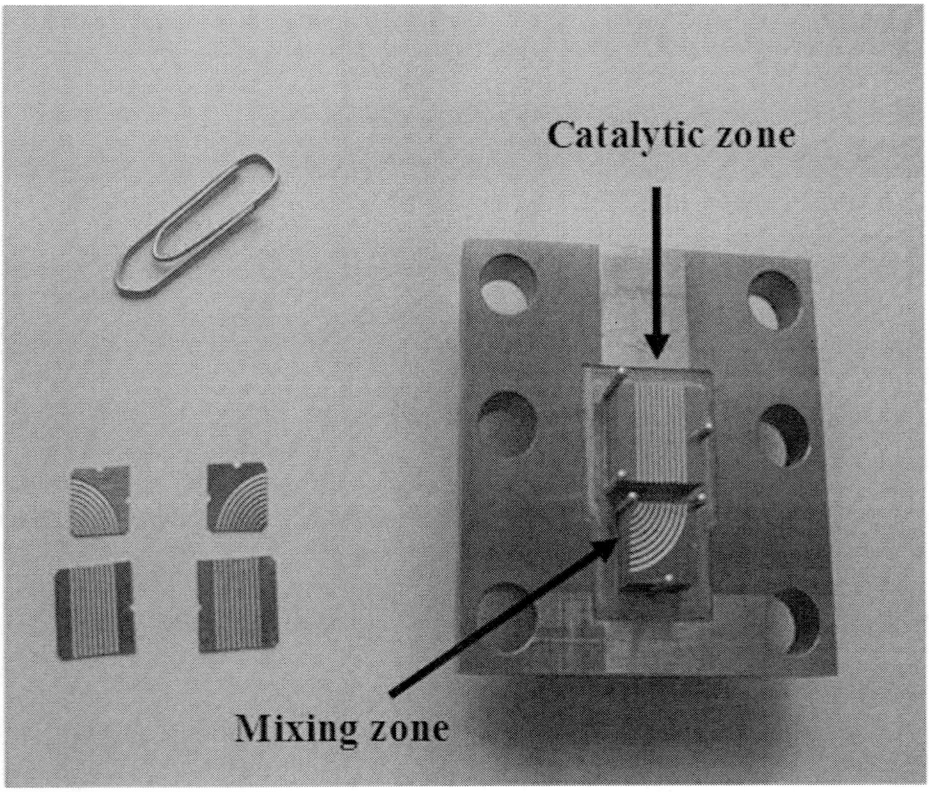

Figure 1. Microreactor chamber (mixing and catalytic zones).

2.1.3. Coating with Cu/(ZSM-5+Al_2O_3) and Cu-ZSM-5

The stainless steel pristine microplates were also coated with Cu/(ZSM-5+Al_2O_3) and Cu-ZSM-5 as follows. In the first method, a mixture containing a ground commercial zeolite powder Na-ZSM-5 (Degussa, Si/Al= 22.6) and ground alumina A-201, La Roche (ratio 1:3) was mixed together and stirred in a 0.9 wt.% aqueous solution of nitric acid at 50 °C. After homogenization, the plates were introduced in the suspension and were left to washcoat for 4 h with mild stirring. The plates were further dried at 100 °C for 1 hour and calcined in air at 550 °C for three hours. The metal (Cu) was introduced by ion exchange using a copper acetate solution (0.05 M) during 24 h with mild stirring. The plates were again dried at 100 °C for 2 h and further calcined at 550 °C for 2 h. The second procedure was done following the guidelines from reference [[16]]. This method consisted in placing 20 g of zeolite (Degussa, Si/Al= 40) in a 0.05 M copper acetate solution during 24 h at room temperature with mild stirring. The catalyst was filtered, washed with distilled water, dried at 110 °C and calcined at 400 °C. 20 g of the dry catalyst were mixed with 75 g of deionized water, 5 g of polyvinyl alcohol (PVA) and 1 g of acetic acid. Firstly, the binder (PVA) was dissolved in water under mild stirring at 60 °C for 2 h and the resulting solution was kept overnight at the same temperature. The zeolite catalyst and acetic acid were added successively, without stirring. The resulting mixture was stirred for 2 h at 60 °C and left successively for five days without heating. The suspension was then drawn with a syringe and used to fill the microchannels (Figure 2), dried during 2 h at 100 °C and further calcined for 3 h at 550 °C in air. The samples are identified as Zeo1 (first preparation procedure) and Zeo2 (second preparation procedure).

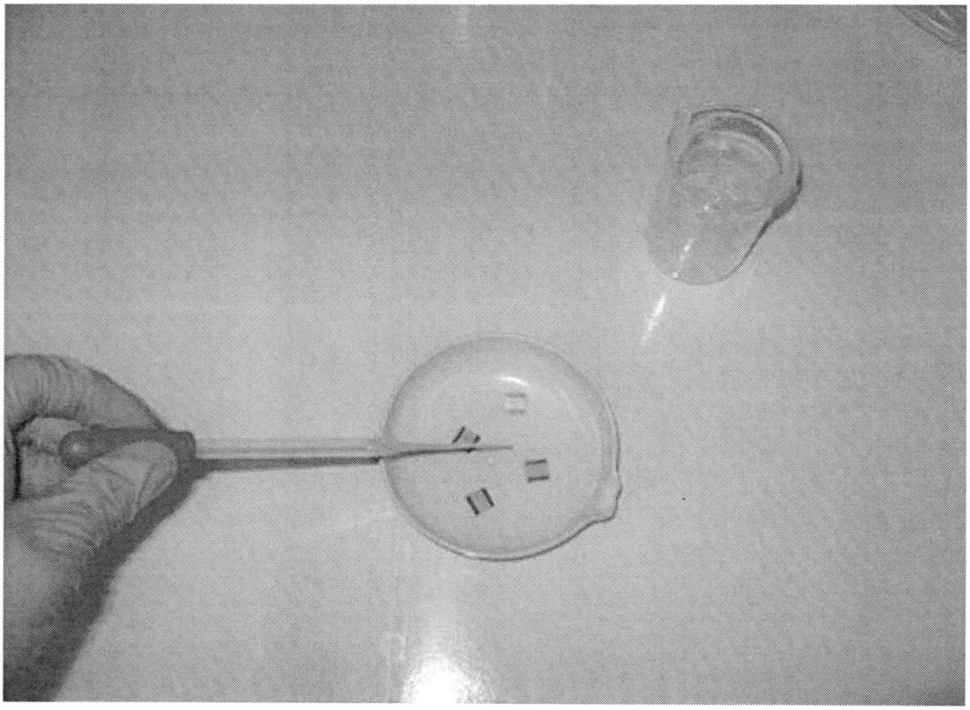

Figure 2. Preparation of the Cu-ZSM-5 coated plates.

2.1.4. Au Deposition

Alumina-coated plates were submerged in one litre of water, tetrachloroaurate and 0.021 M urea solution, elevating the temperature slowly to 81 °C. The temperature was kept overnight, with mild mechanical stirring. Gold nanoparticles were formed on the alumina due to the decomposition of urea. The plates were removed and washed to neutrality. They were further dried overnight at room temperature, followed by calcination at 200 °C for 2 h.

The initial complexes formed on alumina consisted of ammonia-gold species, which upon calcination transformed to Au. Colourless solution indicates full decomposition, since even small amount of $HAuCl_4$ in water gives a yellowish colour.

2.1.5. Ag Supported on Alumina and Ionic Liquids (IL)

Alumina-coated plates were coated with an ionic liquid-silver layer using 1-butyl-3-methyl-imidazolium trifluoromethanesulfonate (Merck, high purity). 1 g of ionic liquid was mixed with 8 g of dry acetone. 0.01 g of silver acetylacetonate (Aldrich, 98%) was successively added to the mixture and was placed in an ultrasound bath for 2 min giving a "milky" suspension.

The suspension was further filtered through a 0.1 μm mesh to capture larger particles. The suspension was deposited in the microplates with a syringe (two droplets per plate) and the plates were exposed to vacuum in a rotary evaporator at approximately 4 mbar at 80 °C for 2 h to vaporize acetone and other volatiles.

2.1.6. Powder Ag/Al_2O_3 Catalyst

For comparison purposes, a conventional silver/alumina catalyst was used in the minireactor tests. 1.91 wt.% Ag/alumina catalyst was prepared by impregnation of a commercial alumina support (A-201 La Roche) with a 0.022 M silver nitrate solution (J.T. Baker) followed by drying and calcination. The crushed and sieved Ag/alumina (particle size 250-500 μm) catalyst was tested in a fixed bed quartz reactor inserted in an oven equipped with a temperature controller. The preparation of the catalyst is described in detail in [[1]].

2.2. Catalyst Characterization

2.2.1. Laser Ablation Inductively Coupled Plasma Mass Spectrometry (LA-ICP-MS)

The characterization of the Ag/alumina catalyst through LA-ICP-MS was carried out by a Merchantek UP-213 (New Wave Research. Inc, USA) LA system coupled to a PerkinElmer SCIEX ICP- Mass Spectrometer (Elam 6100 DRC Plus). The sample surface was irradiated with deep-UV (213 nm) output from a frequency-quintupled Nd: YAG (neodymium doped yttrium aluminium garnet crystal) laser. The laser light couples to the surface of the sample, causing very rapid heating, which, in turn, causes the matrix to be volatilized or ablated. This ablated material is then carried to the ICP-MS in an argon carrier gas stream for analysis. The scan speed used during the analysis was 20 μm/sec, the laser energy output was 50%, and the spot size was 8 μm.

2.2.2. Scanning Electron Microscopy with Energy Dispersive X-ray Analysis (SEM- EDS)

The SEM images were taken with the scanning electron microscope system Leo 1530 Gemini manufactured by Zeiss. The Leo 1530 presents an acceleration voltage of Uacc= 0.2 - 30 kV and a field emission gun (FEG) resolution of 1.0 nm at 20 kV, 2.5 nm at 1 kV, and 5 nm at 0.2 kV. The analyses were carried out by an energy dispersive X-ray detector (EDX, Thermo Noran VANTAG-ESI, 120 kV).

2.2.3. X-ray Photoelectron Spectroscopy (XPS)

A Perkin-Elmer 5400 ESCA spectrometer was used with monochromatized Al K□ radiation (photon energy 1486.6 eV) and a pass energy value of 35 eV. Samples were in contact with ambient air prior to analysis and the analysis was made from the surface of the microplate without any additional chemical or physical treatments. A low energy electron gun (flood gun) was used to stabilize the charging that arises from loss of photoelectrons during X-ray bombardment. To calibrate the binding energy axis accurately, carbon 1s line at 284.6 eV was used [[22]].

2.3. Experimental Setup

The gas-phase microreactor was designed at Åbo Akademi and the parts were purchased from IMM. The device consists of a two-piece cubic chamber with two inlets and one outlet, each with a tube diameter of 710 μm. The lower part of the chamber has two recesses, each filled with a stack of ten microstructured platelets, which are connected by a diffusion tunnel. The first stack contains a total of ten mixing platelets with nine semicircular channels of different radii but with a common centre. They are arranged in the stack in a way that they meet the two inlets in alternation. The second stack consists of ten rectangular platelets with nine parallel shallow microchannels (Ø = 460 μm length 9.5 mm; and depth = 75 μm) coated with Ag/Al_2O_3, $Cu/(Al_2O_3$+ZSM-5), Cu-ZSM-5, Au/Al_2O_3, $Ag/(Al_2O_3$+IL), or a combination between the optimum Ag/Al_2O_3 and Cu-zeolite catalysts. A maximum gas flow of 80 mL/min was permitted. Pressures and temperatures up to 50 bar and 600 °C, respectively, can be achieved. The technical data of the microreactor is presented in Table 1.

The experiments in the microreactors were carried out under atmospheric pressure at 150-550 °C. The catalyst-coated plates were pre-treated with a 6 vol.% O_2/He mixture at 400 °C for 30 minutes. The reaction mixture simulated the exhaust of a diesel engine. In most of the cases, concentrations of 500 ppm NO, 12 vol.% H_2O, 6 vol.% vol. O_2 were used, keeping the C_1/NO ratio between 4 and 6 (e.g., for C_1/NO=6, 375 ppm C_8H_{18} or 187.5 ppm $C_{16}H_{34}$). Helium was used as balance. The total flow rate was set to 50 ml/min. The activity experiments were carried out in the microreactor and the concentrations of the species were determined by a Hewlett Packard GC System 6890 series with TC and FC detectors, an Agilent 3000A MicroGC with four parallel columns and TC detectors and a PPM Systems Chemiluminescent NOx Analyzer model 200AH.

Table 1. Technical data of the gas-phase microreactor with mixing and internal heating

Property	Values
Size (L x W x H, mm)	40 x 40 x 30
Connectors (Inlet/Outlet)	1/4¯
Standard material	Inconell 600 (2.4816) for housing and top plate 1.4571 for mixing and catalyst plates
Number of mixing plates	10
Size of mixing plates (mm)	7.5 x 7.5
Channel geometry of mixing plates (μm)	180 – 490 x 100
Number of catalyst plates	10
Size of catalyst plates (mm)	9.5 x 9.5
Channel geometry of the catalyst plates (μm)	460 x 125

Water and hexadecane were pumped with a syringe pump (CMA/102 Microdialysis). Brooks Mass Flow Controllers 5878 were used to pump the gases. The scheme of the process is presented in Figure 3, while the actual picture of the system is presented in Figure 4.

3. Results and Discussion

3.1. Characterization of the Coated Microplates

The metal uptake by the alumina in the plates was determined by LA-ICP-MS (New Wave Research UP-213 Laser Ablation system with PerkinElmer SCIEX ICP Mass Spectrometer Elam 6100 DRC Plus) measuring the silver content with respect to Al_2O_3. The results of these measurements are presented in Table 2 for different silver nitrate concentrations and impregnation times.

The results obtained by the laser ablation technique suggested that the silver distribution along the channels was uniform and that the concentration variations in different channels in the same plate were negligible. The same silver loadings were achieved for the washcoating/impregnation or only impregnation if the plates were treated under the same conditions (impregnation time and concentration of the $AgNO_3$ solution), thus they are treated indistinctly in this manuscript.

Overall, longer impregnation times resulted in higher metal loadings when the same solution concentration was used. Dependence on solution concentration was less prominent, although higher silver nitrate concentrations were also yielding higher Ag content (on wt.% basis with respect to the alumina support) on the plates (Table 2). The silver content was also investigated by SEM-EDS yielding similar results as presented in Table 2 for the investigated areas. SEM pictures of an uncoated plate are shown in Figure 5. Furthermore, Figure 6 shows a similar picture after the washcoating with alumina and further impregnation with silver. Large spots can be seen at some places in the microchannels. They were further analyzed by XPS and the results are presented below. The washcoats thickness was between 15 and 20 μm (Figure 6, right) with a uniform distribution along the channel radius.

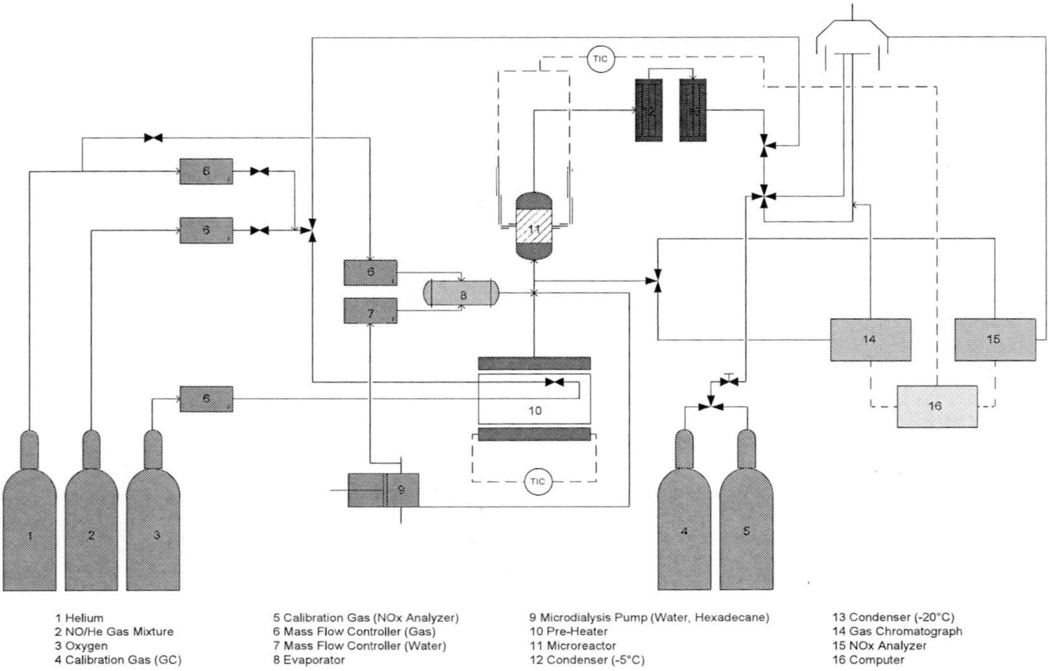

Figure 3. Scheme of the experimental set-up.

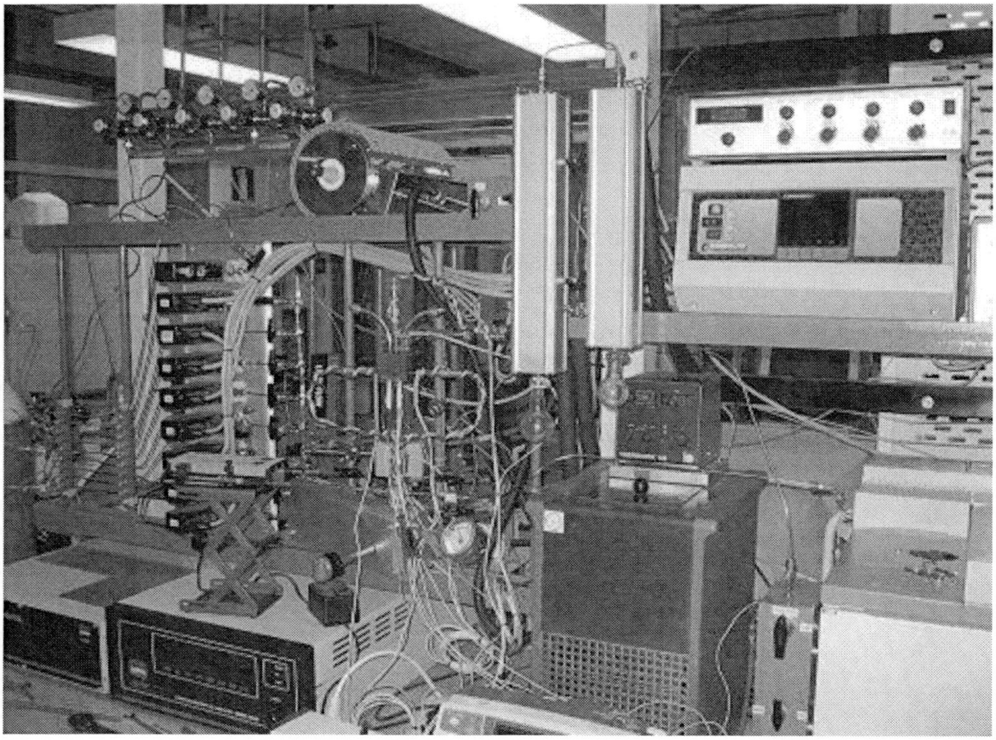

Figure 4. Actual picture of the used system.

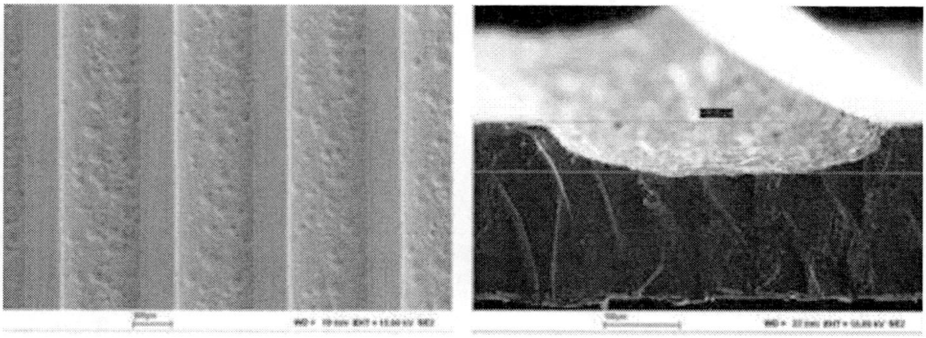

Figure 5. Scanning electron microscope of an uncoated plate (50X, left) and (200X, right).

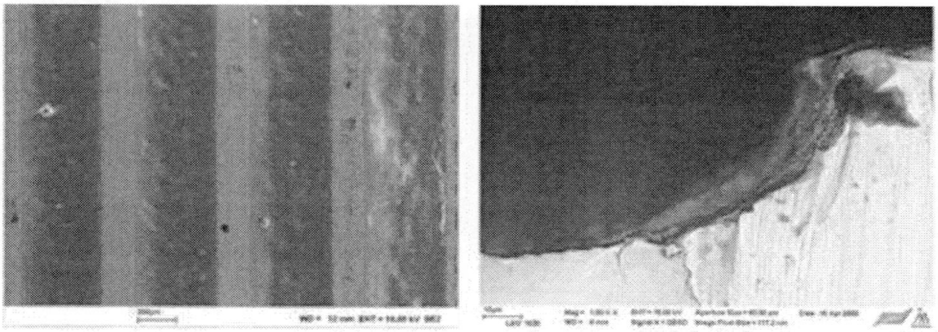

Figure 6. Scanning electron microscope of an Ag/Al_2O_3 coated plate (50X, left) and (1kX, right).

SEM pictures were also taken for the Cu-ZSM-5 coated plates and Figure 7 presents the zeolite structure on the microchannel after the catalytic experiments, indicating that the zeolite structure was preserved (Section 3.3.4).

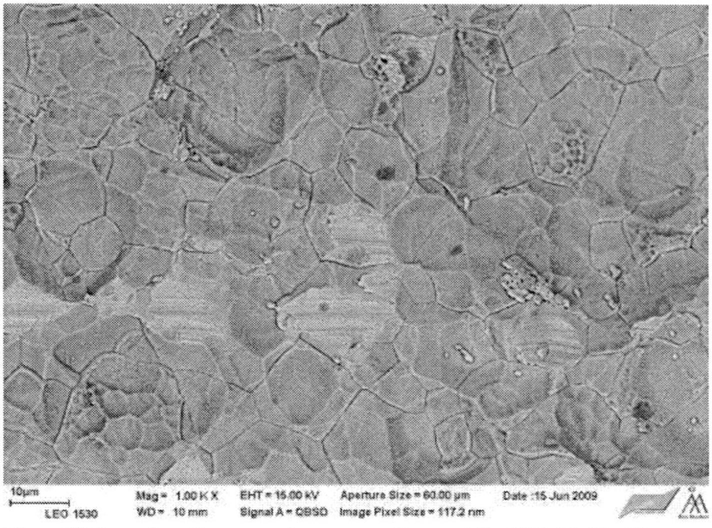

Figure 7. Scanning electron microscope of a used Cu-ZSM-5 (Zeo2) coated plate (1kX).

Table 2. Different Ag loadings from different impregnation times and $AgNO_3$ solution concentrations (LA-ICP-MS and SEM-EDXA)

$AgNO_3$ Solution concentration (M)	Impregnation time (h)	Ag wt.%
0.011	48	5.1 – 6.2
0.022	12	1.2 – 1.5
0.022	24	1.5 – 2.0
0.022	48	3.0 – 3.6
0.04	12	4.5 – 5.2
0.04	24	6.8 – 7.2
0.04	48	7.4 – 7.6

Table 3. Nomenclature for the XPS analysis

Catalyst	Description
A	Ag/alumina (0.04 M, 24h)
B	Ag/alumina (0.04 M, 12h)
C	Ag/(alumina + ionic liquid)
D	Ag/alumina (0.022 M, 24h)
E	Ag/alumina (0.022 M, 12h)
F	Au/alumina (deposition)
G	Ag/alumina (0.04 M, 48h)
H	Ag/alumina (0.011 M, 48h)
I	Ag/alumina (0.022 M, 48h)

Table 4. Binding energies (where applicable) and estimated oxidation state of metal

Catalyst	Al 2p (eV)	C 1s (eV)	O 1s (eV)	Ag $3d_{5/2}$ (eV)	State	Au $4f_{7/2}$	State
Ag/alumina (0.04 M, 24h) - **A**	74.6	285.0	531.1	368.0	Ag^0	–	–
Ag/alumina (0.04 M, 12h) - **B**	74.6	284.8	532.3	368.2	Ag^0	–	–
Ag/(alumina + ionic liquid - **C**	74.0	284.6^{ref}	531.1	368.1	Ag^0	–	–
Ag/alumina (0.022 M, 24h) - **D**	74.6	284.7	530.6	366.4 (58 %) 368.0 (42 %)	Ag^+ Ag^0	–	–
Ag/alumina (0.022 M, 12h) - **E**	74.6	284.7	531.7	368.6	Ag^0	–	–
Au/alumina (deposition) - **F**	74.6	284.8	530.7	–	–	84.2	Au^0
Au/alumina (deposition) ion etch - **F**	74.6	–	530.6	–	–	–	–
Ag/alumina (0.04 M, 48h) - **G**	74.6	284.5	530.6	366.5 (14 %) 368.0 (86 %)	Ag^+ Ag^0	–	–
Ag/alumina (0.011 M, 48h) - **H**	73.8	284.6^{ref}	531.1	368.1	Ag^0	–	–
Ag/alumina (0.011 M, 48h) ion etch - **H**	74.6	284.1	530.3	366.9 (63 %) 368.0 (37 %)	Ag^+ Ag^0	–	–
Ag/alumina (0.022 M, 48h) - **I**	74.6	285.2	531.3	368.3	Ag^0	–	–

^{Ref}Carbon 1s at 284. 6 eV was used as energy reference to compensate for electric charging.

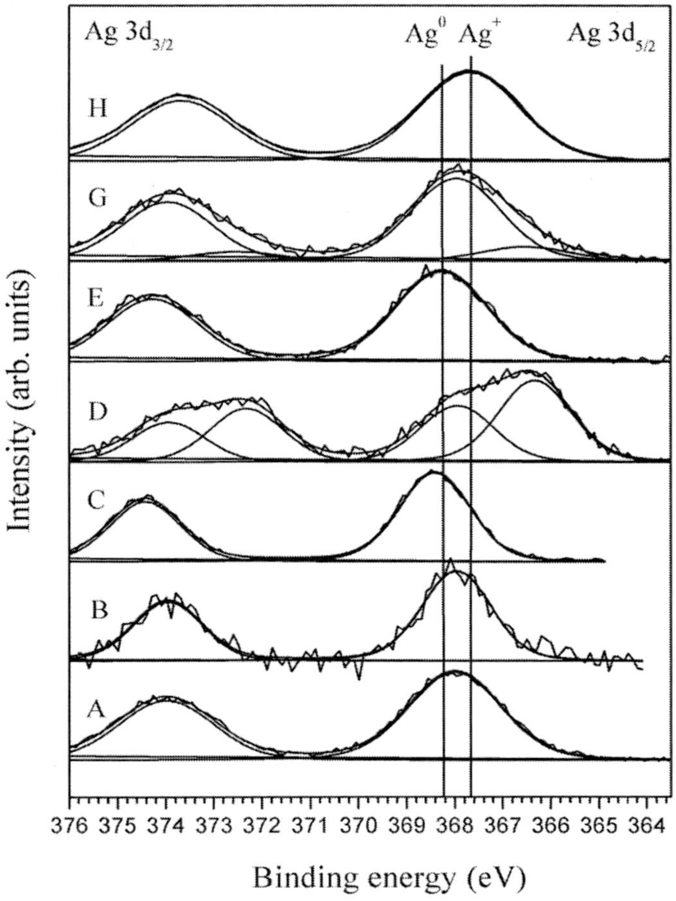

Figure 8. XPS lines from Ag 3d, normalized to peak height prior to background subtraction.

X-ray photoelectron spectroscopy (XPS) was also used for characterization of the prepared plates. The different catalytic plates and their description, are presented in Table 3.

Examples of the XPS analysis for different catalysts are shown in Table 4.

In the line fitting procedure, intensity ratios of 2p3/2:2p1/2, 3d5/2:3d3/2 and 4f7/2:4f5/2 lines were kept fixed at their theoretical values (2:1; 3:2; 4:3) and a mixture of Gaussian and Lorentzian line shapes was used. In silver catalysts with high loadings Doniach-Šunjić line shape was used to fit slightly asymmetric Ag 3d lines. Sensitivity factors used in determining atomic concentration ratios for Ag 3d, Al 2p, Au 4f, C 1s, Cu 2p and Si 2p were 5.198, 0.193, 5.240, 0.296, 4.798 and 0.283, respectively [[22]]. Silver was found to be in a slightly oxidized state in most cases. Figure 8 shows an example of the Ag 3d5/2 lines for different Ag catalysts (see Table 3 for nomenclature). The results obtained by XPS were similar with the other characterization techniques performed in this study.

The observed XPS binding energies indicate that silver was in metallic state in samples A, B, C, E, and H (Table 4). Silver was in two states, metallic and oxidized, in samples D and G. Sample F contained gold in Au^0 state.

Table 5. Atomic concentration ratios

Catalyst	Ag/O	Ag/Al	Ag/C	Au/Al
Ag/alumina (0.04 M, 24h) - **A**	0.016	0.034	0.062	-
Ag/alumina (0.04 M, 12h) - **B**	0.004	0.024	0.006	-
Ag/(alumina + ionic liquid - **C**	0.016	0.035	0.064	-
Ag/alumina (0.022 M, 24h) - **D**	0.004	0.009	0.019	-
Ag/alumina (0.022 M, 12h) - **E**	0.003	0.006	0.027	-
Au/alumina (deposition) - **F**	-	-	-	0.047
Au/alumina (deposition) ion etch - **F**	0.048	0.111	31	-
Ag/alumina (0.04 M, 48h) - **G**	0.008	0.017	0.045	-
Ag/alumina (0.011 M, 48h) - **H**	0.015	0.030	0.098	-
Ag/alumina (0.011 M, 48h) ion etch - **H**	0.002	0.004	N/A	-
Ag/alumina (0.022 M, 48h) - **I**	0.015	0.033	0.059	-

N/A: very low surface atomic carbon concentration.

There were small black spots (ca. 2 mm^2) present on the surface of sample H, which were analyzed in more detail. Analysis area was confined with a mask made of indium metal and thus the black spots on sample H could be analyzed separately. The spots had high carbon content. When ion etching was applied, silver/carbon surface atomic concentration ratio increased from 0.005 to 0.025. Prior to Ar ion etching, silver existed in two oxidation states, at 386.6 eV (Ag^0, 73 %) and 365.4 eV (Ag 27 %). After etching Ag still existed in the two oxidation states, but the fraction of Ag^0 had increased to 80 %. Values tabulated in Tables 4 and 5 are not representative of the black spots. All the obtained atomic concentration ratios are depicted in Table 5. On supported catalysts, metal loadings and the type of metals affect reaction activity; lower metal loadings will result in higher percentage of metal dispersion and smaller metal particle sizes, increasing the NOx reduction activity in the HC-SCR.

Table 6. Binding energies and atomic concentration ratio of the Cu-ZSM-5 catalyst (Zeo2)

Al	Si	C	Cu	O	Cu:Si
76.1[a]	103.1	284.6	932.8 (71 % Cu^0) 935.3 (29 % Cu^{2+})	532.7	0.032

[a] Low signal-to-noise ratio.

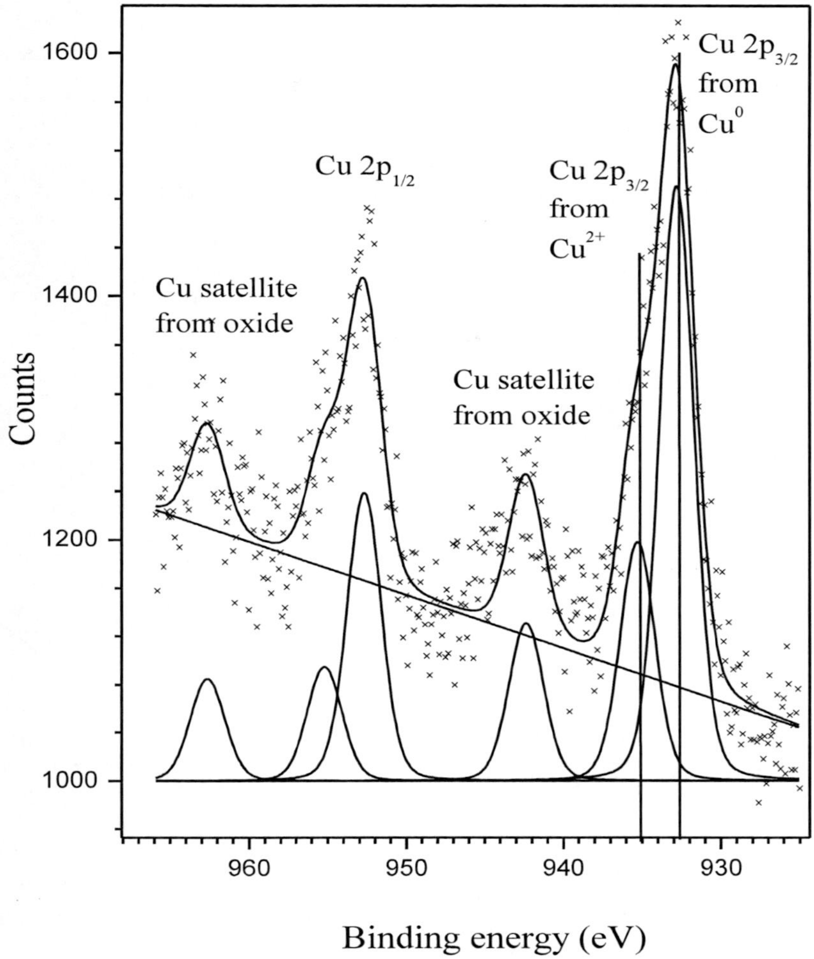

Figure 9. Cu 2p line from a Cu-ZSM-5 (Zeo2) coated plate.

For the case of the Cu-ZSM-5, Figure 9 shows Cu 2p line from the Cu-ZSM-5-coated plate, while Table 6 demonstrates the binding energies and the atomic concentration ratio between Cu and Si for the same catalyst. The metal content in the zeolite (Cu) was found to be ca. 3 wt.%.

3.2. Catalyst Testing

As mentioned before, Ag supported on alumina has exhibited high and stable conversions towards NOx reduction in HC-SCR systems. As a comparison, gold catalysts supported on alumina as well as a silver/alumina catalyst supported on ionic liquids were tested. Hexadecane, a paraffinic high-quality diesel fuel [[23],[24]] was used as a reductant for preliminary experiments, in order to detect the best metal/support combination from the coated microplates. Ag/alumina, Au/alumina and Ag/(alumina+IL) on the microplates were tested. The activity of these catalysts presented as NO-to-N_2 conversion as a function of the reactor temperature can be observed in Figure 10. Silver supported on alumina prepared by the impregnation method showed the higher NO conversion in a temperature window of 150-550 ºC, where the maximum peak was obtained at higher temperatures. Similar behaviour with respect to temperature was observed for Au/Al_2O_3 and Ag/(alumina+IL). Nevertheless, the NOx reduction with the last two catalysts was considerable lower. Figure 11 shows a SEM micrograph of the Au/Al_2O_3 plate that reveals the presence of gold as quite large clusters over the alumina surface. Very large Au particles which are visible are not selective and most likely not active as it was demonstrated by the activity experiments (Figure 10).

According to the mechanism proposed by Eränen et al [[25]], the NO conversion to N_2 is attained by the reaction of NO with a partially oxidized hydrocarbon, and it is presumed that the oxidation of NO is the rate-determining reaction. It has been demonstrated by Kung et al. [[26]] that Au/Al_2O_3 is a very active catalyst for the oxidation of hydrocarbons given that the gold is in the right particle size.

The oxidation of the hydrocarbon possibly takes place before the NOx reduction in this case. The experiments of Ueda et al. [[12]] on NOx reduction over gold supported on alumina evidenced that the oxidation of NO occurs mainly with O_2, which could also confirm the poor selectivity towards NO reduction. Additionally, a lower Au loading was applied in [[12]] with a conclusion that the optimum Au content for NO reduction should be between 0.1 and 0.2 wt% [[12]]. The XPS results from our gold catalyst, showed a metal content of 18 wt.%, much higher than the optimum found by Ueda and collaborators [[12]]. The large Au clusters revealed by SEM images (Figure 11) might also explain the poor activity of our Au/Al_2O_3 catalyst since the gold catalytic activity is very sensitive to its crystal size. Moreover, because of the still low selectivity of Au/alumina for the NO reduction, further research was carried out by Ueda et al., where the gold catalyst was mixed mechanically with Mn_2O_3 revealing higher NO conversions since the added functional catalyst is active for the oxidation of NO [[12]].

On the contrary, Ag/Al_2O_3 is considered less active for hydrocarbon oxidation and consequently, as the oxidation of the organic compounds is promoted, the conversion of NO over the Ag/alumina is increased, mainly due to the promoting of the reduction reaction.

The results obtained during these experiments are in good agreement with numerous studies on HC-SCR carried out during the recent years and overviewed by Burch [[27]].

Even if Au/Al_2O_3 and $Ag/(Al_2O_3$+IL) catalysts were characterized and tested, their potential was by far the least exploited and could still be further investigated in future contributions. Since the conversions of NO-to-N_2 for the cases of Au/Al_2O_3 and $Ag/(Al_2O_3$+IL) were very poor compared with the ones obtained with Ag/alumina, the latter was investigated more in detail.

3.2.1. Impregnation Time and Solution Concentration

Table 2 shows the different metal loadings obtained in the silver/alumina catalysts for different silver nitrate concentrations and different impregnation times. Hexadecane was again used as a probe molecule in this case, and an example of the activity results for the $C_{16}H_{34}$-SCR of NOx is depicted in Figure 12. Three sets of catalysts were investigated, using the same silver nitrate solution (0.022 M) but different impregnation times.

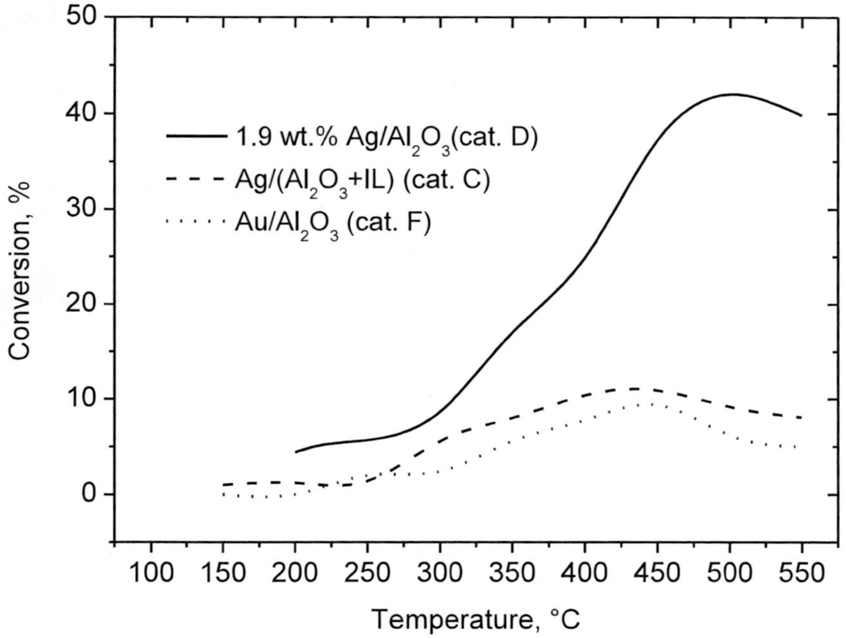

Figure 10. NO-to-N_2 conversion as a function of temperature on the SCR with octane for different catalysts; 6 vol.% O_2; 0.5 vol.% NO; 0.3125 vol.% C_8H_{18}.

The silver loading affected considerable the activity of the catalyst on the NOx reduction with hexadecane as a reducing agent (Figure 12). In some cases, the metal uptake by the coated plate was proportional to the amount of time that the plates had been in contact with the solution, but a clear correlation between the impregnation time and the silver loading in the catalyst was not found (refer to Table 2). After several tests, the optimal silver loading was found to be ca. 2 wt.%, yielding the highest NOx reduction over the broader temperature range. The impregnation with the solution of 0.022 M gave the closest to optimal metal loading, while the impregnation time that resulted in the optimum catalyst was, in general, 24 hours. Having the plates in contact with the solution for longer times resulted in a decrease of the catalytic activity (Figure 12).

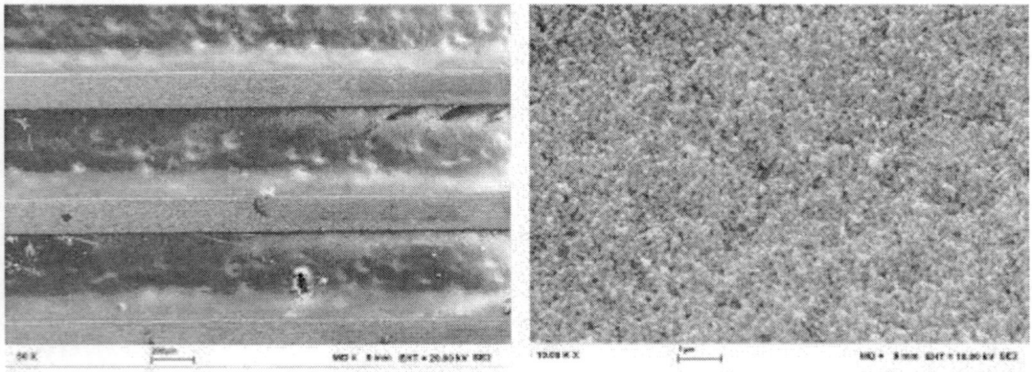

Figure 11. Scanning electron microscope pictures of Au/Alumina plates (50X and 10 kX).

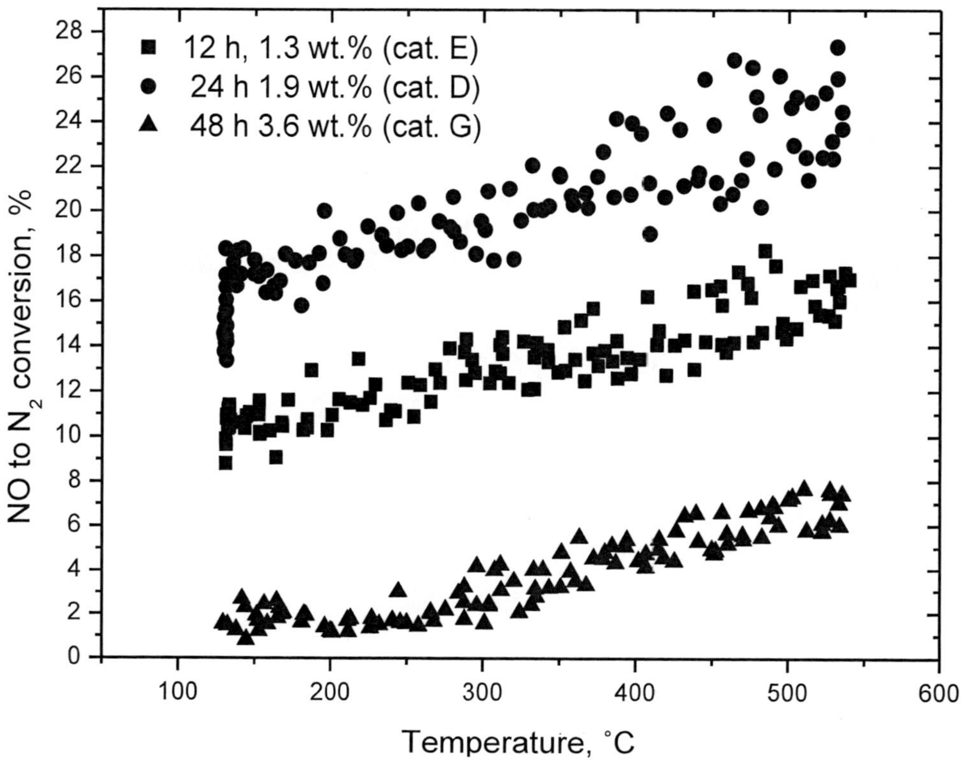

Figure 12. NO-to-N_2 conversion as a function of temperature for the SCR with hexadecane over Ag/Alumina (0.022 M $AgNO_3$) for different impregnation times; 6 vol.% O_2; 0.5 vol.% NO; 0.156 vol.% $C_{16}H_{34}$.

The catalyst that exhibited the highest level of NOx reduction for the SCR with octane and hexadecane was the one impregnated for 24 h in a 0.022 M $AgNO_3$ solution with a silver content close to 2 wt.% (see Table 2). This silver loading has been found to be optimal for conversion and selectivity towards molecular nitrogen from NOx [[1],[28]]. Too high silver loadings are associated with complete oxidation, hence more production of NO_2, a non-desirable pollutant [[27]]. The different patterns of activity are associated with different reaction mechanisms. Burch et al. [[27]] suggested that a mechanism of NO decomposition is operative in the case of high metal loadings on the catalyst, where the big particles of metallic silver are placed on alumina. This mechanism involves the dissociation of NO and the recombination of adsorbed nitrogen atoms to form N_2, or a reaction of N with adsorbed NO to produce N_2O. On the other hand, in the case of low silver loadings, the Ag^+ cations or small clusters of silver oxide prevail with strong interaction with the alumina. It is presumed that the N_2 is formed through a series of parallel and consecutive reactions that involve the formation of numerous intermediaries over silver and alumina. Probably, the oxidized species of nitrogen react with the reduced ones of it to produce N_2 [[27]].

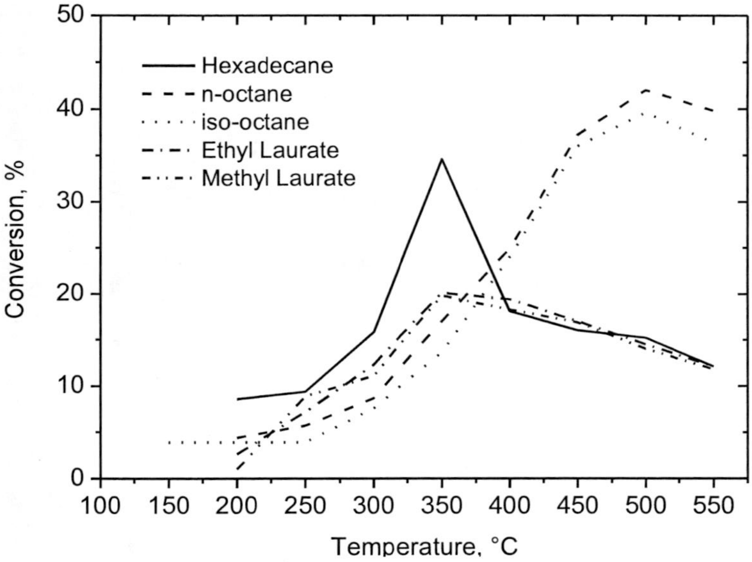

Figure 13. NO conversion as a function of temperature for different hydrocarbons over 1.9 wt.% Ag/Al_2O_3-coated microplates - Comparison with n-octane and iso-octane. 6 vol.% O_2, 0.5 vol.% NO. C_1/NO=5.

3.2.2. HC/Biofuel - SCR of NOx over 1.6 wt.% Ag/Al_2O_3

We have previously tested Ag/Al_2O_3-coated microplates for the SCR of NOx with ethene, propane, and n-octane as reducing agents [[29]]. The reduction showed to be sensitive to the reaction temperature and the number of carbons present in the molecule. For lower hydrocarbons (less than C8), higher carbon atoms present in the molecule, afforded a better reduction. This behaviour was expected, since longer chain hydrocarbons are easier to activate, being in agreement with the previous findings [[2]].

The aim in the present study was to find suitable catalysts for SCR of NOx using bio-derived fuels. The results for different renewable fuels are shown in Figure 13.

First generation biodiesel refers to such components, which are produced by transesterification of fat and oils. Methyl and ethyl laurate are important components that are derived from fatty acids present in vegetable oils. Particularly, coconut and babassu oils, contain a high amount (54%) of C12 saturated fatty acids (lauric acid) [[30]]. Methyl and ethyl laurate were used as test molecules for the SCR of NOx. The NO-to-N_2 conversions were higher at lower temperatures (below 300 °C) compared to the conversion obtained for two reference components representing fossil fuels, n-octane and iso-octane. The reduction potentials of the ethyl and the methyl laurate were very much the same (Figure 13).

A novel and more interesting way of producing bio-derived diesel is by decarboxylation and/or decarbonylation of natural oils and fats, yielding long-chained hydrocarbons (alkanes) [[24]]. Hexadecane was used as a model component in this case. Due to its high cetane number (100), hexadecane is regarded as a high-quality diesel. Similarly as when methyl- and ethyl laurate were used as reducing agents, the NOx reduction with hexadecane, was higher at low temperatures (< 400 °C, Figure 13) than with octane. The hydrocarbon activation at lower temperatures could be the reason for this behaviour. In the high-temperature domain (> 400 °C), this activity was surprisingly sharply lowered. The high catalytic activity at low temperatures could also be the cause of this prominent decrease of activity, since some coke formation could be assumed to occur at higher temperatures (above 350 °C).

Theoretical studies suggest [[31]] that heavier hydrocarbons, such as the octane isomers, exhibit a high enthalpy of adsorption compared to lower hydrocarbon chains. Hence, they can be strongly adsorbed over the oxidized surface and helping capturing the nitric oxide molecule thus trigging its reactivity [[27]]. This case could also be true for hexadecane, possible explaining its high reactivity al lower temperatures. At higher temperatures, due to the presence of large amounts of oxygen, the unselective reaction between O_2 and the hydrocarbon proceeds in a faster rate than the reaction with NO, hence, the reduction of NOx is negatively effected. Similar behaviour was not detected with lower alkanes and alkenes [[29],[32]]. Total oxidation of the small amount of hydrocarbons at higher temperatures is also possible, leaving the system with no reducing agent for the NOx reduction reaction to occur.

3.2.3. Variation of the Carbon to Nitric Oxide Ratio

One of the important aspects about the selective catalytic reduction of NOx using hydrocarbons is the amount of the reducing agent that should be used to optimize the reaction. Such amounts of hydrocarbons are calculated on the basis of the relationship between the number of carbon atoms and the percentage of NO in the feed flow, giving the proportion C_1/NO. Different C_1/NO ratios were used for finding an optimum.

Figure 14 represents the NO conversion curve obtained when changing the concentration of n-octane in the reactor feed.

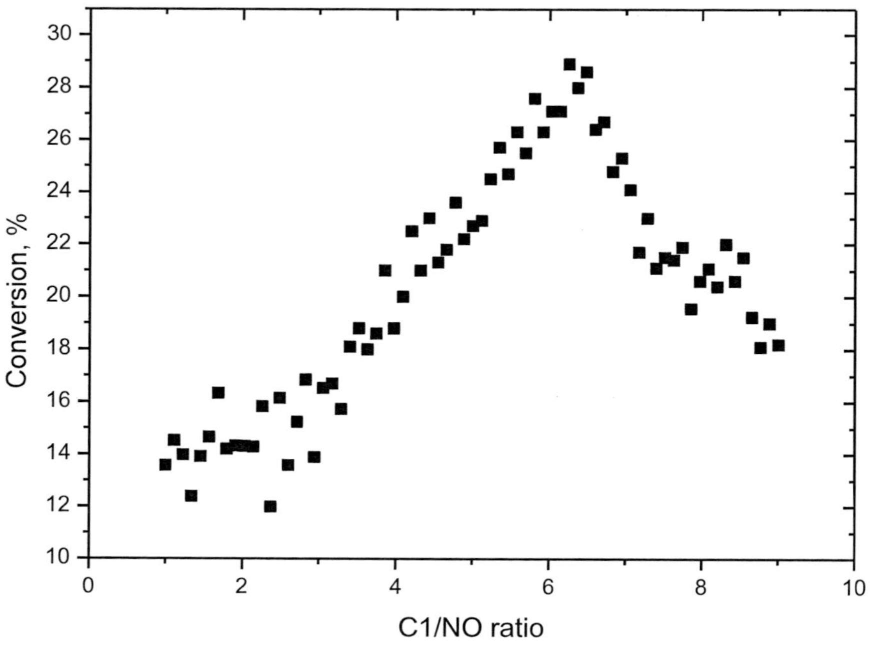

Figure 14. Octane-SCR of NOx with different C_1/NO ratios over 1.9 wt.% Ag/Al_2O_3-coated microplates at 400 °C; 6 vol.% O_2, 0.5 vol.% NO.

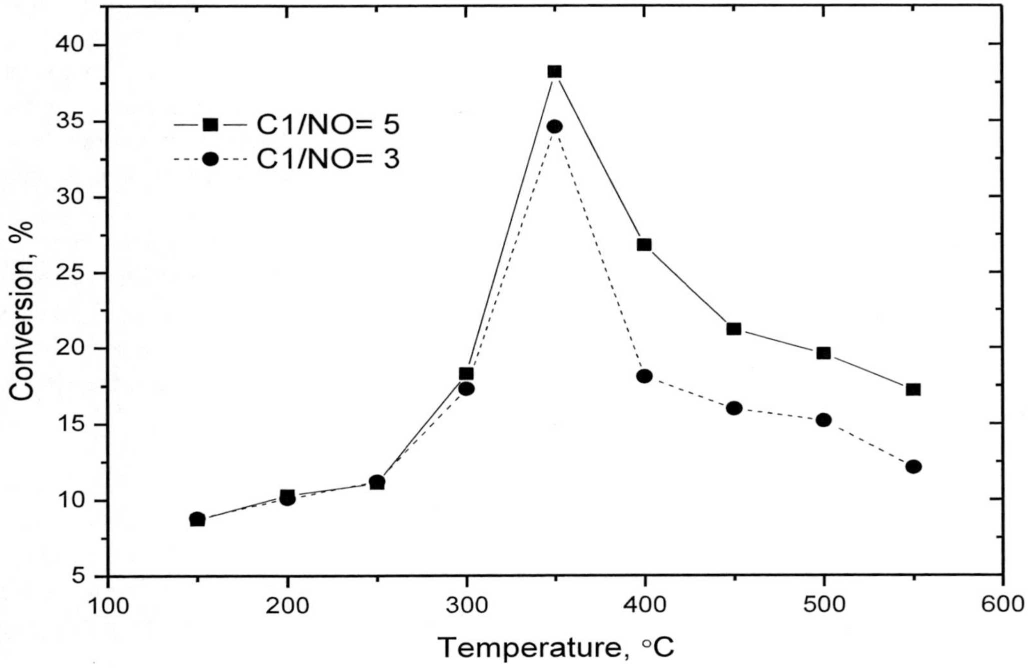

Figure 15. Hexadecane-SCR of NOx with different C_1/NO ratios over 1.9 wt.% Ag/Al_2O_3-coated microplates; 6 vol.% O_2, 0.5 vol.% NO, 0.156 vol.% $C_{16}H_{34}$ (-■-), 0.094 vol.% $C_{16}H_{34}$ (-●-).

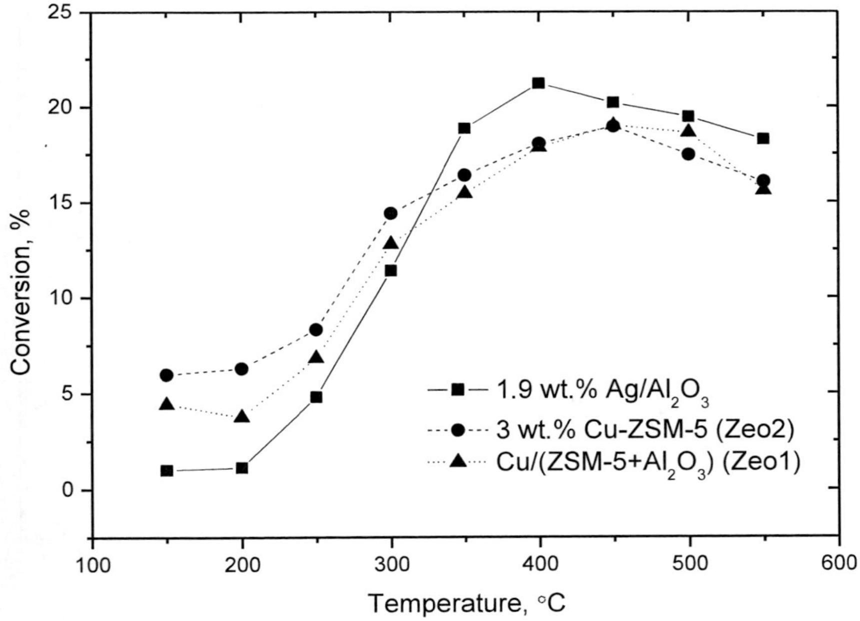

Figure 16. NO-to-N_2 conversion in the hexadecane-SCR of NOx with different reactor configurations. 500 ppm NO, 188 ppm $C_{16}H_{34}$, 6 vol.% O_2, 12 vol.% H_2O (zeolites catalysts without water).

The highest NOx reduction values were observed when the C_1/NO ratio was set between 6 and 7. At higher octane concentrations, the activity decayed considerably, presumably due to a competitive adsorption between the reactive species involved in the reaction. As hexadecane was used as the reducing agent, the effect of two concentrations was evaluated over a larger temperature range. The results obtained for two different ratios of C_1/NO for hexadecane are presented in Figure 15.

The conversion does not show an important variation when increasing the concentration of the hydrocarbon at lower temperatures. However, as the temperature continues to increase (> 350 °C), the positive effect of higher hexadecane concentration in the feed is visible, enhancing the NOx reduction.

3.2.4. Improvement of Low Activity NOx Reduction

Low-temperature SCR (< 300 °C) is challenging and is just emerging as a potentially viable technology, as it still needs to be developed. Usually, NOx reduction conversions are obtained when the temperature is above 350 °C. However, the reduction results for temperatures below 300 °C are not satisfactory and more research is still needed. During the cold start the engine operates in this low temperature range and an effective NOx reduction is mandatory. Some authors suggested that copper-exchanged zeolites could be used as a remedy for the low activity al low temperatures and it has been found that these zeolites provide a fairly good reduction in the low temperature range [[10],[11]]. This was the motivation in the present study, coating the plates with a catalytically active zeolite and testing them in the SCR with hexadecane. The preparation of the active phase is presented in the Experimental section. The versatility of the microreactor allowed the implementation of the two-catalysts system only by changing the plate distribution in the microchamber. Five plates coated with the optimized Ag/alumina catalyst (0.022 M and 24 h impregnation time) and five Cu-ZSM-5 or Cu/(ZSM-5+Al_2O_3) coated plates were placed in the microreactor and tested for the SCR of NOx, using a simulated diesel engine exhaust through the microreactor inlet. Two different preparation methods for the zeolitic coatings were employed, which were essentially impregnation and spraying of the microplates with zeolite containing solutions. When only the zeolite containing plates were used in the reactor, the latter method resulted in slightly better conversions as shown in Figure 16. Both zeolite catalysts exhibited a better activity at low temperature (< 300 °C) than 10 plates of silver/alumina alone.

The zeolite catalyst that exhibited better conversions was used for further comparisons.

When the two catalysts were combined, the reduction efficiency at lower temperature tended to increase. Moreover, the system proved to successfully reduce NOx over the temperature range 150-550 °C with the maximum NOx reduction just 1% lower compared to the case when only Ag/Al_2O_3 was used as a catalyst. Figure 17 shows these results, for the case that water was absent in the system. The most prominent effect was the broadening of the temperature reduction window towards the low temperature range. A NOx reduction up to seven times higher was achieved at temperatures below 350 °C.

The effect of water in the feed was also investigated. Water may poison the zeolites catalyst since the hydroxyl ions could build chemical bridges between the aluminium present in the catalytic structure, making the zeolite structure to collapse. The effect of water is further elucidated in Figure 18. The negative effect of water in the mixture is well observed in the plot, especially for the case when only Cu/(ZSM-5+Al_2O_3) or Cu-ZSM-5 coated plates were used. When a combination of the Ag/alumina and either Zeo1 or Zeo2 catalysts was tested, the water seemed to have a negative effect on the activity at temperatures exceeding 300 °C. Houel et al. [[33]] have found similar behaviour for Cu-ZSM-5 catalyst, proposing as well the dealumination of the zeolite at high temperatures, although their findings suggested a similar activity window for both the Cu-ZSM-5 and the Ag/Al_2O_3 catalysts in minireactors.

The negative effect observed at higher temperatures does not appear to be particularly strong at lower temperatures, as the reduction activity remains at reasonable levels, even with the addition of water to the stream. Presumably, temperatures less than 300 °C are insufficient to produce the aforementioned effect, keeping the zeolite structure intact. The combination of bifunctional zeolitic catalysts with propene as a reducing agent in minireactors has been previously investigated demonstrating a lowering of the optimal reduction temperature by as much as 150 °C [[14],[15]]. Moreover, the presence of water resulted in an enhancement of the zeolitic catalytic activity rather than a decrease, which could be explained by the interaction of the metal structure with the zeolite structure.

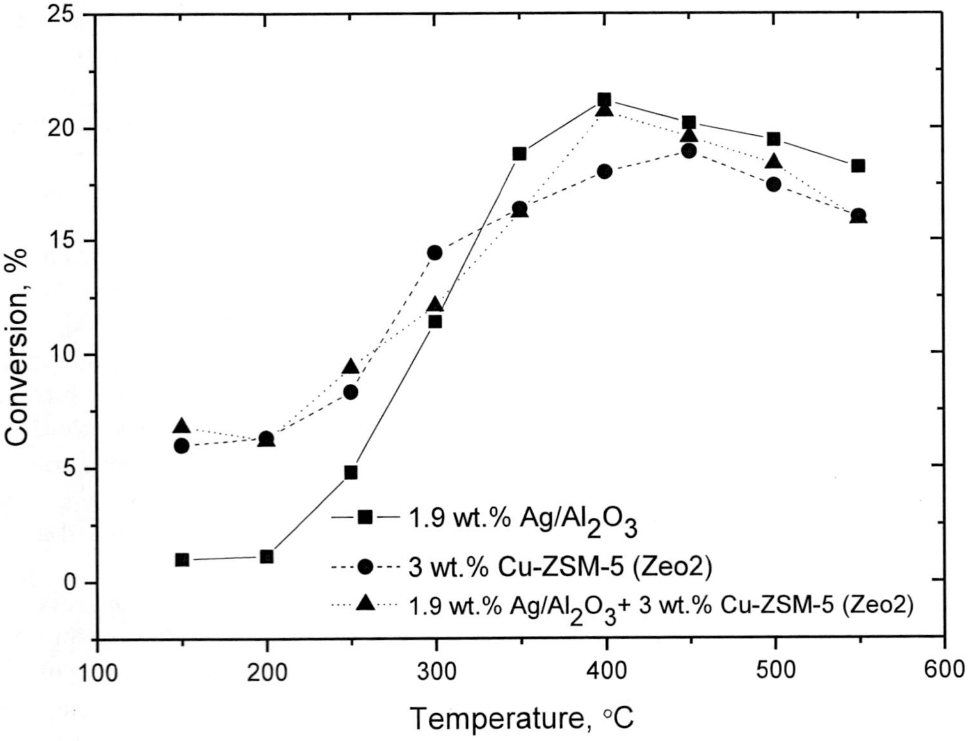

Figure 17. NO-to-N_2 conversion on the SCR of NOx with hexadecane, with different catalyst configurations; 500 ppm NO, 188 ppm $C_{16}H_{34}$, 6 vol.% O_2, 12 vol.% H_2O (for 1.9 wt.% Ag/Al_2O_3 only).

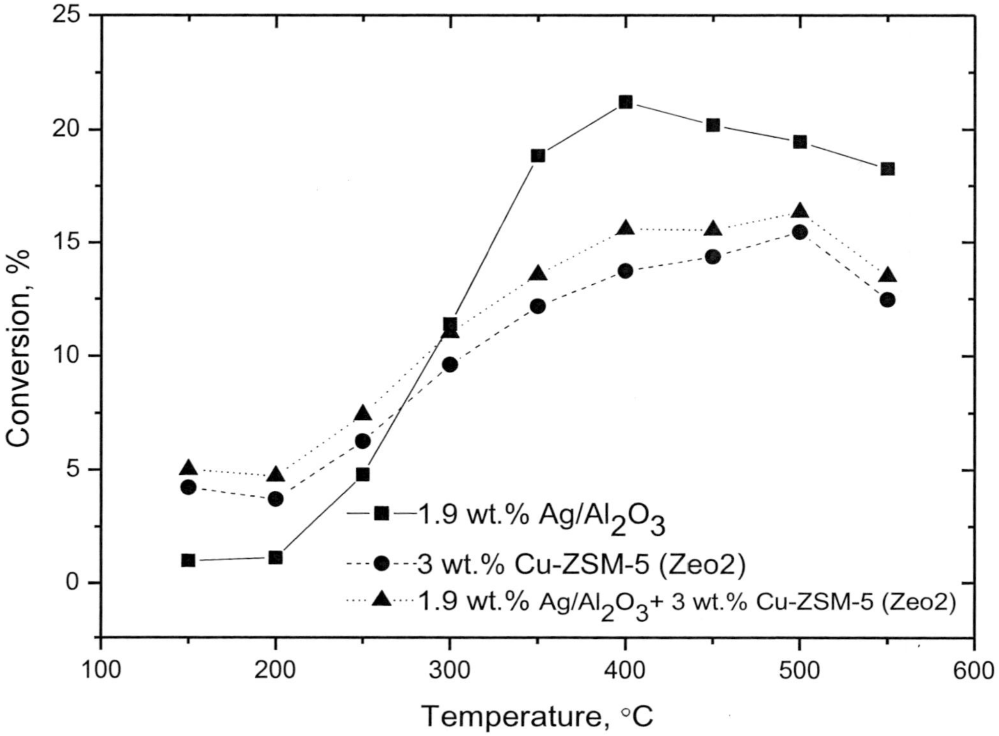

Figure 18. NO-to-N_2 conversion on the SCR of NOx with hexadecane, with different catalyst configurations; 500 ppm NO, 188 ppm $C_{16}H_{34}$, 6 vol.% O_2, 12 vol.% H_2O.

Additionally, the effect of the amount of hydrocarbon (hexadecane) introduced at different temperatures for the optimized system (Figure 19) was studied. It was found that by varying the C_1/NO ratio it was possible to achieve higher NOx conversions. In all the comparison experiments (Figures 16-18) the HC/NO ratio was set to 6. However, changes of this ratio resulted in an improved NO-to-N_2 conversion, especially at the higher temperature range (350-550 °C), where the enhancement of the activity was up to 30%. Figure 19 shows the results of the hexadecane concentration tuning for the combination between the Ag/alumina and the Cu-ZSM-5 catalysts. At lower temperatures, the C_1/NO ratio of 6 (corresponding to 188 ppm hexadecane) resulted in the highest NOx reduction. In this temperature range, the NO reduction was not very much affected by the amount of hydrocarbon introduced on the feed. The low temperature does not seem to be enough for partially oxidizing the hydrocarbon to form oxygenates, a key step for the reduction to take place [[34]]. As the temperature increases and higher reduction levels are achieved, the amount of hydrocarbon that can be introduced to the system tends to be higher. At high temperatures, the oxidation of the hydrocarbon increases, and small amounts of the reductant could not be enough for starting the reduction process. An enhancement response of the reduction behaviour is observed when introducing larger amounts of hydrocarbons, as the temperature raises (Figure 19).

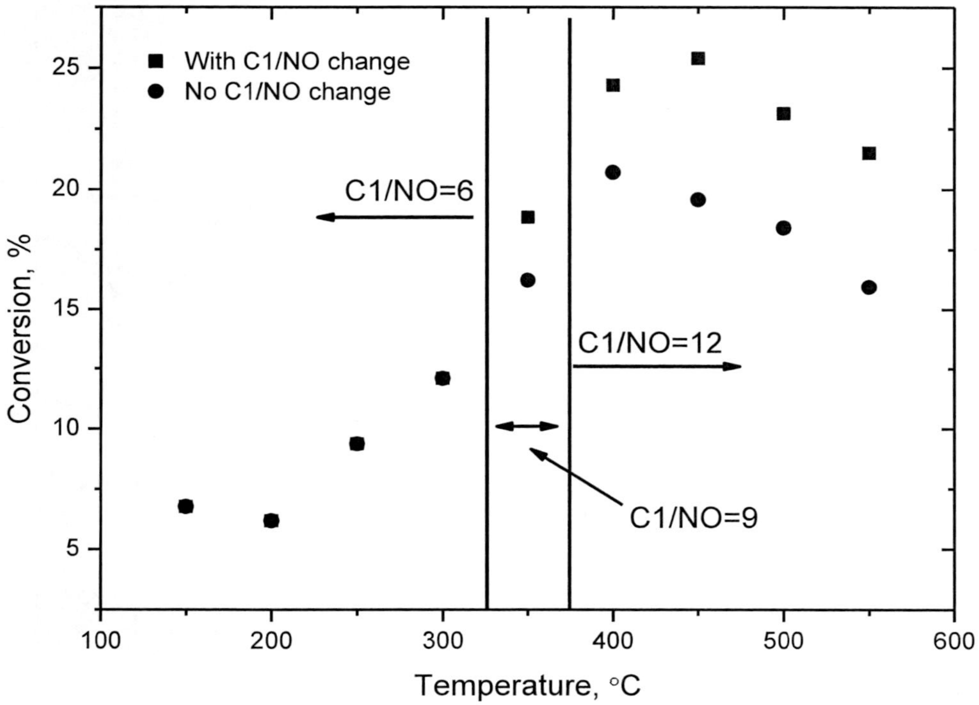

Figure 19. NO-to-N_2 conversion vs. temperature on the SCR of NOx with hexadecane over 1.9 wt.% Ag/Al_2O_3 and 3 wt.% Cu-ZSM-5 (Zeo2) microchannels; varying the C_1/NO ratio; 500 ppm NO, 6 vol.% O_2, 188 ppm $C_{16}H_{34}$ (C1/NO=6), 281.3 ppm $C_{16}H_{34}$ (C1/NO=9), 375 ppm $C_{16}H_{34}$.(C1/NO=12)..

3.2.5. Comparison between the Microreactor and Minireactor Systems

Experiments under similar conditions were conducted in a conventional laboratory reactor, minireactor (Ø= 10 mm). Representative components from conventional diesel fuels have been previously investigated by our group [[29]]. For the case of bio-derived diesel, a comparison between the $C_{16}H_{34}$-assisted-SCR of NOx in a microreactor and a minireactor is presented below.

In order to establish a valid comparison, the ratio between the total volumetric flow and the catalyst amount in both reactors was kept constant. A total flow rate of 700 ml/min was used for the minireactor, with a catalyst mass of 0.1 g. For the microreactor, the total catalytic mass was 7 mg, while the total flow rate was set to 50 ml/min. Figure 20 shows the comparison between the two reactors with hexadecane as reducing agent. The behaviour of both reactors was almost identical and the results are similar to those obtained with fossil-derived fuels by our group [[29]].

In addition to the experimental observations showing very similar results in the mini- and the microreactor, the inherent advantages of the latter, e.g., completely isothermal flows, no hot spots in the reactor, as well as intensified mixing and mass transfer, allow the acquisition of accurate experimental data. Even if the catalyst coating is not straightforward, the enormous amount of data possible to gather with microdevices opens new perspectives in catalytic screening and reaction engineering as an effective tool for the study of catalytic reactions of industrial relevance.

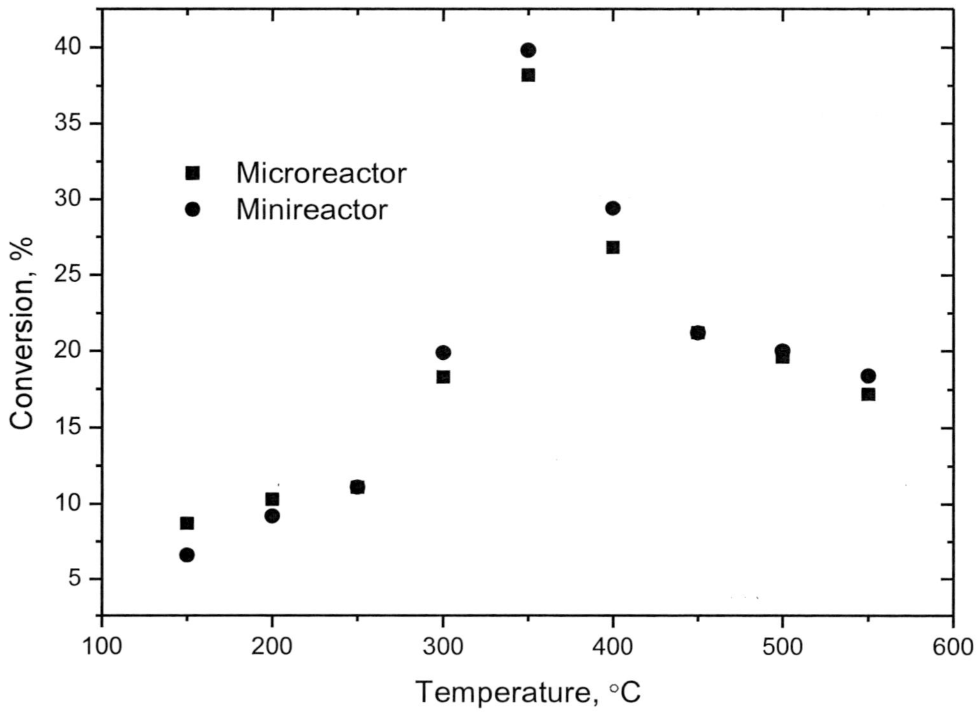

Figure 20. NO-to-N_2 conversion over a 1.9 wt.% Ag/Al_2O_3-coated microchannels (■) and 1.9 wt.% Ag/Al_2O_3 powder (●) for the $C_{16}H_{34}$-SCR; 6 vol.% O_2, 0.5 vol.% NO. C_1/NO=5, 0.156% $C_{16}H_{34}$.

CONCLUSIONS

Continuous reduction of NO with bio-derived fuels in a microreactor (Ø=460 μm) was investigated under an excess of oxygen over Au/Al_2O_3, Ag/(Al_2O_3+IL) and Ag/Al_2O_3. Silver supported on alumina showed the best NO-to-N_2 conversion and the metal loading was further optimized by varying the precursor concentration ($AgNO_3$) and impregnation time. The highest activity for the hexadecane-SCR of NOx was exhibited by an Ag/alumina catalyst with ca. 2 wt.% of silver as shown by XPS spectra, SEM-EDS and LA-ICP-MS. The silver distribution along the microchannels was found to be uniform.

An improvement of activity of the $C_{16}H_{34}$-HC-SCR in the low temperature domain was reached by using a combination of catalysts active in low and high temperature domain, i.e., Cu-ZSM-5 and Ag/Al_2O_3 respectively. Although the overall reduction across the whole temperature range with the combination of catalysts was almost the same, the improvement of conversion for temperatures below 350 °C was up to seven-fold.

The C_1/NO ratio was optimized for the two-catalyst system, obtaining that a progressive increase on the hydrocarbon concentration as the temperature rises, is favourable for the NOx reduction.

The results in the microchannels were also compared to those obtained in a conventional laboratory fixed-bed reactor. The results were very similar, but the intrinsic benefits provided by the micro- compared with the minireactor, present an excellent choice for catalytic testing.

Favourable conditions for working in the microchannels were obtained for further development of kinetic and reactor models for HC-SCR applications.

ACKNOWLEDGEMENTS

Paul Ek at the Laboratory of Analytic Chemistry, Åbo Akademi University is gratefully acknowledged for the LA-ICP-MS analysis. Linus Silvander is also acknowledged for the SEM-EDXA pictures and measurements. This work is part of the activities at Åbo Akademi Process Chemistry Centre within the Finnish Centre of Excellence Program (2000-2011) by the Academy of Finland. Financial support from the Graduate School in Chemical Engineering to J.R. Hernández Carucci is gratefully acknowledged.

REFERENCES

[1] K. Arve, L. Čapek, F. Klingstedt, K. Eränen, L.-E. Lindfors, D.Yu. Murzin, J. Dědecĕk, Z. Sobalik, B. Wichterlová, *Top. Catal.* 30/31 (2004) 91.

[2] K. Eränen, L-E. Lindfors, A. Niemi, P. Elfving, L. Cider, *SAE paper* 2000-01- 2813 (2000).

[3] T.E. Hoost, R.J. Kudla, K.M. Collins, M.S. Chattha, *Appl. Catal. B* 13 (1997) 59.

[4] A. Martínez-Arias, M. Fernández-García, A. Iglesias-Juez, J.A. Anderson, J.C. Conesca, *J. Soria, Appl. Catal. B* 28, (2000) 29.

[5] W. Ehrfeld, V. Hessel, H. Löwe, *Microreactors – new technology for modern chemistry,* Wiley-VCH, Weinheim, 2000.

[6] K. Jensen, *Chem. Eng. Science* 56 (2001) 293.

[7] G. Kolb, V. Hessel, Micro-structured reactors for gas phase reactions, *Chem. Eng. J.* 98 (2004) 1.

[8] K. Eränen, L-E. Lindfors, F. Klingstedt, D. Yu. Murzin, *J. Catal.* 219 (2003) 25.

[9] F. Klingstedt, K. Eränen, L.-E. Lindfors, S. Andersson, L. Cider, C. Landberg, E. Jobson, L. Eriksson, T. Ilkenhans, D. Webster, *Top. Catal.* 30/31 (2004) 27.

[10] L. Čapek, K. Novoveská, Z. Sobalík, B. Wichterlová, L. Cider, E. Jobson, *Appl.Catal.* B 60 (2005) 201.

[11] L. Čapek, J. Dědeček, B. Wichterlová, L. Cider, E. Jobson, V. Tokarová, *Appl.Catal. B* 60 (2005) 147.

[12] A. Ueda, T. Oshima, M. Haruta, *Appl. Catal. B* 12 (1997) 81.

[13] L. Q. Nguyen, C. Salim, H. Hinode, *Appl. Catal. A* 347 (2008) 94.

[14] M. K. Neylon, M. J. Castagnola, N. B. Castagnola, C. L. Marshall, *Catal. Today* 96 (2004) 53.

[15] M. J. Castagnola, M. K. Neylon, C. L. Marshall, *Catal. Today* 96 (2004) 53.

[16] V. Sebastián, O. de la Iglesia, R. Mallada, L. Casado, G. Kolb, V. Hessel, J. Santamaría, *Microporous Mesoporous Mater.* 115 (2008) 147.

[17] E. V. Rebrov, G. B. F. Seijger, H. P. A. Calis, M. H. J. M. de Croon, C. M. van den Bleek, J. C. Schouten, *Appl. Catal. A* 206 (2001) 125.

[18] J. L. Hang Chau, Y. S. Susanna Wan, A. Gavriilidis, K. L. Yeung, *Chem. Eng. J.* 88 (2002) 187.

[19] R. Zapf, C. Becker-Willinger, K. Berresheim, H. Bolz, H. Gnaser, V. Hessel, G. Kolb, P. Loeb, A.-K. Pannwitt, A. Ziogas, *Trans IChemE* 81 (2003) 721.

[20] G. Germani, A. Stefanescu, Y. Schuurman, A.C. van Veen, *Chem. Eng. Sci.* 62 (2007) 5084.

[21] R. S. Howell, M. K. Hatalis, *J. Electrochem. Soc.* 149 (2002) G143.

[22] J. F. Moulder, W. F. Stickle, P. E. Sobol, and K. D. Bomben, Handbook of X-ray photoelectron spectroscopy, Perkin Elmer Corp., *Physical Electronics Division, USA* (1992).

[23] D. Yu. Murzin, I. Kubickova, M. Snåre, P. Mäki-Arvela, *European patent* 05075068.6 (2005).

[24] I. Kubičková, M. Snåre, K. Eränen, P.-M. Arvela, D. Yu. Murzin, *Catal. Today* 106 (2005) 197.

[25] K. Eränen, F. Klingstedt, K. Arve, L. E. Lindfors, D. Yu. Murzin, *J. Catal.* 227 (2004) 328.

[26] M. C. Kung, K.A. Bethke, J. Yan, J.-H. Lee, H.H Kung, *Appl. Surf. Sci.* 121/122 (1997) 261.

[27] R. Burch, J.P. Breen, F.C. Meunier, *Appl. Catal. B* 39 (2002) 283.

[28] K. Arve, K. Svennerberg, F. Klingstedt, K. Eränen, L. R. Wallenberg, J.-O. Bovin, L. Čapek, D. Yu. Murzin, *J. Phys. Chem. B* 110 (2006) 420.

[29] J. R. Hernández Carucci, K. Arve, K. Eränen, D. Yu. Murzin, T. Salmi, *Catal. Today* 133-135 (2008) 448.

[30] G. Knothe, R. O. Dunn, M. O. Bagby, Fuels and Chemicals from biomass, *American Chemical Society,* Washington D.C., 1997.

[31] S. M. Wetterer, D. J. Lavrich, T. Cummings, S. L. Bernasek, G. Scoles, *J. Phys. Chem.* B 102 (1998) 9266.

[32] K. Miyadera, *Appl. Catal.* B 2 (1993) 199.

[33] V. Houel, D. James, P. Millington, S. Pollington, S. Poulston, R. Rajaram, R. Torbati, *J. Catal.* 230 (2005) 150.

[34] K. Arve, J. R. Hernández Carucci, K. Eränen, A. Aho, D. Yu. Murzin, *Appl. Catal. B* 40 (2009) 603.

In: Focus on Catalysis Research: New Developments
Editors: Minjae Ghang and Bjørn Ramel
ISBN: 978-1-62100-455-4

Chapter 9

EMERGING CATALYSTS FOR WET AIR OXIDATION PROCESS

Asuncion Quintanilla, Carmen M. Dominguez, Jose A. Casas and Juan J. Rodriguez
Chemical Engineering Section, Universidad Autónoma de Madrid, Campus de Cantoblanco, Madrid, Spain

ABSTRACT

Wastewater treatment has reached a maturity state but the growing industrialization along with the more stringent environmental regulations demand an increasing dynamism in short and medium term. Accordingly, the existing technologies should improve in both versatility and efficiency and, in this sense, catalysts can play a prominent role. In this chapter, the authors offer a critical review of the catalysts currently investigated for the industrial wastewater decontamination by wet air oxidation. A survey of catalysts industrially implemented and academically investigated is presented and their nature and competitive features in activity and durability remarked. The current trend in the exploration of nanomaterials in the wet air oxidation field is highlighted. Updated research on nanotechnology-based catalysts, specially, carbon nanostructures and gold nanoparticles, is summarized and thorough discussed. Our recent results involving nanoscale gold particles are also included.

FUNDAMENTS

Wet air oxidation (WAO) is a well established technology for the removal of both organic and some inorganic pollutants from industrial process water and wastewater. It involves the oxidation of dissolved or suspended substances in aqueous phase using a gaseous source of oxygen, usually air, at high temperatures and pressures. The elevated temperatures are convenient to assure rapid oxidation and mineralization rates and the elevated pressures are required to carry out the reaction in liquid phase and to enhance the solubility of oxygen in water (Mishra et al.,1995).

The WAO units typically run at 0.5-20 MPa and 400-700 K, with residence time from 10 to 120 min. Conversion of organics in these conditions ranges from 80 to 99%, and chemical oxygen demand (COD) reduction is completed by 60-90%. WAO is a technology with expertise in refineries, coke ovens, organic compounds production plants, pharmaceutical factories, pulp and paper mills, and textile and surface treatments plants (Cybulski, 2007). This wide array of industrial process water and wastewater illustrates the versatility of this technology, which is still more relevant, in the current context of frenetic industrial dynamism and stringent environmental regulations.

The flow diagram of a typical WAO plant is depicted in Figure 1. The wastewater is brought to the system pressure using a high pressure pump. Air is fed to the reactor using a compressor. Both streams are brought together and preheated to raise the operating temperature, by heat recovering from the effluent stream. The pressurized and hot inlet stream enters in the reactor where oxygen diffuses from the gas to the liquid and reacts with the pollutants in water. The bi-phasic effluent is cooled down first by the feed and then by cooling water, and subsequently disengaged in a separator. The off-gases can be expanded in a turbine to recover energy. The liquid phase can be directly discharged or most often subjected to a biological treatment.

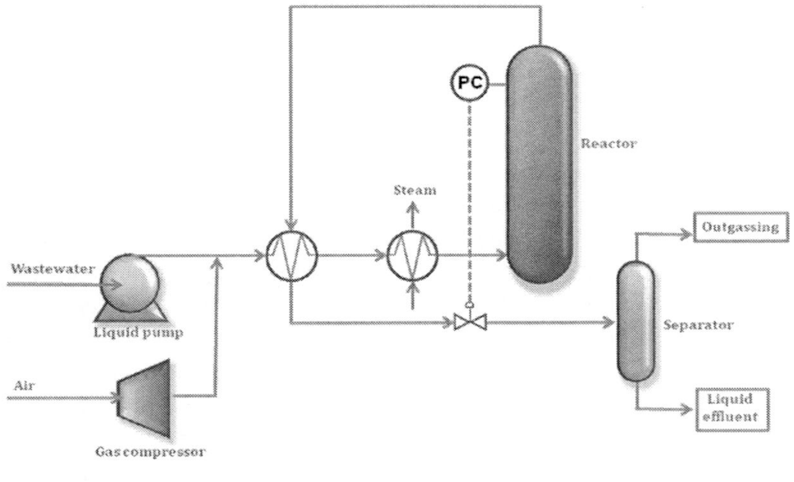

Figure 1. Schematic diagram of a wet air oxidation unit.

The oxidation of organic pollutants proceeds through a free-radical mechanism. Initially, the cleavage of the C-H bond occurs with formation of a free radical R·. This radical reacts with molecular oxygen to give an alkylperoxyl radical (ROO·). This reacts with another organic molecule (RH) producing the alkylhydroperoxyl species (ROOH) which is unstable and decomposes to intermediates with lower number of carbons (Cybulski et al., 2007).

The degree of oxidation depends on the temperature, oxygen partial pressure, residence time and the reactivity of the specific pollutant. The organic contaminants are either partially degraded into low molecular weight oxygenated compounds (*i.e.* maleic, fumaric, malonic, propionic oxalic, acetic and formic acids as well as methanol, ethanol and acethaldehyde) or mineralized into CO_2. Nitrogen is converted mostly to ammonia and also to nitrate and elemental nitrogen. Halogens and sulphur are oxidized to inorganic halides or sulfates. Phosphorous oxidize to phosphates and chlorine compounds to HCl. Debellefontaine and Foussard (2000) reported the following mass balance with a heat value of 435 kJ/mol of oxygen consumed:

$$C_m H_n O_k Cl_w N_x S_y P_z + (m + 0.25\,(n - 3x) - 0.5k + 2(y + z))\,O_2 \rightarrow$$

$$\rightarrow mCO_2 + 0.5(n - 3x)H_2O + xNH_3 + wCl^- + ySO_4^{2-} + zPO_4^{3-} + Q$$

In contrast to other thermal processes, WAO does not produce toxic gases *viz.* NO_x, SO_2, dioxins or furans.

WAO does not guarantee the complete destruction of the pollutants to carbon dioxide and water or to nitrogen. The mineralization occurs to large extent, but some innocuous by-products such as low molecular weight oxygenated compounds, *viz.* acetic acid, as well as ammonia are refractory to the oxidation. Extreme temperatures and rather long residence times would be required to increase the mineralization, factors that would make the process economically unsustainable. Therefore, WAO is most commonly conceived as a first step of a combined strategy in which the effluent from the oxidation is subsequently discharged into a conventional biological treatment plant (Scott and Ollis, 1995; Mantzavinos and Psillakis, 2004). Thus, the biodegradability of the exit stream along with the oxidation degree are the determining factors regarding the efficiency of the WAO process. Surprisingly, biodegradability and ecotoxicity tests of the WAO effluents have only been studied in the recent years (Santos et al., 2004a and b; Pintar et al., 2004; Mantzavinos and Psillakis, 2004; Rubalcaba 2007; Suarez-Ojeda 2007; Santos et al, 2009).

The constrain of WAO processes is the high capital costs associated to the high-pressure equipments and constructions materials (most oxidation by-products are corrosive under the reaction conditions, and then the use of special alloys is demanded) and the energy cost derived from operation at high temperatures and pressures. For this reason, the applicability of WAO is, in practice, limited to wastewater with an organic load content that makes the process autothermal. In general, WAO becomes practical for wastewaters with COD of at least 20.000 mg/L. The use of oxygen instead of air can improve the profitability, despite its higher cost (Prasad and Materi, 1990).

Nowadays, WAO is a proven technology with more than 200 industrial installations in operation, with the first commercial unit for the treatment of sulfite liquors being started in the late 1950s (Cybulski et al., 2007). Maugans and Ellis (2002) presented an extensive review of industrial WAO processes.

The efficiency of WAO (conversions and biodegradability of the exit stream) can be increased and the process energy demand simultaneously decreased by the incorporation of a catalyst. The catalyst is believed to participate in the activation of reactants by promoting the formation of the organic radical. Organic pollutant (RH) are adsorbed on the catalytic active sites and easily transformed to radical species (R·) by abstraction of the hydrogen through a redox reaction with the metal (Sadana and Katzer, 1974; Pintar and Leves, 1994).

As it happens with WAO, catalytic wet air oxidation (CWAO) is considered,for partial oxidation of pollutants into more biodegradably amenable intermediates, but now the presence of catalyst fairly minimizes the emission and allows the operation at significantly milder operating conditions and lower residence times. Hence, catalyst positively affects the economics of the process oxidation. According to Levec (1997), the operating costs of the CWAO unit commercialized by Nippon Shokubai, are 1.5-3 times lower than those for a typical WAO.

CONVENTIONAL CATALYSTS IN WAO

The CWAO technology started in the mid-fifties in the United States (Moses, US 2690425). It implied the development of adequate catalysts and new reactor configurations. Several Japanese companies developed CWAO processes based on heterogeneous catalysis whereas in Europe, the focus was on the homogeneous. Table 1 summarizes reactor types, operating conditions, residence times and common catalysts of the current CWAO commercial units. Extensive description of these units has been published in several reviews (Kolaczkowski et al.,1999; Luck, 1996 and 1999, Silva 2009).

The commercial processes are mostly based on homogeneous catalysis (Cu^{2+}, Fe^{2+}). The few heterogeneous processes involve noble metal catalysts (Ru, Pd, Pt). The plants running with homogeneous catalysts need additional processing steps for the recovery and reuse of the metal, *i.e.* by some precipitation/separation technique (Ciba Geigy), which increases the operating cost. The supported noble metals exhibit higher activities but suffer from deactivation. They are vulnerable to poisons including polymers formed upon CWAO conditions and also lose activity because of the metal oxidation. This sensitivity to deactivation can be reduced by the adequate choice of the support (Cybulski et al., 2007).

Table 1. Reactors, operating conditions, residence times and catalysts in commercial catalytic wet air oxidation processes.

Commercial Units	Reactor	T (K)	P_T (MPa)	t_R (h)	Catalyst
Loprox	Bubble column	313-473	0.5-2	1-3 h	Fe^{2+} and H_2O_2 promoter
IT Enviroscience	Stirred Tank	438 - 548	1.2-7	0.5-2	Br^-, NO_3^-, Mg^{2+}
Ciba-Geigy	Bubble column	573	20	3	Cu^{2+}
Orcan	-	120	0.3	-	Fe^{2+} and H_2O_2 promoter
Kurita	-	170	-	-	Supported Pd
ATHOS[§]	-	235	4	-	Cu^{2+}
NS-LC	Trickle-bed	433-543	0.9-8	1	Pt-Pd/TiO_2-ZrO_2 (monolith), Ru/CeO_2 (pellets)
Osaka Gas	Trickle-bed	523	7	0.5	Fe, Cu, Co or Ni/ Ru, Pd, Pt or Au /TiO_2 or ZrO_2

[§] treatment plant for sludge.

Investigation stimulated to search for active and stable catalysts capable of perform the oxidation of a particular kind of wastewater at milder operating conditions has resulted in the proliferation of a vast literature in the CWAO field since the seventies. More than 700 entries can be found searching with the topic *catalytic wet air oxidation* in the ISI Web of Knowledge. This literature survey is classified in Figure 2 attending to different general aspects: sort of document in which the information is provided (a), model pollutants and real wastewater used as inflowing streams (b) and nature of the investigated catalysts (c).

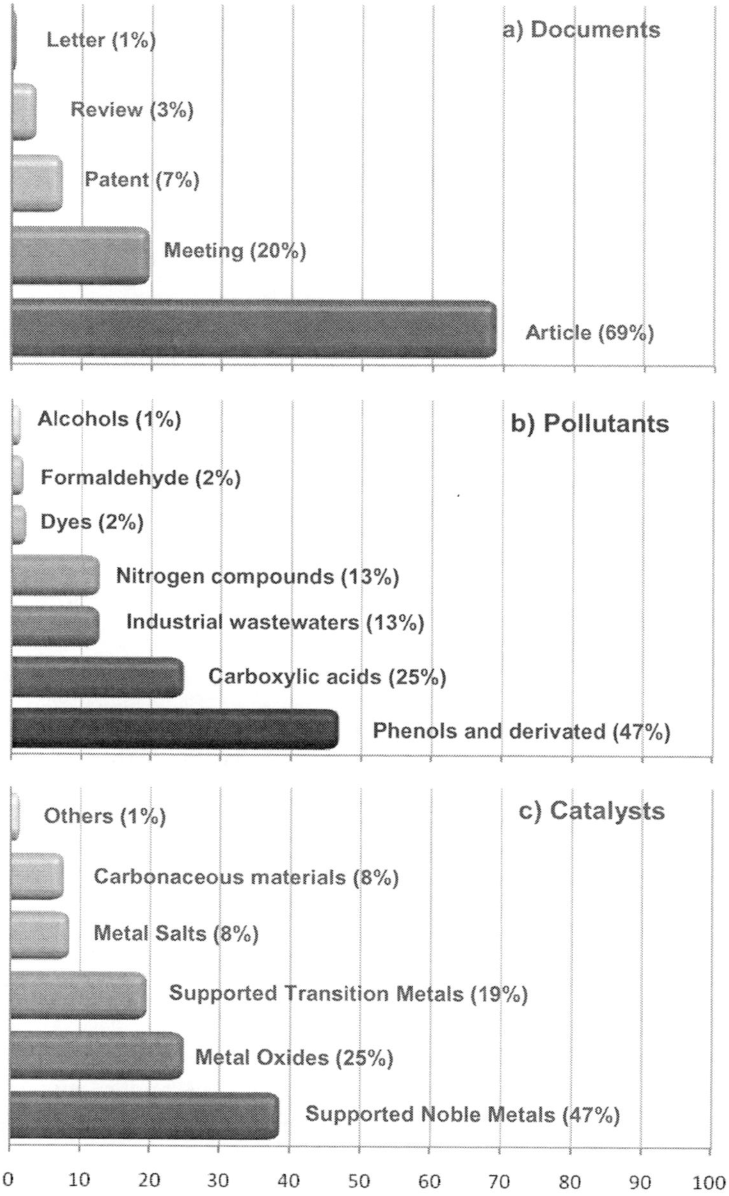

Figure 2. Literature generated in the CWAO field distributed by sort of documents (a), pollutants (b) and catalysts (c) studied. Source: *ISI Web of Knoweldge*.

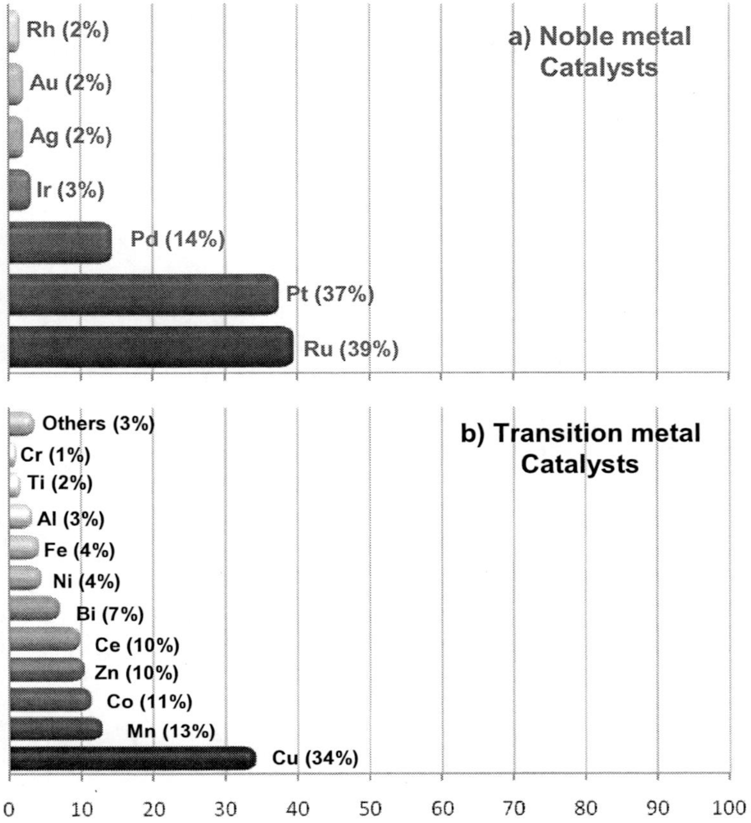

Figure 3. Distribution of papers devoted to noble (a) and transition metal (b) catalysts.

The information deduced from this Figure 2a reveals that only 7% of the documents are patents, in spite of the important proliferation of divulgative documents (articles, meetings, reviews and letters). The patents basically claim for catalysts and operating conditions for the oxidation of model pollutants. These figures manifest the difficulty in the development of suitable catalysts.

Phenols and derivates (chlorophenols and nitrophenols) have been preferentially used as model pollutants followed by carboxylic acids (Figure 2b), the most resistant compounds to the wet oxidation and also by-products in the oxidation of most organic compounds. Nitrogen compounds raised researcher's interest coinciding with the application of CWAO for the treatment of agricultural wastewater polluted by ammonia. An extensive review about the removal of ammonia and nitrogen-containing compounds has been published by Oliviero et al., (2003). Dedication to real wastewater started from the nineties and due to the employ of wastewater of different origin (paper, petrochemical, dye industries, olive processing, wine distilleries etc.) has been possible to be more concise about catalyst durability.

Four types of catalysts tested: (supported) noble metals, (supported and unsupported) metal oxides, metal salts and carbon materials (Figure 2c). The distribution of papers devoted to the different noble and transition metals investigated is complied in Figure 3. Ruthenium, platinum, palladium and copper are the most intensively tested because of their high activity. A comprehensive survey of investigations on catalysts for CWAO processes is provided in the reviews of Matatov-Meytal and Sheintuch (1998), Imamura (1999), Yang et al. (2002), Pirkanniemi and Sillanpää (2002), Bhargava et al. (2006), Levec and Pintar (2007) and Cybulski (2007). In these works, lists of the catalysts used in the oxidation of different compounds along with the operating conditions and reactor types are provided.

Noble Metals

Supported noble metals (ruthenium, platinum, palladium, iridium, silver, and rhodium) are the most studied because of their exhibited high activities, particularly with carboxylic acids. As a consequence, the mineralization to CO_2 and H_2O of high weight molecules such as phenol, *p*-coumaric acid and aniline is usually higher. Different rankings of activity have been reported depending on the model pollutant tested and the metal/support combination. For instance, Okitsu et al., (1995) studied the oxidation of *p*-chlorophenol with noble metals on alumina or titania and found the following activity order: Pt >> Pd > Ru > Rh >Ag. Similar conclusion was later obtained by Qin et al. (2001) when using activated carbon, alumina and ceria as supports. Immaura (1988), compared the activity of the noble metals on alumina, ceria, titania and zirconia in the oxidation of PEG-200, and the activity in terms of TOC followed: Ru = Rh = Pt > Ir> Pd. Barbier et al. (2005), reported for the oxidation of phenol and with ceria as support: Ru > Pd > Pt. Conversely, Trawczýnski (2003) found also for the oxidation of phenol but with carbon black composites (CBC) supports, a different ranking: Pt > Pd > Ru >> CBC. These results evidence the importance of some supports such as ceria and carbon, which results to be catalytically active in the oxidation (Figure 2c and Figure 3b).

In general, noble metal catalysts are comparably more expensive than the rest of catalysts reported, and specially at the high metal loads (1-5wt%) used in the CWAO. In this context, the durability of the catalyst gains relevance. Precious metals are resistant to leaching but more sensitive to poisoning when halogen-, sulphur- and phosphorus-containing compounds are the pollutants. They also deactivate by the fouling originated by the deposition of polymers formed upon reaction. Metal oxidation can also take place leading to a loss of active sites. Different strategies have been studied to increase their longevity: *i)* the use of the appropriate wastewater, *ii)* the use of fixed-bed reactors instead of tanks to employ low liquid-solid ratios in order to reduce the formation of the polymers in the liquid phase (Pintar and Levec, 1992) and *iii)* the addition of small amounts of hydrogen to the gas stream with the aim of maintaining the metallic active sites and reducing the formation of polymer (Kim et al., 2003 and 2005).

Metal Oxides

Supported and unsupported metal oxides (copper, manganese, cobalt, zinc, cerium, and bismuth) have been also tested. The following ranking in the activity was reported by Kochetkova et al., (1992) in the oxidation of phenol: $CuO > CoO > Cr_2O_3 > NiO > MnO_2 > Fe_2O_3 > YO_2 > Cd_2O_3 > ZnO > TiO_2 > Bi_2O_3$. Copper oxide was found to be the most active one. The intensive literature devoted to this catalyst and the main goals achieved has been recently reviewed by Cybulski (2007). Its principal limitation comes from the insufficient durability caused by the acidic pH of the media, inherited to the CWAO, which provokes the metal leaching. Nevertheless, the low cost of these catalysts compared to noble metals has stimulated their investigation even at the cost of lower activity. The investigation on the improvement of the catalytic performance of the metal oxides led to the development of composite oxide catalysts (Ce/Al, Mn/Ni, Mn/Ce, Co/Bi, Co/Mo, Cu/Zn, Cu/Fe, etc.). Copper catalysts with basic active sites like bismuth, cobalt and manganese exhibit relative high activities in the oxidation of carboxylic acids and also reduced the copper leaching (Cybulski, 2007).

The most effective means to prevent leaching accounts for modifications on catalyst preparation. As an example, Hocevar et al. (1999) prepared the mixed $Ce_{1-x}Cu_xO_{2-d}$ by precipitation and sol-gel methods and copper was much less soluble than the copper oxide dispersed on ceria and prepared by impregnation. Alejandre et al. (1998) selected the appropriate metallic phase, *viz.* spinel phase, which resulted to be stable by modifying the calcination temperature. The operating conditions, in particular the pH of the inlet wastewater, can be also manipulated to avoid leaching. The work of Santos et al. (2005a) demonstrated that the strategy of buffering phenolic wastewater to basic values (pH=8 with bicarbonate) succeeds in maintaining the stability of the CuO/Cr_2O_3/graphite catalyst (copper was neither leached nor fouled), though, in addition, copper selectivity is affected. Bicarbonate anions are scavengers of phenol peroxide radicals and consequently different aromatic by-products to those usually detected at acidic conditions are formed. Also, lower reaction rates were observed which was explained by the absence of the homogeneous contribution to the overall oxidation reaction.

Metal Salts

Copper, as well as, niquel, iron, cobalt, chromium and manganese salts were first investigated. The supremacy of the copper activity is exemplified in CIBA unit that uses $CuSO_4$ as catalyst. Due to the oxidizing conditions, the active metals are usually in their highest oxidation states. Their activity is affected by the counteranions (Immamura, 1982), the pH of the reaction media and the nature of the pollutant and by-products formed (Santos et al., 2005b). All these factors ultimately affect the metal solubility.

Carbon Materials

Carbon materials such as activated carbons, graphites and carbon blacks are active catalysts. Stuber et al. (2005) reviewed their application for CWAO processes. The supremacy in the number of papers using activated carbons over the rest of carbon materials was evidenced in this work. Activated carbons are relevant catalysts because of their high adsorption capacity towards several compounds, in particular aromatic compounds and polymers, which are non-biodegradable and ecotoxic. The adsorption contribution to the removal of these compounds increases the efficiency of the treatment when activated carbons are used. The activity of the activated carbon is attributed to the concentration of oxygen bearing functional groups present on the surface. In general, the mild conditions required in order to avoid their partial combustion (T< 433 K and p_{O2}<0.9 MPa, Mundale et al. 1996) invites to use metal active phases, such as noble metals (Ru, Pd, Pt) and metal oxides (Fe_2O_3 and CuO_2) or to use a promoter such as hydrogen peroxide (Quintanilla et al. 2010 and Rubalcaba, 2007) to enhance the activity. Quintanilla et al. (2010) demonstrated the better performance (initial rates, oxidation and mineralization degrees and ecotoxicity removal) in presence of that promoter than iron. Figure 4 shows those results by representing the removal curves of phenol, TOC and ecotoxicity, the latter expressed in toxicity units (TU).

Activated carbons are physical and chemically modified upon reaction. Their surface evolves as a consequence of the oxidation treatment and the pollutant adsorption. The specific surface area significantly reduces and the number of oxygenated surface groups importantly increases but these changes do not affect their catalytic activity (Quintanilla et al., 2007). As an example, the activity of Fe/activated carbon catalyst was unperturbed during 200 h of time on stream in CWAO of phenol at 400 K and 0.8 MPa (Quintanilla et al., 2007). This adequate durability makes them very attractive catalysts for CWAO.

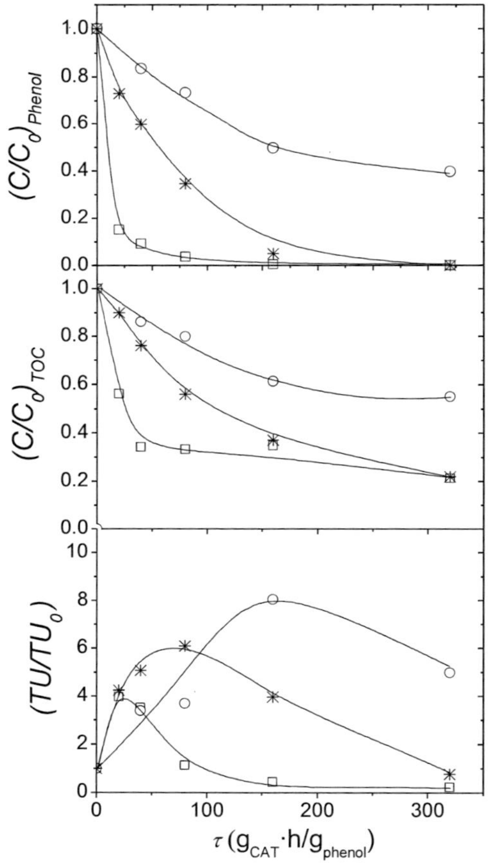

Figure 4. Catalytic activity and ecotoxicity of the effluent in the CWAO of phenol in a trickle-bed reactor over activated carbon (○), 2.5wt% Fe/activated carbon (*) and activated carbon plus H_2O_2 promoter (□). Reaction conditions: C^{0}_{phenol} =1 g/L, Q_{O2}=91.6 NmL/min, w_{cat}=2.5 g, T=400 K, p_{O2}=8 atm. Lines to guide the eye.

Pirkanniemi and Sillanpää (2002) reviewed the supports tested in CWAO. The most common are metal oxides, *viz.* TiO_2, Al_2O_3, ZrO_2 and CeO_2, and activated carbons (Duprez et al., 1996; Cao et al., 2003; Oliviero et al., 2000; Gomes et al., 2000, 2002a, 2002b; Quintanilla et al., 2006). As aforementioned, ceria and carbon are the most interesting supports since improve the catalyst activity. Ceria-containing catalysts are stable and enhanced textural and catalyst redox properties at low temperatures (Trovalrelli et al., 1999; Oliviero et al., 1999; Cylbuski, 2007). Activated carbons are low cost materials and interesting candidates when the wastewater (sort and amount of pollutant) admits mild oxidation conditions.

The experience gathered on the application of the described catalysts, compiled in Figures 2c and Figure 3, in the remediation of a wide array of wastewater (Figure 2b) points out that the selection of the most adequate catalyst is ultimately decided by the nature and amount of the pollutant. The starting hazardous compound and the reaction by-products are decisive for the catalyst durability for several reasons: they can be prone to polymerize giving rise to catalyst fouling and they can contribute either to decrease the pH or form complexes with the metal (Santos et al., 2005; Perathoner and Centi, 2005), promoting, in both cases, the leaching of the active phase in heterogeneous systems or modifying the solubility of the metal salt in homogeneous systems, thus negatively affecting the activity. As an example, two different works came out with a different conclusion regarding the stability of a particular catalyst, CuO-ZnO/Al_2O_3. This catalyst was applied in the CWAO of *p*-coumaric acid (Mantzavinos et al., 1996) and formaldehyde (Silva et al., 2003) at not very different operating conditions. Only in the former, an obvious elution of copper, zinc and aluminium was detected and the homogeneous contribution to the overall oxidation reaction could not be discarded. Therefore, the final selection of catalysts will be dictated by the nature of the particular industrial wastewater to decontaminate.

With the above background, one can deduce that researches investigating in the CWAO field still need to face significant challenges: *i)* searching for economic and environmentally friendly catalysts and *ii)* the optimization of catalyst activity while maintaining a high stability. The persecution of these goals has resulted in the exploration of nanomaterials for catalysts in the CWAO technology.

Nanotechnology in Wet Air Oxidation

The current trends in CWAO explore the use of nanomaterials. The application of nanotechnology in the water field is expected to produce environmental benefits in terms of water management and treatment by improving filtering, decontamination, desalination, conservation, recycling, analysis and monitoring of sewerage systems. The investigation about the application of nanotechology in water treatments can be found in recent reviews (Theron et al. 2008; Derhmendra et al., 2008; Ong et al., 2010). There, filtration and adsorption are evidenced as the dominating application so far. In addition, dechlorination and disinfection by photocatalysis are also processes gaining relevance. Chemical oxidation was not mentioned which reveals the incipient interest of nanomaterials in environmental oxidation technologies.

Table 2. Emerging catalysts in catalytic wet air oxidation treatments.

Reference	Catalyst	Pollutant	poll./cat. (mol:mol)	*T* (K)	*P* (MPa)	Results
Gomes et al. 2004	1% wt. Pt/MWCNT	Aniline	39.3	473	0.7	$X_{aniline}$ = 77% (1 h), 98.8% (2 h)
	1% wt. Pt/AC		39.3			$X_{aniline}$ = 95% (1 h), 100% (2 h)
	1% wt. Pt/CX		39.3			$X_{aniline}$ = 98% (1 h), 100% (2 h)
García et al. 2005	1% wt. Pt/MWCNT	Aniline	39.3	473	0.7	$X_{aniline}$ = 99.4 %, X_{TOC} = 78.3% (2 h)
	1% wt. Pt/AC	Aniline	39.3			$X_{aniline}$ = 100 %, X_{TOC} = 94.5% (2 h)
	1% wt. Pt/MWCNT	Azo dye SG	2.8			X_{dye} = 99.5%, X_{TOC} = 21.2% (2 h)
	1% wt. Pt/MWCNT	Azo dye ERB	8.1			X_{dye} = 99.5%, X_{TOC} = 78.1% (2 h)
	1% wt. Pt/MWCNT	Azo dye C2R	8.4			X_{dye} = 100%, X_{TOC} = 63.5% (2 h)
	1% wt. Pt/MWCNT	Textile wastewater	-			X_{TOC} = 51.2%, X_{COLOR}=100% (2 h)
García et al.2006	1% wt. Ru/MWCNT	Aniline	20.3	473	0.7	$X_{aniline}$ = 100%, X_{TOC} = 80.8% (45min)
	1% wt. Ru/AC		20.3			$X_{aniline}$ = 100%, X_{TOC} = 96.4% (45min)
Ovejero et al. 2006	1% wt. Pt/N-MWCNT	Aniline	-	473	0.42	$X_{aniline}$ = 62.4%, X_{TOC} = 55.3% (0.03h^{-1})
	1% wt.Pt/AC		-			$X_{aniline}$ = 97%, X_{TOC} = 85% (0.03h^{-1})
	1% wt. Cu/N-MWCNT		-			$X_{aniline}$ = 60.1%, X_{TOC} = 57.5% (0.03h^{-1})
	1% wt. Ru/N-MWCNT		-			$X_{aniline}$ = 45.6%, X_{TOC} = 43.2% (0.03h^{-1})
Yang et al. 2008	MWCNT	Phenol	0.0796[§]	433	2	X_{Ph} = 100%, X_{TOC} = 100% (2 h)
Rodríguez et al. 2010	3% wt. Fe/MWCNT	Methylene Blue	0.0031	413	0.88	X_{TOC} = 35%, X_{COLOR} = 82% (2.5 h)
	3% wt. Cu/MWCNT		0.0035			X_{TOC} = 40%, X_{COLOR} = 90% (2.5 h)
Rodríguez et al. 2008	3% wt. Cu/CNF	Textil Wastewaters	157.8	413	0.88	X_{TOC} = 74.1%, X_{COLOR} = 97% (3 h)
	3% wt. Cu/AC		157.8			X_{TOC} = 71.5%, X_{COLOR} = 90% (3 h)
Taboada et al. 2009	2% wt. Pt/CNF	Phenol	-	453-513	1	Catalyst deactivation
	2% wt.Pd/CNF		-			
	2% wt.Ru/CNF		-			
Soria Sánchez et al. 2009	1% wt. Fe acetylacetonate/CNF	Phenol	27.9	413	2	X_{Ph} = 100 %, X_{TOC} = 90% (2.5 h)
Sousa et al. 2010	CNF	Aniline	0.0901[§]	473	0.7	$X_{aniline}$ = 95 %, X_{TOC} = 90% (5 h)
	Carbon foam		157.8			$X_{aniline}$ = 78 %, X_{TOC} = 77% (5 h)
Gomes et al. 2008	CX	Anilina	0.0288[§]	473	6.9	$X_{aniline}$ = 99%, X_{TOC} = 86% (5 h)
Apolinário et al. 2008	CX	MNB,	0.0006[§]	473	0.69	$X_{DNT, TNP}$ > 90%, X_{TOC} = 80%(2 h)
		DNP	0.0020[§]			
		TNP	0.0006[§]			
Besson et al. 2003	5% wt. Au/TiO_2	Succinic acid	50.0	463	0.69	$X_{succinic}$ = 88% (6 h)
Milone et al. 2006	3.8% wt. Au/CeO_2	*p*- Cumaric acid	4.7	353	0.42	X_{TOC} = 85% (2.5 h)
Levi et al. 2008	Mn-Ce (Nanocasted)	Aniline		413	1	$X_{aniline}$ = 100%, X_{TOC} = 90% (5 h^{-1})
Yang et al. 2007	$LaCoO_3$ perovskite	Salycilic acid	0.3	383	0.5	X_{COD} = 85% (2 h)
Royer et al., 2008	$LaFeO_3$ and $LaMnO_3$	Stearic acid	1.4	473	2	Pollutant: X_{LaFeO3} = 65%, X_{LaMnO3} = 60% (3 h)
Zhang et al., 2009	$Zn_{1.5}PMo_{12}O_{40}$ nanotubes	Safranin-T	0.01	298	0.1	$X_{Safranin}$ = 90% (35 min)
Zhao et al. 2010	$[(C_nH_{2n+1})N(CH_3)_3]^{3+}{}_xPV_xMo_{12-x}O_{40}$	Phenol	0.8	298	1	X_{Ph} = 95,3 %, X_{COD} = 98,5 % , X_{TOC} = 93% (1.5 h)

MWCNT: multi-walled carbon nanotube; AC: activated carbon; CX: carbon xerogel. [§] per mol of carbon.

Based on the literature, carbon nanostructures, *viz.* nanotubes, nanofibers and xerogels, and gold nanoparticles can be benchmarked as the most studied nanomaterials. By taking advantages of their extraordinary physical, chemical and electronic properties, they represent, along with the zero-valent iron nanoparticles (Theron et al., 2008), the first generation of nanoscale environmental technologies. The exciting results provided by carbon nanostructures (Ong et al., 2010) and gold (Bond et al., 2006) when they are exploited as catalysts in the hydrodechloration technology, *i.e.* removal of trichloroethene, one of the most common hazardous groundwater pollutants, has recently opened interesting opportunities in water decontamination technologies..

The recent research carried out on the catalytic applications of both carbon nanostructures and gold in WAO processes is collected in Table 2. As can be seen, carbon nanostructure studies have proliferated faster than those based on gold. The reasons point out to *i)* the existing large-scale production of the carbon nanostructures, which allows using commercial ones, while gold catalysts are usually home-made, *ii)* the modest number of factors affecting the activity of carbon nanostructures, which is preferentially ascribed to the surface chemistry, *iii)* the better knowledge of the carbon than gold chemistry. The activity of gold is still unclear and many aspects such as the preparation method, particle size, gold-support interaction and the nature of the supports are crucial for the gold nanoparticle activity, and finally, *iv)* the longer durability of carbon nanotubes than gold nanoparticles during oxidation.

The major aspects of carbon nanostructures and gold catalysts gathered along the short experience on their application in CWAO processes (Table 2) are next discussed.

Carbon Nanostructures

The potential characteristic that has encouraged the application of carbon nanostructured materials in CWAOs is their high external surface areas. The mesoporous surface provides an efficient surface contact between the reactants and the active sites, which is especially important in liquid-phase reactions where the mass transfer phenomenon becomes very significant. The carbon nanostructures are mostly used as catalytic supports, though they exhibit activity after activation (Yang et al., 2008). The activation consists in the chemical modification of the surface nanotubes/nanofibers usually by wet chemistry methods in which HNO_3 and HNO_3/H_2SO_4 are used as oxidants. As a result, functional groups such as carboxylic (-COOH), carbonyl (-CO) and hydroxylic (-OH) are formed on the surface of nanostructure. These oxygenated groups favor the adsorption of metal precursor and act as nucleating sites during the formation of the metal sites. Functionalization also enriches the reactivity of the nanostructures and modifies their wetting characteristics. The proper wetness is vital to avoid direct exposition of the carbon surface to oxygen in the gas stream and the consequent gasification of the carbon nanostructures. Taboada et al., (2009) experienced that the gasification in non-functionalized carbon nanofibers supporting noble metals (Ru, Pt and Pd) in the oxidation of phenol in a trickle-bed reactor.

The work of Yang et al., (2008) is the only one so far studying the applicability of bare carbon nanotubes as catalysts. The functionalized multi-walled carbon nanotubes (MWCNT) exhibited both high activity (97.2% and 77% phenol and TOC conversion, respectively, in 2 h of reaction) and good stability in several running cycles in the CWAO of phenol. The absence of some toxic by-products typically found in the phenol oxidation route, such as catechol and *p*-benzoquinone, assists the potentialities of carbon nanotubes in the CWAO treatment of phenolic wastewaters.

As was previously commented, carbon nanostructures are usually explored as supports for preparing efficient solid catalysts. As shows Table 2, the active phases are noble metals, Ru, Pt and Pd (Gomes et al. 2004; García et al., 2005; Taboada et al. 2009) and metal oxides, Cu and Fe (Ovejero et al. 2006; Rodríguez et al., 2010). A global analysis of these works, fundamentally applied to the oxidation of aniline, induces that CWAO is efficiently performed using metal/carbon nanostructures catalysts. However, outstanding catalytic properties comparing to conventional catalysts, in particular activated carbon-based catalysts, are not observed (Garcia et al., 2005 and 2006; Gomes et al., 2004; Rodriguez et al., 2008). In fact, the mineralization degree achieved with activated carbons is usually higher because of their higher adsorption capacity that contributes to overestimate the activity. Durability of the carbon nanostructures is also an aspect to be refined. Leaching of Ru and Cu has been reported (Garcia et al., 2006; Rodriguez et al., 2008) suggesting a weak metal-support interaction. On this regard, the recent work of Soria-Sanchez et al. (2009) reports about an elegant alternative to the usual wet impregnation method selected for the metal immobilization on the nanotubes/fibers. The novel method consists in the preparation of Fe/carbon nanofiber catalysts by the anchoring of iron through complexation of iron ions on the acetylacetonate functional groups created on the functionalized carbon nanofibers. These catalysts demonstrated their good activity and stability in successive runs.

In addition to carbon nanotubes and nanofibers, xerogels have also been used as support (Gomes et al., 2004) and catalyst (Apolinario et al., 2008). The novel work of Gomes et al., (2004) comparing the activity of platinum supported on xerogel, activated carbon and nanotubes demonstrated that Pt/xerogel provides higher aniline removal and similar mineralization than Pt/carbon in contrast to the lower performance of Pt/nanotubes. This work opened the door to the exploitation of theses materials in CWAO. Xerogels have also been used as catalysts for the oxidation of nitrophenols in water at low concentrations (Apolinario et al., 2008). The activity was high (almost complete conversion of nitrophenols and 78% TOC conversion in 30 min) but with a contribution of adsorption higher than 50% in the pollutant removal. The xerogel stability upon oxidation is promising (MM) but long-term experiments are still required.

In conclusion, the results reported confirm the potentialities of the new carbon materials for the development of catalysts for CWAO but there is still a significant scope for further research and innovation.In fact, conclusive results on the application with real wastewaters, deactivation vulnerability and biodegradability of the effluents are required for a cost-efficiency evaluation. Efforts should be made to optimize catalyst preparation and reaction conditions to enhance efficiencies and prevent deactivation. The use of these carbon nanostructures as catalysts without a metal active phase can be a reasonable solution to gain in stability even at the expenses of lower activity.

Gold Nanoparticles

The work carried out with supported gold catalysts is gathered in a few papers (Besson et al., 2003; Milone et al. 2006; Trana et al., 2008). Wastewater containing different organic pollutants has been tested in these studies (see Table 2). Nitrogen-bearing pollutants are also good candidates as demonstrated by the Nippon Shokubai patent (JP 06142660 A) for the treatment of wastewater containing this type of compounds with an oxide catalyst on which gold and optionally platinum are supported.

Besson et al. (2003) were the first to show the applicability of gold in CWAO. They evaluated the performance of Au/TiO_2 catalysts in the oxidation of succinic acid at 463 K and 2 MPa. In this work, several catalysts were prepared by deposition-precipitation following different protocols with the aim of synthesis selected-size nanoparticles. It was demonstrated that the gold activity was higher as the particle size decreased. The 2 nm particles were the most active, exhibiting a TOF value of 16 h^{-1} and complete conversion after 6 h of reaction. Nevertheless, the activity of Ru/TiO_2, the noble metal catalyst used for comparison, was superior, achieving complete conversion of succinic acid after 3 h at a higher pollutant/gold molar ratio. In addition, the gold activity decreased in a second run, the initial rate being reduced from 9.7 to 2.3 $mol_{succ}/mol_{Au}/h$. No leaching of gold was detected but sintering of the nanoparticles, from 2 to 4 nm, was observed. Analysis of the carbonaceous deposits on gold would be required for further conclusive knowledge of the causes of deactivation. Further studies carried out by the same research group (Trana et al., 2008) proved the activity of Au/CeO_2 catalysts in the oxidation of acetic acid. The catalyst was prepared by deposition-precipitation with urea and subsequently subjected to two alternative thermal treatments, calcination at 773 K in air or reduction at 573 K in H_2 atmosphere. The different distribution of gold species resulting on the ceria supports combined with the activity measurements demonstrated that Au^0 species is required to achieve a high catalytic performance. The catalyst showed aging due to sintering.

Milone et al. (2006) did not succeed in the preparation of Au/CeO_2 catalysts by coprecipitation and deposition-precipitation. According to XRD analyses, gold nanoparticles were only present in the catalysts prepared by coprecipitation, and their average size was 18.6 nm. The presence of gold in none case increased the own activity observed with ceria relative to *p*-cumaric acid conversion. The slight increase in the effluent TOC conversion was explained by the carbonaceous deposits formed on the catalyst surface induced by the presence of gold.

The above results state the activity of immobilized gold nanoparticles in the oxidation of carboxylic acids and also their vulnerability to deactivation. Therefore, CWAO is a fertile area for the application of gold catalysts. In this line and considering the huge variety of wastewater, the previous concern of our research group about the application of gold in CWAO was related to the catalyst versatility. For this, we selected alcohols, phenols and anilines as pollutants in our study. They are less refractory to the oxidation than carboxylic acids, fact that allows the use of milder operation conditions, which is expected to prevent to some extend, the gold sintering.

Two commercial catalysts with different average particle size were tested: 0.9wt%Au/Al_2O_3 (Au_{dp}=1.48±0.02 nm; Mintek, batch: BC17) and 0.8wt%Au/C (Au_{dp}=10.5 nm; World Gold Council). The supports were selected based on the previous works of Prati and Rossi (1998) in which these catalysts were successfully tested in air-oxidation of polyhydroxylated molecules in aqueous basic solutions. The pollutants selected were all aromatic compounds: benzyl alcohol, phenol and aniline. The oxidation experiments were carried out in a slurry reactor at 398 K and 0.6 MPa. The reactor was initially charged with 0.5 g of solid and 200 mL of 1g/L of pollutant solution at pH=10 (adjusted with NaOH) and a continuous flow of oxygen fed to the reactor.

Figure 5a provides the time-evolution of target pollutant, TOC and COD conversions in presence of Au/Al_2O_3 catalyst. As observed, the 1.48 nm particle sizes on alumina are active in the oxidation of both benzyl alcohol and phenol. Results with respect to aniline are not provided in the figure because gold was revealed as slightly active catalyst in the oxidation of this compound since similar conversions as in the blank experiment were obtained ($X_{aniline.}$,$X_{TOC,}$ X_{COD} = 24, 22, 11% with Au/Al_2O_3 after 3 h of reaction whereas 19, 16, 8% were observed in the blank experiment). Therefore, gold activity varies attending to the nature of the target pollutant. The benzyl alcohol-gold system provides better results that the phenol-gold whereas aniline-gold system is scarcely active. The lower reactivity of phenol is in line with the results of Tsunoyama et al. (2006) for the air-oxidation of *p*-hydroxybenzyl alcohol on gold nanoparticles in solution, where this alcohol was oxidized to the corresponding aldehyde without altering the phenol group. Complete conversion was achieved for benzyl alcohol whereas only a 25% of phenol was converted after 3 h of reaction. Similar differences are also observed in TOC and COD conversions.

The different gold activity can be induced by the different pollutant-gold interaction. Oxygen atoms coordinate more predominantly than nitrogen on gold (Bonet et al., 2000). We do not discard that the site occupation by water molecules (Yang et al., 2008) could also contribute to the poor affinity between gold and aniline. In the case of alcohols and phenols though both molecules containes a hydroxyl group a different interaction with gold is expected.

Reviewing the extended literature on the oxidation of benzyl alcohol over gold catalysts in organic solvents, it is generally accepted the formation of the gold-benzyl alcoholate species which evolves to more oxidized products, typically benzaldehyde, via β-hydrogen abstraction and without the participation of the oxidant, which eventually will regenerate the active site (Mallat and Baiker, 2004; Abad et al., 2008; Brink et al., 2000). In the case of phenols, the above mechanism cannot be assumed because of the absence of the β-hydrogen in the phenol molecule. By electrochemical analysis, phenol has been demonstrated to be covalently adsorbed on gold electrodes by the oxygen atom and weakly adsorbed by the aromatic ring (Lezna et al., 1991), a similar situation could be ascribed in the case of gold nanoparticles. Therefore, we speculate that the phenol chemisorption on gold *vs.* the gold-alcoholate complex formation dictates the lower oxidation rate observed when phenols instead of alcohols are the pollutants in CWAO processes over gold catalysts. To obtain experimental evidences on the different interaction between the target organic molecules and gold, experiments under inert atmosphere were performed for each individual compound and Au/Al_2O_3. The results show that the removal of benzyl alcohol was only slightly favored than that of phenol (40 g of benzyl alcohol *vs.* 35 g of phenol per gram of gold disappeared after 24 of pollutant-gold contact) but, however, in the former, benzoic acid was progressively detected through time, reaching up to 40 mg/L after 24 h of contact. These results confirm the reactivity between alcohol and gold and also show the participation of water on the gold selectivity since benzoic acid instead of benzaldehyde is the major by-product obtained (Hou et al., 2008). During the phenol experiment, by-products were never detected in the liquid media and TOC and phenol conversions were coincident. Thus, only adsorption, no reaction, was the contribution to the phenol disappearance. In conclusion, the different reactivity of phenols and alcohols with gold invokes the different gold activity in the CWAO.

Figure 5b shows the results obtained over both Au/C model catalyst and a bared activated carbon (Merck, ref.: 102514), in order to elucidate the activity manifested by gold. Both solids were previously saturated in the corresponding pollutant by mixing the pollutant solution with the catalyst in the same ratio as in the reaction in order to avoid the adsorption contribution to the disappearance of the pollutants. In general, Au/C (Au_{dp}=10.5 nm) exhibited lower activity than Au/Al_2O_3 (Au_{dp}=1.48±0.02 nm) catalyst, which shows that the WAO of organic compounds is a size-dependence reaction as concluded by Besson et al., (2003). The difference in the reactivity between the tested target pollutants and gold is again observed and even more acute now because Au/C does not provide activity with phenol (Figure 5b).

In order to study the durability of Au/Al_2O_3 catalysts, the spent sample after the oxidation of benzyl alcohol was re-used in a second cycle. The results are also depicted in Figure 5a. As can be observed, TOC and COD conversions were half reduced whereas benzyl alcohol conversion was still maintained after 4 h of reaction though a decrease from 71 to 47 h^{-1} is obtained in the initial TOF values. The analyses subjected to the fresh and spent catalysts reveal that gold content and particle size remain over use but carbonaceous material is deposited on the catalyst surface (the carbon weigth percentage reached 4% after two reaction cycles, as confimed by elemental analyses). The deposits are decomposed between 773 and 873 K (according to thermogravimetric analyses under air and inert atmosphere). Therefore, it is reasonable to assume the polymeric nature of the carbonaceous material, more considering the susceptibility of aromatic pollutants to polymerize at the CWAO conditions.

In view of the above discussion, the application of gold in CWAO must be carefully considered attending to the nature of the pollutant. Investigation must be addressed in order to enhance gold activity and increase durability, in particular when dealing with pollutants prone to polymerize. Based on the noble metal catalysts studies, the use of appropriate supports, such as ceria and mixtures of ceria with others oxides, could help in these duties. Also, carbon supports may be interesting when the gold particle size is small (< 5 nm). The use of gold bimetallic catalysts can be also an alternative. As shows in fine-chemistry reactions, gold activity can be enhanced by their collaboration with other noble metals (Hutchings, 2007; Sankar et al., 2009). The investigation about the nature of active sites and the reaction mechanism may provide criteria for optimizing the preparation of gold catalysts and the selection of the support. Obviously, the way towards the industrial application of gold requires experimental work on real wastewater.

Other Emerging Catalysts

More sophisticated catalysts are being developed looking particularly for supreme stability. Most of those novel catalysts are listed in Table 2. They are metals oxide nanostructures, perovskites and polyoxometalates.

A novel nanocasted Mn-Ce-oxide catalyst was tested in the CWAO of aniline in a trickle-bed reactor (Levi et al., 2008). In this case, opposite to the observations with metal oxides, acidification of the solution with HCl was necessary to avoid colloidization and leaching of the nanoparticulate catalyst components by complexation with aniline. This catalyst showed stable performance for over 200 h on stream.

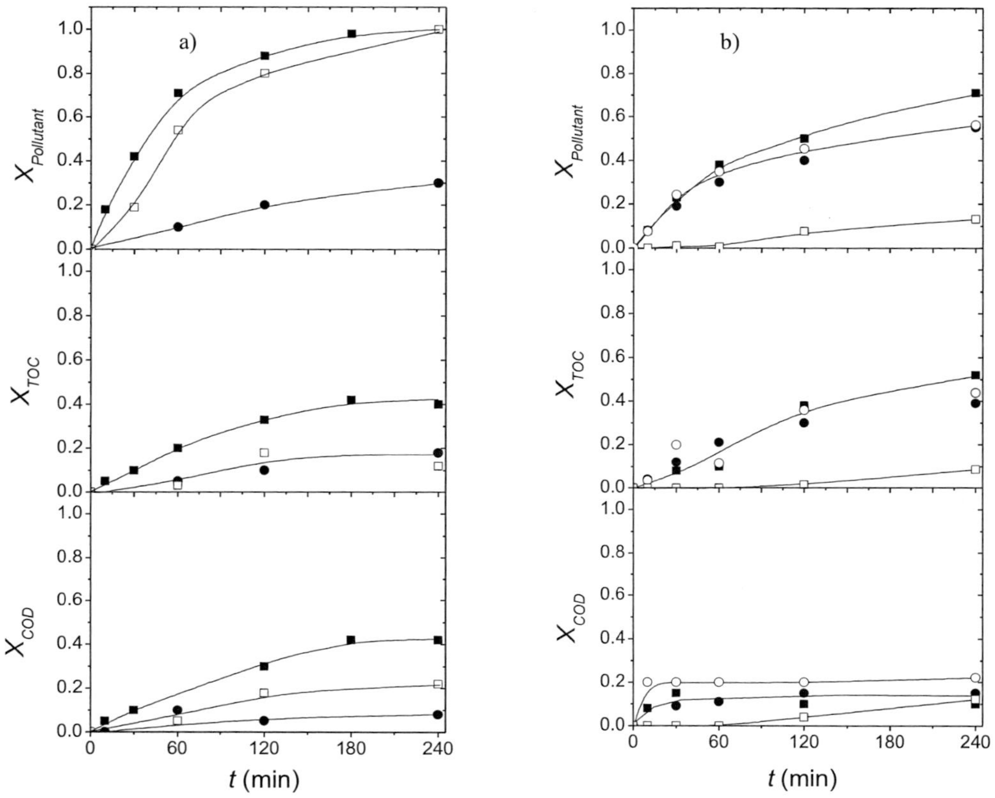

Figure 5. Activity of Au/Al_2O_3 (a) and Au/C and C (b) catalysts in the CWAO of benzyl alcohol (squares) and phenol (circles) expressed in terms of conversions. Reaction conditions: $C^0_{pollutant}$ =1 g/L, Q_{O2}=91.6 NmL/min, V_L=200 mL, pH=10, w_{cat}=0.5 g, T=398 K, p_{O2}=6 atm. Key: (a) ■ fresh catalyst; □ re-used catalyst; (b) closed symbols for Au/C and opened symbols for C. Lines to guide the eye.

The perovskite-type catalysts $LaFeO_3$, $LaCoO_3$ and $LaMnO_3$ prepared by the sol-gel method, have been also presented as novel catalysts (Yang et al., 2007; Royer et al., 2008). Their stability must be studied in each particular case. For instance, Yang et al., (2007) reported activity and structure stability of $LaFeO_3$ in the CWAO of salicylic acid and sulfonic salicylic acid only at T<413 K and Royer et al. (2008) observed the collapse of the $LaCoO_3$ and $LaMnO_3$ structures in the oxidation of stearic acid because of the formation of stable bulk carbonates with lanthanum.

Polyoxometalates, also named heteropolyacids, are oxo-clusters of early transition metals (groups 5 and 6) in their highest oxidation states, namely molybdenum, wolfram or vanadium. They represent an increasingly important class of environmentally benign catalysts that can be used at room temperature for the abatement of water pollutants at low concentrations (COD<200 mg/L). For instance, $Zn_{1.5}PMo_{12}O_{40}$ nanotubes have been synthesized for the wet oxidation of Safranin-T, a hazardous textile dye (Zhang et al., 2009), and micellar catalysts $[(C_nH_{2n+1})N(CH_3)_3]^{3+}{}_xPV_xMo_{12-x}O_{40}$ (x=1, 2, 3; n = 8–18) for phenol (Zhao et al., 2010).

Conclusions

Catalytic wet air oxidation is a developed technology with demonstrated industrial applications. The search for active and stable solid catalysts capable of working at milder operation conditions is an ongoing challenge. In this scenario, wet air oxidation opens new opportunities for the commercial exploitation of nanomaterials such as carbon nanostructures and gold nanoparticles as catalysts. The better knowledge about the chemistry of carbon along with the existing large-scale manufacturing of carbon nanostructures have led to a faster proliferation of research on these materials than on gold. Xerogels are presented as powerful catalysts. Good activity and adequate stability are reported in the few papers existing so far but long-term experiments in continuous mode are claimed for a better knowledge on their stability. Research on the potential application of gold in CWAO is still incipient but deactivation is identified as an important limitation, which opens the door to the use of gold bimetallic catalysts, as occurs for some fine-chemistry reactions. Cost-effective manufacturing and outstanding catalytic performance must be still demonstrated by the nanocatalysts to allow their widespread application in wet oxidation.

Acknowledgments

The authors acknowledge the Ministerio de Ciencia e Innovación (MICINN) of Spain for the financial support through the projects CTQ2008-03988 and CTQ-2010-14807.

References

Abad, A.; Corma A.; Garcia, H. Catalyst Parameters Determining Activity and Selectivity of Supported Gold Nanoparticles for the Aerobic Oxidation of Alcohols: The Molecular Reaction Mechanism. *Chem. Eur. J.* 2008, 14, 212-222.

Alejandre, A.; Medina, Fortuny, A.; Salagre, P.; Sueiras, J. E. Characterisation of copper catalysts and activity for the oxidation of phenol aqueous solutions. *Appl. Catal. B* 1998, 16 (1) 53-67.

Apolinário, A. C.; Silva, A. M. T.; Machado, B. F.; Gomez, H. T.; Araújo, P. P.; Figueiredo, J. L.; J. L. Faria. Wet air oxidation of nitro-aromatic compounds: reactivity on single and multi-component systems and surface chemistry studies with a carbon xerogel. *Appl. Catal. B Environ.* 2008, 24, 75-86.

Barbier, J., Jr; Oliviero, L.; Renard, B.; Duprez, D. Role of ceria supported noble metal catalysts (Ru, Pd, Pt) in wet air oxidation of nitrogen and oxygen containing compounds. *Top. Catal.* 2005, 33 (1-2-3-4), 77.

Besson, M.; Gallezot, P. Deactivation of metal catalysts in liquid phase organic reactions. *Catal. Today* 2003, 81 (4), 547-559.

Bhargava, S. K.; Tardio, J.; Prasad, J.; Föger, K.; Akolekar, D. B.; Grocott, S. C. Wet oxidation and catalytic wet oxidation. *Ind. Eng. Chem. Res.* 2006, 45, 1221-1258.

Bond, G. C.; Louis, C.; Thompson, D. T. *Catalysis by Gold*, Imperial College Press, London, 2006, Chapter 11, page 305.

Bonet, K.; Teikaia-Elhsissen, K.; Vijaya Sarathy, K. Study of interaction of ethylene glycol/PVP phase on noble metal powders prepared by polyol process. *Bull. Mater. Sci.* 2000, 23, 165.

Brink, G.; Arends, I. W. C. E.; Sheldon, R. A. Green, catalytic oxidation of alcohols in water. *Science*, 2000, 287 1636-1639.

Cao, S.; Chen, G.; Hu, X.; Yue, P. L. Catalytic wet air oxidation of wastewater containing ammonia and phenol over activated carbon supported Pt catalysts. *Catal. Today* 2003, 88 (1-2), 37-47.

Cybulski, A. Catalytic Wet Air Oxidation: Are Monolithic Catalysts and Reactors Feasible?. *Ind. Eng. Chem. Res.* 2007, 46, 4007-4033.

Debellefontaine, H.; Foussard, J.N. Wet air Oxidation for the treatment of industrial wastes. Chemical aspects, reactor design and industrial applications in Europe. *Waste Mang.* 2000, 20, 15-25.

Duprez, D.; Delanoe, F.; Barbier, J., Jr.; Isnard, P.; Blanchard, G. Catalytic oxidation of organic compounds in aqueous media. *Catal. Today* 1996, 29 (1-4), 317-322.

Garcia, J.; Gomes, H. T.; Serp, Ph.; Kalck, Ph.; Figueiredo J. L.; Faria, J. L. Platinum catalysts supported on MWNT for catalytic wet air oxidation of nitrogen contaning compounds. *Catal. Today*, 2005, 102-103, 101-109.

Garcia, J.; Gomes, H. T.; Serp, Ph.; Kalck, Ph.; Figueiredo, J. L.; Faria, J. L. Carbon nanotube supported ruthenium catalysts for the treatment of high strength wastewater with aniline using wet air oxidation. *Carbon*, 44, 2006, 2384-2391.

Gomes, H. T.; Figueiredo, J. L.; Faria, J. L. Catalytic wet air oxidation of low molecular weight carboxylic acids using a carbon supported platinum catalyst. *Appl. Catal. B Environ.* 2000, 27 (4), L217-L223.

Gomes, H. T.; Figueiredo, J. L.; Faria, J. L. Catalytic wet air oxidation of butyric acid solutions using carbon-supported iridium catalysts. *Catal. Today* 2002, 75 (1-4), 23-28.

Gomes, H. T.; Figueiredo, J. L.; Faria, J. L.; Serp, Ph.; Kalck, Ph. Carbon-supported iridium catalysts in the catalytic wet air oxidation of carboxylic acids: Kinetics and mechanistic interpretation. *J. Mol. Catal,* A 2002, 182-183, 47-60.

Gomes, H. T.; Samant, P. V.; Serp, Ph.; Kalck, Ph.; Figueiredo, J. L.; Faria, J. L. Carbon nanotubes and xerogels as supports of wel-dispersed Pt catalysts for environmental applications. *Appl. Catal. B Environ.* 2004, 54, 175-182.

Gomes, H. T.; Machado, B. F.; Ribeiro A.; Moreira, I.; Rosario M.; Silva, A. M. T.; Figueiredo J. L.; Faria, J. L. Catalytic properties of carbon materials for wet oxidation of anilina. *J. Hazar. Mat.* 2008, 159, 420–426.

Hocevar, S.; Batista, J; Levec, J. Wet Oxidation of Phenol on $Ce_{1-x}Cu_xO_2$ Catalyst. *J. Catal.* 1999, 184 (1), 39-48.

Hutchings, G. J. Nanocrystalline gold and gold palladium alloy catalysts for chemical síntesis. Chem. Commu. 2008, 1148–1164.

Imamura, S. Catalytic and Noncatalytic Wet Oxidation. *Ind. Eng. Chem. Res.* 1999, 38 (5), 1743-1753.

Imamura, S.; Fukuda, I.; Ishida, S. Wet oxidation catalyzed by ruthenium supported on cerium (IV) oxides. *Ind. Eng. Chem. Res.* 1988, 27 (4), 718-721.

Imamura, S.; Sakai, T.; Ikuyama, T. Wet-oxidation of acetic acid catalyzed by copper salts. *J. Jpn. Petrol. Inst.* 1982, 25 (2), 74.

Kim, S. C.; Park, H. H.; Lee, D. K. Pd-Pt/Al_2O_3 bimetallic catalysts for the advanced oxidation of reactive dye solutions. Catal. Today, 2003, 87 (1-4), 51-57.

Kim, S. C.; Jeong, B. Y.; Lee, D. K. Catalytic wet oxidation of reactive dyes in water. *Top. Catal.* 2005, 33 (1-4), 149-154.

Kochetkova, R. P.; Babikov, A. F.; Shplerskaya, L.I.; Eppel, S. A.; Shmidt, F. K. *Liquid-phase catalytic oxidation of phenol.* Khim. Tekhnol. Topl. Masel 1992, 4, 31.

Kolaczkowski, S. T.; Plucinski, P.; Beltran, F. J.; Rivas, F. J.; McLurgh, D. B. Wet air oxidation: A review of process technologies and aspects in reactor design. *Chem. Eng. J.* 1999, 73 (2), 143-160.

Levec, J. Oxidation technologies for treating industrial wastewaters. In *Proceedings of the 1st European Congress on Chemical Engineering (ECCE-1)*, Florence, Italy, May 4-7, 1997; ERS C.T.: Milan, Italy, 1997; pp 513-516.

Levec, J.; Pintar A. Catalytic wet-air oxidation processes: a review. *Catal. Today,* 2007, 124, 172-184.

Levi, R.; Milman, M.; Landau, M.V.; Brenner A.; Herskowitz, M. Catalytic wet air oxidation of aniline with nanocasted Mn-Ce-oxide catalyst. *Environ. Sci. Technol.*, 2008, 42 (14), 5165-5170.

Lezna, R. O.; Tacconi, N. R.; Centeno, S. A.; Arvia, A. J. Adsorption of phenol on gold as studied by capacitance and reflectance measurements. *Langmuir*, 1991, 7, 1241-246.

Luck, F. A review of industrial catalytic wet air oxidation processes. *Catal. Today.* 1996, 27 (1-2), 195-202.

Luck, F. Wet air oxidation: Past, present and future. *Catal. Today*, 1999, 53 (1), 81-91.

Mallat, T.; Baiker A. Oxidation of Alcohols with Molecular Oxygen on Solid Catalysts. *Chem. Rev.* 2004, 104, 3037-3058.

Mantzavinos, D.; Hellenbrand R.; Livingston A. G.; Metcalfe I. S. Catalytic Wet Oxidation of p-Coumaric Acid: Partial Oxidation Intermediates, Reaction Pathways and Catalyst Leaching. *Appl. Catal. B: Environ.* 1996, 7, 379-96.

Mantzavinos, D.; Psillakis, E. Enhacement of biodegradability of industrial wastewater by chemical oxidation pre-treatment. *J. Chem. Technol.Biotechnol.* 2004, 79, 431-454.

Mantzavinos, D.; Sahibzada, M.; Livingston, A. G.; Metcalfe, I. S.; Hellgardt, K. Wastewater treatment: Wet air oxidation as a precursor to biological treatment. *Catal. Today*, 1999, 53 (1), 93-106.

Matatov-Meytal, Y. I.; Sheintuch, M. Catalytic Abatement of Water Pollutants. *Ind. Eng. Chem. Res.* 1998, 37 (2), 309-326.

Maugans, C. B.; Ellis, C. *Wet air oxidation: A review of commercial sub-critical hydrothermal treatment.* Presented at IT3'02 Conference, New Orleans, LA, May 13-17, 2002.

Milone, C.; Fazio M.; Pistone A.; Galvagno S. Catalytic wet air oxidation of p-coumaric acid on CeO_2, platinum and gold supported on CeO_2 catalysts. *Appl. Catal. B Environ.* 2006, 68, 28-37.

Mishra, V. S.; Mahajani, V. V.; Joshi, J. B. Wet Air Oxidation. *Ind. Eng. Chem. Res.* 1995, 34 (1), 2-48.

Mundale, V. D.; Joglekar, H. S.; Kalam, A.; Joshi, J. B. Regeneration of spent activated carbon by wet air oxidation, *Canadian J. Chem. Eng.* 1991, 69, 1149–1159.

Nutt, M. O.; Heck, K. N.; Alvarez, P.; Wong, M. S. Improved Pd-on-Au bimetallic nanoparticle catalysts for aqueous-phase trichloroethene hydrodechlorination. Appl. *Catal. B Environ.* 2006, 69, 115–125.

Okitsu, K.; Higashi, K.; Nagata, Y.; Dohmaru, T.; Taakenaka, N.; Bandow, H.; Maeda, Y. Decomposition of p-chlorophenol by wet oxidation in the presence of supported noble metal catalysts. Nippon Kagaku Kaishi 1995, 3, 202; *Chem. Abstr.* 1995, 122, 221805.

Oliviero, L.; Barbier, J., Jr.; Labruquere, S.; Duprez, D. Role of the metal-support interface in the total oxidation of carboxylic acids over Ru/CeO_2 catalysts. *Catal. Lett.* 1999, 60 (1-2), 15-19.

Oliviero, L.; Barbier, J., Jr.; Duprez, D.; Guerrero-Ruiz, A.; Bachiller-Baeza, B.; Rodríguez-Ramos, I. Catalytic wet air oxidation of phenol and acrylic acid over Ru/C and Ru-CeO_2/C catalysts. *Appl. Catal. B Environ.* 2000, 25 (4), 267-275.

Oliviero, L.; Barbier, J., Jr.; Duprez, D. Wet Air Oxidation of nitrogen-containing organic compounds and ammonia in aqueous media, *Appl. Catal. B: Environ.* 2003, 40 (3), 163-184.

Ong, Y. T.; Ahmad, A. L.; Zein, S. H. S.; Tan, S. H. A review on carbon nanotubes in an environmental protection and green engineering perspective. *Braz. J. Chem. Eng.* 2010, 27(2), 227-242.

Ovejero, G.; Sotelo, J. L.; Romero, M. D.; Rodriguez, A.; Ocaña, M. A.; Rodriguez, G., García J. Multiwalled carbon nanotubes for liquid-phase oxidation. Functionalization, characterization, and catalytic activity. *Ind. Eng. Chem. Res.* 2006, 45, 2206-2212.

Prati, L.; Rossi, M. Gold on Carbon as a New Catalyst for Selective Liquid Phase Oxidation of Diols. *J. Catal.* 1998, 176(2), 552-560.

Perathoner, S.; Centi, G. Wet hydrogen peroxide catalytic oxidation (WHPCO) of organic waste in agro-food and industrial streams. *Topics Catal.* 2005, 33 (1-4), 207-224.

Pintar, A.; Levec, J. Catalytic Liquid-Phase Oxidation of Refractory Organics in Waste Water. *Chem. Eng. Sci.* 1992, 47(1-2), 2395-2400.

Pintar, A.; Levec J. Catalytic Liquid-Phase Oxidation of Phenol Aqueous Solutions. A Kinetic Investigation. *Ind. Eng. Chem. Res.* 1994, 33, 3070-77.

Pintar, A.; Besson, M.; Gallezot, P.; Gibert, J.; Martin, D. Toxicity to Daphnia magna and Vibrio fischeri of Kraft bleach plant effluents treated by catalytic wet-air oxidation. *Water Res.* 2004, 38 (2), 289-300.

Pirkanniemi, K.; Sillanpää, M. Heterogeneous water phase catalysis as an environmental application: A review. *Chemosphere* 2002, 48 (10), 1047-1060.

Prasad, J.; Materi, G. E. Comparative Study of Air- and Oxygen-Based Wet Oxidation Systems. In *Proceedings of the 7th National Conference on Hazardous Wastes and Hazardous Materials*, Silver Spring, MD, 1990.

Qin, J.; Zhang, Q.; Chuang, K. T. Catalytic Wet Oxidation of p-Chlorophenol over Supported Noble Metal Catalysts. *Appl. Catal. B: Environ.* 2001, 29, 115-23.

Quintanilla, A.; Casas, J. A.; Zazo, J. A.; Mohedano, A. F.; Rodriguez, J. J. Wet air oxidation of phenol at mild conditions with an Fe/activated carbon catalyst. *Appl. Catal. B Environ.* 2006, 62 (1), 115-120.

Quintanilla, A.; Menéndez, N.; Tornero, J.; Casas, J. A.; Rodríguez, J. J. Changes of surface properties of carbon-supported iron catalyst during the wet air oxidation of phenol. *Appl. Catal. B Environ.* 2007, 76, 135-145.

Quintanilla, A.; Casas, J. A.; Rodriguez, J. J. Hydrogen peroxide-promoted-CWAO of phenol with activated carbon. *Appl. Catal. B. Environ*. 2010, 93, 339-345.

Rodríguez, A.; Ovejero, G.; Romero, M. D.; Díaz, C.; Barreiro, M.; García, J. Catalytic wet air oxidation of textile industrial wastewater using metal support on carbon nanofibers. *J. Supercrit. Fluids,* 2008, 46(2), 163-172.

Rodríguez, A.; Ovejero, G.; Mestanza, M.; Callejo, V.; García, J. Degradation of Methylene Blue by Catalytic Wet Air Oxidation with Fe and Cu Catalyst Supported on Multiwalled Carbon Nanotubes. *Chemg. Eng. Trans*. 2009, 17, 145-150.

Royer, S.; Levasseur, B.; Alamdari, H.; Barbier, J.; Duprez, D.; Kaliaguine, S. Mechanism of stearic acid oxidation over nanocrystalline $La_{1-x}ABO_3$ (A= Sr, Ce; B = Co, Mn): The role of oxygen mobility. *Appl. Catal. B Environ*. 2008, 80(1-2), 51-61.

Rubalcaba, A.; Suarez-Ojeda, M. E.; Carrera, J.; Font, J.; Stuber, F.; Bengoa, C.; Fortuny, A.; Fabregat, A. Biodegradability enhancement of phenolic compounds by Hydrogen Peroxide Promoted Catalytic Wet Air Oxidation. *Catal. Today*, 2007, 124(3-4), 191–197.

Sadana, A.; Katzer, J. R. Catalytic Oxidation of Phenol in Aqueous Solution over Copper Oxide. *Ind. Eng. Chem. Fundam*. 1974, 13(2), 127-134.

Sankar, M.; Dimitratos, N.; Knight, D.W.; Carley, A.F.; Tiruvalem, R.; Kiely, C.J.; Thomas, D.; Hutchings, G.J. Oxidation of Glyercol of Glycolate by using Supported Gold and Palladium Nanoparticles. *ChemSusChem,* 2009, 2, 1145-1151.

Santos, A; Yustos, P.; Quintanilla, A.; Garcia-Ochoa, F; Casas J. A.; Rodriguez J. J. Evolution of Toxicity upon Wet Catalytic Oxidation of Phenol. Environ. *Sci. Technol.* 2004a, 38, 133-138.

Santos, A.; Yustos, P.; Quintanilla, A.; Garcia-Ochoa, F. Lower toxicity route in catalytic wet oxidation of phenol at basic pH by using bicarbonate media. *Appl. Catal. B Environ.* 2004b, 53, 181–194.

Santos, A.; Yustos, P.; Quintanilla, A.; García-Ochoa, F. Influence of pH on the wet oxidation of phenol with copper catalyst. *Topics Catal*. 2005a, 33(1-4), 181-192.

Santos, A.; Yustos, P.; Quintanilla, A.; Ruiz, G.; García-Ochoa, F. Study of the copper leaching in the wet oxidation of phenol with CuO based catalysts: Causes and effects. *Appl. Catal. B,* 2005b, 61 (3-4), 323-333.

Santos, A.; Yustos, P.; Rodriguez, S.; Vicente, F.; Romero, A. Kinetic Modeling of Toxicity Evolution during Phenol Oxidation. *Ind. Eng. Chem. Res*. 2009, 48(6), 2844–2850.

Scott, J. P.; Ollis D. F. Integration of chemical and biological oxidation processes for water treatment: review and recommendations. *Environ. Prog*. 1995, 14(2), 88-103.

Silva, A. M. T.; Quinta-Ferreira, R. M; Levec, J. Catalytic and non-catalytic wet oxidation of formaldehyde. A novel kinetic novel. *Ind. Eng. Chem. Res*. 2003, 42, 5099-5108.

Silva, A. M. T. Environmental Catalysis from nano- to macro-scale. *Mat. Tech*. 2009, 43(3), 113-121.

Soria-Sánchez, M.; Maroto-Valiente, A.; Álvarez-Rodríguez, J.; Rodríguez-Ramos I.; Guerrero-Ruiz A. Efficient catalytic wet oxidation of phenol using iron acetylacetonate complexes anchored on carbon nanofibres. *Carbon*, 2009, 47, 2095-2102.

Sousa, J. P. S.; Silva, A. M. T.; Pereira M. F. R.; Figueiredo, J. L. Wet air oxidation of anilie using carbon foams and fibers enriched with nitrogen. *Sep. Sci. Technol*. 2010, 45, 1546-1554.

Stüber, F.; Font, J.; Fortuny, A.; Bengoa, C.; Eftaxias, A.; Fabregat, A. Carbon materials and catalytic wet air oxidation of organic pollutants in wastewater. *Top. Catal.* 2005, 33 (1-4), 3-50.

Suarez-Ojeda, M. E.; Guisasola, A.; Baeza, J. A.; Fabregat, A.; Stuber, F.; Fortuna, A.; Font, J.; Carrera, J. Integrated catalytic wet air oxidation and aerobic biological treatment in a municipal WWTP of a high-strength o-cresol wastewater. *Chemosphere*, 2007, 66, 2096–2105.

Taboada, C. D.; Batista, J.; Pintar, A.; Levec, J. Preparation, characterization and catalytci properties of carbon nanofiber-supported Pt, Pr, Ru monometallic particles in aqueous-phase reactions. *Appl. Catal. B Environ.* 2009, 89, 375-382.

Theron, J.; Walker, J. A; Cloete, T. E. Nanotechnology and water treatment: applications and emerging opportunities. *Crit. Rev. Microbiol.* 2008, 34, 43-69.

Tiwari, D. K.; Behari J.; Sen P. Application of nanoparticles in waste water treatment. World App. Sci. J. 2008, 3(3), 417-433.

Trana, N. D.; Besson, M.; Descormea, C.; Fajerwerg, K.; Louis C.; Méthivierb C. Ceria-supported gold catalysts for wastewater treatment: influence of the pre-treatment conditions. 2008, In *Proceedings of the 5th International Conference in Environmental Catalysis, Belfast, Ireland*, 31 st August to 3rd September, 2008, AW-8.

Trawczýnski, J. Noble metals supported on carbon black composites as catalysts for the wet-air oxidation of phenol. *Carbon*, 2003, 41(8), 1515-1523.

Trovarelli, A.; de Leitenburg, C.; Boaro, M.; Dolcetti, G. The utilization of ceria in industrial catalysis. *Catal. Today*, 1999, 50 (2), 353-367.

Tsunoyama, H.; Sakurai, H.; Tsukuda, T. Size effect on the catalysis of gold clusters dispersed in water for aerobic oxidation of alcohol. *Chem. Phys. Letters*, 2006, 429 528–532.

Yang, M.; Xu, A.; Du, H.; Sun, C.; Li, C. Removal of salicylic acid on perovskite-type oxide $LaFeO_3$ catalyst in catalytic wet air oxidation process. *J. Hazar. Mat.* 2007, 139(1), 86–92.

Yang, S.; Feng, Y.; Wan, J.; Cai, W. *Catalytic wet air oxidation*. Harbin Gongye Daxue Xuebao. 2002, 34 (4), 540.

Yang, S.; Li, X.; Zhu, W.; Wang, J.; Descorme C. Catalytic activity, stability and structure of multi-walled carbón nanotubes in the wet air oxidation of phenol. *Carbon*, 2008, 46, 445-452.

Zhang, Y.; Li, D.; Chen, Y.; Wang, X.; Wang, S. Catalytic wet oxidation of dye pollutants by polyoxomolybdate nanotubes under room conditions. *Appl. Catal. B Environ.* 2009, 86, 182-189.

Zhao, S.; Wang, X.; Huo, M. Catalytic wet air oxidation of phenol with air and micellar molydovanadophosphoric polyoxometalates under room conditions. *Appl. Catal. B Environ.* 2010, 97, 127-134.

In: Focus on Catalysis Research: New Developments
Editors: Minjae Ghang and Bjørn Ramel
ISBN: 978-1-62100-455-4

Chapter 10

HETEROPOLYANION (HPA) CATALYZED OZONE DELIGNIFICATION IN ORGANIC SOLVENTS: EXTREMELY EFFECTIVE AND SELECTIVE PULP BLEACHING APPROACH

Anatoly A. Shatalov
Forest Research Center, Technical University of Lisbon, Institute of Agronomy
Tapada da Ajuda, 1349-017 Lisbon, Portugal

ABSTRACT

The highly efficient and selective environmentally benign bleaching approach using polyoxometalate (POMs) catalyzed ozonation of chemical pulps in organic solvent reaction media has been developed. A number of low-boiling polar aprotic and protic organic solvents showed a well-defined capacity for ozonation improvement in the presence of α-Keggin-type mixed-addenda heteropolyanions (HPAs), particularly molybdo-vanado-phosphate HPAs of series $[PMo_{(12-n)}V_nO_{40}]^{(3+n)-}$. Aqueous acetone solution was found to be the best suited reaction media for ozone delignification catalyzed by $[PMo_7V_5O_{40}]^{8-}$ (HPA-5) polyanion. The effect of solvent and catalyst concentration, pH and ionic strength on ozone bleaching of commercial eucalypt kraft pulp has been examined. The solvent proportion in solution and medium acidity were the principal factors affecting catalytic efficiency of HPA during ozonation. Under optimized conditions, the brightness improvement of bleached pulp by ca. 15% ISO, with additional lignin removal by ca. 39% and increase in intrinsic viscosity by ca. 3% was observed in comparison with conventional (solvent/catalyst free) ozonation. The elimination of the Donnan effect by increase in ionic strength (cation concentration) of pulp suspension substantially improved delignification capacity of HPA/O_3 bleaching system. The catalyst recovery/re-oxidation by ozone has been examined using partially reduced polyanion species (HPA-5_{red} or heteropoly blue). It was shown that the bulk portion (ca. 90%) of HPA-5_{red} is readily re-oxidized by ozone during first reaction minutes thereby confirming the closure of redox catalyst cycle, as a required condition for successful HPA-catalyzed oxidative delignification. Addition of organic solvent had favorable effect on HPA-5_{red} re-oxidation by ozone (k of 0.29 min^{-1} vs. 0.43 min^{-1} for ozonation, respectively, in water and 10% v/v acetone solution; pH 2), being evidently the principal reason for high efficiency of solvent-assisted HPA/O_3 bleaching. The bleaching effect of HPA/O_3 system was substantially improved by enzymatic (xylanase) pulp pre-treatment before ozonation.

INTRODUCTION

Over the last two decades, the increased market demand for TCF (totally chlorine-free) and low AOX (absorbable organic halogens) cellulosic pulps produced by worldwide pulp and paper industry caused renewed great interest in non-chlorine oxidative bleaching chemicals such as oxygen, hydrogen peroxide and ozone. Ozone is the most powerful and particularly potential oxidation agent in pulp bleaching technology [1]. Despite the considerable research to elucidate the bleaching potential of ozone completed already 30-40 years ago, its commercial application became feasible only in the 1990s under the pressure of environmental issues in an attempt to minimize discharges of chlorinated compounds in bleach plant effluents [2-4]. The ozone bleaching stage, as an integrated part of more advanced TCF bleaching sequences, was recently applied to different types of pulps from a wide spectrum of wood and non-woody (agro-based) fiber sources [5,6]. Despite the extremely high ozone reactivity (oxidizing potential of +2.07 eV [7]), the low selectivity of ozone treatment towards lignin, due to unwanted radical-induced reactions with cellulose leading to a deterioration of pulp quality, restricts the delignification capacity of ozone and limits its commercialization in pulp bleaching technology as a whole. A variety of different chemicals have been tested as the additives for reaction solution with the aim to improve ozonation performance [8-11]. The selectivity of ozone bleaching still remains to be solved.

The change in red-ox properties of oxidation bleaching system by application of POM catalysis can be a feasible way to increase selectivity and efficiency of pulp ozonation. Heteropoly compounds are known as soluble polyoxoion salts of anions of general formula $[X_xM_mO_y]^{q-}$ ($x \leq m$), where X is heteroatom (P, Si, B, etc.) and M is addenda atom (early transition metal in high oxidation state, e.g., Mo^{VI}, W^{VI}) [12]. The α-Keggin-type mixed-addenda heteropolyanions $[X_xM'_nM_{12-n}O_{40}]^{q-}$ (Figure 1), particularly molybdo-vanado-phosphate heteropolyanions (HPAs) of series $[PMo_{(12-n)}V_nO_{40}]^{(3+n)-}$, have found a wide application as versatile green catalysts for homogeneous liquid-phase oxidations of different organic compounds [13-15], including phenolics [16]. Substitution of Mo^{VI} by V^V in the polyanion molecule tends to increase the oxidation potential of HPAs, as a result of extra-electrons trapping by vanadium atom [13]. Such remarkable features of HPAs as structural and functional mobility, that can be easily controlled during rather simple and low-cost synthesis, high solubility in water and various oxygen-containing organic solvents, high stability over a wide range of reaction conditions and, finally, easy regeneration (re-oxidation) after utilization, made HPAs (particularly polymolybdovanadates) very attractive catalysts for oxidative delignification of lignocellulosics [17,18], particularly for dioxygen delignification (bleaching) of wood kraft pulps [19].

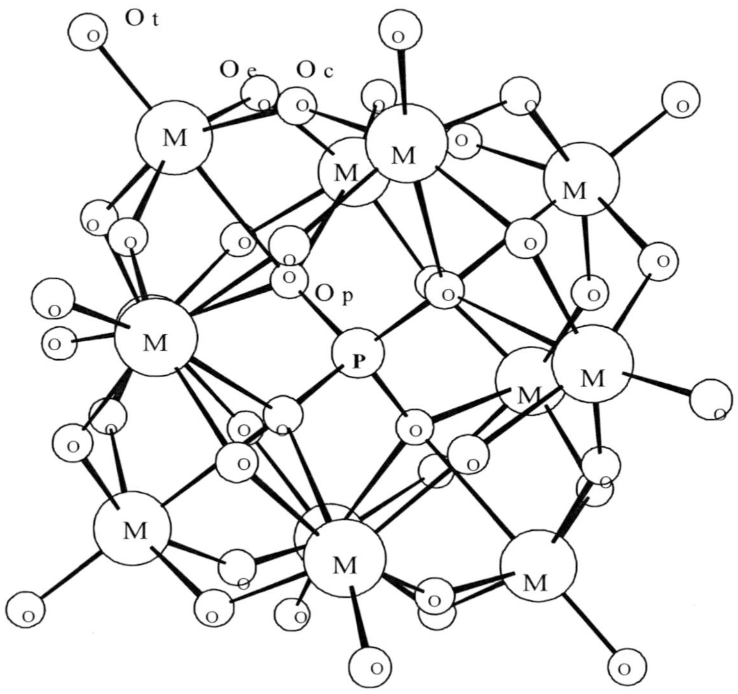

Figure 1. The Keggin structure of $[PM_{12}O_{40}]^{3-}$ anion in ball-and-stick representation (the central PO_4 tetrahedron is surrounded by 12 MO_6 octaedra). The terminal (O_t), corner-sharing (O_c), edge-sharing (O_e) and phosphorous-linked (O_p) oxygen atoms and the metal atoms (primarily W, Mo, V) are depicted by small and big colorless balls, respectively [28].

Oxidation catalysis by polymolybdovanadates occurs due to the ability of V^V to accept electrons from the substrate via a Mars-van Krevelen mechanism [20,21]:

$$\text{HPA-}n + \text{Sub} + m\text{H}^+ \rightarrow \text{H}_m(\text{HPA-}n) + \text{Sub}_{\text{ox}} \quad (1)$$

where $H_m(\text{HPA-}n) = H_m[PMo_{12-n}V^V{}_{n-m}V^{IV}{}_mO_{40}]^{(3+n)-}$ is the partially reduced form of catalyst with m atoms of vanadium (IV). To maintain the charge of the polyanion, the HPA-n reduction in solution is accompanied by its protonation [12].

The reduced V^{IV} can be oxidized back to V^V via reaction with an appropriate oxidant (e.g., with dioxygen), closing thereby the redox cycle by catalyst:

$$\text{H}_m(\text{HPA-}n) + m/4\text{O}_2 \rightarrow \text{HPA-}n + m/2\text{H}_2\text{O} \quad (2)$$

The (Mo-V-P)-HPAs have therefore the properties of easy reversible multi-electron oxidants. The effective HPAs re-oxidation (recuperation) in the separate reactor (two-stage oxidation) or in the same reactor (one-stage oxidation) provides selectivity of the overall catalytic process and opens the possibility to introduce the so-called TEF (totally effluent-free) technology of pulp bleaching with continuous reuse (recirculation) of catalytic solution in a close-loop mode and with carbon dioxide and water as the only byproducts after lignin oxidation [22].

Table 1. HPA-5 catalyzed ozone bleaching of eucalypt (*E. globulus* L.) kraft pulp (3% pulp consistency; $[O_3]$=0.8%; [HPA-5]=0.5mM; 6% (w/w) solvent; pH=2)

Reaction media	Brightness (% ISO)	Lignin (% odp)	Viscosity (ml/g)	CHO (mmol/100g)
Control (water)	53.2	1.83	920	11.12[a]
HPA/water	53.4	1.75	951	8.80
HPA/methanol	55.8	1.54	970	7.08
HPA/ethanol	55.2	1.58	1035	7.53
HPA/acetone	56.8	1.42	950	8.40
HPA/dioxane	55.8	1.59	1026	6.60
HPA/1-propanol	54.1	1.81	1028	6.72
HPA/isopropanol	51.7	1.78	1017	7.44
Ethanol	53.7	1.79	995	8.30
Unbleached pulp[b]	41.6	2.39	1320	0.97

[a] mmol of CHO groups per 100 grams of dry pulp.
[b] properties of initial untreated (unbleached) pulp.

Recently, the new potential method of POM catalyzed oxidative delignification has been developed [23-27]. The HPA catalyzed ozonation of chemical pulps in organic solvent media was found to be particularly effective and selective environmentally benign bleaching approach providing a way for substantial increase in pulp brightness, viscosity and degree of delignification in comparison with any other existing ozone-based bleaching techniques. The basic principals of this bleaching approach are reviewed in this chapter.

SELECTION OF REACTION MEDIA FOR HPA CATALYZED OZONATION

Of all the (Mo-V-P)-HPAs tested for oxygen (dioxygen) delignification of chemical (kraft) pulps, the heteropolyanion $[PMo_7V_5O_{40}]^{8-}$ or HPA-5 showed the best results in terms of delignification [19]. However, the selectivity of process was poor due to undesirable polysaccharide degradation under high medium acidity (pH 1-2) required for structural stability and catalytic activity of HPA-5 [28]. The low pH limits HPA-5 utilization for oxygen delignification, but makes it very attractive for ozone bleaching operating under the same pH range. The extremely high oxidation potential of ozone can substantially reduce the red-ox cycle of catalyst regeneration in comparison with oxygen or peroxide delignification, thereby providing an accelerated rate of bleaching reactions and higher efficiency of the bleaching process as a whole. The process selectivity is provided by stoichiometric electron-transfer mechanism of HPA catalyzed lignin oxidation, in contrast to unselective and uncontrolled radical-induced ozone delignification inevitably accompanied by substantial carbohydrate degradation. The early reported experiments on HPA-5 catalyzed ozone bleaching in water showed, however, substantial loss in pulp viscosity, indicating still low suppression of radical reactions. The addition of organic solvents, as radical scavengers, was expected to improve ozonation performance. To meet economical and ecological concerns related to chemical recovery, a few common low-boiling polar aprotic and protic organic solvents (i.e., methanol, ethanol, acetone, dioxane, 1-propanol and iso-propanol) were tested as potential reaction media for HPA-5 catalyzed pulp ozonation. For solvent screening experiments, the industrial eucalypt (*E. globulus*) kraft pulp was ozonated under fixed conditions of ozone charge and catalyst (HPA-5) concentration in the presence of 6% (w/w) of organic solvent. The brightness development as well as the extent of lignin and carbohydrate degradation during ozonation was examined and compared with control (solvent/catalyst free) test.

The experimental data summarized in Table 1 show that even moderate solvent proportion in reaction mixture can substantially improve HPA catalyzed ozone bleaching, causing notable simultaneous gain in brightness, degree of delignification and viscosity in comparison with HPA/water and, particularly, with conventional control ozonation. Evidently, the bleaching effect depends on the nature of organic solvent used. The highest ozonation effectiveness in terms of pulp delignification (or lignin removal), numerically expressed as lignin decrease per unit of ozone applied (Figure 2), was shown in HPA/acetone medium (50.53% over 33.13% for HPA/water), followed by methanol (44.45%), ethanol (42.45%) and dioxane solutions (41.51%). The ozonation effectiveness in terms of pulp brightening (or brightness improvement), commonly used as a measure of relative pulp bleachability and expressed as increase in brightness per unit of applied active bleaching chemical (Figure 3), was also the highest in HPA/acetone solution (45.79%), followed by dioxane (42.61%), methanol (42.55%) and ethanol (40.93%), over 35.51% for HPA/water. Finally, the HPA/ethanol solution showed the highest ozonation selectivity, as a measure of polysaccharide degradation and expressed as lignin decrease per unit of viscosity decrease (Figure 4), followed by dioxane, acetone and methanol solutions. The ozonation results in *n*-propanol and isopropanol solutions were somewhat poorer and comparable with solvent-free (HPA/water) system.

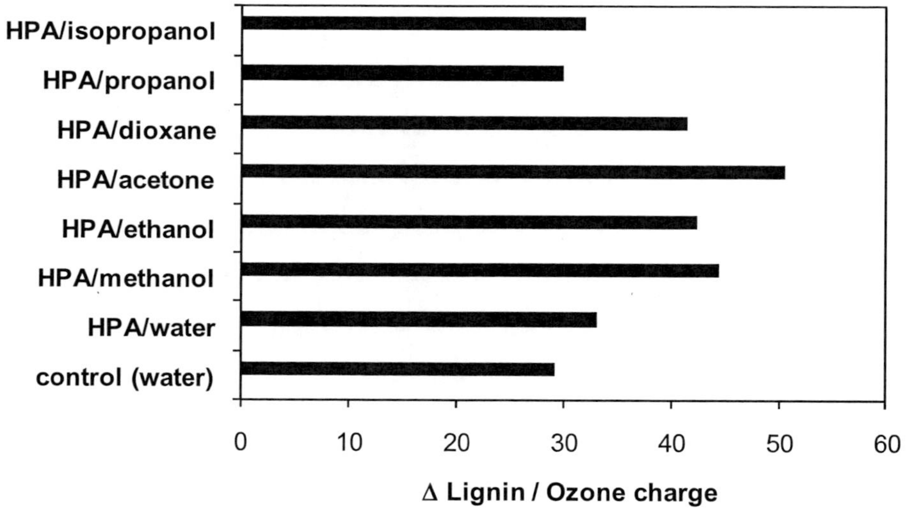

Figure 2. Delignification efficiency of HPA catalyzed pulp ozonation in organic solvent media: $[O_3]$=0.8%; [HPA-5]=0.5 mM; 6% (w/w) solvent; pH=2.

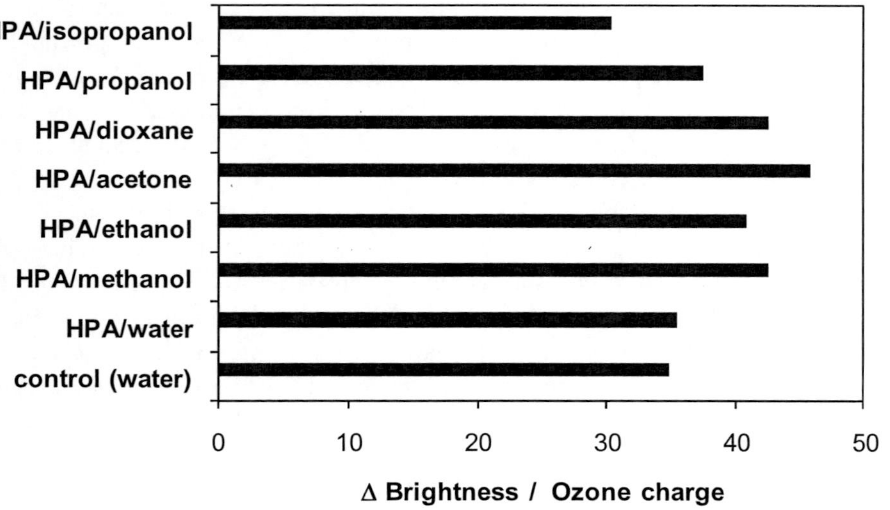

Figure 3. Brightening efficiency of HPA catalyzed pulp ozonation in organic solvent media: $[O_3]$=0.8%; [HPA-5]=0.5 mM; 6% (w/w) solvent; pH=2.

The content of aldehyde (CHO) groups in ozonated pulp can be used as important indicator the degree of polysaccharide oxidative degradation [29]. As can be seen from Table 1, the presence of HPA inhibits oxidation reactions leading to decrease in CHO-groups by ca. 20% in comparison with control, presumably due to reported radical scavenging capacity of HPAs [20]. In HPA/solvent media, the CHO content of ozonated pulps is 4.5-25% less than in HPA/water and 24-40 % less than in control (water) ozonation, indicating substantial suppression of carbohydrate oxidative degradation in the presence of solvent.

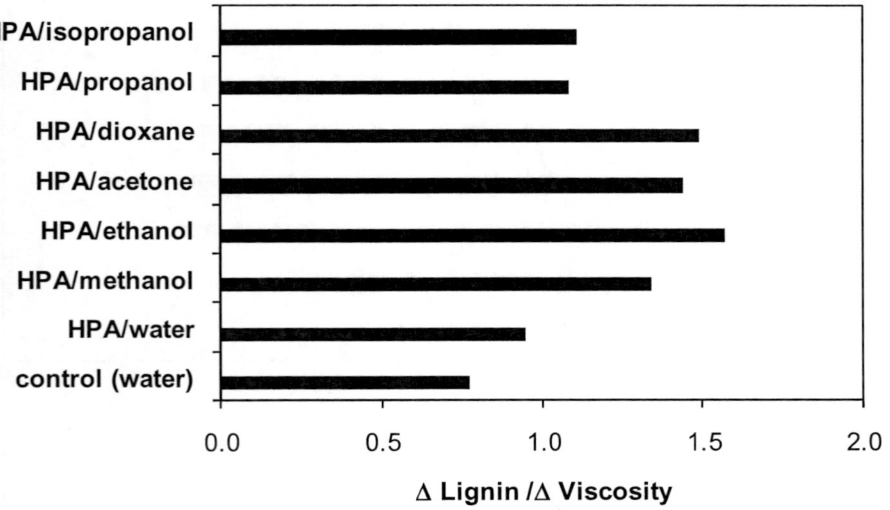

Figure 4. Selectivity of HPA catalyzed pulp ozonation in organic solvent media: $[O_3]$=0.8%; [HPA-5]=0.5 mM; 6% (w/w) solvent; pH=2.

Table 2. Physical properties of HPA-5/O_3 bleached eucalypt (*E. globulus* L.) kraft pulp (3% pulp consistency; [O_3]=0.8%; [HPA-5]=0.5 mM; 6% (w/w) solvent; pH=2)

Reaction media	Tear index ($mN\ m^2/g$)	Burst index ($kPa\ m^2/g$)	Tensile index ($N\ m/g$)
Control (water)	3.86	0.12	3.02
HPA/water	4.14	0.15	3.51
HPA/methanol	4.43	0.16	3.98
HPA/ethanol	4.43	0.15	4.21
HPA/acetone	4.16	0.18	4.58
HPA/dioxane	4.52	0.21	4.44
HPA/1-propanol	4.27	0.19	4.08
HPA/isopropanol	4.08	0.17	4.31
Ethanol	4.23	0.15	3.85
Unbleached pulp[a]	4.08	0.20	4.68

[a] Properties of initial untreated (unbleached) pulp.

The higher intrinsic viscosity of pulps ozonated in HPA/solvent media, as a measure of cellulose preservation, is directly related to better strength (physical) properties of these pulps (Table 2). The found tear, burst and tensile indexes are notably higher than those found for control and HPA/water ozonation and are near the strength properties of untreated (unbleached) pulp. The pulps ozonated in dioxane solution showed the highest tearing and bursting strength, while the ozonation in acetone solution gave the highest tensile strength pulps.

To underline the catalytic effect of HPA-5, the catalyst-free pulp ozonation in ethanol solution, reported to have favorable action on ozone bleaching [30], was carried out and compared with HPA/ethanol ozonation (Table 1). After HPA-5 addition, the notable increase in pulp brightness by 1.6% ISO, degree of delignification by ca. 12% and pulp viscosity by ca. 4% was observed in comparison with non-catalyzed system, indicating the critical role (contribution) of catalyst for successful pulp ozonation.

To summarize, the four tested HPA-5 catalyzed solvent-based ozone bleaching systems, i.e., methanol, ethanol, acetone and dioxane-based, showed significant advantage over the conventional ozone pulp bleaching in water as well as HPA-5/water ozonation. The ozonation efficiency in terms of brightening and delignification, as the principal objectives of pulp bleaching, was found higher in acetone solution, pointing to this reaction medium as the best suited for HPA catalyzed pulp ozonation.

HPA Catalyzed Ozonation in Acetone/water Solution

The effect of solvent and catalyst concentration, pH and ionic strength on HPA-5 catalyzed ozone bleaching of commercial eucalypt kraft pulp in aqueous acetone solution has been examined with the aim of process optimization.

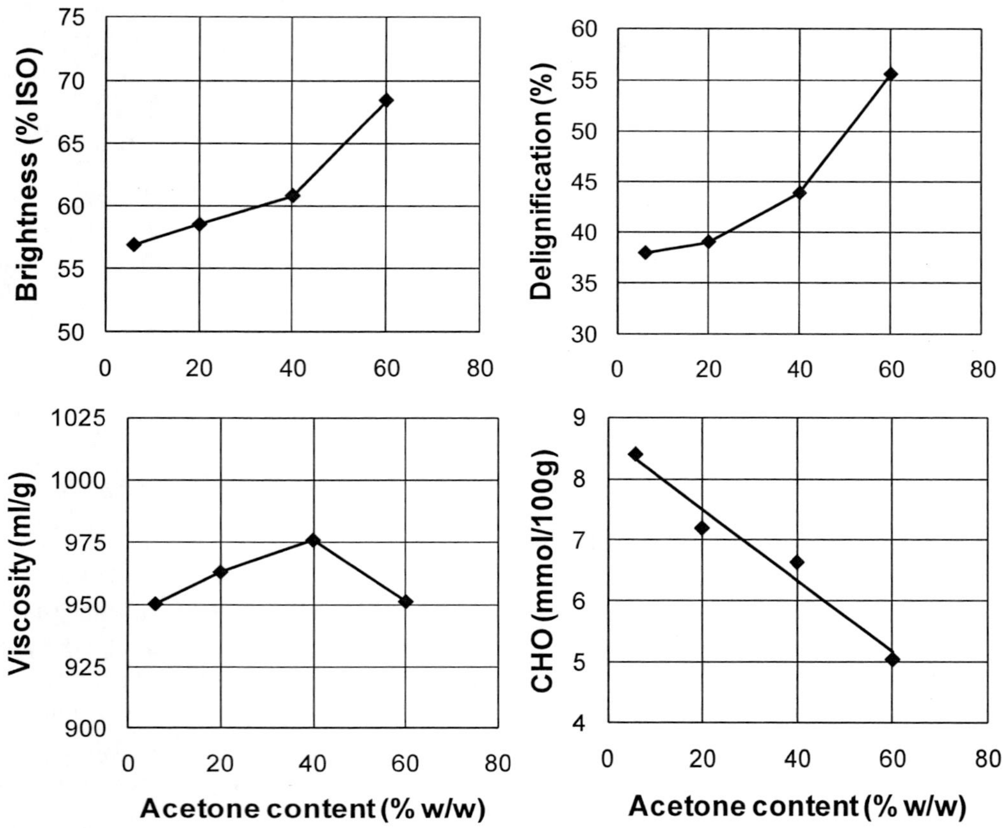

Figure 5. Effect of acetone content in reaction solution on HPA-5 catalyzed ozone bleaching of eucalypt (*E. globulus*) kraft pulp: 3% pulp consistency; $[O_3]$=0.8%; [HPA-5]=0.5 mM; pH=2.

Solvent Content

As illustrated in Figure 5, the progressive increase in solvent proportion in the reaction mixture causes drastic increase in pulp brightness and delignification. A brightness gain by 11.6% ISO and additional lignin removal by 17.8% was observed with increase in solvent content in the reaction mixture from 6% to 60% (w/w), offering substantial bleaching improvement (by 15.1% and 39.4%, respectively, for brightening and delignification) in comparison with HPA/water ozonation and, particularly, with control (water) ozonation. The accelerated intensity of bleaching reactions was particularly notable with acetone proportion exceeded 40% (w/w), pointing to marginal concentration of solvent required for successful ozonation.

The observed strong positive effect of acetone can be explained by better ozone solubility in organic solvents in comparison with water, as a result of reduced interfacial tension of the liquid phase, leading to facilitated ozone mass transfer to the solvent solution [11]. The higher concentration of dissolved (active) ozone in solution enhances the catalyst (HPA) regeneration (re-oxidation) during bleaching (according to reaction scheme *(2)*), accelerating in its turn the rate of delignification reaction as a whole (reaction scheme *(1)*). Obviously, the concentration of dissolved ozone will be directly related to solvent proportion in bleaching solution, explaining substantial acceleration of ozonation activity under elevated acetone concentrations.

The enhanced solubility of lignin (or lignin degradation products) in organic solvents can also contribute to better pulp delignification during solvent-assisted ozonation in comparison with water [31]. The dissolution effect varies for different solvents and depends on their hydrogen-bonding capacity (δ–value), which normally increases in solvent mixtures with water [32].

The increase in solvent content up to 40% (w/w) has favorable (protective) effect on carbohydrate complex of ozonated pulp (Figure 5), increasing pulp intrinsic viscosity and thereby improving selectivity of ozonation as a whole, as a result of scavenging by solvent of harmful oxygen-centered radical species, basically hydroxyl and perhydroxyl radicals formed during ozonation [7]. The decreased content of CHO groups in ozonated pulps confirms effective suppression of oxidative degradation reactions. The subsequent increase in solvent proportion in reaction solution (above 40% w/w) causes substantial drop in pulp viscosity, indicating intensive carbohydrate degradation and loss in ozonation selectivity (Figure 5). The similar effect was previously reported for ethanol-assisted HPA/oxygen delignification and was partially associated with increased intensity of acid-catalyzed hydrolytic (solvolytic) degradation reactions of polysaccharides under elevated solvent concentrations [28].

The change in solvent concentration has some effect on physical (mechanical) properties of ozonated pulps (Table 3). The bursting and, particularly, tensile pulp strength was improved almost twice in comparison with control and HPA/water ozonation after increase in acetone concentration up to 60%. At the same time, the tearing pulp strength was not affected by solvent addition.

Catalyst Concentration

The effect of HPA-5 concentration on ozonation performance is shown in Figure 6.

As can be seen, the maximal efficiency of HPA-5 catalyzed ozonation, i.e., the maximal catalytic effect of HPA-5, can be achieved under some marginal catalyst concentration of 1mM. Within the tested economically reasonable range of HPA concentration of 0.5-4.0 mM and keeping fixed acetone content of 40% w/w, the additional lignin removal by ca. 16% and gain in brightness by 1.2% ISO were observed after change in HPA-5 concentration from 0.5 mM to 1.0 mM. The following increase in HPA concentration led to substantial loss in pulp brightening and delignification, thereby greatly decreasing the ozonation efficiency as a whole.

The observed concentration dependence of HPA-5 catalytic activity can be understood through the structural and redox changes of catalyst in the reaction solution under ozonation conditions. It was first reported by Souchay et al. in 1968 [33] that in diluted acid solutions (the case of ozone bleaching) the Mo-V-phosphoric polyanions HPA-n (n=2-6) undergo a reversible dissociation with release of VO_2^+ ions and formation of so-called defect (lacunary) heteropoly species:

$$(13\text{-}x)\ [H_{x-1}PV_x^VMo_{12-x}O_{40}]^{4-} + 16H^+ \leftrightarrow (12\text{-}x)\ [H_{x-2}PV^V_{x-1}Mo_{13-x}O_{40}]^{4-} + 12VO_2^+ + H_3PO_4 + 12H_2O \quad (3)$$

Table 3. Effect of solvent content on physical properties of HPA-5/O_3 bleached eucalypt (*E. globulus* L.) kraft pulp (3% pulp consistency; [O_3]=0.8%; [HPA-5]=0.5 mM; pH=2)

Reaction media	Tear index ($mN\ m^2/g$)	Burst index ($kPa\ m^2/g$)	Tensile index (N m/g)
Control (water)	3.86	0.12	3.02
HPA/water	4.14	0.15	3.51
HPA/6% w/w acetone	4.16	0.18	4.58
HPA/20% w/w acetone	3.95	0.19	4.83
HPA/40% w/w acetone	3.91	0.22	5.75
HPA/60% w/w acetone	4.09	0.22	6.09

The VO_2^+ ions, having higher oxidation potential than the parent HPA (0.87 V vs 0.71 V, respectively), were shown to be the principal active species in catalyzed aerobic (dioxygen) oxidation reactions [13], causing degradation of lignin and (to some extent) of carbohydrates during pulp bleaching [28].

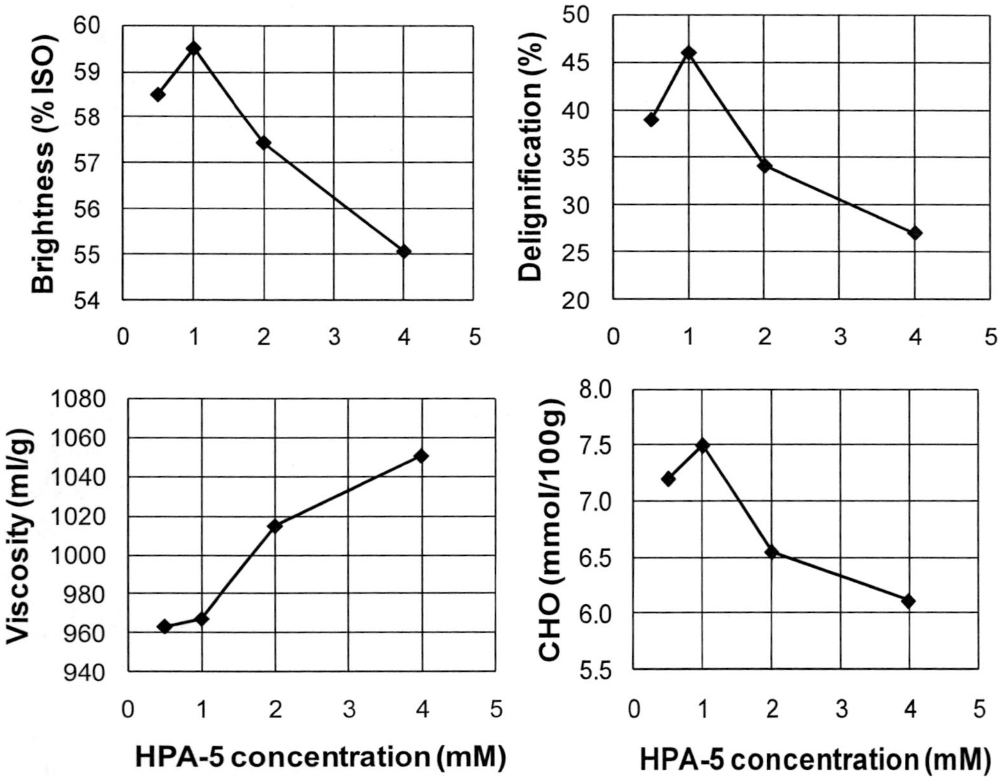

Figure 6. Effect of catalyst (HPA-5) concentration on ozone bleaching of eucalypt (*E. globulus*) kraft pulp in aqueous acetone solution: 3% pulp consistency; [O_3]=0.8%; 40% (w/w) acetone; pH=2.

Table 4. Effect of catalyst concentration on physical properties of HPA-5/O_3 bleached eucalypt (*E. globulus* L.) kraft pulp (3% pulp consistency; [O_3]=0.8%; 40% w/w acetone; pH=2)

Reaction media	Tear index ($mN\ m^2/g$)	Burst index ($kPa\ m^2/g$)	Tensile index (N m/g)
0.5 mM HPA	3.91	0.22	5.75
1.0 mM HPA	3.87	0.22	5.61
2.0 mM HPA	3.84	0.24	5.02
4.0 mM HPA	3.82	0.25	4.63

Counting also the increased HPA resistance to degradative dissociation in concentrated solutions [13], the highest bleaching efficiency of ozonation noted at 1 mM HPA-5 can be therefore associated with maximal release of free VO_2^+ ions from the coordination sphere of HPA. With subsequent increase in catalyst concentrations (above 1 mM), the retarded HPA dissociation and correspondently reduced content of free VO_2^+ ions in solution cause substantial decrease in pulp delignification and brightening. This assumption was supported by analysis of CHO-groups in the ozonated pulps. The highest degree of pulp oxidation revealed at 1 mM HPA (Figure 6) confirmed the maximal intensity of oxidative bleaching reactions under this catalyst concentration. An excellent agreement in HPA-dependent ozonation profiles of CHO-groups, lignin removal and brightness improvement was observed.

The limited dissociation of HPA-5 and reduced release of VO_2^+ ions is an obvious reason of the observed increase in intrinsic viscosity of pulps ozonated under elevated catalyst concentrations (Figure 6). The suppression of polysaccharide degradation leads to viscosity improvement up to 9.1% with change in HPA-5 concentration in reaction solution from 0.5 mM to 4.0 mM. Contrary to the expectations, no appreciable drop in pulp viscosity was noted in 1 mM HPA-5 solution, exhibiting maximal oxidation activity. That allowed keeping the highest level of ozonation selectivity towards lignin at this critical catalyst concentration, being obviously optimal for this bleaching system.

The pulp mechanical properties were only slightly affected by change in catalyst concentration on ozonation (Table 4). The loss of ca. 19% in tensile pulp strength was observed with increase in HPA concentration in the range of 0.5-4.0 mM (most likely due to increased proportion of residual lignin); while the change in bursting and tearing pulp strength had no any statistical significance.

Table 5. Effect of medium acidity (pH) on HPA-5 catalyzed ozone bleaching of eucalypt (*E. globulus* L.) kraft pulp (3% pulp consistency; [O_3]=0.8%; [HPA-5]=1.0 mM; 40% (w/w) acetone)]

Medium acidity	Brightness (% ISO)	Lignin (% odp)	Viscosity (ml/g)	CHO (mmol/100g)
pH 1.0	59.1	1.66	1053	5.32[a]
pH 1.5	60.5	1.47	1002	6.94
pH 2.0	63.6	1.36	989	7.48

[a] mmol of CHO groups per 100 grams of dry pulp.

Medium Acidity (pH)

Medium acidity (pH) has a critical importance for the discussed HPA/O_3 bleaching technology. As it was mentioned above, the synergetic bleaching effect of HPA/O_3 system is feasible first of all due to unique possibility to apply the same pH range (pH 1-2) required for optimal operation performance of both the HPA as catalyst and the ozone as oxidant. The careful pH control during HPA/O_3 bleaching makes it possible to regulate formation and reactivity of active oxidation species in ozonation solution thereby directly affecting the course of bleaching performance and the properties of bleached pulps. As can be seen from Table 5, within the tested pH 1-2 range and keeping fixed catalyst concentration of 1.0 mM and solvent content of 40% w/w, the decrease in medium acidity from pH 1 to pH 2 leads to substantial increase in pulp brightening (by 4.5% ISO) and delignification (by ca. 18%) thus notably improving the bleaching efficiency of ozonation.

Taking into account the higher structural stability (resistance to dissociation by *(3)*) of HPA-5 at increased pH [20] and the highest reported bleaching activity of ozone at pH 2 [1], the observed remarkable bleaching improvement can be associated with joint oxidative effect of catalyst (with parent heteropolyanion and VO_2^+ ions as reactive catalyst species) and ozone. The highest activity of oxidation reactions at pH 2 was confirmed by the elevated content of CHO-groups. Despite the drop in pulp viscosity, due to intensified oxidative degradation of polysaccharides, the selectivity of ozonation towards delignification reaction was even improved after acidity decrease from pH 1 to pH 2, pointing to pH 2 as a preferable pH for HPA-5 ozonation.

Ionic Strength of Bleaching Solution

The regulation of salt molar ratio (ionic strength) of bleaching solution can be effective method to control the pH gradient between fiber walls and outer solution, i.e., so-called the Donnan effect [34], induced by acidic functional groups linked to fiber matrix and affecting the distribution of ionic species in pulp suspension [35]. Due to strong repelling between POM polyanion and the negatively charged fiber, the Donnan effect can completely prevent penetration of POM catalyst into the fiber wall during pulp bleaching, excluding a direct contact between reactive species in the bleaching solution and pulp lignin [36]. The elimination of the Donnan effect could play therefore an essential role in delignification efficiency improvement of POMs.

Increase in ion (basically alkali metal cation) concentration in reaction solution was shown to have notable effect on HPA-5 catalyzed pulp ozonation in acetone solution (Table 6). Improvement in pulp brightness by 0.7% ISO and delignification by ca. 19% was observed after addition (up to 0.25 mol/L) of sodium sulphate salt to the reaction mixture. The concentration of mobile ionic species in bleaching solution (mostly Na^+ and SO_4^{2-} ions derived from sodium sulphate, HPA and sulfuric acid media) was sufficient to screen effectively the negative charges of fibers and polyanions thereby providing facilitated access of active lignin sites to catalyst through elimination of the Donnan effect. The possible formation of ion pairs of HPA with alkali metal cations (Na^+) in solution, leading to substantial increase in HPA oxidation potential and reactivity [37], can be another reason of enhanced pulp delignification during HPA-ozonation at high ionic strength.

Table 6. Effect of ionic strength on HPA-5 catalyzed ozone bleaching of eucalypt (*E. globulus* L.) kraft pulp (3% pulp consistency; $[O_3]$=0.8%; [HPA-5]=1.0 mM; 40% (w/w) acetone)

Salt concentration	Brightness (% ISO)	Lignin (% odp)	Viscosity (ml/g)	CHO (mmol/100g)
Control (no salt)	60.5	1.47	1002	6.94[a]
0.1 M Na_2SO_4	61.0	1.36	1017	5.71
0.25 M Na_2SO_4	61.2	1.21	1040	5.48

[a] mmol of CHO groups per 100 grams of dry pulp.

The suppressed oxidative degradation of carbohydrates under high ionic strength, confirmed by CHO-group analysis (Table 6) and resulted from increased HPA-5 stability and inhibited release of VO_2^+ ions [28, 38], caused substantial increase in pulp viscosity and improvement of ozonation selectivity as a whole.

HPA-5 RE-OXIDATION (RECOVERY) DURING OZONATION

The HPA-*n* solutions are the complex systems containing multiple reactive forms of catalyst. As it was noted above, under acidic conditions, the parent (Mo-V-P)-HPAs undergo reversible dissociation with release of VO_2^+ ions and formation of so-called defect (lacunary) heteropoly species according to reaction scheme *(3)*. The partial reduction of HPAs, with formation of heteropoly blues or mixed-valence complexes containing V(IV) $[PMo_{12-n}V^{V}_{n-m}V^{IV}_{m}O_{40}]^{(3+n)-}$, causes, in addition, a reversible elimination of VO^{2+} ions from the coordination sphere of the polyanion [20]. In reaction solution, the released VO_2^+ and VO^{2+} ions stay in equilibrium with the heteropolyanion species and with one another:

$$VO^{2+} + H_2O \leftrightarrow VO_2^+ + 2H^+ + \bar{e} \quad (4)$$

The simultaneous presence of multiple interconvertable catalytically competent reactive forms complicates significantly quantitative species-specific description of HPA reactivity, particularly kinetic, during catalyst re-oxidation reaction [39].

In the one-stage HPA-5 catalyzed ozone bleaching (the case of the present study), both the stoichiometric oxidation of pulp lignin (reaction scheme *(1)*) and the re-oxidation of reduced catalyst (reaction step *(2)*) take place simultaneously at the same reactor. The thermodynamic condition for HPA-5 catalyzed delignification by ozone is the following:

$$E^o\,(\text{Lignin}) < E^o\,(\text{POM}) < E^o\,(O_3) \quad (5)$$

where E^o (Lignin), E^o (POM) and E^o (O_3) are the oxidation potentials, respectively, of lignin (0.4-0.6 eV vs. NHE at pH 1) [40], HPA (0.68-0.71 eV vs. NHE at pH 1) [41] and ozone (2.07 eV at pH 2) [7].

The bleaching efficiency of HPA-5 (reaction scheme *(1)*) is therefore completely controlled by intensity (efficiency) of HPA-5 regeneration along the bleaching (reaction step

(2)). Simulating the mechanism of HPA catalysis with the *m* electrons transfer from the substrate (lignin) to catalyst, the original HPA-5 was partially reduced by hydrazine monohydrate to give heteropoly blue ($H_mHPA\text{-}5$; $HPA\text{-}5_{red}$) species by reaction scheme:

$$N_2H_4{\cdot}H_2O + 4/m\ HPA\text{-}5 \rightarrow N_2\uparrow + 4/m\ H_mHPA\text{-}5 + H_2O \quad (6)$$

To examine the heteropoly blue capacity for re-oxidation by ozone during pulp bleaching, the diluted $H_mHPA\text{-}5$ solution in water or aqueous acetone (10% w/w) was ozonated under variable conditions of ozone charge and pH, followed by spectrophotometric determination of $H_mHPA\text{-}5$ concentration, as a total content of all V^{IV} (poly- and mononuclear) species in solution.

The essentially identical time-depended profiles of $H_mHPA\text{-}5$ oxidation, consisting of two sharply-defined oxidation stages, were observed in both reaction media (Figure 7). The major part (bulk portion) of partially reduced HPA-5 was readily re-oxidized back during a few first minutes of ozonation, with almost twice higher rate in solvent media. The discussed before enhanced ozone solubility in organic solvents, leading to elevated concentration of active ozone species in reaction solution [11, 42], is a most probable reason of higher intensity of HPA re-oxidation in acetone media. The small residual portion of $H_mHPA\text{-}5$ reacted very slowly with ozone in water or even remained non-oxidative in acetone solution, similar to observation reported for (Mo-V-P)-heteropoly blues oxidation by dioxygen [43]. The slow formation of active intermediate complexes of the [$HPA_{red} \cdot O_2$] type under low degrees of catalyst reduction ($m<1.5$) at the end of the oxidation process was found to be a rate-determined reaction step responsible for incomplete HPA recovery [44]. The formation of similar intermediate complexes between $HPA\text{-}5_{red}$ and ozone can explain the incomplete $HPA\text{-}5_{red}$ re-oxidation by ozone observed in the present study.

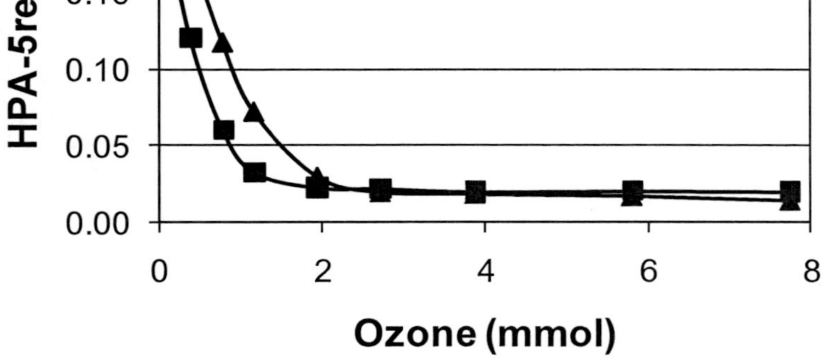

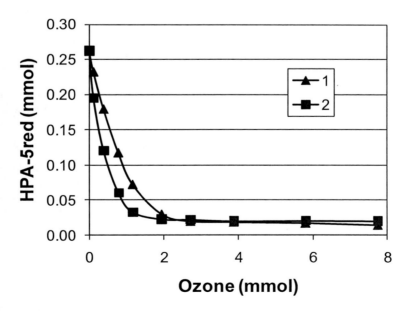

Figure 7. $HPA\text{-}5_{red}$ recovery (re-oxidation) by ozone in water (*1*) and 10% w/w aqueous acetone solution (*2*): 0.39 mmol O_3/min; pH 2.

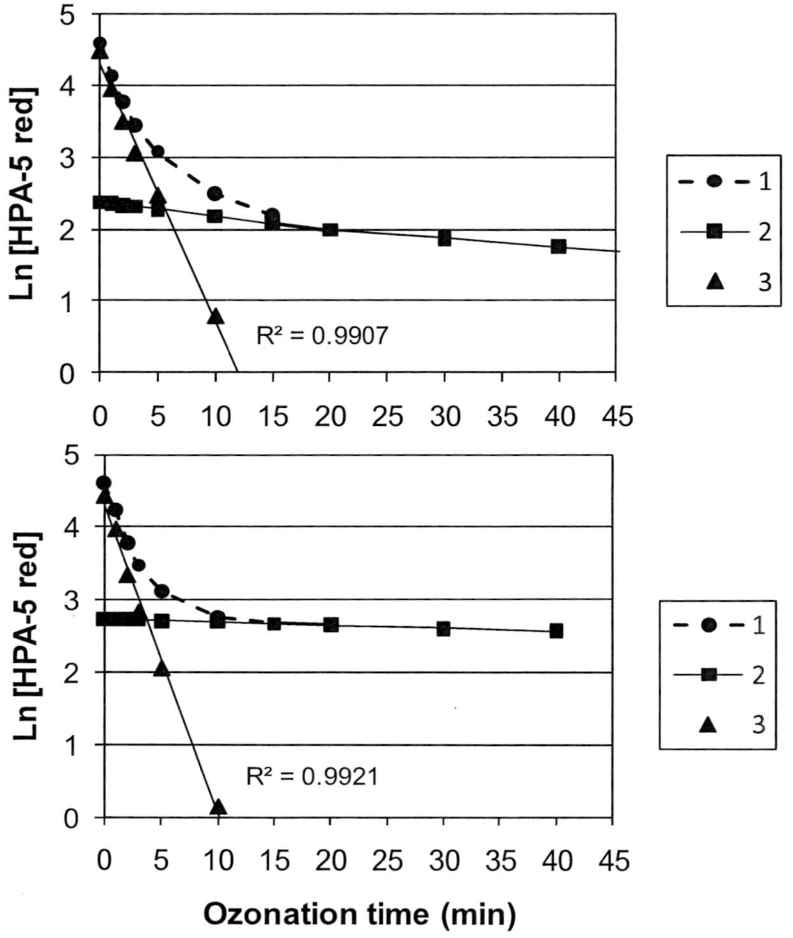

Figure 8. Kinetics HPA-5_{red} re-oxidation by ozone in water (top) and 10% w/w aqueous acetone solution (bottom) at 0.39 mmol O_3/min and pH 2: *1*- experimental kinetic curve $\ln[HPA_{red}]=\ln([HPA_{red}]_1+[HPA_{red}]_2)=f(t)$; *2*- calculated kinetic curve $\ln[HPA_{red}]_2=f(t)$; *3*- calculated kinetic curve $\ln[HPA_{red}]_1= \ln([HPA_{red}]-[HPA_{red}]_2)=f(t)$; $[HPA_{red}]_1$ and $[HPA_{red}]_2$ – reactive catalyst fractions.

The observed kinetic heterogeneity of HPA_{red} re-oxidation can be accurately described by original analytical approach developed for multi-component reaction systems [45]. At this case, the heteropoly blue solution is assumed to consist of a few (*i*) catalyst fractions (states), corresponding to (*i*) active catalyst species (or groups of species) with distinct reactivity to ozonation, as the kinetically homogeneous systems. Consequently, the heteropoly blue oxidation by ozone can be considered as a complex of (*i*) irreversible first-order reactions with similar final product. The employed method of graphical differentiation of kinetic curves [45] makes it possible to quantify accurately the content and reactivity of each separate kinetically homogeneous catalyst fraction (state) present in the reaction system, giving thereby the most complete kinetic description of HPA_{red} re-oxidation by ozone (Figure 8).

Table 7. Kinetic data (content of catalyst fractions and oxidation rate constants) on re-oxidation of partially reduced HPA-5 by ozone in water and aqueous acetone solution (0.5 mM HPA; 18.6 mg O_3/min)

Medium/pH	$HPA\text{-}5_{red,1}$		$HPA\text{-}5_{red,2}$	
	$[HPA]_{1,0}$ (%)	k_1 (min^{-1})	$[HPA]_{2,0}$ (%)	$k_2 \times 10^2$ (min^{-1})
Water				
pH 1	88.19	0.97	11.81	0.64
pH 1.5	88.42	0.64	11.58	0.84
pH 2	89.25	0.29	10.75	1.93
Acetone (10% w/w)				
pH 1	89.75	1.11	10.25	0
pH 1.5	90.25	0.86	9.75	0
pH 2	87.62	0.43	12.38	0.45

The performed analysis showed that the complex oxidation kinetics of $HPA\text{-}5_{red}$ by ozone can be accurately described in terms of two irreversible first order reactions corresponding to two different kinetically homogeneous catalyst fractions present in ozonation solution:

$$[HPA_{red}]_t = [HPA_{red}]_{1,0}\,(1\text{-}\exp(-k_1 t)) + [HPA_{red}]_{2,0}\,(1\text{-}\exp(-k_2 t)) \qquad (7)$$

where $[HPA_{red}]_t$ is a current concentration of $HPA\text{-}5_{red}$ at ozonation time (t); $[HPA_{red}]_{1,0}$ and $[HPA_{red}]_{2,0}$ are the contents of catalyst fractions at $t=0$; k_1 and k_2 are the effective rate constants of respective fractions oxidation.

It can be seen from Table 7 that found proportion of different reactive forms of partially reduced catalyst is very similar in water and acetone solution. In both reaction medium, the highly reactive (easily oxidizing) catalyst fraction comprises ca. 89% of total $HPA\text{-}5_{red}$. The rest ca. 11% of catalyst is re-oxidized with two-order lower rate in water or not re-oxidized at all (excepting at pH 2) in acetone solution. The pH of reaction media, having notable effect on catalyst reactivity, generally, does not affect the proportion of different catalyst forms in reaction solution.

Table 8. Effect of enzymatic pre-treatment of eucalypt (*E. globulus* L.) kraft pulp with commercial xylanase preparation (EC 3.2.1.8) on HPA-5 catalyzed ozone bleaching in acetone solution

	X – stage[a]		Z_{HPA} – stage[b]		E_R – stage[c]	
	Control[d]	Enzyme	Control	Enzyme	Control	Enzyme
Brightness (% ISO)	43.5	46.5	64.3	71.8	64.8	73.9
Kappa number	11.1	9.6	3.1	1.9	2.9	1.2
Viscosity (ml/g)	1320	1330	982	996	992	1003

[a] Enzymatic pre-treatment (Ecopulp TX-200A; 65°C; pH 7.0).
[b] HPA catalyzed ozone bleaching ($[O_3]$=0.8%; [HPA-5]=1.0 mM; 40% (w/w) acetone).
[c] Alkaline extraction in the presence of reducing agent (1% NaOH; 0.1% $NaBH_4$; 60°C).
[d] Enzyme-free treatment.

Strong correlation (R^2=0.99) was observed between experimental data on HPA-5 recovery and simulated data calculated by Eq. 7 (sum of square residuals SQR=0.0073), providing the validation test for applied kinetic description.

COMBINATION OF HPA CATALYZED OZONATION WITH ENZYMATIC (XYLANASE) PRE-TERATMENT

The enzymatic pulp pre-treatment with highly specific hemicellulolytic enzymes (particularly xylanases) before chemical bleaching (so-called bio-bleaching) has attracted considerable recent interest as very potential ecologically friendly biotechnological approach [46,47]. The key function of xylanase pre-treatment is to enhance the bleaching effect of other chemicals in subsequent bleaching stages, i.e., bleach boosting. The limited hydrolysis of xylan network improves fibre permeability thereby increasing the accessibility of lignin to bleaching reagents and facilitating the removal of lignin degradation products into solution [48].

To assess effect of enzymatic pulp pre-bleaching on HPA catalyzed ozonation, the industrial eucalypt kraft pulp was treated with commercial xylanase preparation Ecopulp TX-200A (endo-1,4-β-xylanase activity, EC 3.2.1.8) and then ozonated in acetone solution in the presence of HPA-5 catalyst, following by alkaline extraction of lignin degradation products. The bleaching results after each stage were compared with control enzyme-free samples (Table 8).

The direct bleaching effect of xylanases, i.e., direct brightening and delignification, was already observed immediately after the enzymatic stage, as a result of enzymatic attack on lignin-carbohydrate complexes (LCC) with removal of some lignin components and lignin-associated chromophoric groups [49]. The intrinsic viscosity was also somewhat increased, evidently, due to partial removal of low-molecular weight xylan fractions.

It can be seen from Table 8 that xylanase pre-treatment substantially promotes the subsequent HPA-5 catalyzed ozone bleaching in acetone media. The brightness, delignification and viscosity of xylanase-treated pulp were increased, respectively, by 11.7%, 38.7% and 1.4% as compared with control (untreated) samples, significantly improving the efficiency and selectivity of ozonation. The bleaching results were further improved by alkaline extraction of lignin degradation products from ozonated pulp into reaction solution. Reduction of ozonated pulps with sodium borohydride in alkaline extraction stage enabled to limit the alkali-induced degradation reactions of carbohydrates, through conversion of alkali-sensitive carbonyl groups formed during ozonation to hydroxyl groups [50], providing thereby the high bleached yield and pulp viscosity.

CONCLUSION

Heteropolyanion (HPA) catalyzed ozonation of chemical pulps in organic solvent media was proved to be very effective and selective environmentally benign bleaching approach providing a way for substantial increase in pulp brightness, viscosity and degree of delignification in comparison with other ozone-based bleaching techniques. The practical (commercial) feasibility of the proposed bleaching approach will be depended on the costs/efficiency of chemicals recovery and, consequently, on the environmental impact of this technology. The confirmed catalyst recovery during pulp ozonation permits to apply the totally effluent-free (TEF) concept of HPA bleaching with repeated recirculation of the reaction solution, minimizing any catalyst and solvent losses with the liquid process streams after pulp ozonation and meeting thereby the economical and ecological concerns related to chemical recovery. The high efficiency of solvent-assisted HPA/ozonation opens the possibility for a development of new short TCF bleaching sequences, e.g., in combination with bio-bleaching, to get desirable bleaching objectives without considerable ecological impact.

REFERENCES

[1] Van Lierop B.; Skothos, A.; Liebergott, N. In *Pulp bleaching. Principles and practice; Dence,* C. W.; Reeve, D. W.; Eds,; Tappi Press: Atlanta, 1996, pp 321-345.

[2] Byrd Jr., M.V.; Gratzl, J.S.; Singh, R.P. *Tappi J.* 1992, 75, 207-213.

[3] Liebergott, N.; Van Lierop, B.; Skothos, A. T*appi J.* 1992a, 75, 145-152.

[4] Liebergott, N.; Van Lierop, B.; Skothos, A. *Tappi J.* 1992b, 75, 117-124.

[5] Bokstrom, M.; Tuomi, A. *Paperi Ja Puu - Paper and Timber* 2001, 83, 124-127.

[6] King, J. E.; Van Heiningen, A. R. P. *Pulp Paper* - Canada 2003, 104, 38-42.

[7] Rice, R. C.; Netzer, A. *Handbook of ozone technology and application*; Ann Arbor Science Publishers: Ann Arbor, 1982.

[8] Kamishima, K.; Fujii, T.; Akkatsu, I. *Japan Tappi* 1977, 31, 699-706.

[9] Ni, Y.; Van Heiningen, A. R. P.; Lora, J.; Magdzinski, L.; Pye, E. K. *J. Wood Chem.* Technol. 1996, 16, 367-380.

[10] Saake, B.; Lehnen, R.; Schmekal, E.; Neubauer, A.; Nimz, H. *H. Holzforschung* 1998, 52, 643-650.

[11] Cogo, E.; Albert, J.; Malmary, G.; Coste, C.; Molinier, J. *Chem. Eng. J.* 1999, 73, 23-28.

[12] Pope, M. T. In *Handbook on the Physics and Chemistry of Rare Earths*; Gschneidner, Jr. K. A.; Bunzli, J.- C.; Pecharsky, V. K.; Eds.; Elsevier; 2007; Vol. 38, Chap. 240, pp 337-382.

[13] Kozhevnikov, I. V. *Catalysis by polyoxometalates;* Wiley & Sons: Chichester, England, 2002.

[14] Neumann, R.; Khenkin, A. M. *Chem. Commun.* 2006, 24, 2529-2538.

[15] Hill, C. L. *J. Mol. Catal. A: Chem.* 2007, 262, 2–6.

[16] Kolesnik, I. G.; Zhizhina, E. G.; Matveev, K. I. *J. Mol. Catal. A: Chem.* 2000, 153, 147–154.

[17] Weinstock, I. A.; Barbuzzi, E. M. G.; Sonnen, D. M.; Hill, C. L. *ACS Symp. Ser.* 2002, 823, 87-100.

[18] Evtuguin, D. V.; Neto, C. P.; Gaspar, A. R. *O Papel* 2005, 66, 28-34.

[19] Gaspar, A. R.; Gamelas, J. A. F.; Evtuguin, D. V.; Neto, C. P. *Green Chem.* 2007, 9, 717-730.

[20] Kozhevnikov, I. V. *Chem. Rev.* 1998, 98, 171-198.

[21] Hill, C. L. In Comprehensive Coordination Chemistry. Part II*: From Biology to Nanotechnology;* Wedd, A. G.; Ed.; Elsevier Ltd.: Oxford, UK, 2004; Vol. 4, pp 679–759.

[22] Weinstock, I. A.; Atalla, R. H.; Reiner, R. S.; Moen, M. A.; Hammel, K. E.; Houtman, C. J.; Hill, C. L. *New J. Chem.* 1996, 20, 269-275.

[23] Shatalov, A. A.; Pereira, H. In 10th Europ. *Workshop Lignocellulosics and Pulp, Proceed.;* Stockholm, Sweden, 2008.

[24] Shatalov, A. A.; Pereira, H. In 15th Int. Symp. *Wood Fibre and Pulping Chem., Proceed.;* Oslo, Norway, 2009.

[25] Shatalov, A. A.; Pereira, H. *Chem. Eng. J.* 2009, 155, 380-387.

[26] Shatalov, A. A.; Pereira, H. *Bioresour. Technol.* 2010, 101, 4625-4630.

[27] Shatalov, A. A.; Pereira, H. *Bioresour. Technol.* 2010, 101, 9330-9334.

[28] Shatalov, A. A.; Evtuguin, D. V.; Pascoal Neto, C. *Carbohydr. Polym.* 2000, 43, 23-32.

[29] Chandra, S.; Gratzl, J. S. In *Int. Pulp Bleach. Conf.;* CPPA: Quebec, Montreal, Canada, 1985; pp 27-30.

[30] Ni, Y.; Van Heiningen, A. R. P.; Lora, J.; Magdzinski, L.; Pye, E. K. *J. Wood Chem. Technol.* 1996, 16, 367-380.

[31] Quesada, J.; Rubio, M.; Gomes, D. *J. Appl. Polym. Sci.* 1998, 68, 1867-1876.

[32] Schuerch, C. *J. Am. Chem. Soc.* 1952, 74, 5061-5067.

[33] Souchay, P.; Chauveau, F.; Courtin, P. *Bull. Soc. Chim. Fr.* 1968, 6, 2384-2389.

[34] Donnan, F. G.; Harris, A. B. J. *Chem. Soc.* 1912, 99, 1554-1577.

[35] Towers, M.; Scallan, A. M. *J. Pulp Pap. Sci.* 1996, 22, J332-J337.

[36] Ruuttunen, K.; Vuorinen, T. *J. Pulp Pap.Sci.* 2004, 30, 9-15.

[37] Grigoriev, V. A.; Cheng, D.; Hill, C. L.; Weinstock, I. A. *J. Am. Chem. Soc.* 2001, 123, 5292-5307.

[38] Pettersson, L.; Andersson, I.; Selling, A.; Grate, J. *H. Inorg. Chem.* 1994, 33, 982-993.

[39] Pope, M. T.; Muller, A. *Polyoxometalate chemistry from topology via self-assembly to applications*; Kluwer Academic Publishers: Dordrecht, 2001.

[40] Weinstock, I. A.; Attala, R. H.; Reiner, R. S.; Moen, M. A.; Hammel, K. E.; Houtman, C. J.; Hill, C. L.; Harrup, M. K. *J. Mol. Catal. A: Chem.* 1997, 116, 59-84.

[41] Odyakov, V. F.; Zhizhina, E. G.; Matveev, K. I. *J. Mol. Catal. A: Chem.* 2000, 158, 453-456.

[42] Bin, A. K. *Ozone: Sci. Eng.* 2006, 28, 67-75.

[43] Zhizhina, E. G.; Odyakov, V. F.; Simonova, M. V.; Matveev, K. I. *React. Kinet. Catal. Lett.* 2003, 78, 373-379.

[44] Kozhevnikov, I. V.; Burov, Yu. V.; Matveev, K. I. Izv. Akad. Nauk SSSR. *Ser. Khim.* 1981, 11, 2428-2435.

[45] Shatalov, A. A.; Pereira, H. Ind. *Crops Prod.* 2005, 21, 203-210.

[46] Viikari, L.; Kantelinen, A.; Sundquist, J.; Linko, M. *FEMS Microbiol. Rev.* 1994, 13, 335-350.

[47] Suurnakki, A.; Tenkanen, M.; Buchert, J.; Viikari, L. In *Biotechnology in the Pulp and Paper Industry*; Eriksson, K.; Ed.; Springer Verlag: Berlin, 1997, pp. 261-287.

[48] Kantelinen, A.; Hortling, B.; Sundquist, J.; Linko, M.; Viikari, L. Holzforschung 1993, 47, 318-324.

[49] Paice, M. G.; Gurnagul, N.; Page, D. H.; Jurasek, L. *Enz. Microb. Technol.* 1992, 14, 272-276.

[50] Lindholm, C.-A. *J. Pulp Pap. Sci.* 1993, 19, 108-113.

In: Focus on Catalysis Research: New Developments
Editors: Minjae Ghang and Bjørn Ramel

ISBN: 978-1-62100-455-4

Chapter 11

MICROWAVE-ASSISTED CATALYTIC ASYMMETRIC TRANSFORMATIONS

Carolina Vargas[1,*], Alina Mariana Balu[2], Juan Manuel Campelo[2], Maria Dolores Gracia[2], Elia Losada[2], Rafael Luque[2], Antonio Pineda[2], Antonio Angel Romero[2]

[1]Departamento de Tecnología Química y Ambiental. Universidad Rey Juan Carlos, C/ Tulipán s/n, 28933, Móstoles, Madrid, Spain

[2]Departamento de Química Orgánica, Universidad de Córdoba, Edificio Marie Curie (C-3), Ctra Nnal IV, Km 396, E14014, Córdoba, Spain

ABSTRACT

Environmental and economical considerations in past decades have urged scientists to redesign commercially important processes towards the use of more environmentally friendly substances avoiding the use of toxic compounds and generation of waste. In this way, the substitution of conventional heating by microwave irradiation as an alternative source of energy is increasingly attractive. Microwave-assisted heating is particularly interesting because it offers improved conversions (and often seletivities to target products) under milder conditions at remarkably reduced times of reaction for a wide range of conventional reactions.

In this chapter we aim to explore several asymmetric syntheses carried out under microwave irradiation, giving a general overview of recently developments in this field. Significant rate enhancements, a decrease of chiral catalyst loading and higher enantioselectivities are the key achievement of microwave activation protocols.

* E-mail: carolina.vargas@urjc.es

1. Introduction

Asymmetric catalysis is currently a field of great importance in the synthesis of useful intermediates for the preparation of optically active organic compounds for the fine chemical industry [1]. Different chiral catalysts based in organic ligand structures with different functional groups (coordinated with a variety of metals in order to control the asymmetric properties of the compounds) have been used under homogeneous conditions since a few decades ago [2]. Homogeneous chiral catalysis is a fruitful research area with plenty of examples: from the relevant and notorious Sharpless catalyst based in titanium isopropoxide and diethyl tartrate [3] (Nobel Prize in Chemistry, 2001) (Figure 1) and Jacobsen catalyst for enantioselective epoxidation of olefins in early 1980s [4], until recent achievements such as the use of alternative two phase systems [5] or the design of new ligands in asymmetric catalysis [6].

Nevertheless, the disadvantages of homogeneous compared to heterogeneous systems are well known in general for all type catalytic processes [7]. Furthermore, the recent developments achieved in solid-phase chemistry have resulted in highly active and stereoselective heterogeneous catalysts systems that benefit from an improved separation and enhanced reusability after several uses [8,9]. These advantages combined to the minimization of the generation of toxic waste and the harsh reaction conditions (in energy intensive protocols) are the consequence of the enormous commercial and environmental interest of the chemistry industry.

Scientists have therefore been recently interested in the development of heterogeneous systems for such homogeneous chiral catalysts. Leaching of the catalyst from the support is an important factor to be taken into account. Consequently heterogeneization alternatives are studied in order to tackle this important issue. Many reports can be found in the literature related to different methods for the heterogeneization of homogeneous catalysts (e.g., by impregnation [10], ion exchange [11], or covalent grafting [12]).

Different supports have also been utilised in order to achieve the immobilization of these homogeneous chiral catalysts. These include polymeric resins [13], clays [14], zeolites [15] and mesostructured silica materials such as MCM-41 and SBA-15 [16]. Particularly, mesoporous solids have emerged as an interesting alternative support due to their remarkably properties.

MCM-41 is one of the most relevant members of the family of ordered mesoporous silicates characterized by its uniform array of mesopores with surface areas typically around 1000 m^2/g (Figure 2). Both the internal and external surfaces areas are covered by silanol groups, which can be potentially useful as anchoring sites for the immobilization of homogeneous complexes [17,18]. The well-known mesostructured SBA-15 materials present higher pore wall thickness and hydrothermal stability and larger pore size than MCM-41 [19]. These textural properties make both types of silica solids interesting supports for the heterogeneization of voluminous catalytic complexes formed by chiral organic ligands coordinated with different metals, as they may overcome diffusional and/or steric hindrances observed in asymmetric transformations where bulky molecules are involved [20].

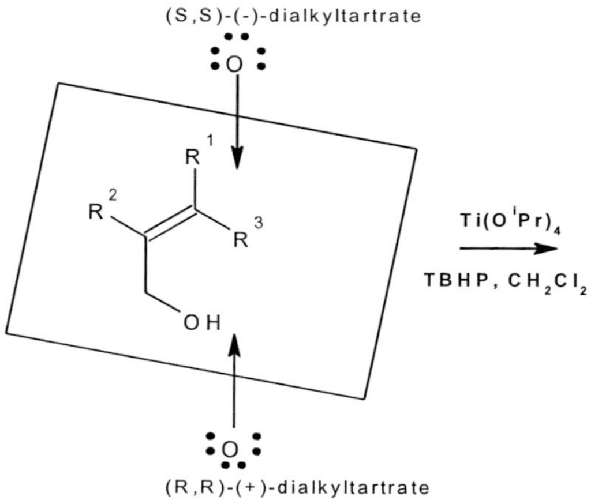

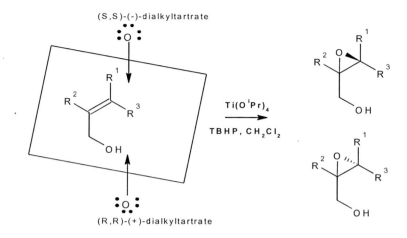

Figure 1. Enantioselective epoxidation of allylic alcohols using the Sharpless catalyst.

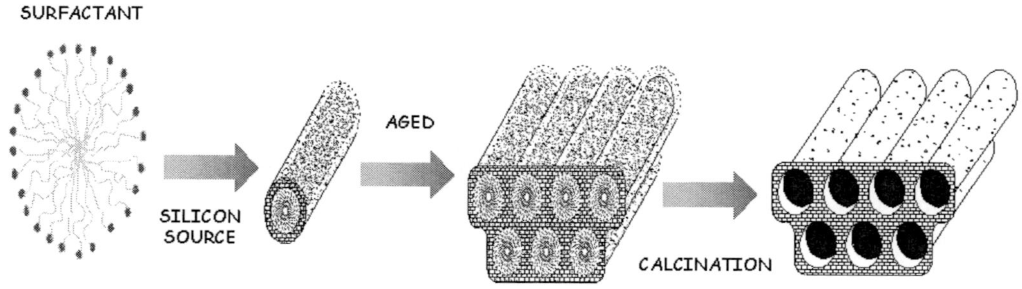

Figure 2. Synthesis method of mesostructured silica supports.

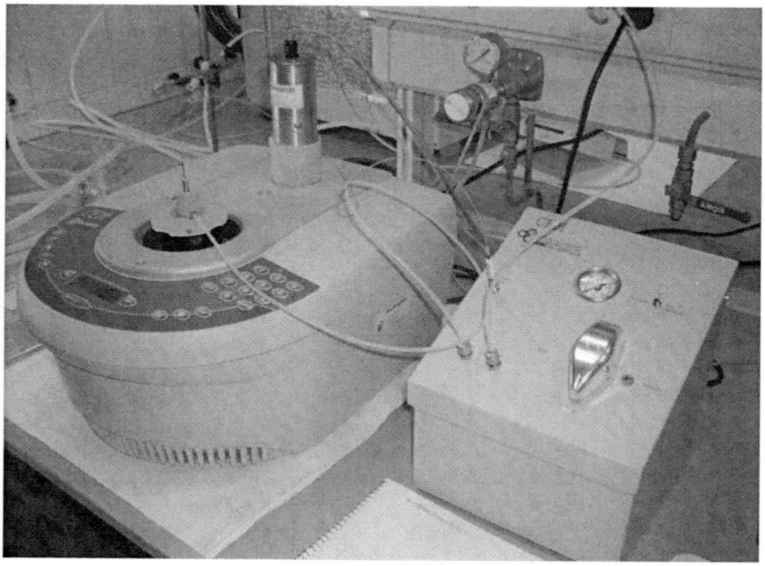

Figure 3. CEM microwave reactor system.

Microwave-assisted chemistry has rapidly emerged as an environmental and economical alternative to conventional heating [21]. The most evident potential benefits of microwaves are based in the remarkably reduced times of irradiation to achieve similar activities to those obtained under conventional heating, with the consequent reduction of waste and energy consumption. These reduced times are related to the selective microwave activation of molecules in which the energy transfer occurs in a fast way instead of being the heating localised at one molecular centre [22].

The equipment used for microwave irradiation have experienced remarkable advances since the first experiments carried out in a domestic household microwave oven without any temperature or pressure measurements control in the 1980s [23,24]. Different groups subsequently studied the use of microwaves in dry-media reactions working on with open-vessel technology in order to avoid potential explosions [25]. Nevertheless, in this case neither the temperature and pressure nor the complete mixing of the reactants could be optimally controlled, in particular when scale-up issues needed to be addressed. To solve these drawbacks, current microwaves reactors (e.g. CEM, Anton Parr, etc.) offer different sets of rotor systems with several open and close vessel designs and sizes for various synthesis applications [26]. Temperature and pressure measurements are controlled by both infrared and electronic sensors respectively and monitorized in an external PC connected to the MW instrument. The system carefully regulates the amount of power being supplied depending on the temperature and pressure of the processes, and if there is risk of too high temperatures or over-pressurised conditions it will automatically shut down. Protocols can also be precisely designed to control of the power during the reaction time. A cooling system is also required in order to restart the room temperature after the end of the experiment (Figure 3).

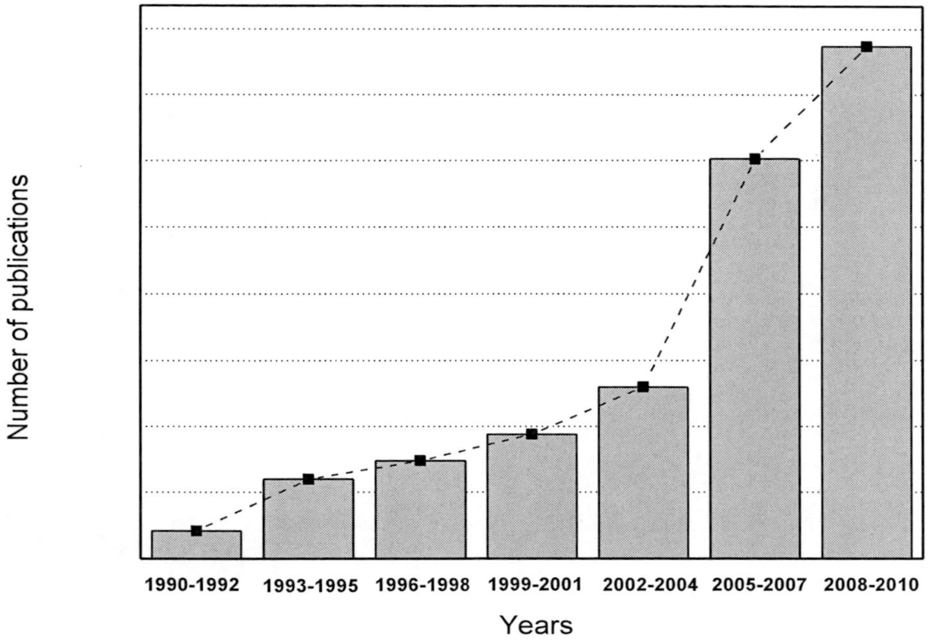

Figure 4. Development of number of publications related to microwave-assisted asymmetric transformations.

Microwave methodologies were proved to provide at least comparable yields and selectivities towards determined products in a wide range of reactions (Figure 4) using milder conditions as compared to those utilised under conventional heating [27,28]. In fact, the development of microwave-assisted protocols in the field of asymmetric organic transformations is currently active finding a variety of chiral organic compounds including natural products [29]. These methodologies can combine a microwave method which allows a reduction of the long reaction times traditionally needed for asymmetric synthesis as well as low temperatures and mild conditions required to achieve high enantiomeric excesses. The fusion of the asymmetric synthesis and microwave irradiation can therefore be considered a major step forward in the development of greener and more efficient strategies in this field.

This manuscript is intended to be a mini-contribution towards the state of the art and future prospects of the design of chiral catalysts in asymmetric reactions under microwave irradiation based on environmental and economical advantages with respect to conventional heating.

2. State-of-the-art in Asymmetric Reactions via Microwave Irradiation

The state of the art in microwave-assisted asymmetric transformations follows the directions of more environmental and economical routes that can improve the green credentials of the process. Chiral catalysts generally employed in past years have been based on chiral metal complexes, biocatalysts and chiral purely organic molecules. Organometallic chiral catalysts, formed between the combination of a chiral organic ligand and the corresponding metal, are generally utilised with a wider substrate scope due to their inherent stability and greener chemistry behaviour compared to the others. These active organocatalysts for asymmetric synthesis are rapidly gaining importance for their application in different processes [30].

A representative number of organocatalytic reactions studied are based on polar intermediates that are affected by microwave irradiation. Indeed, enamine or iminium catalysis promotes aldol and Mannich reactions as well as Michael additions and/or related processes with good to excellent results [31]. The aim of this section is to revise some selected and most relevant microwave-assisted transformations including reported asymmetric synthesis using chiral catalysts as compared to those performed under conventional heating.

2.1. Asymmetric Reduction of Imines

Palladium asymmetric reduction of imines mediated by chiral auxiliaries and assisted by microwave irradiation was recently reported by Espinoza-Moraga *et al.* [32]. Particularly, the reduction of different imines was carried out under microwave-assisted irradiation by using 8-phenylmenthylchloroformate as chiral auxiliary and $PdCl_2/Et_3SiH$ as reducing agent in CH_2Cl_2 (Table 1). Compared to conventional heating conditions, reaction times were significantly reduced to nine minutes achieving yields ranged from 65-90 % (Table 1).

Table 1. Microwave-assisted irradiation (90 W) and thermal reduction of imine 2 to 6 by using chloroformates of 8-phenylmenthyl (1a) as chiral auxiliary and $PdCl_2/Et_3SiH$ as reducing agent in CH_2Cl_2.

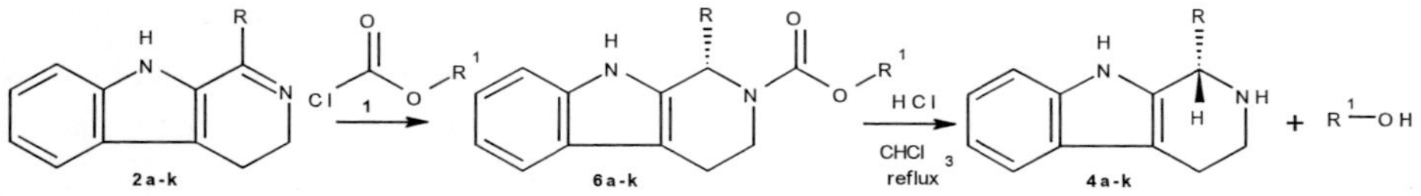

R^1: 8-phenylmenthyl **1a**
trans-phenylcyclohexyl **1b**

Entry	Imine X (R)	MWA reaction		Thermal reaction	
		Time and Temperature	Compound, Diastereomeric ratio (dr)[a] and yield[b]	Time and Temperature	Compound, Diastereomeric ratio (dr)[a] and yield[b]
1	Me, 2a	9 min (60 °C)	6a, 5:1 (90%)	0.5 (−78 °C)	6a, 6:1 (87%)
2	Et, 2b	9 min (60 °C)	6b, 7:1 (85%)	1 (−78 °C)	6b, 6:1 (87%)
3	*iso*-Pr, 2c	9 min (60 °C)	6c, 9:1 (82%)	3 (−78 °C)	6c, 12:1(88%)[1c]
4	1-Pentenyl, 2d	9 min (60 °C)	6d, 11:1 (78%)	3 h (−78 °C)	6d, 10:1 (85%)
5	Ph, 2e	9 min (60 °C)	6e, 1:1 (65%)	1 h (−78 °C)	6e, 1.5:1 (85%)
6	Bn, 2f	9 min (60 °C)	6f, 4:1 (75%)	1 h (−78 °C)	6f, 3.5:1 (90%)
7	$(CH_2)_2CO_2Me$, 2g	9 min (60 °C)	6g, 5:1 (77%)	2 h (−78 °C)	6g, 5:1 (91%)
8	$(CH_2)_3CO_2Me$, 2h	9 min (60 °C)	6h, 5.5:1 (80%)	2 h (−78 °C)	6h, 5:1 (89%)
9	$(CH_2)_7CH{=}CH(CH_2)_7Me$, (*Z*)-2i	9 min (60 °C)	6i, 7:1 (89%)	2 h (−78 °C)	6i, 8:1 (70%)
10	$(CH_2)_7(CH{=}CHCH_2)_4(CH_2)_3Me$, (*Z,Z*)-2j	9 min (60 °C)	6j, 9:1 (88%)	2 h (−78 °C)	6j, 10:1 (75%)
11	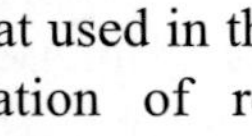 , 2k	9 min (60 °C)	6k, 12:1 (88%)	1 h (−78 °C)	6k, 13:1(85%)[2b]

[a] Diastereomeric ratio (dr) calculated based on HPLC of product *6a–k*.
[b] Isolated yields.

Applying a microwave-assisted protocol, the temperature of the reaction mixture inside the reaction vessel reached a value of 60ºC at 90 W. Although this temperature was higher than that used in the thermal conditions, times of reaction were shorter preventing in this way degradation of reactants and products. These microwave-assisted conditions can be considered an example of a green chemistry approach which features short times of reaction, reduced energy consumption, and increased efficiencies.

2.2. Oxidative Coupling of Amines

A K-10 montmorillonite catalyzed microwave-assisted oxidative coupling of amines to imines was reported by Lange *et al.* [33]. Imines are versatile starting materials for chiral amine synthesis [34]. The direct oxidative coupling of amines to imines without the use of any carbonyl compound can serve as a novel alternative for this reaction [35]. Over the past two decades, microwave-assisted organic synthesis has emerged as an important area because is a convenient and time saving method, which promoted the application of environmentally benign approaches such as solvent-free and heterogeneous catalytic reaction conditions [36, 37]. The major advantages of the highlighted oxidative coupling of amines are the use of a readily available and economic catalyst, a solvent-free system as well as short times of reaction.

K-10 is a strong solid acid prepared from natural montmorillonite by mineral acid treatment also used in bifunctional catalysis. Originally the oxidative self-coupling of substituted benzylamines using traditional conductive heating (external oil bath) was observed, but the application of microwave irradiation significantly increased the reaction rates and yields (3 min as compared to 24 h under conventional heating). Table 2 shows the corresponding substituted benzylidene-benzylamines could be obtained in good to excellent yields at very short reaction times.

The microwave-assisted oxidative coupling of substituted benzylamines with anilines (Table 3) and aliphatic amines (Table 4) using K-10 montmorillonite as catalyst was also studied. Although anilines and aliphatic amines cannot undergo self-coupling by themselves (only in low yields), they readily react with benzylamines as shown in Tables 3 and 4.

Reactions are however relatively slow and the yields are moderate. It appears that steric effects play little role in these reactions, cyclic amines react similarly to open chain analogs. Further increase in reaction times resulted in a significant amount of byproducts.

Table 2. Microwave-assited oxidative self-coupling of substituted benzylamines on K-10 montmorillonite

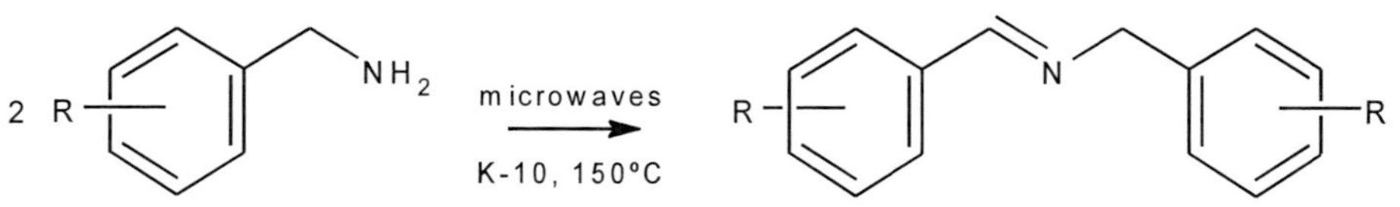

Entry	R	Time (min)	Yield[a] (%)
1	H	3	80
2	*p*-CH_3	3	88
3	*m*-CH_3	1	80
4	*o*-F	1	86
5	*p*-F	1	98
6	*p*-Cl	2	96
7	*m*-CF_3	6	76
8	*o*-OCH_3	2	87[b]

Reaction conditions: benzylamine (1.0–4.6 mmol), 1 g catalyst, 150°C, 1 bar air.
[a]Based on benzylamine, determined by GC and GC–MS.[b]At 140°C.

Table 3. Microwave-assisted oxidative coupling of substituted benzylamines with anilines on K-10 montmorillonite

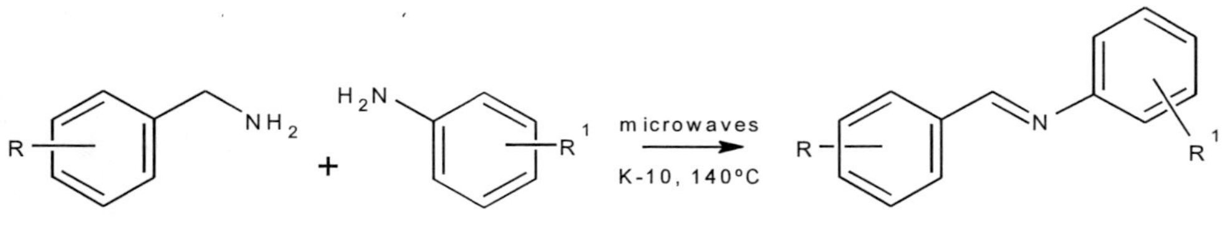

Entry	R	R_1	Time (min)	Yield[a] (%)
1	H	*p*-F	18	90
2	H	*m*-CF_3	33	79
3	*p*-CH_3	H	45	84
4	*p*-CH_3	*p*-F	21	83
5	*m*-CH_3	H	35	85
6	*m*-CH_3	*p*-F	18	93
7	*o*-OCH_3	*p*-F	30	78
8	*p*-F	*p*-F	23	81
9	*p*-Cl	*p*-F	18	86
10	*m*-CF_3	*p*-F	18	98
11	*m*-CF_3	*m*-CF_3	28	79

Reaction conditions: benzylamine (2.75 mmol), aniline (2.75 mmol), 1 g catalyst, 140°C, 1 bar air.
[a]Based on aniline, determined by GC and GC–MS.

Table 4. Microwave-assisted oxidative coupling of substituted benzylamines with aliphatic amines on K-10 montmorillonite

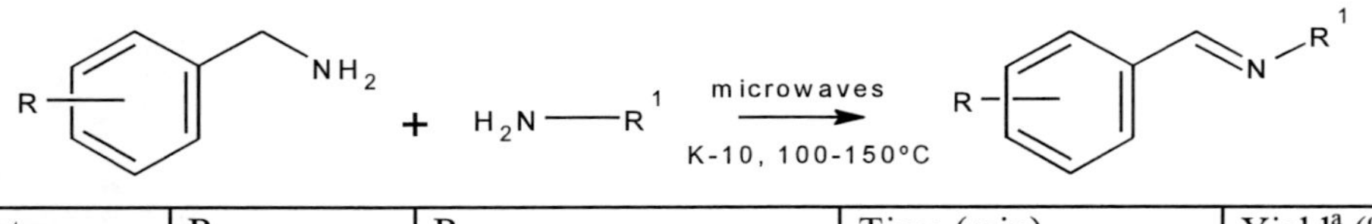

Entry	R	R_1	Time (min)	Yield[a] (%)
1	H	Cyclopentyl[b]	19	72
2	H	*n*-Hexyl[c]	30	50
3	*p*-CH_3	Cyclopentyl[b]	60	34
4	*p*-CH_3	*n*-Hexyl[c]	30	28
5	*p*-F	Cyclopentyl[b]	21	64
6	*p*-F	*n*-Hexyl[c]	40	34
7	*p*-Cl	*n*-Hexyl[c]	20	28
8	*p*-Cl	Cyclopentyl[b]	30	33

Reaction conditions: benzylamine (2.75 mmol), alkylamine (5.5 mmol), 1 g catalyst, 100°C, 1 bar air; after half reaction time extra 0.5 equiv alkylamine was added.
[a]Based on alkylamine, determined by GC and GC–MS.
[b]140°C.
[c]150°C.

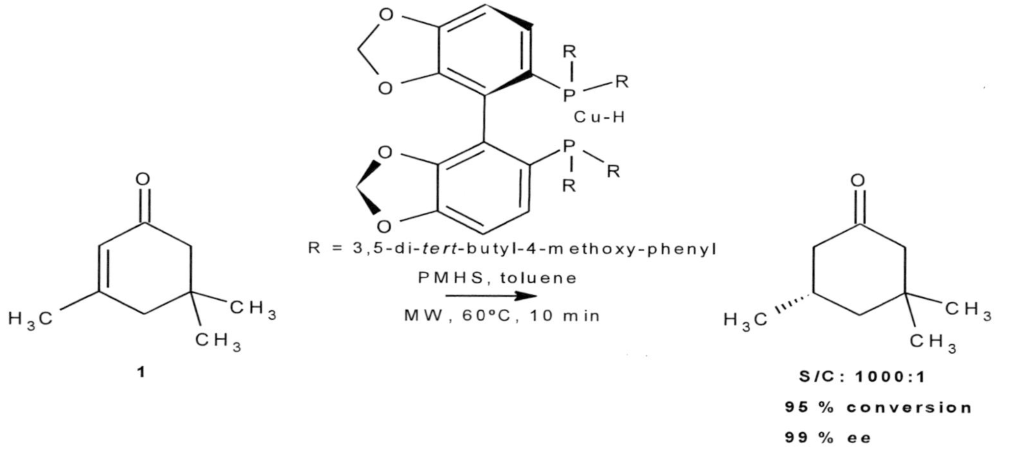

Scheme 1. Microwave asymmetric reduction of enones.

2.3. Asymmetric Hydrosilylations and Reductions

Excellent ee´s and good conversions were obtained in several asymmetric hydrosilylations at room temperature, but also microwave heating could be applied for the reduction of enone **1** (Scheme 1). Times of reaction could be reduced from several hours to only 10 minutes by using a substrate/catalyst ratio (S/C) of 1000:1 [38].

The enantioselective ethylation of aryl aldehydes to 1-aryl-1-propanols could also be efficiently performed under microwave irradiation, with greatly accelerated rates of reaction and high ee´s (up to 92% ee).

2.4. Synthesis and Polymerization of Chiral Acrylamide

Microwave irradiation accelerates considerably the process of acrylamide formation with exceptional selectivity. In contrast, the reaction mixture heated in oil bath was completely polymerized after only 2 min [39]. Microwave activation performed in the presence of AIBN afforded, in a single step, optically active polymers or a blend of polymers of variable composition and structure as a function of the applied microwave power.

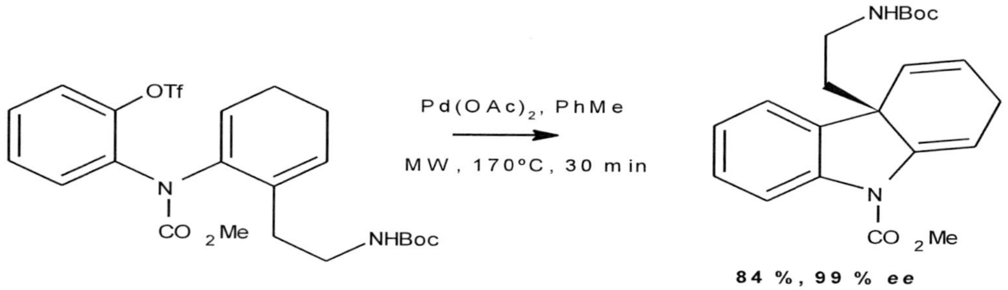

Scheme 2. Palladium catalysed Heck reaction.

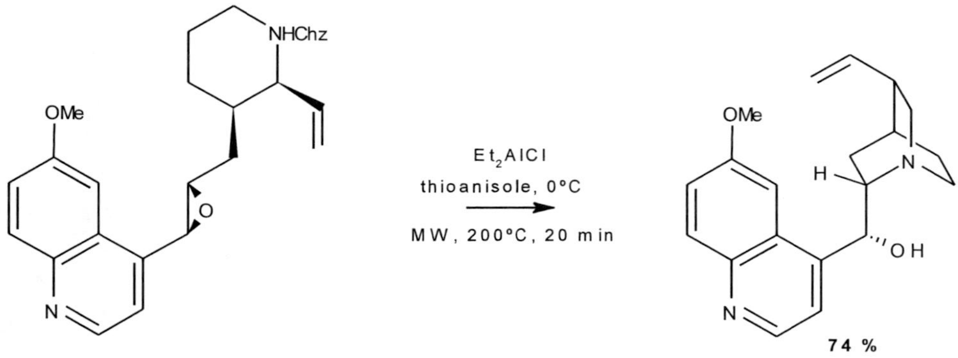

Scheme 3. Microwave enhanced deprotection/nucleophilic substitution sequence.

3. Application of Microwave Irradiation Protocols to Natural Products

One of the greatest tests of any synthetic methodology is the application into complex molecules, in particular natural products, and in recent times it has become clear that many complex systems can tolerate the high temperatures associated with microwave heating at least for short periods of time [40]. Moreover, it is also clear that microwave heating can rescue a synthetic route and be an important part of a successful target synthesis endeavour [41].

Douney *et al.* have reported in past years an asymmetric palladium catalysed Heck cyclidation as part of a synthetic approach to the strychnos alkaloid minfiensine (Scheme 2) [42]. Under conventional heating, the reaction was slow, but it could be accelerated using microwave heating from 100 h to 30 min, with no loss of enantioselectivity compared to the conventionally heated system using a system based on $Pd(OAc)_2$, PhMe and PMP as ligand.

In a similar way, Jacobsen used a microwave enhanced deprotection/nucleophilic substitution sequence in order to complete the total synthesis of the cinchona alkaloids quinine and quinidine (Scheme 3). Whereas the reported microwave protocol reached completion within 20 min, all other approaches led to poor yields and/or longer reaction times [43].

Balu *et al.* have also recently described a one-step methodology for the diastereoselective production of enantiomerically pure menthols [with (-)-menthol in high selectivites] from (+)-citronellal [44].

Supported metal nanoparticles (e.g. Pd, Pt and Cu) on a range of mild acidic supports (e.g. Al- and Ga-MCM41 and SBA-15) were employed in a novel cyclisation/hydrogenation tandem process in which the acid sites of the support were claimed to be responsible of the cyclisation of citronellal to isopulegol (1st step) and then the final transfer hydrogenation process (22) takes places on the metallic sites [45].

In this way, unprecedent selectivities close to 75% to (-)-menthol could be obtained at conversion values of starting material superior to 90% only when chiral modifiers (e.g. cinchonidine) were introduced into the reaction. These observations imply that the use of chiral modifiers has a promising effect in this relevant process [44].

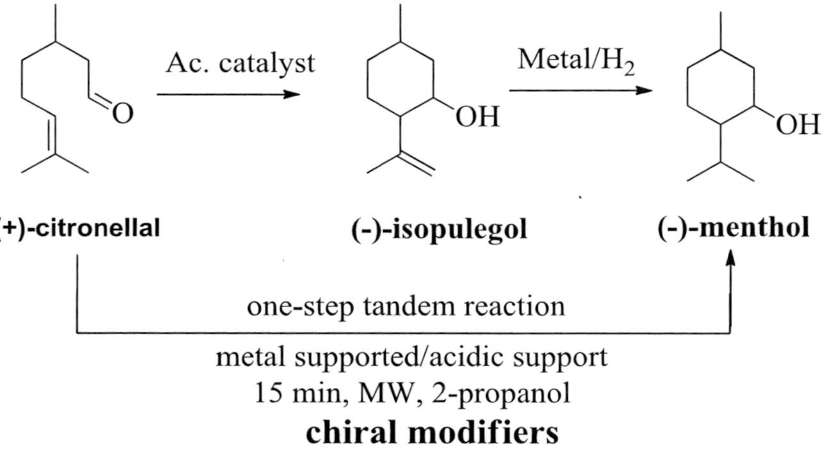

Scheme 4. Production of (-)menthol from (+)-citronellal using heterogeneous catalsyts and chiral modifiers.

4. Conclusions

Through selected elegant examples, we aimed to provide a brief overview of the usefulness of microwave technologies in the catalysed asymmetric transformation of a wide range of compounds, from simple systems (e.g. oxidative couplings) to total synthetic protocols in which complex molecules such as alkaloids are involved. In all cases, microwaves were able to grant access to a series of reactions that could not be otherwise performed using conventional methodologies. With these reports, we want to highlight the promising future of microwave-assisted catalysed asymmetric processes in the selective synthesis of natural products and/or intermediates to them that find many applications in medicine, fragrances, and pharmaceuticals.

Acknowledgments

RL gratefully acknowledges Ministerio de Ciencia e Innovacion, Gobierno de España for the concession of a Ramon y Cajal contract (RYC-2009-04199). Funding from projects P09-FQM-4781 and FQM-02695 (Consejería de Educación y Ciencia de la Junta de Andalucía) and Ministerio de Ciencia e Innovación (Project CTQ2008-01330/BQU) are also gratefully acknowledged.

References

[1] Blaser, H.U.; Spindler, F.; Studer, M. *Appl. Catal. A. Gen.* 2001, *221*, 119.
[2] Consiglio, G.; Waymouth, R.M. *Chem. Rev.* 1989, *89*, 257.
[3] Katsuki, T; Sharpless, K.B. *J. Am. Chem. Soc.* 1980, *102*, 5974.

[4] Zhang, W.; Loebach, J.L.; Wilson, S.R.; Jacobsen, E.N. *J. Am. Chem. Soc.* 1990, *112*, 2801.

[5] Yoo, M-S.; Kim, D-G; Ha, M-W.; Jew, S; Park, H.; Jeong, B-S. *Tetrahedron Lett.* 2010, *51*, 5601.

[6] Li, Y-M.; Kwong, F-Y.; Yu, W-Y; Chan, A.S.C. *Coordination Chemistry Reviews* 2007, *251*, 2119.

[7] Serrano, D.P.; Aguado, J.; García, R.A.; Vargas, C. *Stud. Surf. Sci.Catal.* 2005, *158*, 1493.

[8] Karjalainen, J.K.; Hormi, O.E.O.; Sherrington, D.C. *Tetrahedron Asymm.* 1998, *9*, 1563.

[9] Serrano, D.P.; Aguado, J.; Vargas, C. *Appl. Catal. A. Gen.* 2008, *335*, 172.

[10] Bein, T; Ogunwumi, S.B. *Chem. Commun.* 1997, 901.

[11] Yin, D.; Liu, J.; Zhang, Y.; Gao, Q.; Yin, D. *Stud. Surf. Sci.Catal.* 2005, *156*, 851.

[12] Kureshy, R.I.; Ahmad, I.; Khan, N.H.; Abdi, S.H.R.; Pathak, K.; Jasra, R.V. *J. Catal.* 2006, *238*, 134.

[13] Disalvo, D.; Dellinger, D.B.; Gohdes, J.W. *React. Funct. Polym.* 2002, *53*, 103. [14] Das, P.; Kuzniarska-Biernacka, I.; Silva, A.R.; Carvalho, A.P.; Pires, J.; Freire, C. *J. Mol. Catal. A Chem.* 2006, *248*, 135.

[15] Choudary, B.M.; Chowdary, N.S.; Kantam, M.L.; Santhi, P.L. *Catal. Lett.* 2001, *3-4*, 76.

[16] Kureshy, R.I.; Ahmad, I.; Khan, N.H.; Abdi, S.H.R.; Pathak, K.; Jasra, R.V. *Tetrahedron Asymm.* 2005, *16*, 3562.

[17] Beck, J.S.; Vartuli, J.C.; Roth, W.J.; Leonowicz, M.E.; Kresge, C.T.; Schmitt, K.D.; Chu, C.T-W.; Olson, D.H.; Sheppard, E.W.; McCullen, S.B.; Higgins, J.B.; Schlenker, J.L. *J. Am. Chem. Soc.* 1992, *114*, 10834.

[18] Galarneau, A.; Gambon, H.; Martin, T.; De Menorval, L-C.; Brunel, D.; Di Renzo, F.; Fajula, F. *Stud. Surf. Sci. Catal.* 2002, *141*, 395.

[19] Davidson, A.; Davidson, P.; Imperor-Clerc, M. *J. Am. Chem. Soc.* 2000, *122*, 11925.

[20] Linssen, T.; Cassiers, K.; Cool, P.; Vansant, E.F. *Advances in Colloid and Interfase Science* 2003, *103*, 121.

[21] Kappe, C.O. *Comprehensive Medicinal Chemistry II* 2007, Elsevier.

[22] Tierney, J.P. & Lidström P. *Microwave assited organic synthesis* 2005, Blackwell publishing.

[23] Gedye, R.; Smith, F.; Westaway, K.; Ali, H.; Baldisera, L. *Tetrahedron Lett.* 1986, *27*, 279.

[24] Giguere, T.L.B.; Scott, M.D.; Majetich, G. *Tetrahedron Lett.* 1986, *27*, 4945.

[25] Feinberg, M.; Suard, C.; Ireland-Ripert, J. *Chemometrics and Intelligent laboratory Systems* 1994, *22*, 37.

[26] Kappe, C.O.; Stadler, A. *Microwaves in Organic and Medicinal Chemistry* 2005, Wiley-VCH.

[27] Kappe, C.O. *Chem. Soc. Rew.* 2008, *37*, 1127.

[28] Polshettiwar, V.; Varma, R.S. *Acc. Chem. Res.* 2008, *41*, 629.

[29] Luque, R.; Balu, A.M.; Campleo, J.M.; Romero, A.A. *9th Congress on Catalysis Applied to Fine Chemicals* 2010, 93.

[30] Hosseini, M; Stiasni, N.; Barbieri, V.; Kappe, C.O. *J. Org. Chem.* 2007, *72*, 1417.

[31] Mossé, S.; Alexakis, A. *Org. Lett.* 2006, *8*, 3577.

[32] Espinoza-Moraga, M.; Caceres, A.G.; Silva, L. *Tetrahedron Lett.* 2009, *50*, 7059.
[33] Landge, S.M.; Atanassova, V.; Thimmaiah, M.; Török, B. *Tetrahedron Lett.* 2007, *48*, 5161.
[34] Enders, D.; Reinhold, V. *Tetrahedron: Asymmetry* 1997, *8*, 1895.
[35] Sheldon, R.A.; Kochi, J.K. *Metal Catalyzed Oxidations of Organic Compounds* 1981, Academic Press: New York.
[36] Loupy, A. *Microwaves in Organic Synthesis* 2002, Wiley-VCH.
[37] Ahn, J.A.; Chang, D.H.; Park, Y.J.; Yon, Y.R.; Loupy, A.; Jun, C.H. *Adv. Synth. Catal.* 2006, *55*, 348.
[38] Lipshutz, B.H.; Frieman, B.A. *Angew. Chem. Int. Ed.* 2005, *44*, 6345.
[39] Iannelli, M.; Alupei, V.; Ritter, H. *Tetrahedron* 2005, *61*, 1509.
[40] Caddick, S.; Fitzmaurice, R. *Tetrahedron* 2009, *65*, 3325.
[41] Kang, Y.; Mei, Y.; Du, Y.G.; Jin, Z.D. *Org. Lett.* 2003, *5*, 4481.
[42] Dounay, A.B.; Overman, L.E.; Wrobleski, A.D. *J. Am. Chem. Soc.* 2005, *127*, 10186.
[43] Raheem, I.T.; Goodman, S.N.; Jacobsen, E.N. *J. Am. Chem. Soc.* 2004, *126*, 706.
[44] Balu, A.M.; Campelo, J.M.; Luque, R.; Romero, A.A. *Org. Biomol. Chem.* 2010, *8*, 2845.
[45] Gracia, M.J.; Campelo, J.M.; Losada, E.; Luque, R.; Marinas, J.M.; Romero, A.A. *Org. Biomol. Chem.*, 2009, *7*, 4821.

In: Focus on Catalysis Research: New Developments
Editors: Minjae Ghang and Bjørn Ramel
ISBN: 978-1-62100-455-4

Chapter 12

METALLOCORROLES AS CATALYST: CURRENT STATUS AND FUTURE DIRECTIONS

Achintesh Narayan Biswas and Pinaki Bandyopadhyay*
Department of Chemistry, University of North Bengal,
Raja Rammohunpur, Siliguri, India

ABSTRACT

Since the discoveries of facile methodologies for the synthesis of triarylcorroles and the corresponding metal complexes, metallocorroles have attracted the interest of chemists immensely because of their ability to catalyze diverse reactions. The huge number of studies performed during the last decade on metallocorrole catalyzed reactions has led to several important catalytic systems encompassing oxidation catalysis, reduction catalysis, group transfer catalysis, etc. The aim of the present chapter is to provide a comprehensive account on the catalytic properties of metallocorroles with an emphasis on recent advances in the area and future directions.

1. INTRODUCTION

Corroles [1] are tetrapyrrolic macrocycles who owe their name to cobalt-chelating corrin of vitamin B_{12} with whom they share an identical skeleton. Corroles are extensively conjugated and more closely related to the iron-chelating porphyrins of heme enzymes and proteins. More precisely, corroles are based on the [18] annulene structural framework, just one *meso* carbon short from the porphyrin skeleton, as shown in Figure 1.

* Correspondence email: pbchem@rediffmail.com

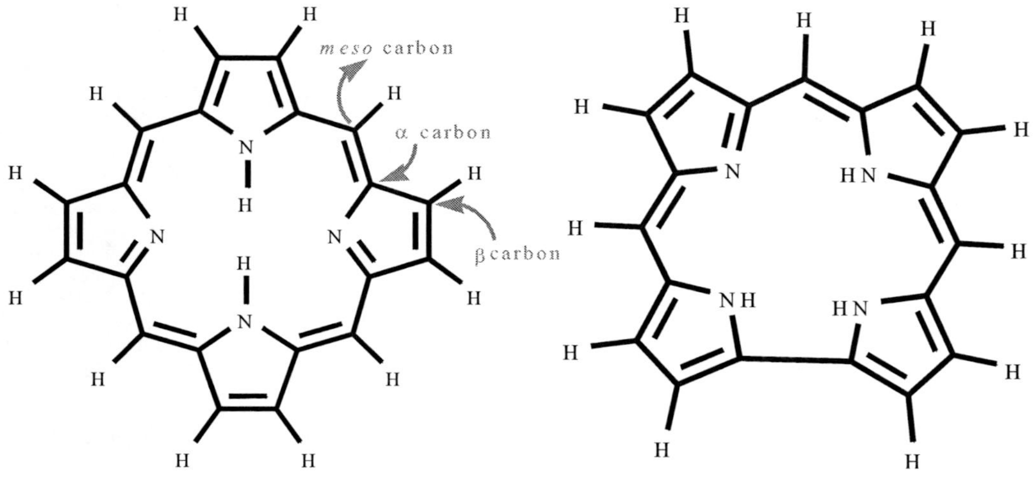

Figure 1. The structure of free base porphyrin (left) and free base corrole (right). The positions of C_{α}, C_{β} and C_{meso} are noted with arrows.

The missing *meso* carbon leads to a smaller central cavity compared to porphyrin and reduces symmetry from D_{4h} to C_{2v}. Free base corroles act as a trianionic ligand owing to the presence of three inner –NH protons. Being a trianionic macrocycle with a small cavity corroles has unique property of stabilizing high metal oxidation states. The most stable oxidation numbers in metallocorroles are often one positive charge higher than in the case of analogous metalloporphyrins. The chemistry of corroles unlike porphyrin remained underdeveloped for a long period, largely due to lack of simple methods for their synthesis [2]. Corroles were first synthesized as early as in 1964 by Johnson and Kay [2a] and the crystallographic characterization of a free base corrole was first reported by Hodgkin *et al.* [3] in 1971. The *meso* substituted corrole was first synthesized in 1993 [4], almost 30 years after the first report of corrole synthesis. All these synthetic processes suffer from the limitations of very low yield, longer reaction time and absence of readily available starting materials. Finally, major breakthrough was made in 1999 by Gross *et al.* [5] and Paolesse *et al.* [6] of one-pot corrole syntheses involving pyrrole-aldehyde condensations. The research group of Paolesse prepared a wide range of free base triaryl corroles under Alder-Longo-type protic acid catalyzed reactions using glacial acetic acid as the solvent and with an aldehyde/pyrrole molar ratio of 1:3 [6]. In contrast, Gross and his research group reported another convenient method which involves an essentially solvent free condensation of pyrrole with aldehydes in the presence of basic alumina under aerobic condition, followed by oxidation with DDQ in dichloromethane [5]. The directed synthesis of corrole macrocycles continues to be an active area of research as evident from the recent contributions from the groups of Gryko [7], Bruckner [8], Collman [9] and Geier [10].

Investigations into the syntheses of free-base and metal corroles have flourished in recent years and a tremendous amount of interest has been generated on the chemistry of this contracted macrocycle [1]. Corroles and their metal derivatives find their application in diverse fields like catalysis [1, 11], medical science [1d, 1i], sensors [12], photophysics [13] and dye-sensitized solar cells (DSSCs) [14].

2. METALLOCORROLES AS CATALYSTS

Past few decades have witnessed a tremendous amount of research activities in metallo-porphyrin catalyzed reactions with biological relevance. It is, thus, not surprising that metallo-derivatives of corrole, a close analogue of porphyrin, are emerging as prospective candidates for catalyzing diverse types of reactions. Corroles have always been considered promising in this regard due to the presence of three inner –NH's (instead of two in porphyrins) and smaller cavity (than porphyrin) that in turn, stabilizes high metal oxidation states. Due to their unique capacity to access high metal oxidation states, various metallocorroles have been synthesized and turned out as successful catalysts for epoxidation [15], hydroxylation [15a, 16], cyclopropanation [15a, 17] and aziridination [18]. In fact, there are several reactions which are catalyzed only by metallocorroles. The examples are iron-based aziridination by Chloramine-T, the facile N-H activation of amines and specially the disproportionation of peroxynitrite which is of immense potential in medicinal chemistry.

The *meso*-substituted corrole derivatives are found to be robust like porphyrin analogues and almost all the catalytic reactions reported so far are associated with *meso*-aryl substituted metallocorroles. Among the *meso*-substituted corroles, 5,10,15-tris(pentafluorophenyl) corrole (abbreviated as H_3tpfc) has emerged as the most important member.

2.1. Oxidation Catalysis

The oxidative functionalization of hydrocarbons to more valuable products under mild conditions is an area of current interest [16]. The selective insertion of one oxygen atom from oxygen donors into various hydrocarbon molecules under mild conditions remains a challenge in chemical science. The biological system has evolved a group of enzymes called monooxygenases, which catalyze monooxygenation reaction under very mild conditions [20]. Extensive effort has been devoted to develop synthetic homogeneous catalysts as functional models for monooxygenases. In this direction metalloporphyrins catalyzed oxidation of hydrocarbons has been extensively studied [21].

2.1.1. Hydroxylation

Gross *et al.* reported the first ever application of metallocorrole as catalyst in hydroxylation of ethylbenzene with iodosylbenzene (PhIO) using iron(IV) corrole [(tpfc)Fe-Cl] (1) [15a]. In benzene medium, ethylbenzene has been oxidized with an overall yield of 10.8% forming 1-phenylethanol (6.6%) and acetophenone (4.2%) as the major products (Scheme 1).

Difluoroantimony(V)corrole has also displayed high activity and selectivity for the photoinduced hydroxylation of allylic and benzylic C-H bonds with dioxygen as the oxidant [16d]. In case of olefinic substrates like cyclohexene and cyclooctene, only allylic oxidations have been noted. Cumene, under the photocatalytic reaction condition employed has been converted solely to cumene hydroperoxide, which has been converted into the corresponding alcohol on addition of triphenylphosphine (Scheme 2). The failure of this oxidizing system to hydroxylate substrates like adamantane or ethylbenzene suggests the involvement of singlet oxygen as the oxidizing species.

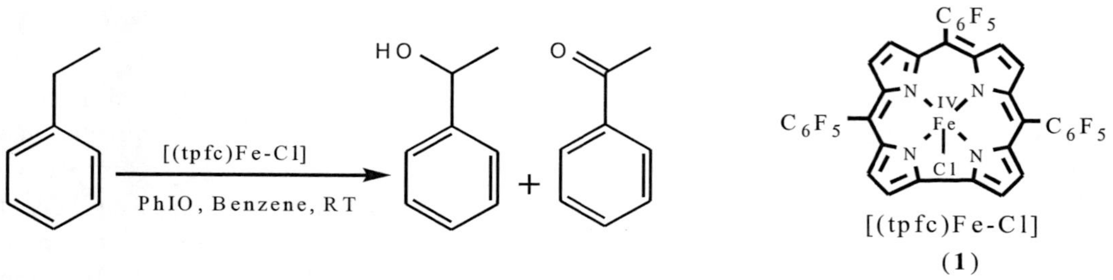

Scheme 1.

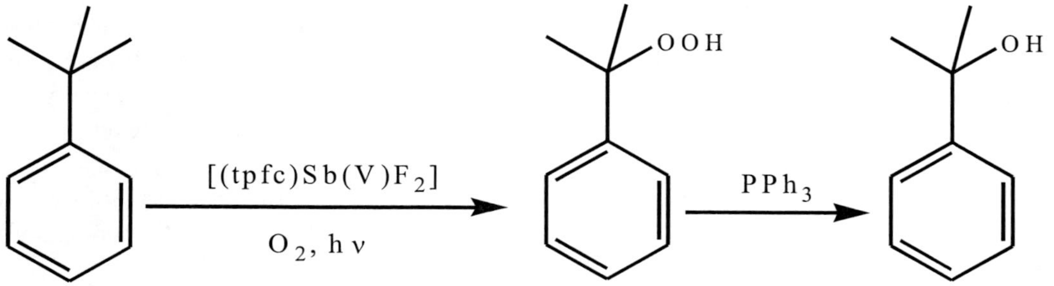

Scheme 2.

Interestingly, there was no follow-up report of the above results on metallocorrole catalyzed hydroxylation of alkanes and alkylbenzenes. However, very recently several iron- and manganese-corrole complexes have been successfully employed as catalysts to hydroxylate alkanes and alkylbenzenes [16a-c]. Selective hydroxylation of cyclohexane and adamantane has been reported for the first time using [(tpfc)Fe-Cl] as catalyst with *m*-choloroperbenzoic acid (*m*-CPBA) as terminal oxidant [16a]. Cyclohexane has been hydroxylated with 100% selectivity producing cyclohexanol in 50% yield. This is also the first report of the hydroxylation of adamantane catalyzed by any metallocorrole. The overall conversion in the hydroxylation of adamantane is achieved up to 75% with 3°/2° ratio of 10.3-10.7 (Scheme 3).

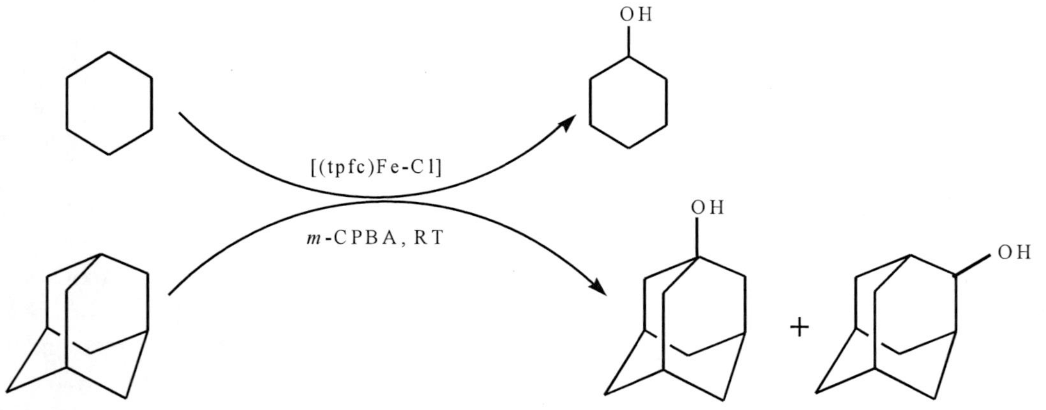

Scheme 3.

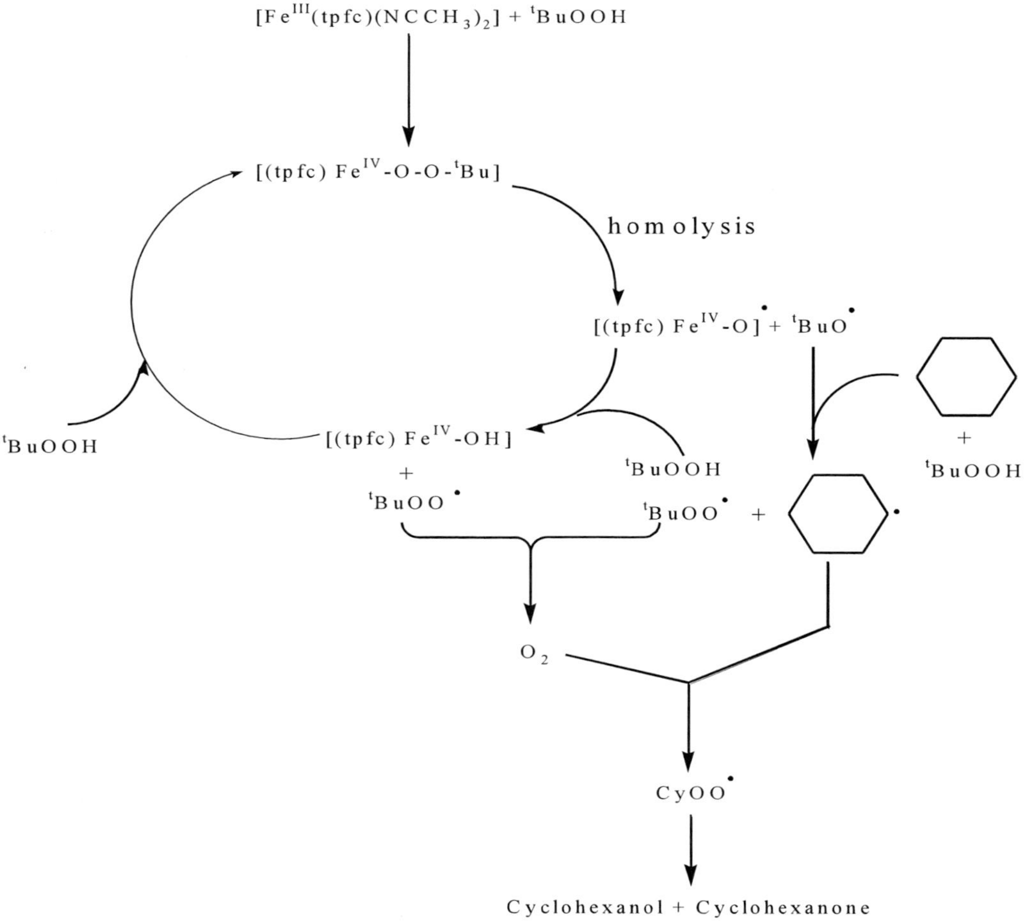

Scheme 4. (adapted from ref. 16b).

The complex [(tpfc)Fe-Cl] has also been found to be catalytically active in hydroxylating alkanes with mild and environmentally benign *tert*-butyl hydroperoxide (*t*-BuOOH) at ambient condition [16b]. Here, the catalytic hydroxylation of alkane proceeds *via* Fenton-type chemistry, with alkoxyl radicals (^{t}BuO• and ^{t}BuOO•) responsible for hydrogen atom abstraction from the alkanes rather than a high-valent iron-oxo species. The proposed mechanism for the hydroxylation reactions is outlined in Scheme 4. The behaviour of the iron corrole towards *t*-BuOOH is in contrast to that of the corresponding iron porphyrins, where oxoiron(IV) cation radical has frequently been invoked as the intermediate [22]. The chemistry of iron corroles with *tert*-butyl hydroperoxide resembles more to that of several non-heme iron catalysts with tripodal pyridylamine ligands [23].

Electron deficient manganese(III) corroles (2-4) (Scheme 5) have also been reported to be active in catalyzing hydroxylation of alkanes and alkylbenzenes with *t*-BuOOH as the oxidant [16c]. The reactivity pattern of the manganese(III) corroles are similar to that of iron(III) corroles as it has been showed that the main role of the catalysts is the activation of alkylhydroperoxide rather than oxygen atom transfer catalysis (OAT).

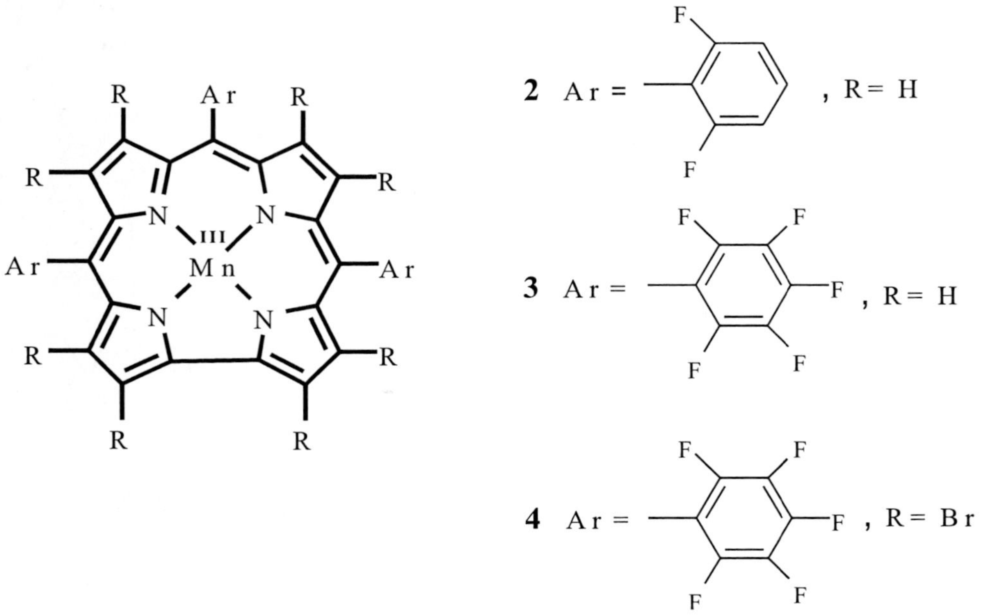

Scheme 5.

The reactions have been shown to proceed with the participation of alkylperoxy radicals ($ROO^{\bullet}$) radicals and organo-hydroperoxide (ROOH) and the proposed mechanistic pathway has been outlined in Scheme 6.

$Mn^{III}(corrole) + t\text{-}BuOOH \longrightarrow Mn^{IV}(corrole)(OH) + t\text{-}BuO^{\bullet}$	(1)
$Mn^{IV}(corrole)(OH) + t\text{-}BuOOH \longrightarrow Mn^{IV}(corrole)(OOBu\text{-}t) + H_2O$	(2)
$Mn^{IV}(corrole)(OOBu\text{-}t) \longrightarrow Mn^{III}(corrole) + t\text{-}BuOO^{\bullet}$	(3)
$2\ t\text{-}BuOO^{\bullet} \longrightarrow 2\ t\text{-}BuO^{\bullet} + O_2$	(4)
$t\text{-}BuO^{\bullet} + RH \longrightarrow R^{\bullet} + t\text{-}BuOH$	(5)
$R^{\bullet} + O_2 \longrightarrow ROO^{\bullet}$	(6)
$2\ ROO^{\bullet} \longrightarrow ROH + R'{=}O + O_2$	(7)
$ROO^{\bullet} + t\text{-}BuOO^{\bullet} \longrightarrow R'{=}O + t\text{-}BuOH + O_2$	(8)
$ROO^{\bullet} + RH \longrightarrow ROOH + R^{\bullet}$	(9)
$2ROOH \longrightarrow RO^{\bullet} + ROO^{\bullet} + H_2O$	(10)
$RO^{\bullet} + RH \longrightarrow ROH + R^{\bullet}$	(11)

Scheme 6. Reaction pathway for the oxidation of hydrocarbons by *t*-BuOOH in the presence of manganese(III) corroles.

2.1.2. Epoxidation

Complex [(tpfc)Fe-Cl] catalyzed epoxidation of styrene with PhIO was the first example of metallocorrole catalyzed epoxidation [15a]. Styrene was converted to a mixture of stryrene oxide and phenylacetaldehyde (3:1) with an overall yield of 87%, which was significantly lower than the corresponding iron porphyrin catalysts. The results stimulated interest among the chemists to develop more effective epoxidation systems based on metallocorroles [15b-g]. The electronegative manganese(III) corrole [(tpfc)Mn] catalyzed epoxidation of styrene with

PhIO was first reported by Gross *et al.* [15b]. The addition of PhIO to the green coloured manganese(III) catalyst produces red coloured authentic oxomanganese(V) corrole which is isolated in pure form. Surprisingly, the oxomanganese(V) corrole fails to transfer its oxygen atom to the olefins under stoichiometric condition. This observation raised ambiguity over the key intermediate associated with the process. Gross *et al.* proposed a high valent Mn(VI)O species, generated by disproportionation of oxomanganese(V) corrole, as the active oxidant [15b].

Collman and coworkers put forward another mechanistic view regarding the intermediates in manganese(III) corrole catalyzed epoxidation. They argued in favour of multiple oxidant forms in competitive epoxidation reactions of styrene and cyclooctene as a function of the different iodosylarenes [15f]. The catalyst-oxidant adduct was considered as an important oxygen transferring intermediate.

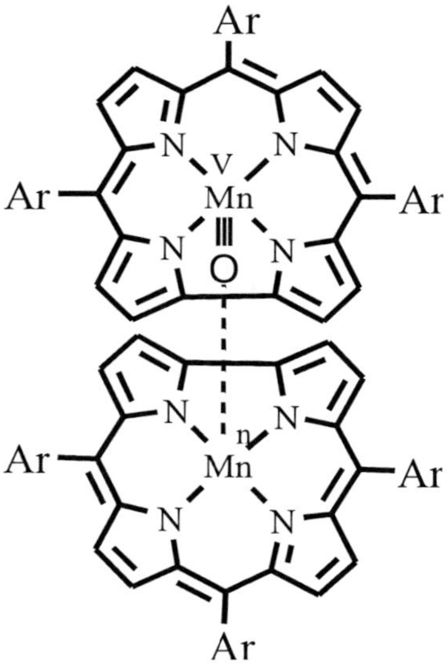

Scheme 7.

Newcomb *et al.* provided insight into the mechanistic pathway for oxygen atom transfer from oxomanganese(V) corrole [15e]. The oxomanganese(V) transient was generated by laser flash photolysis of the corresponding Mn(IV) chlorates and its decay as well as its reactivity with organic reductants were monitored. The results obtained by kinetic measurements can be interpreted by either one of the following two mechanisms: (i) oxidation by Mn(VI) corrole species formed *via* disproportionation of Mn(V)O species and (ii) oxidation by free Mn(V)O species that equilibrates with inactive "sequestered" form (Scheme 7). The involvement of single oxidant form in photochemical reaction has been proposed whereas the presence of multiple oxidant forms is known in normal catalytic turnover reactions [15f].

The perhalogenation of manganese(III) corroles with bromine [24] or fluorine [25] shows enhanced catalytic activity with PhIO, which is reflected from higher turnover numbers. Catalytic activity of the oxomanganese(V) corroles with axial coordination of imidazole show enhanced reactivity towards olefins [24]. Theoretical calculations have suggested the weakening of Mn=O bond due to the presence of coordinated axial ligand. Moreover, axial ligation also raises the energy of HOMO and LUMO thereby making the oxomanganese(V) transients more reactive towards olefins. A bulky bis-pocket manganese(V)oxo complex of 5,10,15-tris (2,4,6-triphenylphenyl) corrole (**5**) catalyze the epoxidation of styrene to styrene oxide with PhIO. This provides strong evidence in favour of OAT in between the Mn(V) oxo species and styrene [25]. The catalyst (**5**) was also found effective in shape selective epoxidation of nonconjugated olefins in presence of N-methyl imidazole as the axial ligand [25]. This catalytic system shows significant higher selectivity toward the less-substituted but more accessible double bond.

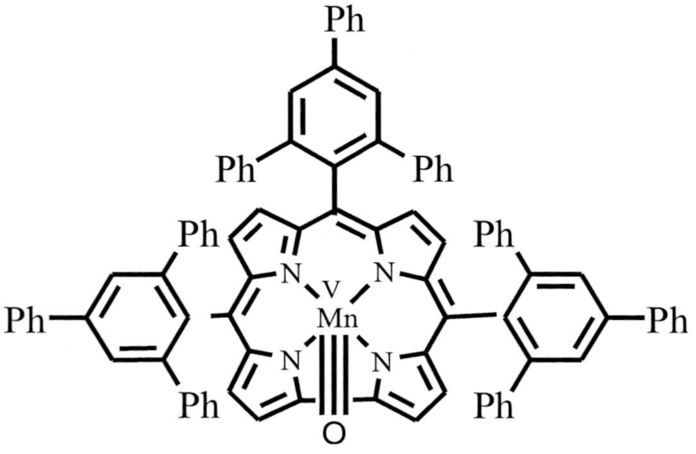

(5)

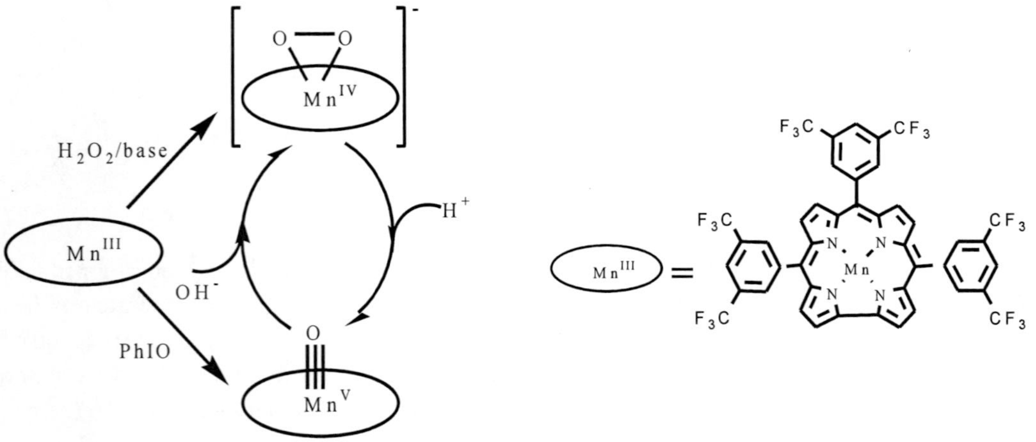

Scheme 8.

The epoxidation reaction in metallocorrole chemistry has been studied mostly with iodosylarenes as the terminal oxidants [15, 24, 25]. The use of mild and inexpensive oxidants like hydrogen peroxide, dioxygen etc. is rare. With *t*-BuOOH as the terminal oxidant, metallocorroles catalyzed oxidation of alkenes mainly afford allylic oxidation products (in case of cyclohexene) or aldehydes (in case of styrene) indicating radical-driven reactions rather than OAT [16a-c]. Recently Nam *et al.* reported the generation of Mn(IV)-peroxo corroles by the action of H_2O_2 on the Mn(III) corrole (Scheme 8) in presence of tetramethylammonium hydroxide in acetonitrile [26]. The catalytic potential of the peroxo species appears to be promising.

2.1.3. Sulfoxidation

The metal complexes of 2,17- or 3,17-bis-sulfonated corroles have been shown to bind tightly with serum albumins [27]. The formation of strong non-covalent conjugates of catalytically active metallocorroles with chiral proteins [28] offers a prospective route to induce asymmetric catalysis. The iron(III) and manganese(III) complexes of these amphiphilic corroles (6 & 7) were employed as catalysts in asymmetric sulfoxidation reactions in phosphate buffer medium (pH 7.0) with hydrogen peroxide as terminal oxidant (Scheme 9) [29]. The albumin source significantly affects the ee's of the process. Again, the formation of the major enantiomer shows dependence on the nature of metal ions. The Fe(III) and Mn(III) corroles, conjugated with the same albumin, produce enantiomers of opposite optical activity. The manganese(III) catalysts have been found to be superior to iron(III) corroles in terms of enantioselectivity, yield and catalyst survival.

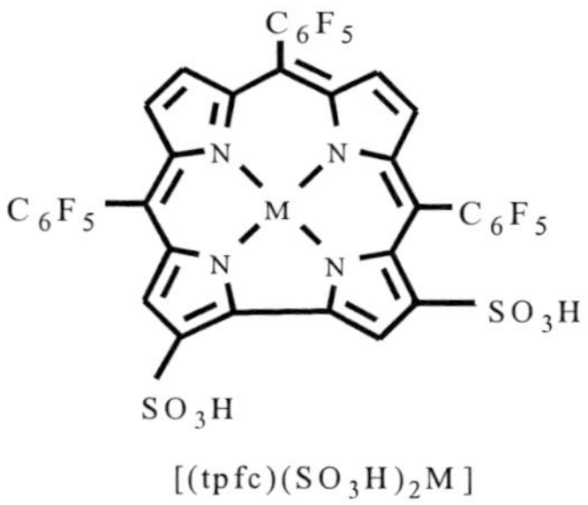

6 M= Fe(III)
7 M= Mn(III)

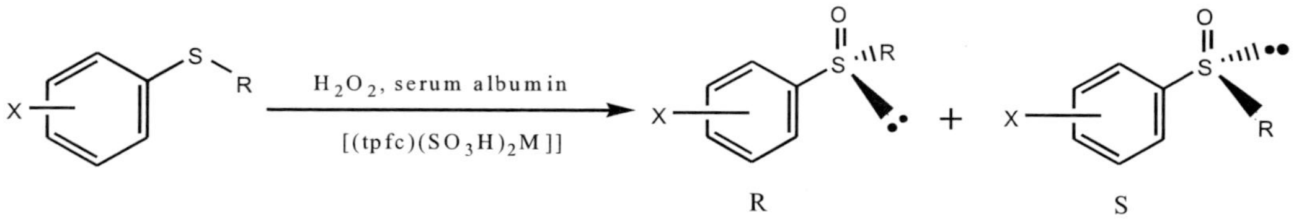

Scheme 9.

Biomimetic asymmetric sulfoxidation reactions catalyzed by water soluble manganese(III) corroles with H_2O_2 has been shown to proceed *via* the formation of coordinated hydroperoxo adduct (Scheme 10).

A close analogue of corrole is corrolazine, which is *meso*-N-substituted corrole. The manganese(III) or iron(III) complex of corrolazine, [(tbp$_8$Cz)M] (tbp8Cz = octakis(4-tert-butylphenyl)corrolazinato) (Scheme 11) [30] have been utilized as catalysts in sulfoxidation. Goldberg *et al.* thoroughly investigated the mechanistic aspects of Mn(III)-corrolazine catalyzed sulfoxidation reactions with PhIO as the active oxidant and provided evidence for a "third oxidant" namely the oxidant coordinated Mn(V)O species which can be thought as a hybrid of oxidant coordinated Mn(III)-corrolazine and the Mn(IV) oxo π-cation radical.

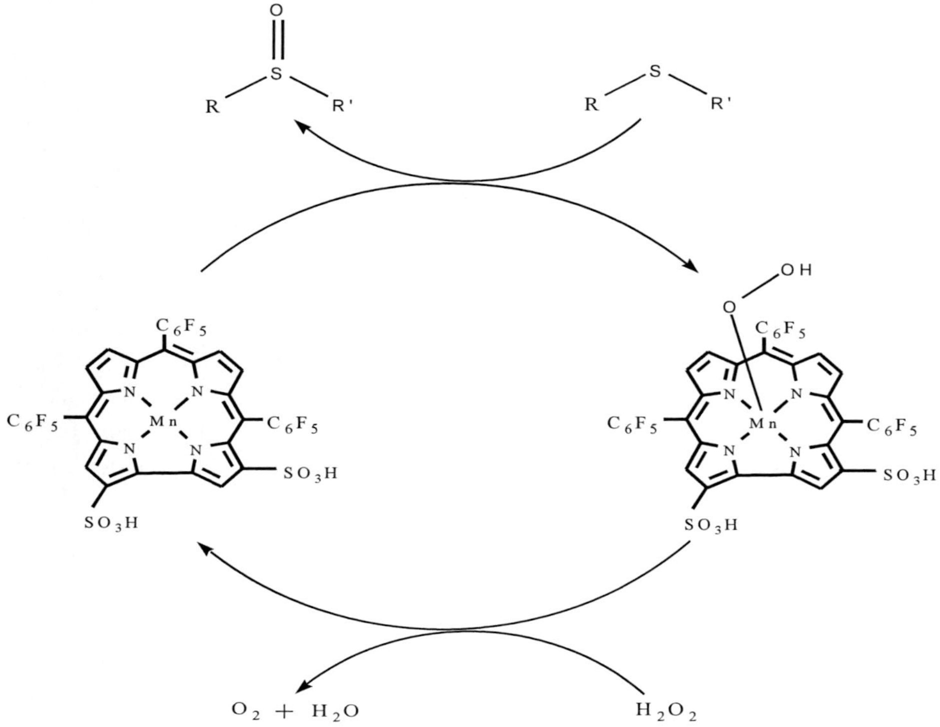

Scheme 10.

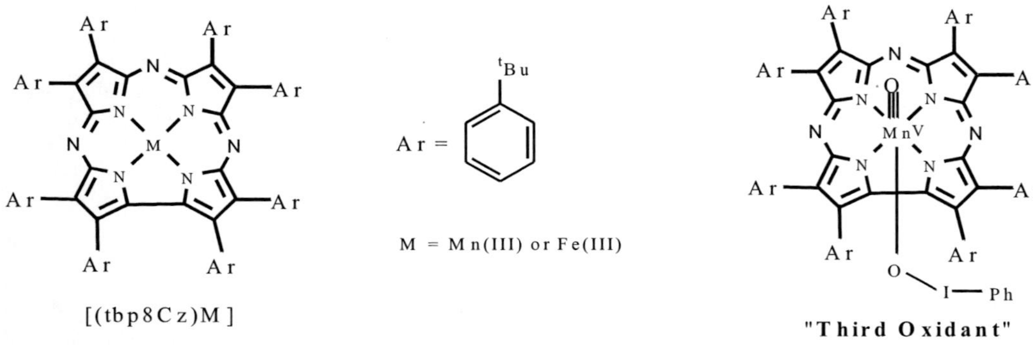

Scheme 11.

The iron(III) complex of corrolazine [$(tbp_8Cz)Fe^{III}$] (tbp8Cz = octakis(4-tert-butylphenyl) corrolazinato) (Scheme 11) has been found to be a potent catalyst in oxidizing thioether substrates with H_2O_2 as the oxidant [31]. However, in this case, low temperature EPR measurements indicated the involvement of an antiferromagnetically coupled oxoiron(IV) π-cation radical as the prime intermediate. Although Ghosh *et al.* [32] earlier concluded that the corrolazine ligand should stabilize the oxoiron(V) transient over the oxoiron(IV) π-cation radical, the former species has not been observed in this case.

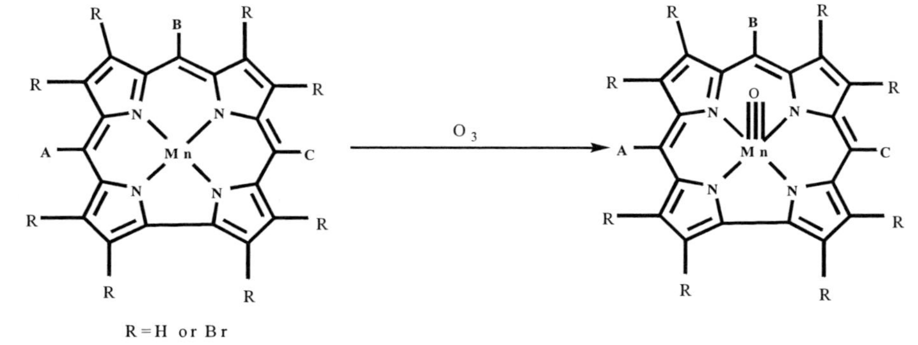

	A	B	C
8	C_6F_5	C_6F_5	C_6F_5
9	C_6F_5	C_7H_7O	C_6F_5
10	$C_6H_3Br_2$	C_6F_5	$C_6H_3Br_2$
11	$C_6H_3F_2$	$C_6H_3F_2$	$C_6H_3F_2$
12	$C_6H_3Cl_2$	$C_6H_3Cl_2$	$C_6H_3Cl_2$

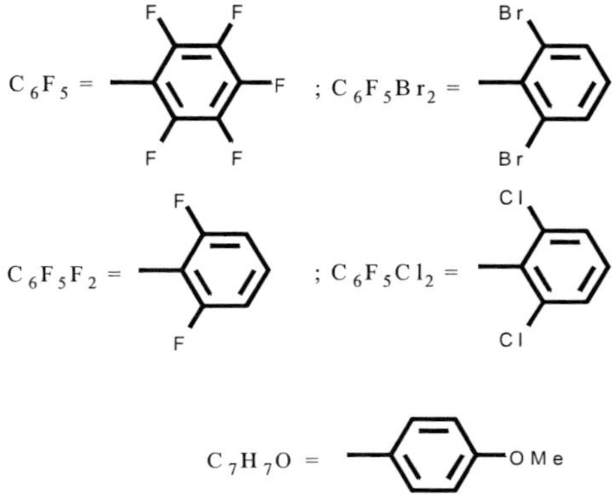

Scheme 12. Isolated (oxo)manganese(V) corroles utilized in the OAT reactions to sulfides.

Very recently, Gross and coworkers have reported the catalytic activity of a series of (oxo)manganese(V) complexes of A_3- and A_2B-corroles (8-12), (Scheme 12) in oxygen atom transfer (OAT) reactions to sulfides [33]. The (oxo)manganese(V) corroles were isolated from the oxidation of Mn(III) corroles by ozone. This eludes the possibility of generation of oxidant-coordinated reaction intermediates. The mechanistic investigation of isolated (oxo) manganese(V) corroles towards a series of *p*-thioanisoles revealed that (oxo)manganese(V) corroles are legitimate OAT agents. Much less negative Hammet ρ values and the non-first order dependence on the concentration of the oxomanganese(V) species have been interpreted in terms of the oxygen atom transfer from $[(oxo)\ manganese(VI)]^+$ generated by the disproportionation of the oxomanganese(V) transient.

2.1.4. Oxidation of Phosphorous Compounds

The chromium corrole catalyzed oxidation of triphenylphosphine is the first example of aerobic oxidation in corrole chemistry [34]. The activation of dioxygen by chromium(III) corrole results in the formation of oxo-chromium(V) corrole, which oxidizes Ph_3P to Ph_3PO followed by reducing itself to chromium(III) corrole. The reaction shows solvent dependence and the best result is obtained in acetonitrile (33 catalytic turnovers) without any external oxidizing or reducing agents.

2.1.5. Alcohol Oxidation

Oxidation of alcohols to aldehydes or ketones is an important functional group transformation in organic synthesis [35]. The carbonyl compounds represent an important group of products and intermediates in the fine chemicals [36]. Thus, several methods have been explored to accomplish selective oxidation of alcohols to more valuable carbonyl compounds. The existing methods [37] suffer from one or more limitations like the use of expensive chemicals, prolonged reaction time, strongly acidic condition, toxic waste and tedious work-up procedure. Thus development of catalytic method for selective oxidation of alcohol using safe, economic, and environmentally benign oxidizing agents remains a critical challenge in organic synthesis [38]. Very recently, electron deficient manganese(III) corrole (**2-4**) have been shown to be effective in catalyzing alcohol oxidation with *t*-BuOOH at room temperature (Scheme 13) [39]. Manganese(III) corroles with greater electron withdrawing substituents have been found to be better catalysts in oxidizing alcohols. The preliminary results of the corresponding iron catalysts in oxidation of alcohols to aldehydes or ketones appear to be impressive [40].

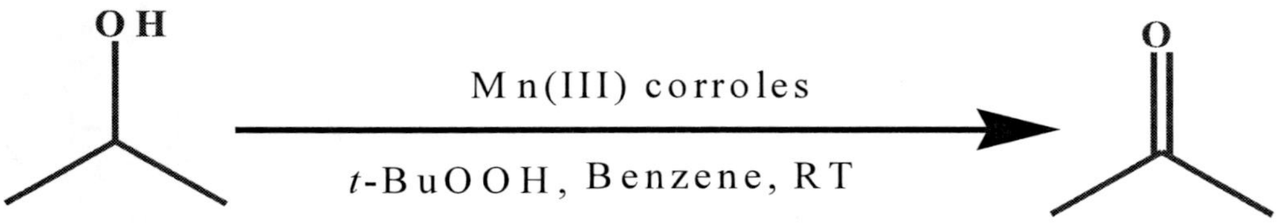

Scheme 13.

2.2. Reduction Catalysis

Due to the superiority of corroles (over the porphyrins) in stabilizing high metal oxidation states, the low-valent metallocorroles are expected to be very reactive. Thus, it is not surprising that various metallocorroles have been utilized as catalysts in the activation of reducible molecules [1d]. Significant development in this regard has been achieved in this area over the past few years. Because of the potential applications metallocorroles as reduction catalysis in diverse areas encompassing environmental chemistry to bio-medical chemistry, this branch is expected to emerge as one of the frontier areas in metallocorrole chemistry.

2.2.1. Reduction of O_2

Considerable efforts have been devoted to develop materials capable of performing four-electron electrocatalytic reduction of dioxygen due to the possible use of these materials in energy conversion technologies such as fuel cells, metal-air batteries, oxygen sensors etc [41]. Catalytic reduction of dioxygen in biology is also of great importance [42]. In aerobic organisms, cytochrome c oxidase and several other heme and copper proteins selectively catalyze four-proton, four-electron conversion of dioxygen to water without the formation of partially reduced species such as peroxides or superoxides [43]. Attempts have been made to utilize various metal macrocycles, such as metalloporphyrins, metallophthalocyanines, dibenzotetraazaannulenes etc., to mimic the active site of these enzymes [44]. However, only a few catalysts have been found to catalyze the four-electron reduction of dioxygen to water with a low overvoltage [44a, 44b, 45]. One notable example is the linked dicobalt bisporphyrins adsorbed on the surface of a graphite electrode, which efficiently catalyze such reduction in acidic solutions [44b]. Although, the electroreduction, in this case has been achieved at positive potentials ($E_{1/2}$ = 0.71 V *vs.* SHE at pH 0), the reversible thermodynamic potential of the reduction of dioxygen is substantially higher (1.23 V vs. SHE at pH 0) indicating the need of optimization of the catalysts.

$$O_2 + 4H^+ + 4e^- \rightleftharpoons 2H_2O$$

$$\Delta E^0 \text{ (pH 7)} = 0.8 \text{ V}; \; \Delta G^0 = 80 \text{ kcal/mole}$$

Kadish and coworkers [45b, 46] first demonstrated the catalytic activity of a series of cobalt corroles towards reduction of dioxygen (Scheme 14). The cyclic voltammetric studies of the catalysts show four-electron reduction of dioxygen in the range of 0.33 to 0.4 V in 1M $HClO_4$. The Co^{III}-Co^{II} corrole-porphyrin dyads have evolved as more effective catalysts than the Co^{III}-Co^{III} corrole-corrole dyads [45b].

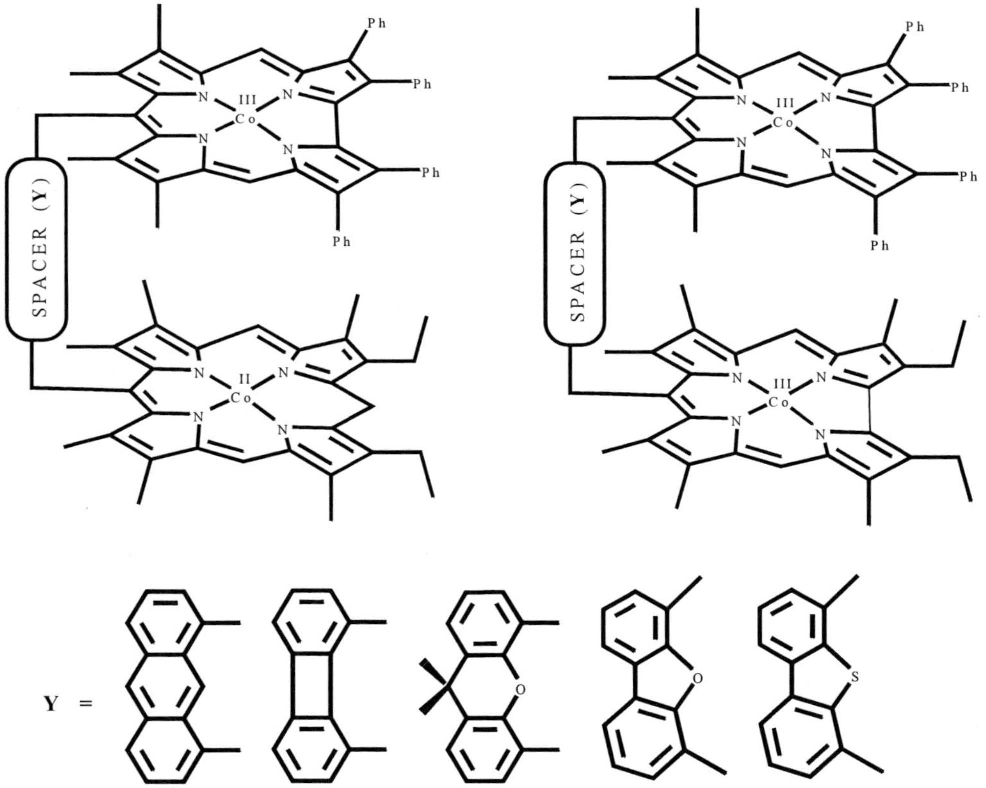

Scheme 14. Biscobalt corrole-porphyrin and biscobalt corrole-corrole dyads utilized as electrocatalysts for reduction of dioxygen.

Recently, they have also demonstrated that the catalytic activity of biscobalt corrole-porphyrin dyads towards electroreduction of dioxygen is largely dependent on the nature and position of the substituents on the corrole macrocycle as well as on the type of spacer (Y) [47]. Quantitative four-electron reduction of dioxygen by the reductant 1,1′-

dimethylferrocene was achieved with the catalysts 13 and 14 (Scheme 15) containing the spacer 9,9-dimethylxanthane.

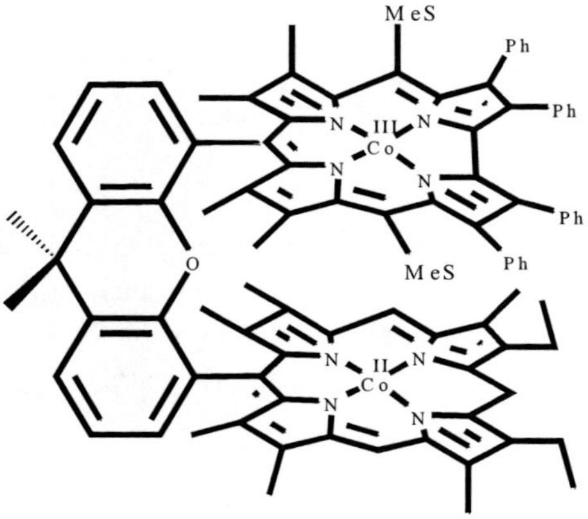

13

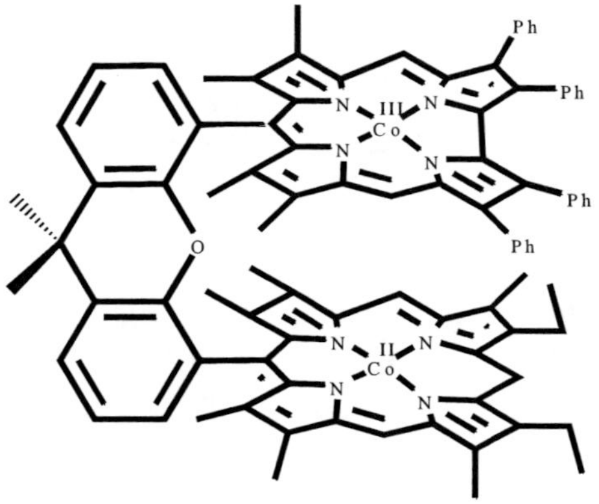

14

Scheme 15.

Collman and coworkers [48] also reported the catalytic activity of metallocorroles for dioxygen reduction using the rotating disc electrode method (Scheme 16). The metallocorroles have been found to be operative in the potential region of 0.5-0.7 V. The catalytic activity has been shown to depend on the nature of metal ions; the relative order of reactivity being Co > Fe >> Mn > free base corrole. Interestingly, iron corroles have been found to catalyze the four-electron reduction of dioxygen to water directly, whereas the corresponding iron porphyrins are known to adopt (2+2) electron pathway. Moreover, iron corroles catalyze the reduction of dioxygen more efficiently than that of hydrogen peroxide, whereas iron porphyrins show comparable reactivity towards both the substrates [45j].

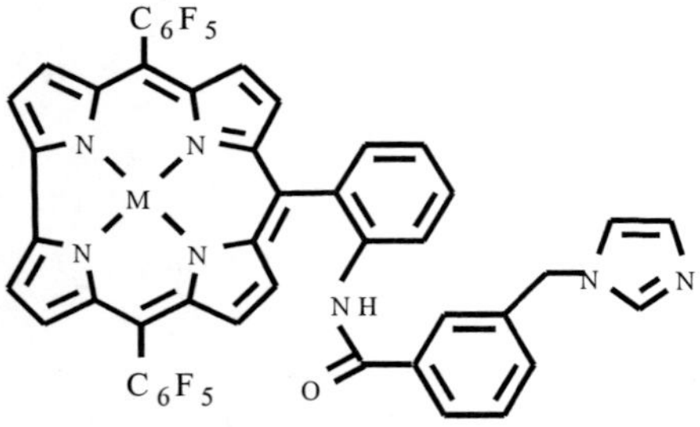

M = Fe, Co

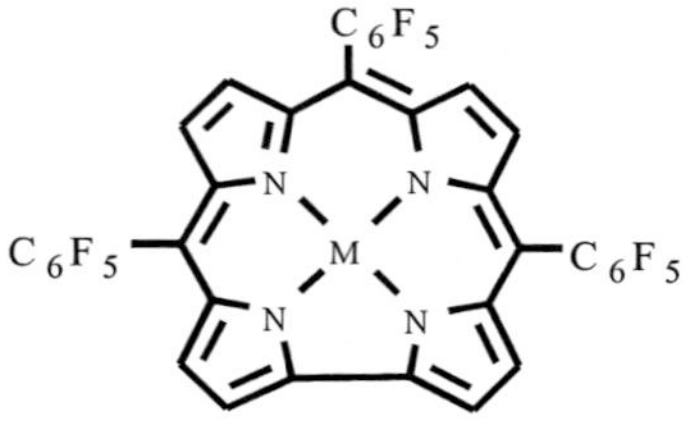

M = Fe-Cl, Mn, Co-PPh3

Scheme 16. Metallocorroles used for the reduction of dioxygen.

In spite of significant progress in this area, it is obvious that more optimization specially the tuning of the electronic properties by suitable modification of the substituents on corroles and porphyrins are to be achieved.

2.2.2. Reduction Of CO_2

Carbon dioxide, the most abundant green house gas, comes from anthropogenic sources such as burning of fossil fuels and its rise in atmosphere has adverse effect on global climate change. Therefore, methods for transforming CO_2 into a source of fuel offer a very attractive way to decrease its atmospheric concentration. This may be achieved through the photolytic or electrocatalytic reduction of CO_2 to various products (Table 1) [49]. Zerovalent metalloporphyrins and monovalent phthalocyanine radical anion catalyze the reduction of CO_2 in homogeneous medium [50].

Table 1. Reduction potentials of CO_2

Reaction	$E^{0'}$ (volt) vs. SCE at pH 7
$CO_2 + 2H^+ + 2e^- \rightarrow HCOOH$	-0.85
$CO_2 + 2H^+ + 2e^- \rightarrow CO + H_2O$	-0.77
$CO_2 + 4H^+ + 4e^- \rightarrow C + 2H_2O$	-0.44
$CO_2 + 4H^+ + 4e^- \rightarrow HCHO + H_2O$	-0.72
$CO_2 + 6H^+ + 6e^- \rightarrow CH_3OH + H_2O$	-0.62
$CO_2 + 8H^+ + 8e^- \rightarrow CH_4 + 2H_2O$	-0.48

The application of metallocorroles for the reduction of CO_2 is least exploited so far. Highly reduced cobalt and iron corroles have been shown to catalyze the reduction of CO_2 [51]. Chemical, electrochemical and photochemical reduction of the metallocorroles has been achieved chemically in acetonitrile medium and monovalent cobalt and iron complexes have been shown to reduce CO_2. The carbon dioxide reduction has been achieved photochemically using CO_2-saturated acetonitrile containing *p*-terphenyl (TP) as the sensitizer and triethylamine (TEA) as the reductant. The reduced iron(I) corrole, $[Fe(I)(tdcc)Cl]^{2-}$ (tdcc= 5,10,15-tris(2,6-dichlorophenyl)corrole) has emerged as the best CO_2-reduction catalyst [51].

The chemical reduction has been shown to proceed in a stepwise manner and the proposed route is outlined in Scheme 17.

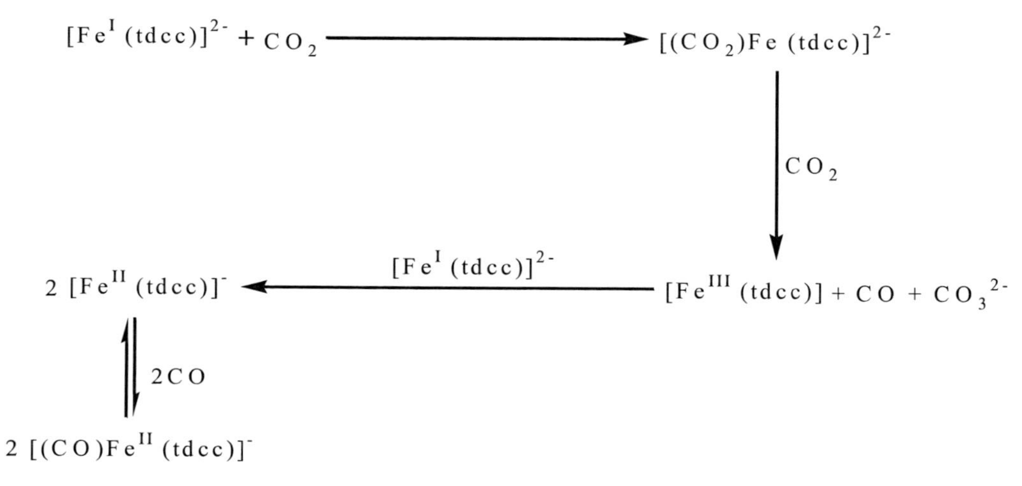

Scheme 17. Reduction of CO_2 by $[Fe(I)(tdcc)Cl]^{2}$.

2.2.3. Decomposition of Peroxynitrite

The chemistry of peroxynitrite decomposition continues to be one of the most important topics in contemporary medicinal chemistry [52]. Peroxynitrite is formed in vivo by the reaction of NO and O_2^- and it decays spontaneously producing radical-decomposition products that damage essential biomolecules such as DNA, proteins, lipids, etc. [53]. Accumulation of peroxynitrite and its decomposition are responsible for biological malfunctions initiated by oxidative stress and also for several neurodegenerative disorders like Alzheimer's, Parkinson's and Huntington's disease [54]. The peroxy linkage in the protonated peroxynitrite (HOONO) is known to cleave homolytically forming •OH and $•NO_2$ (equation 12) [55]. These strong oxidizing and nitrating species are primarily responsible for damaging the biomolecules [56]. Unfortunately, there is no biological defense mechanism to prevent the spontaneous decay of peroxynitrite. Intense research in the area of peroxynitrite decomposition to benign products has indicated two distinct pathways for detoxification of peroxynitrite [57]. They are (i) isomerization of peroxynitrite to nitrate (equation 13) and (ii) disproportionation to nitrite and molecular oxygen (equation 14).

$$HOONO \longrightarrow \dot{O}H + \dot{N}O_2 \qquad (12)$$

$$HOONO \longrightarrow H^+ + NO_3^- \qquad (13)$$

$$HOONO \longrightarrow 2H^+ + 2NO_2^- + O_2 \qquad (14)$$

Endogenous reducing agents like ascorbate and glutathionate react with peroxynitrite but the processes are too slow to be effective [58]. Similarly, dissolved carbon dioxide reacts with peroxynitrite but this decomposition pathway is also slow with respect to transmembrane diffusion and reactions with metal centers [59]. In the mid-nineties, research groups of Groves [60] and Stern [61] showed that water-soluble iron(III) porphyrins are potent catalysts in decomposing peroxynitrite. On the contrary, manganese(III) porphyrins have been shown to decompose peroxynitrite only in presence of stoichiometric amounts of sacrificial antioxidants [62].

Over the last few years, iron(III) and manganese(III) complexes of the amphiphilic corrole $[(tpfc)(SO_3H)_2]$ have emerged as excellent catalysts for peroxynitrite decomposition [63]. The iron(III) complexes catalyze the decomposition with a very fast rate; the half life period of peroxynitrite have been found to be reduced to 0.025 s from 1.8 s. The catalytic activity of the complex is second only to a most optimized positively-charged iron porphyrin [64]. Mechanistic studies have revealed that the catalytic activity of this amphiphilic iron(III) corrole is similar to the corresponding iron(III) porphyrins. Both induce isomerization of peroxynitrite and form nitric acid (equation 13) [64, 65].

On the other hand, the rate at which the amphiphilic manganese(III) corrole catalyzes peroxynitrite is significantly lower than their iron(III) counterpart. But the decomposition pathway in this regard deserves special mention (Scheme 18). It has been found that manganese(III) corrole disproportionate peroxynitrite yielding nitrites (equation 14) [63]. It is also noteworthy that this is the first manganese catalyst, which can decompose peroxynitrites without any sacrificial reducing agent.

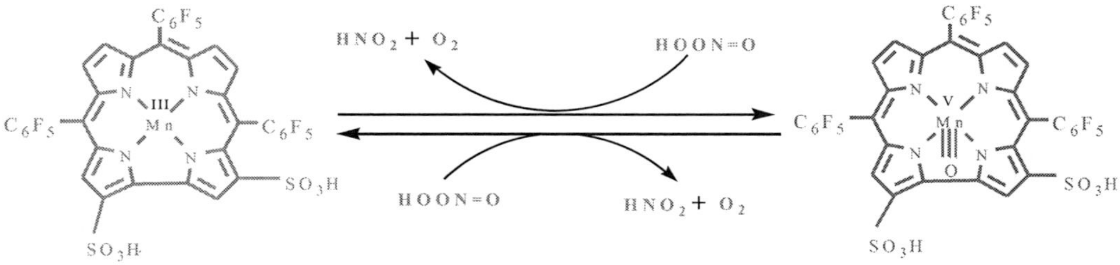

Scheme 18. Proposed catalytic cycle for disproportionation of peroxynitrite catalyzed by amphiphilic manganese(III) corrole.

Another water-soluble manganese(III) corrole (15) has been synthesized and used successfully as catalyst in peroxynitrite decomposition [66]. The cationic Mn(III) corrole (15) catalyzes the decomposition of peroxynitrite at a much faster rate (almost ten times) than [tpfc(SO_3H)$_2$Mn] (7).

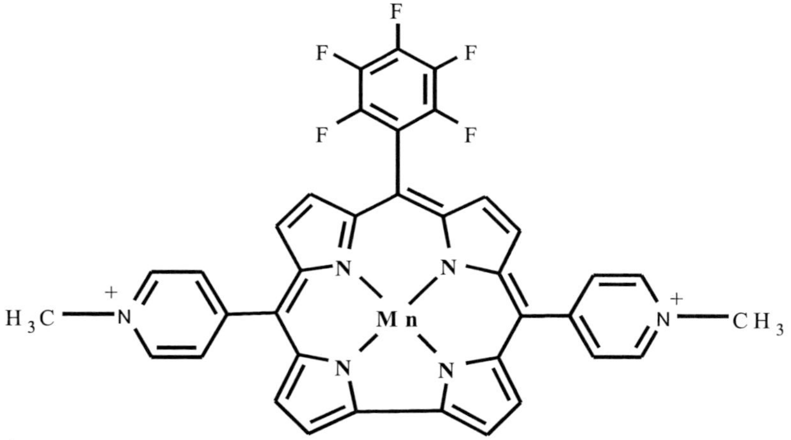

15

These chemical results have simulated significant interest among the scientists and metallocorrole has started to find its place in various medicinal applications. The iron(III) complex [tpfc(SO_3H)$_2$Fe] (**6**) have been found particularly effective in eliminating intermediates of lipid peroxidation [67]. This result is extremely encouraging since oxidation of lipoproteins is one of the main causes behind the heart disease 'atherosclerosis'. The corresponding Mn(III) corrole, [tpfc(SO_3H)$_2$Mn] (**7**), on the other hand acts as pro- rather than anti-oxidant under similar condition. Another distinct advantage of metallocorroles in this regard is their preferential binding to lipoproteins at the expense of all serum proteins thereby providing easy transport of the complexes to their site of action, i.e., the arterial wall [67].

2.3. Group Transfer Catalysis

2.3.1. Carbene Transfer to Olefins (Cyclopropanation Reactions)

The cyclopropanation of olefins using the transition metal catalyzed decomposition of diazoalkanes is one of the most widely studied reactions in organic chemistry (Scheme 19) [68].

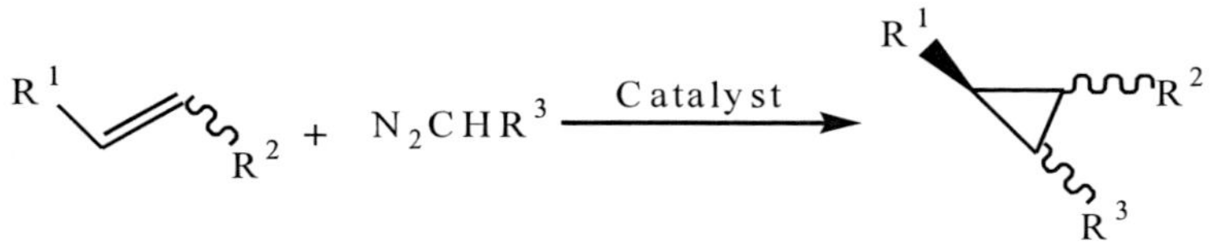

Scheme 19.

Gross and coworkers first noted the catalytic property of the iron(IV) corrole [Fe(tpfc)Cl] for the cyclopropanation of styrene by ethyl diazoacetate and also by a chiral carbenoid (Scheme 20) [15a].

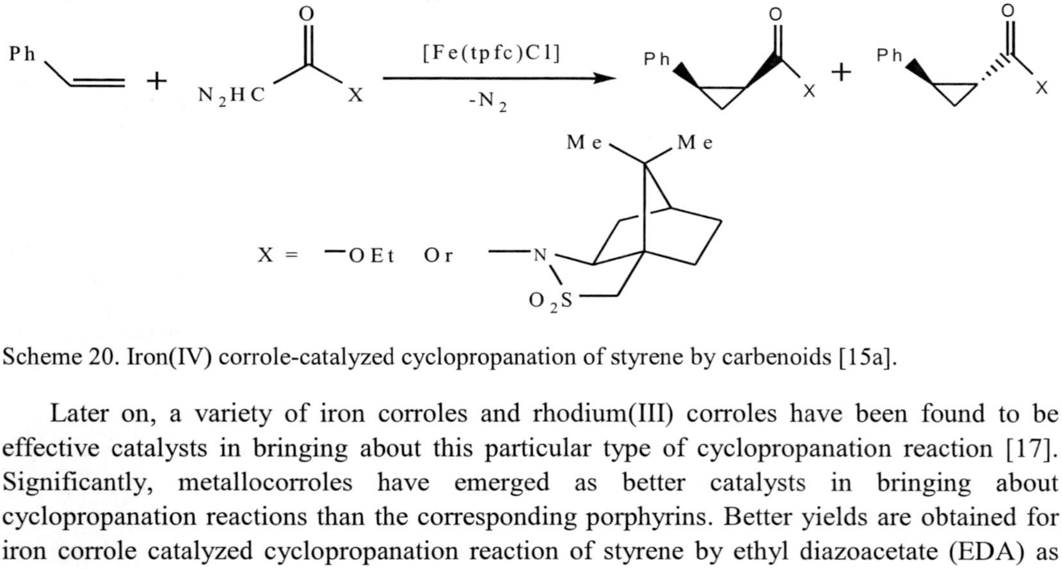

Scheme 20. Iron(IV) corrole-catalyzed cyclopropanation of styrene by carbenoids [15a].

Later on, a variety of iron corroles and rhodium(III) corroles have been found to be effective catalysts in bringing about this particular type of cyclopropanation reaction [17]. Significantly, metallocorroles have emerged as better catalysts in bringing about cyclopropanation reactions than the corresponding porphyrins. Better yields are obtained for iron corrole catalyzed cyclopropanation reaction of styrene by ethyl diazoacetate (EDA) as well as by the much larger unichiral diazene compound (Scheme 20) (71 and 41% respectively), whereas only 24 and 10% chemical yields were obtained in case of 5,10,15,20-tetrakis(pentafluorophenyl)porphyrinatoiron(III) chloride catalyzed cyclopropanation reactions [17b]. Another significant difference of reactivity between these two sets of catalysts is the smaller *trans*:*cis* ratio of the cyclopropanation products in iron corrole catalyzed reactions than that observed in the iron porphyrin catalyzed reactions. These differences have been attributed to the absence of the fourth meso aryl group in corroles, which in turn, favour the formation of the relatively large metal-carbene intermediates (Scheme 21)

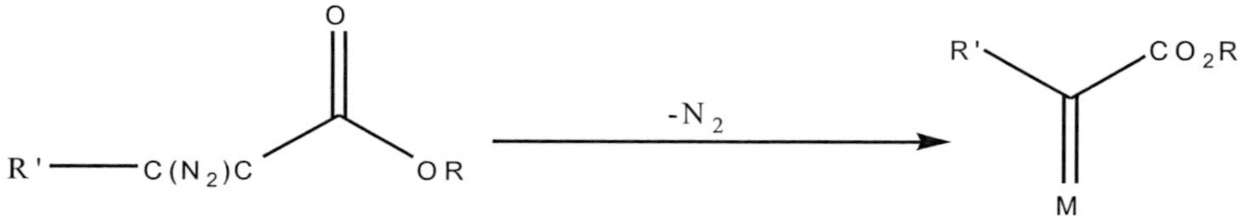

Scheme 21. Proposed metal-carbene intermediate in metal-catalyzed carbene transfer reactions.

Moreover, whereas iron(III) porphyrins first reduce to iron(II) porphyrins, iron(III) corroles are found to be the true catalysts not requiring any reducing agent for their catalytic activity. However, for the iron(IV) corrole (**1**), EDA is proposed to act as the reducing agent like iron(III) porphyrin catalyzed carbene transfer reactions.

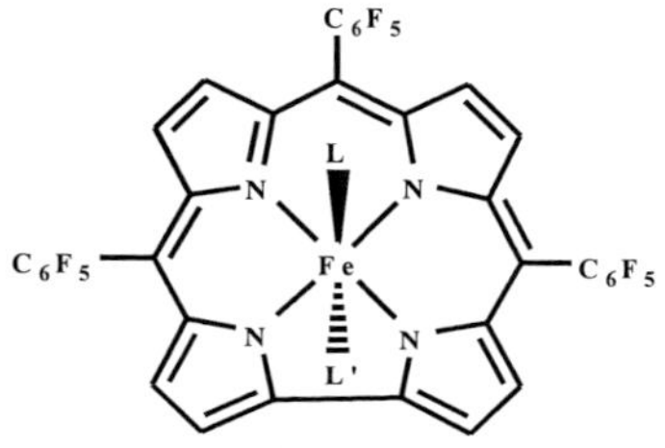

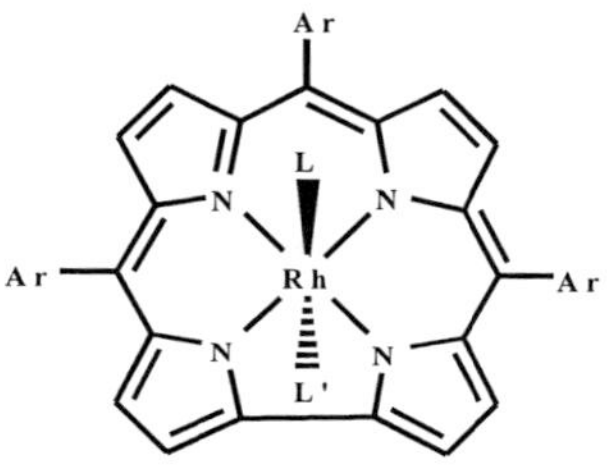

16a: L=L'=OEt_2 [Fe(tpfc)$(OEt_2)_2$]

16b: L=L'=Py [Fe(tpfc)$(Py)_2$]

16c: L=NO [Fe(tpfc)(NO)]

17a: L=L'=PPh_3, Ar=C_6F_5 [Rh(tpfc)$(PPh_3)_2$]

17b: L=L'=$P(C_6H_{11})_3$, Ar=C_6F_5 Rh(tpfc)$[P(C_6H_{11})_2]$

17c: L=L'=Py, Ar=C_6F_5 [Rh(tpfc)$(Py)_2$]

17d: L=L'=PPh_3, Ar=2,6-$Cl_2C_6H_3$ [Rh(tdcc)$(PPh_3)_2$]

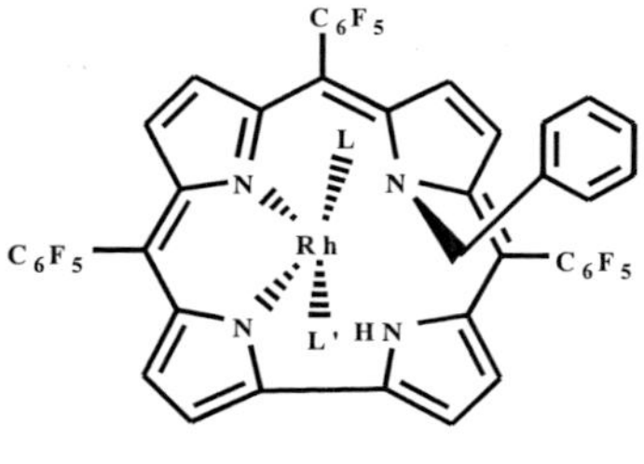

18:L=CO, L'=PPh_3

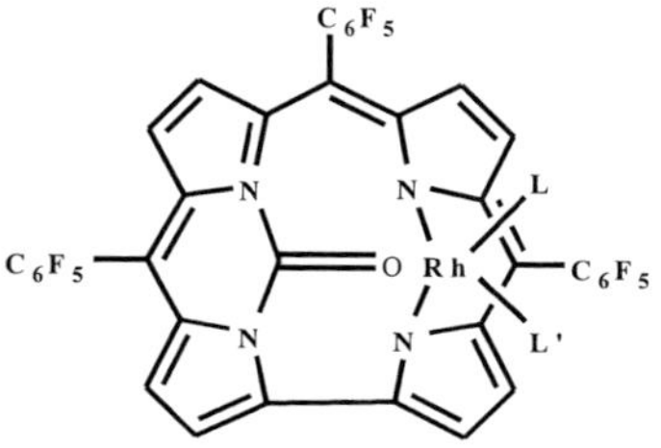

19a: L=L'=CO

19b: L=CO, L'=PPh_3

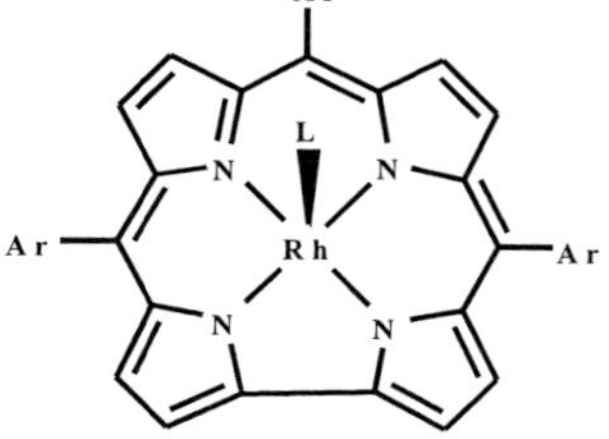

20: L=

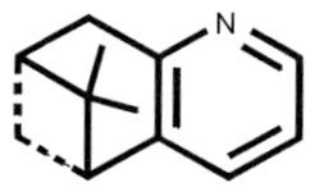

Scheme 22. Metallocorroles utilized in cyclopropanation reaction.

The selectivity for carbene transfer in iron corrole catalyzed cyclopropanation reactions, is somewhat low due to the formation of diethyl maleate *via* EDA coupling. The rhodium corrole catalyzed cyclopropanation reactions show better selectivity [17]. The reaction yields significant amount of the less stable *syn*-isomer (*anti*/*syn* ratio: 1.5-2.7). Different metallocorroles utilized in carbene transfer reactions have been shown in Scheme 22. Solvent effect in rhodium corrole catalyzed reactions is less pronounced. Moreover, the rhodium catalysts have been shown to be poisoned by external amine (20) probably due to the coordination of the amine to the metal. Asymmetric catalysis has also been observed for one of the chiral catalysts (19b).

2.3.2. C-H Insertion of Carbenes

Driven by the superior reactivity of metallocorroles in cyclopropanation reaction, Gross and coworkers utilized iron and rhodium corroles as catalysts for the insertion of carbene moiety of EDA into the allylic C-H bonds of cycohexene [18b]. Using the iron corrole catalysts [Fe(tpfc)Cl] (1) and [Fe(tpfc)(OEt_2)$_2$] (16a), the C-H bond of the cyclohexene has been achieved with EDA (Scheme 23).

In case of rhodium corrole (12a) as catalyst, preferential C=C activation was observed instead of C-H activation [18b]. This particular field in metallocorrole chemistry has not been explored completely as evident from the number of publications in this regard [17, 18], it certainly promises to usher in several new ventures to prepare complex organic molecules in coming years.

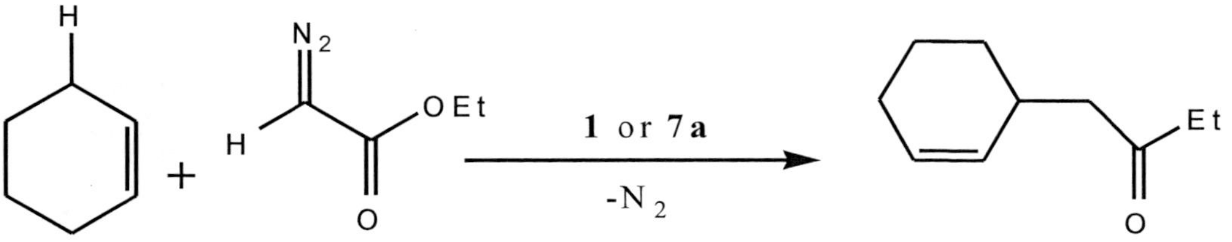

Scheme 23. Catalytic insertion of carbene of EDA into allylic C-H bond of cyclohexene.

2.3.3. Nitrene Transfer to Olefins (Aziridination Reactions)

The transition-metal-catalyzed addition of a nitrene to an olefin or the addition of a carbene to an imine provides aziridines [69]. Aziridines are important building blocks for the synthesis of various nitrogen-containing compounds, mainly due to their highly regio- and stereo-selective ring opening reactions [70]. Most of the initial activity in this area was based on the transfer of the NTs moiety of PhI=NTs (Ts=tosylate) into C-H and C=C bonds [71]. Despite these notable progresses, catalytic aziridination reactions based on PhI=NTs did not enter the realm of synthetic organic synthesis, due to several drawbacks [72]. This classical aziridination reagent is an expensive and inconvenient nitrogen source and it generates iodobenzene in equimolar amounts. Moreover, oxygenated hydrocarbons are generally formed in high amounts rendering its synthetic applicability minimal [72]. In this regard, chloramines-T, utilized first by Sharpless in 1998 in bromine-catalyzed aziridination of olefins [73], offer several advantages as the aziridination reagent; it is commercially available, cost effective and forms only NaCl as the byproduct.

Iron corroles (1 and 11a-11c) appeared as efficient catalysts for aziridination of styrene with PhI=NTs (scheme 24) [18].

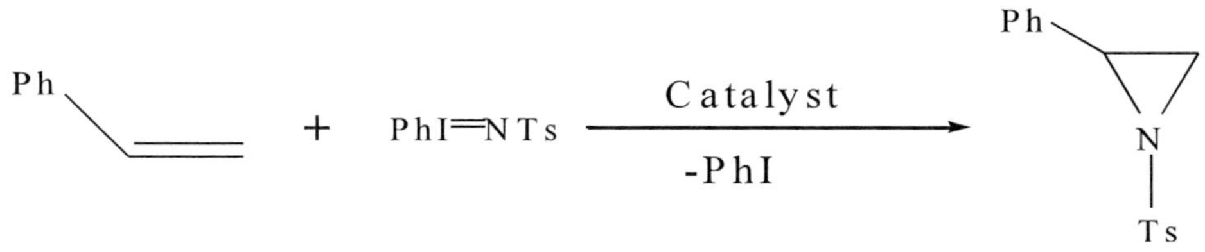

Scheme 24. Catalytic aziridination of styrene by with PhI=NTs

For all the iron corrole catalysts, the desired aziridine was obtained together with the oxygenated hydrocarbons (styrene oxide and benzaldehyde). In comparison with the corresponding porphyrin catalysts, better selectivity was obtained in corrole-based reactions [18]. The iron(III) corrole catalyst [Fe(tpfc)(OEt_2)$_2$] (11a) emerged as the best catalyst providing the 84% desired aziridine with the largest selectivity (aziridine/oxygenated products = 11.4) [18a]. For these reactions, an iron-nitrene intermediate has been proposed. The intermediate can either reacts with styrene to produce the desired aziridine or it may hydrolyse to the oxoiron complex subsequently oxidizing styrene to styrene oxide and phenyl acetaldehyde (Scheme 25). The superior reactivity of the corrole catalysts than that of the porphyrin catalysts in aziridination reaction may be rationalized in terms of the formation of the metal-nitrene intermediate; corroles, being less sterically crowded can stabilize the metal-nitrene intermediate over the metal-oxo intermediate.

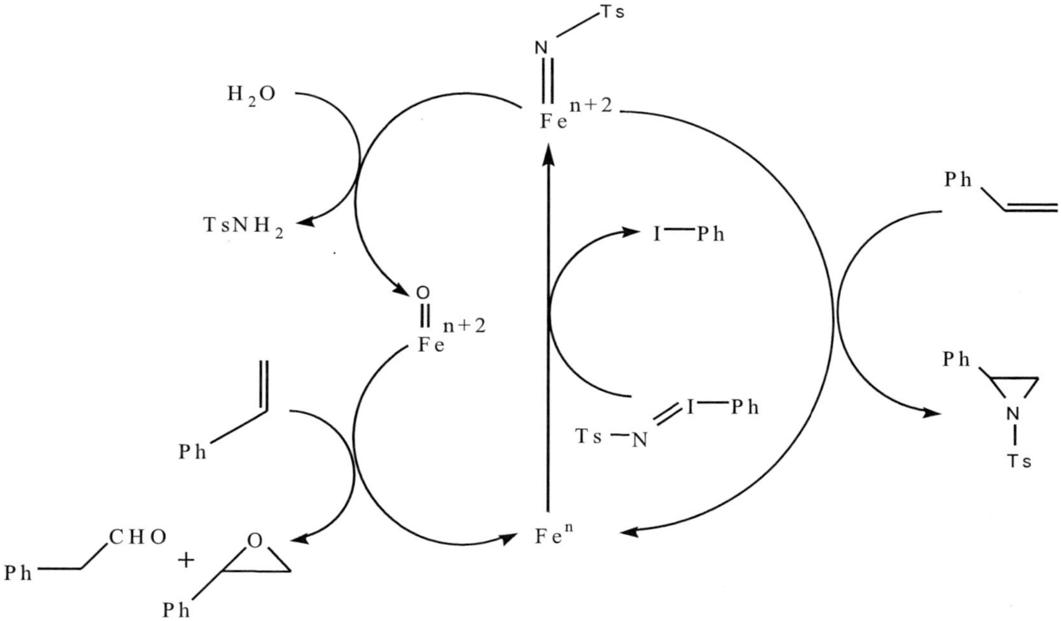

Scheme 25. Proposed catalytic routes for iron complexes catalyzed aziridination of styrene with PhI=NTs (adapted from ref. 18a).

Among the different iron corroles, the iron(IV) corrole catalyst **(1)** has been found to be effective catalyst for the more 'synthetically important' aziridination reactions of styrene with Chloramine-T [18a]. In this case, no oxygenated byproducts of styrene are obtained and the isolated yields of aziridines are in the range of 45-60%. Proposed activation of Chloramine-T by 1 is shown in Scheme 26. Upon coordination to the iron(IV) corrole, the polarity of the N-Na bond is inverted allowing it to react as an electrophile for the olefins. Strongly polarizable high valent iron (as in **1**) seems to be an absolute necessity for this mechanism to occur. This explains the failure of other iron corroles to activate Chloramine-T.

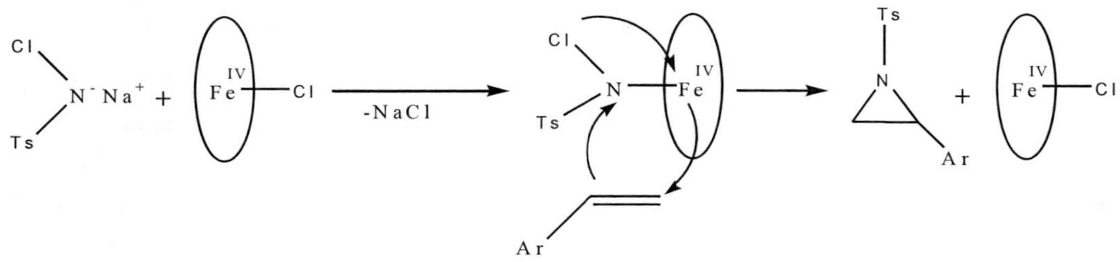

Scheme 26. Proposed catalytic routes for **1** catalyzed aziridination of styrene with Chloramine-T (adapted from ref. 18a).

The corresponding manganese(III) complex [(tpfc)Mn] (2) has also been shown to be active in catalyzing aziridination reactions of olefins with PhI=NTs. However, in this case, authors have elegantly shown that the often-postulated high-valent Mn(V) imido species is not the active group transfer reagent. They have synthesized the manganese(V) tosylimide (16) by reacting the five-coordinate [(tpfc)Mn(EtoAc)] [18b] with PhI=NTs and characterized it thoroughly. Through labeling experiments it has been shown that while 2 serves as an active catalyst for aziridination of styrene substrates with PhI=NTs, [(tpfc)Mn^V=NTs] (21) does not transfer [NTs] to styrene substrates to give aziridine [18b]. In this case, an iodoimine (ArI=NTs) adduct of [(tpfc)Mn^V=NTs] (21) has been proposed to be the active oxidant, analogous to the "third oxidant" (Scheme 11) proposed by Goldberg *et al.* [30].

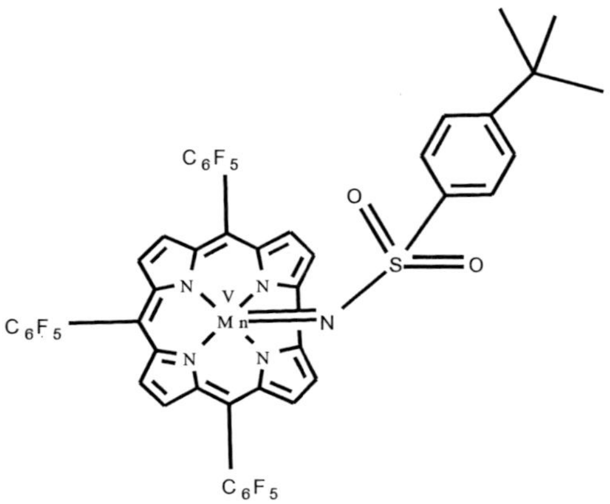

21

2.3.4. N-H Insertion of Nitrenes

Analogous to carbene insertion into the allylic C-H bond of cyclohexene, [Fe(tpfc)Cl] (**1**) and [Fe(tpfc)(OEt_2)$_2$] (11a) have also shown to be active in catalyzing the insertion of EDA into the amine NH bonds (Scheme 27) [18b].

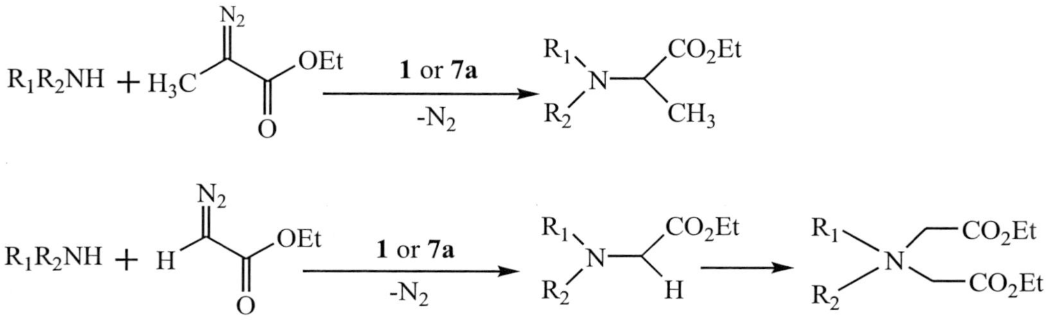

Scheme 27. Catalytic insertion of nitrene of EDA into the N-H bonds of ammines.

Iron corroles have appeared as the most promising catalyst in this regard. With only 0.1 mol% of the iron(III) catalyst, the products (N-aryl-glycines) are obtained in very high yields (>90%) [74]. The rhodium corrole (22) has also been found to be active in catalyzing the N-H insertion reactions [18b]. However, in this case greater catalyst loading (2 mol%) is required and the overall reaction completes in one hour (in contrast to the corresponding iron complexes where the reactions complete within 3-5 min) yielding the desired product (92%) together with a trace amount of diethyl maleate and diethyl fumarate.

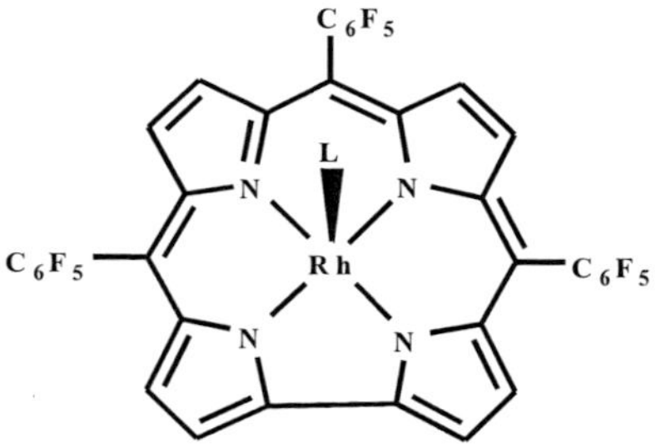

22: L=PPh$_3$

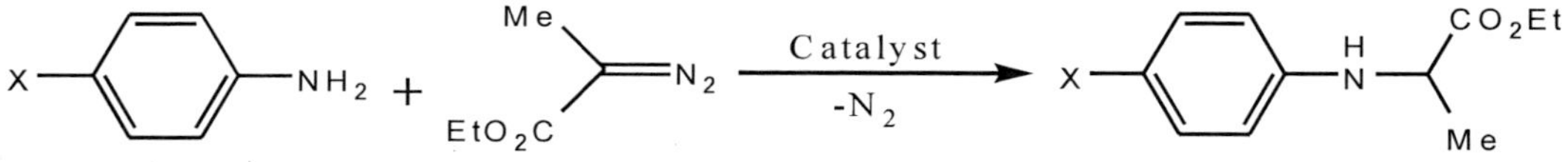

X= H, Cl, OMe

Scheme 28.

In an attempt to develop non-racemic alanine derivatives, ethyl diazopropionate (EDP) has been used instead of EDA (scheme 28) [75]. However, racemic products are obtained in all the cases. Racemic mixtures have also been obtained using sulfonato iron(III) corroles [(tpfc)(SO_3H)$_2$Fe] together with myoglobin or serum albumin in aqueous medium [18b]. Interestingly CD spectrum of the protein before and after the reaction showed absolutely no change disapproving the hypothesis that the failure to obtain effective chiral induction due to damage to the chirality inducing host (myoglobin or serum albumin). Since chiral induction in the sulfoxidation of thioanisole by H_2O_2 (scheme 9) [29] was achieved successfully, the failure in this case indicated a different mechanistic pathway for the N-H insertion reactions.

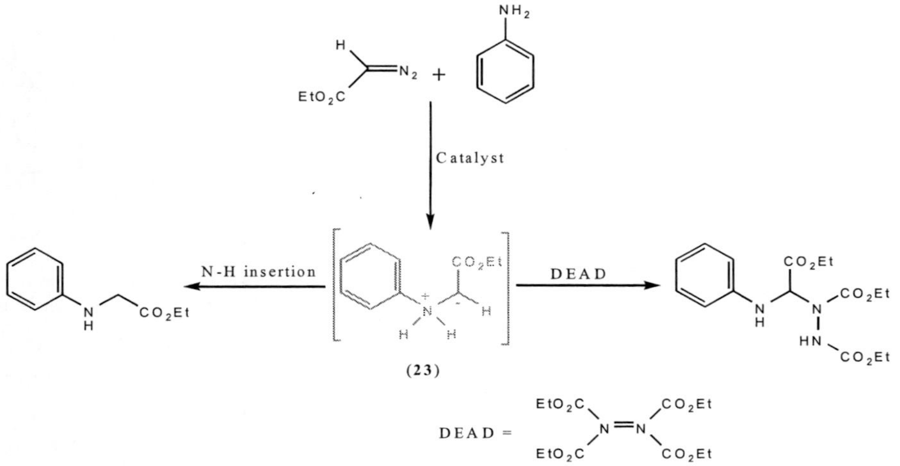

Scheme 29. Trapping of the ylide intermediate by DEAD [18b].

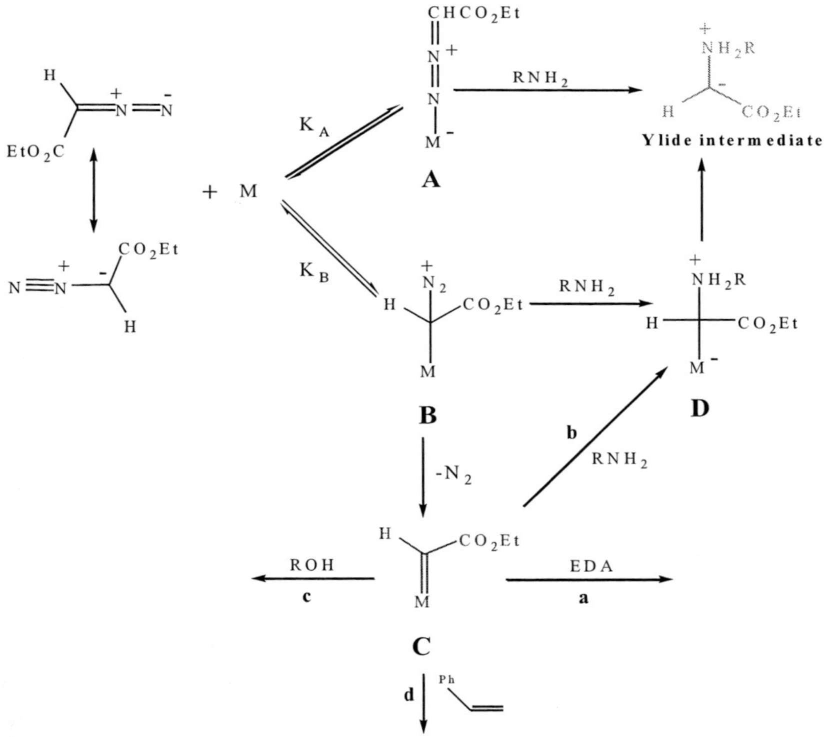

Scheme 30. Plausible mechanistic pathways leading to metal-carbene intermediate and N-ylide in catalytic N-H insertion reaction. (M= Metal ion) [adapted from ref. 18b].

The mechanistic investigations have indicated rapid formation of N-ylides (23) (Scheme 29). The ylide intermediate has been efficiently trapped by diethyl azodicarboxylate (DEAD) [18b]. The proposed mechanistic pathway is presented in Scheme 30.

Theoretical calculations showed that the formation of the carbene intermediate (C→D) is the rate-determining step having activation energy of about 17 kcal mol^{-1}. The intermediate, then rapidly reacts with the substrates (routes a-d) forming the products. The metal-carbene intermediate has been proposed to be formed by the reaction of the diazoalkane with the metal (B→C). The intermediate C is also the precursor of the nitrogen ylide transient.

Metallocorroles have been shown to be excellent catalysts in N-H insertion reactions and various novel corrole based catalytic systems are expected to emerge leading to the synthesis of important molecules such as α-amino esters, N-heterocycles *etc.*

3. Summary and Future Directions

Intensive research over the last decade in the area of metallocorrole catalyzed reactions resulted in the development of a number of processes. Some of them, particularly disproportionation of peroxynitrite and aziridination by chloramines-T, are unique. Metallocorroles as catalysts also enjoy distinctive superiority to metalloporphyrins in the N-H activation of ammonia, cyclopropanation reactions, four electron reduction of dioxygen to water etc.

The exceptional pattern of reactivity of metallocorroles towards decomposition of active oxygen species (ROS) and active nitrogen species (RON) coupled with their easy transport (binding with lipoproteins) and sometimes, their fluorescence property is believed to play a vital role in several biomedical application such as treatment of diseases caused by nitrooxidative stress, heart-diseases like atherosclerosis and even cancer.

Major emphasis was given so far on the catalytic processes involving high-valent metallocorroles, but little attention was devoted to the low-valent metallocorrole catalysts. The low-valent metallocorroles appear prospective for reductive activation of small molecules particularly CO and CO_2, which are of great import from the view point of environmental science. Therefore, the area of reductive activation of small molecules by low-valent metallocorroles desrves more attention.

Asymmetric catalysis by metallocorroles is yet to gain momentum although they enjoy the advantage of their intrinsic low symmetry. In this direction, water soluble metallocorroles associated with proteins provided a breakthrough in catalytic asymmetric oxidation reaction.

Photocatalysis by metallocorroles is another important area which needs more attention. Antimony corroles have been found to be successful photocatalysts in aerobic oxidation reactions. Recent spurt in photophysical and photochemical investigations on corroles and metallocorroles may lead to the development of efficient photocatalysts in future.

Despite the impressive achievements of metallocorroles as versatile catalysts, in general, they suffer from low turnover numbers. It may be expected that surface bound or solid supported metallocorroles will be increasingly utilized in the near future to attain more efficiency.

ACKNOWLEDGMENT

The financial support (SR/S1/IC-08/2007) from DST, Government of India, is gratefully acknowledged.

REFERENCES

[1] (a) Paolesse, R. in *The Porphyrin Handbook*, Eds. Kadish, K.M.; Smith, K.M.; Guilard, R; Academic Press, New York, 2000, vol. 2, ch. 11, pp. 201–232; (b) Erben, C; Will, S; Kadish, K.M. in *The Porphyrin Handbook*, ed. Kadish, K.M; Smith, K.M; Guilard, R; Academic Press, New York, 2000, vol. 2, ch. 12, pp. 233–300; (c) Sessler, J. L.; Weghorn, S. J. *Expanded, Contacted, and Isomeric Porphyrins*; Pergamon: Oxford, 1997; (d) Aviv, I.; Gross, Z. *Chem. Commun.*, 2007, 1987; (e) Nardis, S.; Monti, D.; Paolesse, R. *Mini-Rev. Org. Chem.*, 2005, *2*, 546; (f) Gryko, D.T.; Fox, J.P.; Goldberg, D.P. *J. Porphyrins Phthalocyanines*, 2004, *8*, 1091; (g) Barbe, J.-M.; Canard, G.; Brandes, S.; Guilard, R. *Chem. Eur. J.*, 2007, *13*, 2118. (h) Paolesse, R. *Synlett.* 2008, 2215. (i) Aviv, I.; Gross, Z. *Chem. Eur. J.* 2009, *15*, 8382.

[2] (a) Johnson, A.W.; Kay, I.T. *Proc. Chem. Soc. London*, 1964, 89; (b) Johnson, A.W.; Kay, I.T. *J. Chem. Soc.*, 1965, 1620; (c) Johnson, A.W.; Kay, I.T. *Proc. R. Soc. London, Ser. A: Math. Phys. Sci.*, 1965, *288*, 334.

[3] Harrison, H.R.; Hodder, O.J.R.; Hodgkin, D.C. *J. Chem. Soc. B*, 1971, 640.

[4] Paolesse, R., Licoccia, S., Fanciullo, M., Morgante, E., Boschi, T., *Inorg. Chim. Acta*, 1993, *203*, 107.

[5] (a) Gross, Z., Galili, N, Saltsman, I. *Angew. Chem. Int. Ed.*, 1999, *38*, 1427; (b) Gross, Z., Galili, N, Simkhovich, L, Saltsman, I., Botoshsnsky, M, Blaser, D, Boese, R., Goldberg, I. *Org. Lett.*, 1999, *1*, 599.

[6] (a) Paolesse, R., Jaquinod, L., Nurco, D. J., Mini, S., Sagone, F., Boschi, T., Smith, K. M. *Chem. Commun.*, 1999, 1307; (b) Paolesse, R., Nardis, S., Sagone, F., Khoury, R.G. *J. Org. Chem.*, 2001, *66*, 550.

[7] (a) Gryko, D.T., Kosazrna, B. *Org. Biomol. Chem.*, 2003, *1*, 350; (b) Gryko, D.T. *Chem. Commun.*, 2000, 2243; (c) Gryko, D.T., Fox, J.P., Goldber, D.P. *J. Porphyrins Phthalocyanines,* 2004, *8*, 1091; (d) Gryko, D.T., *Eur. J. Org. Chem.*, 2002, 1735.

[8] Brinas, R.P., Bruckner, C., *Synlett*, 2001, 442.

[9] Collman, J.P., Decreau, R.A., *Tetrahedron Lett.*, 2003, *44*, 1207.

[10] Geier, G.R., III, Chick, J.F.B., Callinan, J.B., Reid, C.G., Auguscinski, W.P., *J. Org. Chem.*, 2004, *69*, 4159.

[11] Gross, Z.; Gray, H. B. *Adv. Synth. Catal.*, 2004, *346*, 165.

[12] (a) Guilard, R.; Gros, C.P.; Bolze, F.; Je´ro^me, F.; Ou, Z.; Shao, J.; Fisher, J.; Weiss, R.; Kadish, K.M. *Inorg. Chem.* 2001, *40*, 4845; (b) Kadish, K. M.; Ou, Z.; Shao, J.; Gros, C.P.; Barbe, J.-M.; Je´ro^me, F.; Bolze, F.; Burdet, F.; Guilard, R. *Inorg. Chem.* 2002,

[13] *41*, 3990.

[14] Flamigni, L.; Gryko, D.T. *Chem. Soc. Rev.* 2009, *38*, 1635.

[15] Walker, D.; Chappel, S.; Mahammed, A.; Weaver, J.J.; Brunschwig, B.S.; Winkler, J.R.; Gray, H.B.; Zaban, A.; Gross, Z. *J. Porphyrins Phthalocyanines,* 2006, *10*, 1259.

[16] (a) Gross, Z.; Simkhovich, L.; Galili, N. *Chem. Commun*., 1999, 599; (b) Golubkov, G.; Bendix, J.; Gray, H. B.; Mohammed, A.; Goldberg, I.; DiBilio, A. J.; Gross, Z. *Angew. Chem. Int. Ed*., 2001, *40*, 2132; (c) Gross, Z.; Golubkov, G.; Simkhovich, L. *Angew. Chem. Int. Ed*., 2001, *39*, 4045; (d) Liu, H.-Y.; Lai, T.-S.; Yeung, L.-L.; Chang, C. K. *Org. Lett*., 2003, *5*, 617; (e) Zhang, R.; Harischandra, D. N.; Newcomb, M. *Chem. Eur. J*., 2005, *11*, 5713; (f) Collman, J. P.; Zeng, L.; Decreau, R. A. *Chem. Commun*., 2003, 2974.

[17] (a) Biswas, A.N.; Das, P; Agarwala, A.; Bandyopadhyay, D.; Bandyopadhyay, P. *J. Mol. Catal. A: Chem.* 2010, *326*, 94; (b) Biswas, A.N.; Pariyar, A.; Bose S.; Das, P.; Bandyopadhyay, P. *Catal. Commun.* 2010, *11*, 1008; (c) Bose S.; Pariyar, A.; Biswas, A.N.; Das, P.; Bandyopadhyay, P. *J. Mol. Catal. A: Chem.* 2010, *332*, 1; (d) Luobeznova, I.; Raizman, M.; Goldberg, I.; Gross, Z. *Inorg. Chem*., 2006, *45*, 386.

[18] 17. (a) Simkhovich, L; Mohammed, A.; Goldberg, I.; Gross, Z. *Chem. Eur. J*., 2001, *7*, 1041; (b) Saltsman, I.; Simkhovich, L; Balazs, Y. S.; Goldberg, I.; Gross, Z. *Inorg. Chim. Acta*., 2004, 357, 3038; (c) Simkhovich, L; Goldberg, I.; Gross, Z. *J. Porphyrins Phthalocyanines*, 2002, *6*, 439; (d) Simkhovich, L; Gross, Z. *Tett. Lett.* 2001, *42*, 8089.

[19] (a) Simkhovich, L; Gross, Z. *Tetrahedron Lett.,* 2001, *42*, 8089; (b) Zdilla, M.J.; Abu-Omar, M.M. *J. Am, Chem. Soc.*, 2006, *128*, 16971; (b) Aviv, I.; Gross, Z.; *Synlett* 2006, *6*, 951; (c) Aviv, I.; Gross, Z.; *Chem. Eur. J*., 2008, *14*, 3995.

[20] (a) Giri, R.; Shi, B.F.; Engle, K.M.; Maugel, N.; Yu, J-Q. *Chem. Soc. Rev.* 2009, *38* 3242; (b) Shul'pin, G.B. *Mini-Rev. Org. Chem.* 2009, *6*, 95; (c) Labinger, J.A.; Bercaw, J.E. *Nature* 2002, *417*, 507; (d) Sen, A. *Acc. Chem. Res.* 1998, *31*, 550.

[21] Liu, K.E.; Lippard, S.J. *Adv. Inorg. Chem.* 1995, *42*, 263.

[22] a) Groves, J.T. In *Cytochrome P450; Structure, Mechanism and Biochemistry*; Ortiz de Montellano, P.R., Ed.; Kluwer Academic/Plenum Publishers: New York, 2005; pp 1-43. (b) Wertz, D.L.; Valentine, J.S. *Struct. Bonding (Berlin)* 2000, *97*, 37-60. (c) Shaik, S.; Kumar, D.; de Visser, S.P.; Altun, A.; Thiel, W. *Chem. Rev.* 2005, *105*, 2279. (d) Nam, W. *Acc. Chem. Res.* 2007, 40, 522.

[23] (a) Agarwala, A.; Bandyopadhyay, D. *Chem. Commun.* 2006, 4823. (b) Denisov, I.G.; Makris, T.M.; Sligar, S.G.; Schlichting, I. *Chem. Rev.* 2005, *105*, 2253.

[24] (a) Singh, B.; Long, J.R.; de Biani, F.F.; Gatteschi, D.; Stavropoulos, P. *J. Am. Chem. Soc.* 1997, *119*, 7030. (b) Barton, D.H.R.; Sawyer, D.T. in *The Activation of Dioxygen and Homogeneous Catalytic Oxidation*; Ed.; Barton, D.H.R.; Martell, A.E.; Sawyer, D.T., Plenum Press, New York, 1993, p 4.

[25] Liu, H. Y.; Zhou, H.; Liu, L. Y.; Ying, X.; Jiang, H. F.; Chang, C. K. *Chem. Lett.* 2007, *36*, 274.

[26] Liu, H. Y.; Yam, F.; Xie, Y.-T; Li, X.-Y.; Chang, C. K. *J. Am. Chem. Soc.* 2009, *131*, 12890.

[27] Kim, S.H.; Park, H.; Seo, M.S.; Kubo, M.; Ogura, T.; Klajn, J.; Gryko, D.T.; Valentine, J.S.; Nam, W. *J. Am. Chem. Soc.* 2010, *132*, 14030.

[28] (a) Saltsman, I.; Mahammed, A.; Goldberg, I.; Tkachenko, E.; Botoshansky, M.; Gross, Z. *J. Am. Chem. Soc.* 2002, *124*, 7411. (b) Mahammed, A.; Goldberg, I.; Gross, Z. *Org. Lett.* 2001, *3*, 3443.

[29] Mahammed, A.; Gray, H.B.; Weaver, J.J.; Sorasaenee, K.; Gross, Z. *Bioconjugate Chem.* 2004, *15*, 738.
[30] Mahammed, A.; Gross, Z. *J. Am. Chem. Soc.* 2005, *127*, 2883.
[31] Wang, S. H.; Mandimutsira, B. S.; Todd, R.; Ramdhanie, B.; Fox, J. P.; Goldberg, D.P. *J. Am. Chem. Soc.*, 2004, *126*, 18.
[32] McGown, A.J.; Kerber, W.D.; Fujii, H.; Goldberg, D.P. *J. Am. Chem. Soc.*, 2009, *131*, 8040.
[33] Wasbotten, I.; Ghosh, A. *Inorg. Chem.* 2006, *45*, 4910.
[34] Kumar, A.; Goldberg, I.; Botoshansky, M.; Buchman, Y.; Gross, Z. *J. Am. Chem. Soc.*, 2010, 132, 15233.
[35] (a) Meier-Callahan, A.E.; DiBilio, A.J.; Simkhovich, L.; Mahammed, A.; Goldberg, I.; Gray, H.B.; Gross, Z. *Inorg. Chem.* 2001, *40*, 6788. (b) Mahammed, A.; Gray, H.B.; Meier-Callahan, A.E.; Gross, Z. *J. Am. Chem. Soc.* 2003, *125*, 1162.
[36] (a) Sheldon, R. A.; Kochi, J. K. *Metal-Catalyzed Oxidation of Organic Compounds*, Academic Press, New York, 1981. (b) Sheldon, R.A.; Arends, I.C.W.E.; Dijksman, A. *Catal. Today* 2000, *57*, 157. (c) Bäckvall, J.-E. *Modern Oxidation Methods*, Wiley-VCH, Weinheim, 2004.
[37] Kockritz, A.; Sebek, M.; Dittmar, A.; Radnik, J.; Bruckner, A.; Bentrup, U.; Pohl, M.M.; Hugl, H.; Magerlein, W. *J. Mol. Catal. A: Chem.* 2006, *246*, 85.
[38] (a) *Organic Synthesis by Oxidation with Metal Compounds*, ed. Mijs, W. J. and De Jonge, C. R. H. I. Plenum, New York, 1986. (b) March, J. *Advanced Organic Chemistry*, Wiley, New York, 1992, 4th edn., p. 1167. (c) Highet, R.J.; Wildman, W.C. *J. Am. Soc. Chem.* 1955, *77*, 4399. (d) Lee, D.G.; Spitzer, U.A. *J. Org. Chem.* 1970, *35*, 3589. (e) Stevens, R.V.; Chapman, K.T.; Weller, H.N. *J. Org. Chem.* 1980, *45*, 2030. (f) Menger, F.M.; Lee, C. *J. Org. Chem.* 1979, *44*, 3446.
[39] Mallat, T.; Baiker, A. *Chem. Rev.* 2004, *104*, 3037.
[40] Bose, S.; Pariyar, A.; Biswas, A.N.; Das, P.; Bandyopadhyay, P. *Catal. Commun.* 2011, *12*, 446.
[41] Pariyar, A.; Bose, S.; Biswas, A.N.; Das, P.; Bandyopadhyay, P. unpublished results.
[42] Winter, M.; Brodd, R. J. *Chem. Rev.* 2004, *104*, 4245.
[43] Collman, J.P.; Boulatov, R.; Sunderland, C.J.; Fu, L. *Chem. Rev.* 2004, *104*, 561.
[44] (a) Barrientos, A.; Barros, M. H.; Valnot, I.; Rotig, A.; Rustin, P.; Tzagoloff, A. *Gene* 2002, *286*, 53. (b) Kitagawa, T. *J. Inorg. Biochem.* 2000, *82*, 9. (c) Yoshikawa, S.; Shinzawa-Itoh, K.; Tsukihara, T. *J. Inorg.Biochem.* 2000, *82*, 1. (d) Wikstrom, M. *Biochim. Biophys. Acta* 2000, *1458*, 188. (e) Mills, D. A.; Florens, L.; Hiser, C.; Qian, J.; Ferguson-Miller, S. *Biochim. Biophys. Acta* 2000, *1458*, 180. (f) Michel, H. *Nature* 1999, *402*, 602. (g) Michel, H. *Biochemistry* 1999, *38*, 15129.
[45] (a) Chang, C. J.; Deng, Y. Q.; Shi, C. N.; Chang, C. K.; Anson, F. C.; Nocera, D. G. *Chem. Commun.* 2000, 1355. (b) Collman, J. P.; Denisevich, P.; Konai, Y.; Marrocco, M.; Koval, C.; Anson, F. C. *J. Am. Chem. Soc.* 1980, *102*, 6027. (c) Cui, H. F.; Ye, J. S.; Liu, X.; Zhang, W. D. *Nanotechnology* 2006, *17*, 2334. (d) Wang, B. J. *Power Sources* 2005, *152*, 1. (e) Abramson, J.; Riistama, S.; Larsson, G.; Jasaitis, A.; Svensson-Ek, M.;
[46] Laakkonen, L.; Puustinen, A.; Iwata S.; Wikstr ¨om, M. *Nat. Struct. Biol.* 2000, *7*, 910. (f) Zagal, J. H. *Coord. Chem. Rev.* 1992, *119*, 89.
[47] (a) Fukuzumi, S.; Okamoto, K.; Gros, C. P.; Guilard, R. *J. Am. Chem. Soc.* 2004, *126*, 10441. (b) Kadish, K. M.; Fremond, L.; Ou, Z. P.; Shao, J. G.; Shi, C. N.; Anson, F. C.;

Burdet, F.; Gros, C. P.; Barbe, J. M.; Guilard, R. *J. Am.Chem. Soc.* 2005, *127*, 5625. (c) LeMest, Y.; Inisan, C.; Laouenan, A.; Lher, M.; Talarmin, J.; ElKhalifa, M.; Saillard, J. Y. *J. Am. Chem. Soc.* 1997, *119*, 6095. (d) Shin, H.; Lee, D. H.; Kang, C.; Karlin, K. D. *Electrochim. Acta* 2003, *48*, 4077. (e) Gojkovic, S. L.; Gupta, S.; Savinell, R. F. *J. Electroanal. Chem.*1999, *462*, 63. (f) Choi, A.; Jeong, H.; Kim, S.; Jo, S.; Jeon, S. *Electrochim. Acta* 2008, *53*, 2579. (g) Zhang, W.; Shaikh, A. U.; Tsui, E. Y.; Swager, T. M. *Chem. Mater.*2009, *21*, 3234. (h) Zhou, Q.; Li, C. M.; Li, J.; Lu, J. T. *J. Phys. Chem. C* 2008, *112*, 18578. (i) Lucero, M.; Ramirez, G.; Riquelme, A.; Azocar, I.; Isaacs, M.;

[48] Armijo, F.; Forster, J. E.; Trollund, E.; Aguirre, M. J.; Lexa, D. *J. Mol. Catal. A: Chem.* 2004, *221*, 71. (j) Chen, W.; Akhigbe, J.; Brükner, C.; Li, C.M.; Leu, Y. *J. Phys. Chem. C.* 2010, *114*, 8633.

[49] (a) Kadish, K. M.; Fre′mond, L.; Burdet, F.; Barbe, J.-M.; Gros, C. P.; Guilard, R. *J. Inorg. Biochem.* 2006, *100*, 858. (b) Kadish, K. M.; Shao, J.; Ou, Z.; Fre′mond, L.; Zhan, R.; Burdet, F.; Barbe, J.-M.; Gros, C. P.; Guilard, R. *Inorg. Chem.* 2005, *44*, 6744– 6754.

[50] Kadish, K. M.; Fr′emond, L.; Shen, J.; Chen, P.; Ohkubo, K.; Fukuzumi, S.; Ojalml, M.E.; Gros, C.P.; J.-M. Barbe; R. Guilard, *Inorg. Chem.* 2009, *48*, 2571.

[51] Collman, J.P.; Kaplun, M.; Decréau, R. *Dalton Trans.* 2006, 554.

[52] Morris, A.J.; Meyer, G.J.; Fujita, E. *Acc. Chem. Res.* 2009, *42*, 1983.

[53] (a) Grodkowski, J.; Behar, D.; Neta, P.; Hambright, P. *J. Phys. Chem. A* 1997, *101*, 248. (b) Behar, D.; Dhanasekaran, T.; Neta, P.; Hosten, C. M.; Ejeh, D.; Hambright, P.; Fujita, E. *J. Phys. Chem. A* 1998, *102*, 2870. (c) Grodkowski, J.; Dhanasekaran, T.; Neta, P.; Hambright, P.; Brunschwig, B. S.; Shinozaki, K.; Fujita, E. *J. Phys. Chem. A* 2000,

[54] *104*, 11332. (d) Grodkowski, J.; Neta, P. *J. Phys.Chem. A* 2000, *104*, 1848–1853.

[55] Grodkowski, J.; Neta, P.; Fujita, E.; Mahammed, A.; Simkhovich, L.; Gross, Z. *J. Phys.Chem. A* 2002, *106*, 4772–4778.

[56] (a) Fukuto, J.M.; Ignarro, L.G. *Acc. Chem. Res.* 1997, *30*, 149. (b) Alvarez, B.; Radi, R. *Amino Acids* 2003, *25*, 295.(c) Masood, A.; Belcastro, R.; Li, j.; Kantores, C.; Jankov, R.P.; Tanswell, A.K.; *Free Radic. Biol. Med.* 2010, *49*, 1182.

[57] (a) SzabJ, C.; Ischiropoulos, H.; Radi, R. *Nat. Rev. Drug Discovery* 2007, *6*, 662. (b) Peluffo, G.; Radi, R. *Cardiovasc. Res.* 2007, *75*, 291.

[58] (a) Wiedau-Pazos, M.; Goto, J. J.; Rabizadeh, S.; Gralla, E. B.; Roe, J. A.; Lee, M. K.; Valentine J. S.; Bredsen, D.E. *Science* 1996, *271*, 515; (b) Beckman, J. S.; Carson, M.; Smith C. D.; Koppenol, W. H. *Nature* 1993, *364*, 584.

[59] Mahoney, L. R. *J. Am. Chem. Soc.* 1970, *92*, 4244.

[60] de la Asuncion, J.G., Millan, A., Pla, R., Bruseghini, L., Esteras, A., Pallardo, F.V., Sastre, J., Vina, J., *FASEB J.* 1996, *10*, 333.

[61] Coddington, J.W.; Hurst, J.K.; Lymar, S.V. *J. Am. Chem. Soc.* 1999, *121*, 2438.

[62] Shi, X.; Rojanasakul, Y.; Gannett, P.; Liu, K.; Mao, Y.; Daniel, L.N.; Ahmed, N.; Saffiotti, U. *J. Inorg. Biochem.* 1994, *56*, 77.

[63] (a) Goldstein, S.; Czapski, G. *J. Am. Chem. Soc.* 1998, *120*, 3458. (b) Houk, K. N.; Condorski, K. R.; Pryor, W. A. *J. Am.Chem. Soc.* 1996, *118*, 13002. (c) Radi, R. *Chem. Res. Toxicol.* 1998, *11*, 720.

[64] Groves, J.T.; Marla, S.S. *J. Am. Chem. Soc.* 1995, *117*, 9578.

[65] Stern, M.K.; Jensen, M.P.; Kramer, K. *J. Am. Chem. Soc.* 1996, *118*, 8735.

[66] Lee, J.; Hunt, J.A.; Groves, J.T. *J. Am. Chem. Soc.* 1998, *120*, 6053.
[67] (a) Mahammed, A.; Gross, Z. *Angew. Chem., Int. Ed.* 2006, *45*, 6544. (b) Mahammed, A.; Gross, Z. *Angew. Chem.* 2006, *118,* 6694.
[68] SzabA, C.; Mabley, J.G.; Moeller, S.M.; Shimanovich, R.; Pacher, P.; VirMg, L.; Soriano, F.G.; Van Duzer, J.H.; Williams, W.; Salzman, A.L.; Groves, J.T. *Mol. Med.* 2002, *8*, 571.
[69] Ferrer-Sueta, G.; Quijano, C.; Alvarez, B.; Radi, R. *Methods Enzymol.* 2002, *349*, 23.
[70] (a) Gershman, Z.; Goldberg, I.; Gross, Z. *Angew. Chem.* 2007, *119*, 4398. (b) Gershman, Z.; Goldberg, I.; Gross, Z. *Angew. Chem. Int. Ed.* 2007, *46*, 4320.
[71] (a) Haber, A.; Mahammed, A.; Fuhrman, B.; Volkova, N.; Coleman, R.; Hayek, T.; Aviram, M.; Gross, Z. *Angew. Chem.* 2008, *120*, 8014. (b) Haber, A.; Mahammed, A.; Fuhrman, B.; Volkova, N.; Coleman, R.; Hayek, T.; Aviram, M.; Gross, Z. *Angew. Chem. Int. Ed.* 2008, *47*, 7896. (c) Okun, Z.; Kupeshmidt, L.; Amit, T.; Mandel, S.; Bar-Am, O.; Youdim, M.B.H.; Gross, Z. *ACS Chem. Biol.* 2009, 4, 910.
[72] Lebel, H.; Marcoux, J-F.; Molinaro, C.; Charette, A.B. *Chem. Rev.* 2003, *103*, 977.
[73] Padwa, A.; Murphree, S.S. *ARKIVOV* 2006, *iii*, 6.
[74] (a) Burtoloso, A.C.B.; Correia, C.R.D. *J. Organomet. Chem.* 2005, *690*, 5636. (b) Huang, H.; Wang, Y.; Chen, Z.; Hu, W. *Adv. Synth. Catal.* 2005, 347, 531. (c) Davis, F.A.; Wu, Y.; Xu, H.; Zhang, J. *Org. Lett.* 2004, 6, 4523. (d) Fructos, M.R.; Belderrain, T.R.; Nicasio, M.C.; Nolan, S.P., Kaur, H.; Diaz-Requejo, M.M.; Pérez, P.J. *J. Am. Chem. Soc.* 2004, *126*, 10846. (e) Singh, G.S. *Curr. Org. Synthesis* 2005, *2*, 377.
[75] (a) Breslow, R.; Gellman, S. H. *J. Chem. Soc., Chem. Commun.* 1982, 1400. (b) Mansuy, D.; Mahy, J.-P.; Dureault, A.; Bedi, G.; Battioni, P. *J. Chem. Soc., Chem. Commun.* 1984, 1161. (c) Mansuy, D.; Mahy, J.-P. In *Metalloporphyrins Catalyzed Oxidations*; Montanari, F.; Casella, L., Eds.; Kluwer: Dordrecht, 1994; pp. 175–206.
[76] Mahy, J.-P.; Bedi, G.; Battioni, P.; Mansuy, D. *J. Chem. Soc., Perkin Trans. 2* 1988, 1517.
[77] Jeong, J. U.; Tao, B.; Sagasser, I.; Henniges, H.; Sharpless, K. B. *J. Am. Chem. Soc.* 1998, *120*, 6844.
[78] Aviv, I.; Gross, Z. *Chem. Commun.* 2006, 4477.
[79] Liu, B.; Zhu, S.-F.; Zhang, W.; Chen, C.; Zhou, Q.-L. *J. Am. Chem. Soc.* 2007, *129*, 5834.

In: Focus on Catalysis Research: New Developments
Editors: Minjae Ghang and Bjørn Ramel
ISBN: 978-1-62100-455-4

Chapter 13

APPLICATIONS OF DENDRIMERS IN CATALYSIS

G. Rajesh Krishnan and Krishnapillai Sreekumar
Department of Applied Chemistry,
Cochin University of Science and Technology, Cochin, Kerala, India

INTRODUCTION

Dendrimers are highly monodisperse organic nanoparticles prepared by an iterative method of synthesis. They are considered as the fourth generation polymer architecture. Due to their particular structure and the properties arising due to this structure, dendrimers have aroused the curiosity of scientists from all disciplines of research ranging from chemistry to medicine. Catalysis is an area of science really blessed by dendrimers. Contrary to their high molecular weight, many dendrimers have shown well defined solubility properties and because of that dendrimers are considered to have filled up the gap between homogeneous and heterogeneous catalysts. Dendrimers offer a number of possibilities in fine tuning the activity and selectivity of many catalysts by properly attaching them to the dendrimers. The present chapter deals with the developments in catalysis after the materialization of dendrimers.

1. DENDRIMERS

Nanoscience is at the forefront of scientific and technological development and is the science of the 21st century. Nanochemistry occupies a place of choice in this discipline and can be regarded as the use of synthetic chemistry to make nanoscale building blocks of different size and shape, composition and surface structure, charge and functionality. These building blocks may form, or can be used to form, even more sophisticated architectures having different properties and particular uses. [1]

Dendrimers, monodisperse nanosized polymeric molecules composed of a large number of perfectly branched monomers that emanate radially from a central core, can be considered as one of the most fascinating molecules arising from this area of research. Dendritic molecules are repeatedly branched species that are characterized by their structure perfection. The latter is based on the evaluation of both symmetry and polydispersity. They are considered as the fourth major macromolecular architecture after linear, cross-linked and branched polymers. The name comes from the Greek "δενδρον"/*dendron*, meaning "tree". The ability to easily tune the size, topology, molecular weight and consequently the properties of these nano-objects has led to their widespread use in a variety of applications from biology to material science, i.e. at the interface of many disciplines. Remarkably, their unique branched topologies lead to properties that differ frequently from those of linear polymers, thus exciting the interest and curiosity of thousands of researchers worldwide. Dendritic architecture is one of the most pervasive topologies observed in nature at the macro- and microdimensional-length scales. At the nanoscale (molecular level), there are relatively few natural examples of this architecture. Most notable are glycogen and amylopectin, the two proteins used for energy storage in many higher organisms. This also adds curiosity to these molecules as a totally new synthetic molecular architecture that is new to the nanoworld, but has so many potentials. [2]

2. THE DENDRITIC STRUCTURE

Dendrimers are highly ordered, regularly branched, globular macromolecules prepared by stepwise iterative approach. A higher generation dendrimer can be considered as a three dimensional sphere with nanoscale dimensions. Their structure is divided into three distinct architectural regions: (i) a core or focal moiety, (ii) layers of branched repeat units emanating from this core and (iii) end groups on the outer layer of repeat units as shown in Figure 1.

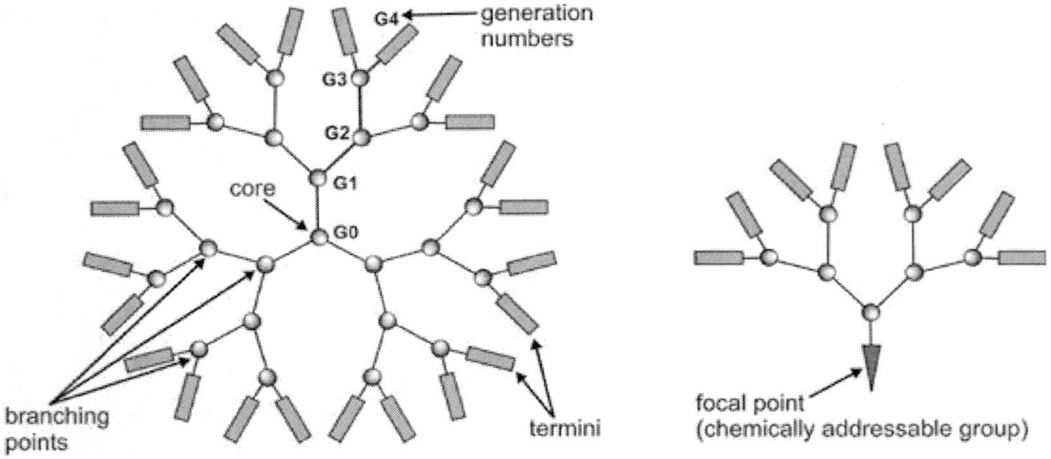

Figure 1. The dendritic structure.

At least three characteristic features of dendrimers are in sharp contrast to those of traditional linear polymers because of their characteristic structure. (i) A dendrimer can be isolated as an essentially monodisperse single compound, unlike most linear polymers whose synthesis affords a range of molecular species differing in molecular weight (MW). Size monodispersity results from a well-designed iterative synthesis that allows reactions to be driven to completion, side-reactions to be avoided, and in some cases, the dendritic products to be purified at intermediate steps during their growth. (ii) As their molecular weight increases, the properties of dendrimers (e.g., solubility, chemical reactivity, glass transition temperature) are dominated by the nature of the end groups. Unlike linear polymers that contain only two end groups, the number of dendrimer end groups increases exponentially with generation, and therefore the end-groups frequently become the primary interface between the dendrimer and its environment. (iii) In contrast to linear polymer growth that, theoretically, can continue ad infinitum barring solubility issues, dendritic growth is mathematically limited. During growth of a dendrimer, the number of monomer units increases exponentially with generation, while the volume available to the dendrimer only grows proportionally to the cube of its radius. As a result of this physical limitation, dendritic molecules develop a more globular conformation as generation increases. At a certain generation, a steric limit to regular growth, known as the De Gennes dense packing [3] is reached. Growth may be continued beyond de Gennes dense packing, but this leads to irregular dendrimers incorporating structural flaws. The most common dendrimer structures are shown in figure 2. In addition to these a number of other dendrimers were reported and interested reader may go through the references given at the end of the chapter.

3. Dendrimers In Catalysis

Among the two area of research consecrated by dendrimers, the most vital one is catalysis and the other is biotechnology and medicine. Dendrimers are considered to build the gap between homogeneous catalysts and heterogeneous catalysts and they show many advantages over conventional catalysts. [4] In principle, dendritic catalysts can provide systems that

1. Show the kinetic behavior and thus the activity and selectivity of a conventional homogeneous catalyst. Catalysts supported on heterogeneous systems show diminished activity compared to the homogeneous analogues, which is because of reduced accessibility.
2. Can easily be removed from the reaction mixture by membrane or nanofiltration technique because of their large size compared to the products (an advantage of heterogeneous catalysts).
3. Allow mechanistic studies, because of the monodisperse, uniform character of their catalytic sites and the symmetry of the molecules (an advantage of homogeneous catalysts).
4. Allow fine-tuning of their catalytic centers by precise ligand design (an advantage of homogeneous catalysts) and
5. Require relatively low metal loading (an advantage of homogeneous catalysts over heterogeneous catalysts).

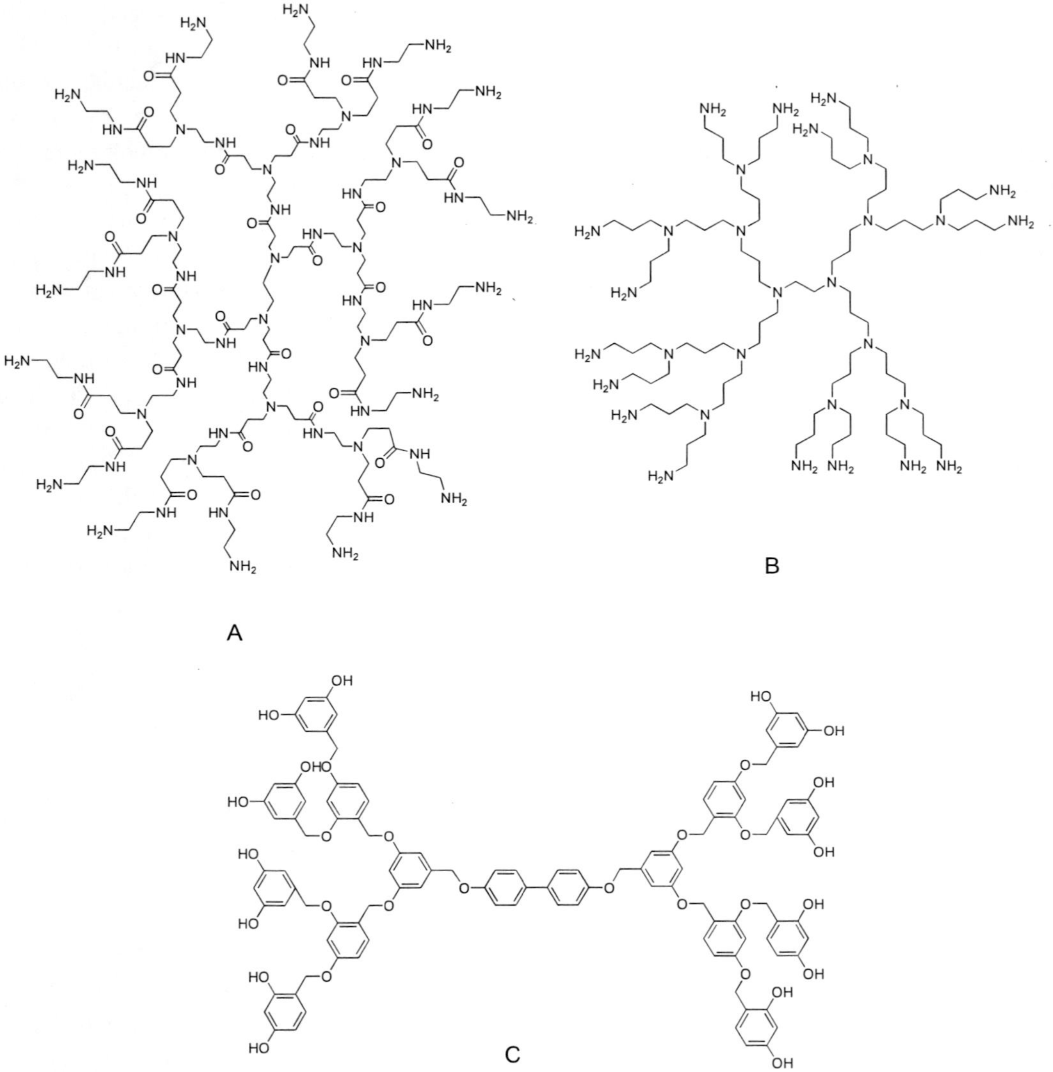

Figure 2. Some common dendrimers (A) poly(amidoamine) (PAMAM) (B) poly(propylimine) (PPI) (C) poly(aryl ether) dendrimers.

The advantages of dendrimer catalysts arise from their peculiar structure and the properties arise due to this structure. Since dendrimers generally contain three distinguishable parts in their structure, one can distinguish periphery functionalized, core functionalized and focal point functionalized catalytic systems. All these systems show their own activity and selectivity. After the introduction of dendrimer-based catalysts, a new term was coined in dendrimer chemistry and it is 'the dendrimer effect'. The special properties that may result from catalyst incorporation onto these macromolecules can be described as a "dendrimer effect." This term has been generally invoked in the literature to explain phenomena that arise as the generation of dendrimer increases. [5] This may be positive or negative depending on the dendrimer platform, the placement of catalysts, the reaction conditions and the mechanism of reaction.

The biggest stumbling block in the wide spread application of dendrimers in catalysis even in the midst of all these advantages is the lengthy and complicated synthetic procedure of dendrimer based catalysts. Any attempt to create a perfect harmony between these advantages and disadvantages are valuable in dendrimer research.

Dendrimer based catalysts can be classified into three. The first one is homogeneous catalysts based on dendrimers, which will be described in detail in the following sections of this chapter. The second is dendrimers attached to some supports like polymers or silica. A discussion on polymer supported dendrimer catalysts can be found in the chapter on polymer supported catalysts. The third type is dendrimer encapsulated nanoparticles. A brief discussion of the same can be found at the end of this chapter.

4. Dendrimers in Homogeneous Catalysis

Dendrimer based homogeneous catalysts is one of the most widely studied catalysts among all dendrimer based catalysts. This is because their synthesis and characterization are more easily compared to many supported dendrimers and since it is in homogeneous condition mechanistic studies can more easily be carried out. These homogeneous catalysts can be classified based on the position where the catalytic site is placed. The catalyst can be placed on either of three points in the dendrimer ie. periphery, core or focal point. Both metal based catalysts and organocatalysts can be placed at any of these points. Among them periphery functionalized dendrimers are more common followed by core functionalized. There are no examples of focal point functionalized dendrimer catalysts.

4.1. Periphery Functionalized Dendrimers in Catalysis

Periphery-functionalized dendrimers have their ligand systems, and thus the metal complexes, at the surface of the dendrimer. The catalytic sites will be directly available for the substrate, in contrast to the other two systems, in which the substrate has to penetrate the dendrimer prior to reaction. This accessibility allows reaction rates that are comparable with homogeneous systems. On the other hand, the periphery functionalized systems contain multiple reaction sites and ligands, which results in extremely high local catalyst and ligand concentration. In reactions where excess ligand is required to stabilize the catalyst, this local concentration effect can indeed result in stable systems. Furthermore, if a reaction proceeds by a bimetallic mechanism, the dendritic catalysts might show better performance than the monomeric species. Periphery functionalized dendrimer catalyst can be represented as in figure 3.

First example of a dendrimer-based catalyst was reported by Keijsper and co-workers. [6] A hexaphosphine catalyst containing a benzene core was used in the palladium catalyzed polyketone formation by alternating polymerization of CO and alkene. By using this catalyst, it was observed that, fouling was reduced to 3% compared to the 50% given by monomeric version under similar conditions. A possible explanation is that, in the dendritic catalysts, the palladium remains attached to the surface of the growing polymer and do not go into solution during the chain-transfer reaction (which may lead to fouling). This is the first example of a dendrimer effect observed in catalysis.

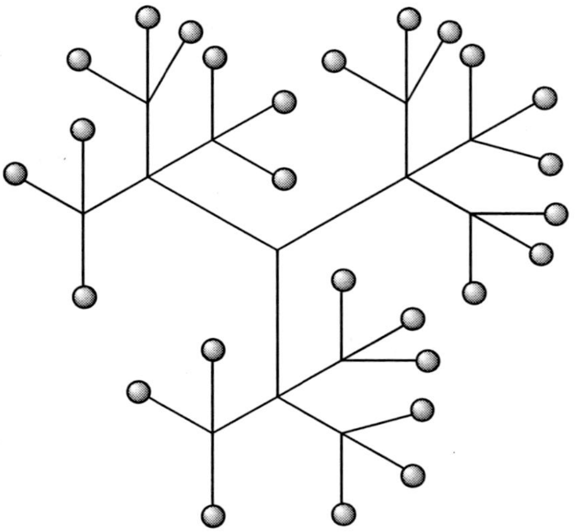

Figure 3. Schematic structure of periphery functionalized dendrimer where the dark spheres represent catalysts.

Another classical example of dendritic catalyst was reported by Knapen *et al* who functionalized generation zero and one carbosilane dendrimers with up to twelve pincer like NCN-nickel(II) groups. [7] These dendrimers were applied as catalysts for the Kharasch addition of CX to alkene (figure 4). The catalytic activity of the dendritic catalysts was slightly lower than that of the monomeric parent compound. A negative dendrimer effect was observed and the activity decreased with increase in generation of the catalyst.

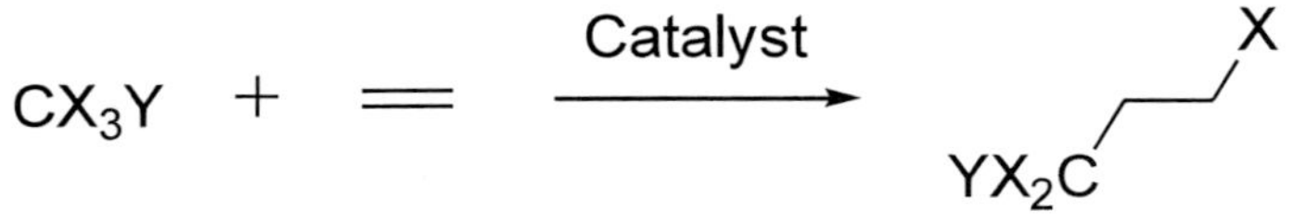

where catalyst is

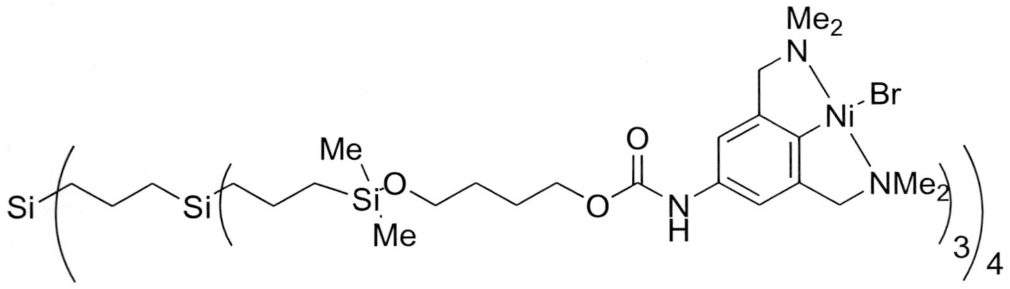

Figure 4. Carbosilane dendrimers carrying pincer like NCN-nickel(II) groups catalyzed Kharash addition.

Molecular modeling showed that the accessibility of the catalytic sites was similar for dendrimers and monomers, and it was proposed that the lower rates were because of the high local concentration of nickel centers and interaction between the neighboring Ni(II) and NI(III) sites. A similar effect was observed by Miedaner *et al.* in palladium-functionalized dendrimers when used as catalysts in the electrochemical reduction of CO_2 to CO. [8]

Catalyst for carbon-carbon bond formation is another area of interest in which periphery functionalized dendrimers have found wide spread applications. Several groups have prepared systems that were functionalized with phosphine ligands at the periphery of the dendrimer. Commercially available DAB dendrimers were equipped with diphenylphosphine groups at the periphery by Reetz *et al* by a double phosphination of the amines with diphenylphosphine and formaldehyde. [9] The palladium complexes of these dendrimers were prepared and used as catalysts in the Heck reaction of bromobenzene and styrene to form stilbene (figure 5). Interestingly, the dendrimers showed larger turnover number than the monomeric parent compounds, which was ascribed to the higher thermal stability of the dendritic palladium complexes. No palladium black formation was observed when dendrimer catalysts were used. After the reaction, the dendritic catalyst was completely precipitated upon addition of diethyl ether. In this way, the catalysts were recycled and a second catalytic run gave similar results with out loss of activity.

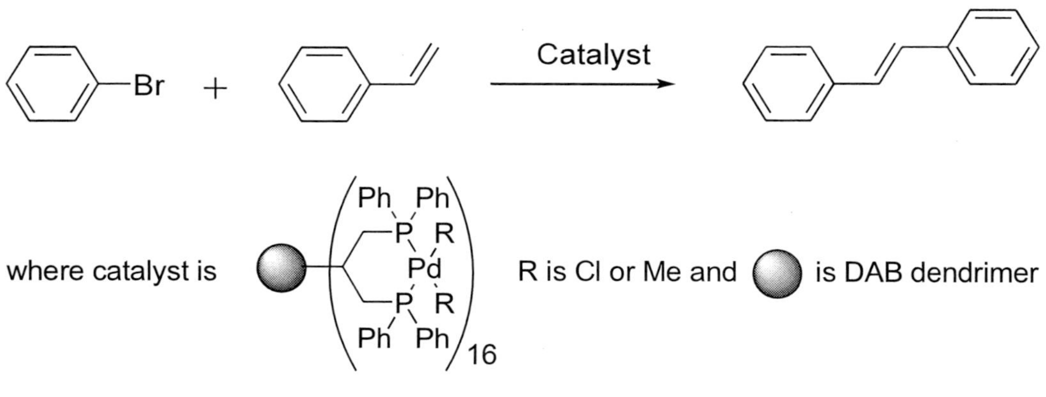

Figure 5. Heck reaction catalyzed by palladium complexes of DAB dendrimer with diphenylphosphine groups at the periphery.

Brinkmann and co-workers also used the palladium-dendrimer complex described above as catalyst in allylic substitution. [10] But the catalyst was not stable under the reaction conditions tested and palladium leaching was observed. The catalyst showed better stability with higher generation dendrimer and fourth generation dendrimer-palladium complex showed the highest activity.

De Groot *et al* prepared a similar catalyst but with a different dendritic backbone and in this case, the DAB dendrimer was replaced by a carbosilane dendrimer. But this catalyst too did not perform well in allylic substitution and metal leeching was observed in considerable amount. [11]

Later, Astruc and co-workers showed that palladium complexes of DAB dendrimer containing bis-phosphines on the periphery are highly efficient catalysts in Sonogashira coupling between aryl halides and terminal alkynes. [12-14] The two catalysts developed were so effective that the reaction proceeded with out the presence of copper whereas conventional processes required both palladium and copper to catalyze the reaction effectively. In addition, the catalyst was effective even at –40 °C. But a negative dendrimer effect was observed and the rate of reaction decreased with increase in dendritic generation.

Another interesting example of dendrimer catalyst used for carbon-carbon bond formation was that reported by Sarkar *et al.* [15] A second generation DAB dendrimer surface functionalized with highly basic bicyclic azido phosphine moiety was used as base catalyst for Michael addition and Henry reaction. The catalyst showed good performance and was considered due to the co-operative effect of sixteen azido phoshine units in each dendrimer. The structure of the catalyst is shown in the Figure 6. This study proved the ability of dendrimers to act as organocatalysts also, in addition to their role as multidentate ligands in metal complex based catalysts.

Sreekumar and co-workers showed that simple poly(aimidoamine) dendrimers carrying primary amino groups are able to act as highly efficient organocatalysts in Knoevenagel condensation between aldehydes and active methylene compounds. [16] The reaction proceeded with 90-100% yield and 100% selectivity towards substituted olefins. They also showed that the same dendrimer was able to catalyze three component Manich reactions. So the same dendrimer is able to catalyze both carbon-carbon and carbon-hetero atom bond formation.

The same group used a half generation PPI dendrimer carrying four nitrile groups in the preparation of catalytically active silver complex. [17] The complex was an efficient catalyst in both three component Mannich reaction and in the synthesis of 1,5-benzodiazepines from o-diaminobenzene and ketones. Both the Mannich adduct and benzodiazepines were obtained in high yield and the catalyst was reusable for four times.

A classical example of dendrimer catalyst is the one reported by Mizugaki et al. [18] A third generation DAB dendrimer was funtionalized with diphenylphosphine and palladium complex of this dendrimer was used as catalyst in the hydrogenation of cyclopentadiene. The catalyst showed better activity than the corresponding monomeric complex and soluble polystyrene supported complex. The dendrimer backbone was proposed to act as a base to capture HCl, there by accelerating the formation of catalytically active (diphosphine)PdHCl species. Gade and Findeis reported another dendrimer catalyst in which tripodal phosphine ligands were attached to a dendrimer periphery and rhodium complex of this dendrimer was used as catalyst in the hydrogenation of alkenes in THF. [19]

Many dendritic catalysts for polymerization were reported by various groups and the important ones are described here. Seyferth et al. synthesized carbosilane dendrimers with 4, 8, and 12 peripheral zirconocene, hafnocene and titanocene groups. [20, 21] The zirconocene containing dendrimer was used as catalyst in the methylaluminoxane (MAO) activated olefin copolymerization and in silane polymerization. The active form of the metallocene catalyst used for α-olefins was cationic and was typically generated by a co-catalyst. Industrially, this is usually methylaluminoxane or perfluorophenylborane. The interaction between the ion pair affects the overall activity, steroregularity, chain transfer, termination rate and lifetime of the metallocene catalyst. Less nucleophilic counter ions are particularly desirable; this has, in the past, been achieved through charge delocalization or steric shielding.

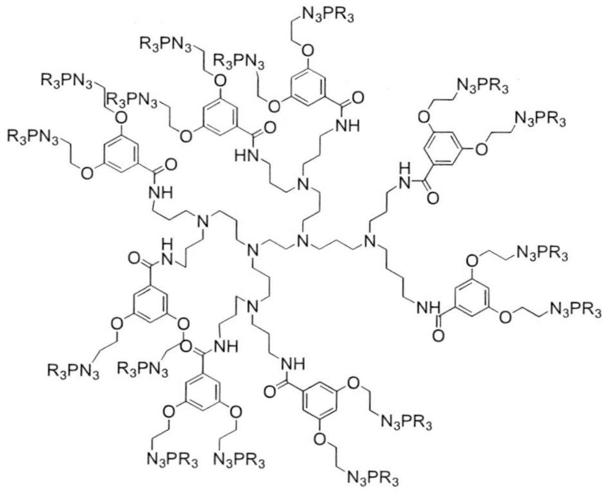

where R_3P is $P(iBuNCH_2CH_3)_3$

Figure 6. DAB dendrimer surface functionalized with highly basic bicyclic azido phosphine moiety.

However, in their work, Mager and co-workers achieved this through steric crowding at the surface of a peripherally modified carbosilane dendrimer. [22] Co-catalysts G0 to G2 were synthesized with 4, 12, and 36 alkyl-tris- (pentafluorophenyl)borates at the periphery (Figure.7). When employed with $[(Ind)_2ZrMe_2]$ metallocene co-catalysts, these non-coordinating dendrimer polyanions were highly active in ethylene polymerization and copolymerization with propylene or 1-hexene, although activity was not generation-dependent. Unlike non-coordinating small molecule anions resulting from $B(C_6F_5)_3$, the dendrimer polyanions gave high activities even in aliphatic solvents, with no apparent loss in activity at reaction times greater than 40 min. This positive dendrimer effect was thought to arise from the specific and unique interaction between the active cationic metallocene co-catalyst and the crowded anionic surface of the dendrimer.

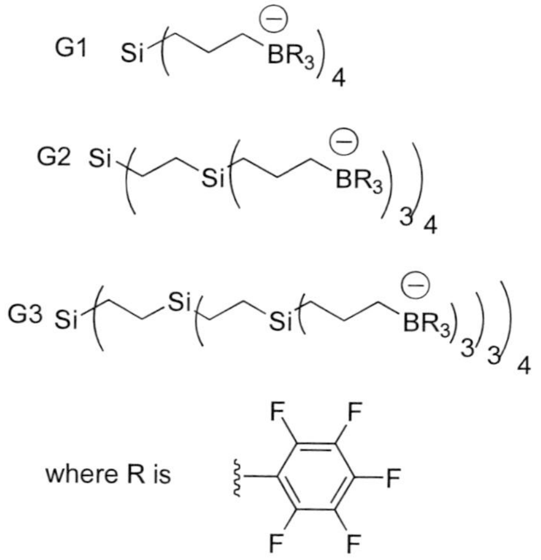

Figure 7. Polycarbosilane dendrimers used as polymerization catalysts.

In a similar study relating the nature of the interaction between a dendritic polymerization catalyst with a conventional anionic co-catalyst, Zheng and coworkers prepared first and second generation carbosilane metallodendrimers bearing bis(imino)pyridyl iron(II) catalyst precursors at their periphery. [23] After activation with modified methylaluminoxane (m-MAO) at a ratio of Al/Fe>1200, the rate of ethylene polymerization for either of the multivalent dendrimer catalyst was comparable to the parent Fe catalyst. However, at lower Al/Fe ratios (<1000), the activities of both dendrimer catalysts were superior to the small molecule catalyst. In addition, the molecular weight and the melting temperature (Tm) of the polyethylene obtained increased with the generation of the dendrimer catalyst: Tm=127.9, 133.9 and 134.1 8 °C for the parent, G1 and G2 catalysts, respectively. Again, this positive dendrimer effect was generally thought to involve a fundamentally different interaction of the dendritic periphery with the polymeric counter ion derived from m-MAO than was experienced with the small molecule catalyst system. Benito et al. synthesized a series of carbosilane dendrimers with surface Ni(II)pyridylimines. [24] These were used as catalysts in the polymerization of α-olefins. They reported a strong generation dependence on the molecular weight and topology of the final product. With an increase in generation of the dendrimer catalyst, there was a strong preference for oligomerization and hence chain transfer, over polymerization.

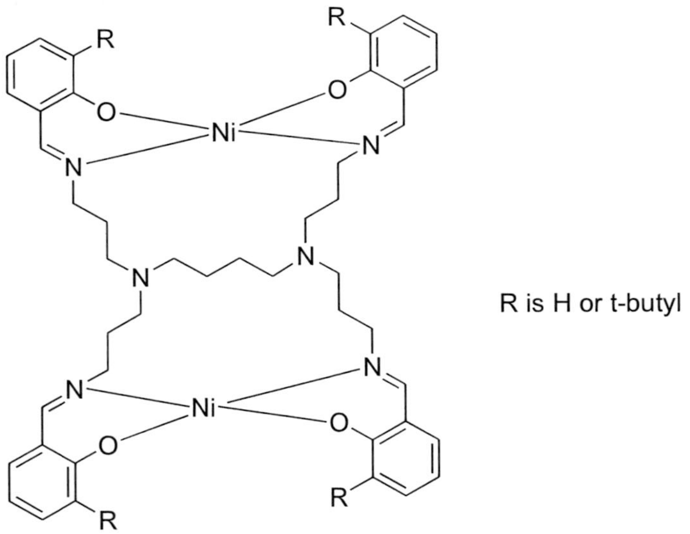

Figure 8. First generation dendrimeric Ni complex used in the polymerization of norbornene.

Nickel complexes of dendrimeric salicylaldimine showed promising results in catalyzing the vinyl addition polymerization of norbornene. [25] First and second generations of diaminobutane dendrimer carrying four and eight amino groups respectively were reacted with salicylaldehyde and the salicylaldimines thuse obtained were used as the macromolecular ligands in the synthesis of dendrimeric nickel complexes. The nickel complexes were able to catalyze the polymerization of norbornene selectively through the vinyl addition polymerization route. Catalyst prepared from higher generation dendrimer showed higher activity compared to that formed from lower generation dendrimer. It was also found that the substituents on the aldehyde part of the ligand also influenced the activity of the catalysts and the activity decreased with bulkier substituents. Another interesting factor regarding this catalyst is that irrespective of the catalyst used, the all polymers obtained showed similar molecular weight distribution. The general structure of the catalyst is shown in figure 8.

Cole-Hamilton and coworkers showed that Rh-phosphine-terminated dendrimers based on polyhedral silsesquioxane cores with 16 PPh_2 arms gave much higher linear selectivities (14: 1) than their small molecular analogues (3.4: 1) in the hydroformylation of cyclooct-1-ene. [26, 27] The scheme of the reaction and the structure of the catalyst used are given in figure 9.

The peptide dendrimers for ester hydrolysis described by Delort show a positive dendrimer effect. [28] Four generations of a dendrimer containing His-Ser in all branches were synthesized by the techniques of solid phase peptide synthesis and after cleavage from the support these dendrimers were screened for their catalytic activity in a ninety-six well plate set up. It was observed that the rate of hydrolysis of pyrene trisulfonate esters increased with increase in generation. The dendrimers displayed enzyme like Michaelis-Menten kinetics.

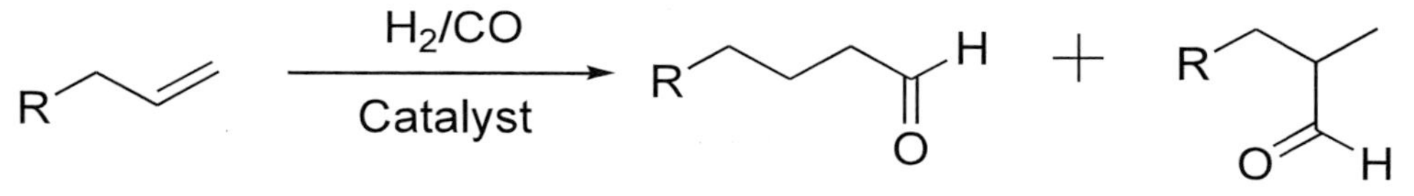

where catalyst is Rh complex of

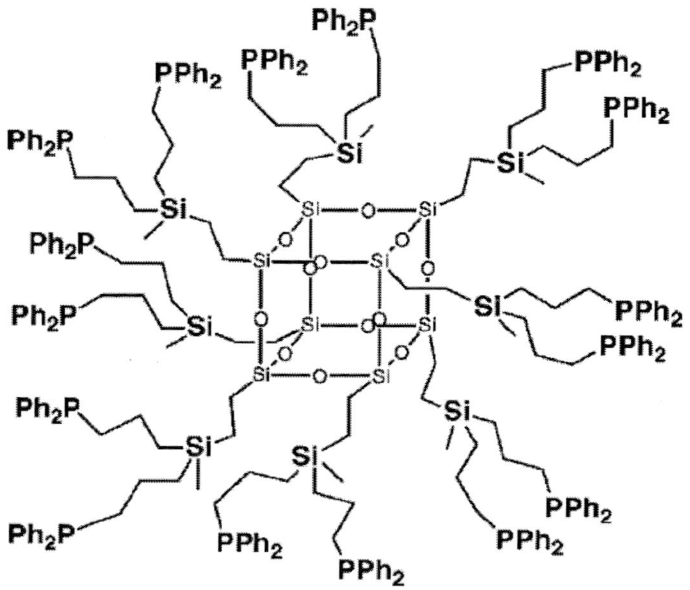

Figure 9. Hydroformylation catalyzed by Rh complex of phosphine-terminated dendrimers based on polyhedral silsesquioxane cores.

Two series of dendritic compounds functionalized at the periphery with polyoxometalate (POM) units were synthesized using two synthetic strategies by Nlate et al. [29] The first series involved the $[CpFe]^{+}$ induced functionalization of polymethylarenes, leading to dendrimers in which the dendronic tripod was directly bonded to the arene core. In the second procedure, dendrimers with a spacer group of six atoms between the arene core and the tripod units were synthesized through a coupling reaction between the phenol dendron and bromobenzyl derivatives. These polyallyl dendrimers were functionalized at the periphery to give quaternary poly-ammonium salts. Reactions of the latter with $H_3PW_{12}O_{40}$ in the presence of hydrogen peroxide led to dendrimers containing $\{PO_4[WO(O_2)_2]_4\}^{3-}$ species at the periphery. These compounds are efficient catalysts for the selective oxidation of alkenes, sulfides and alcohols in an aqueous–$CDCl_3$ biphasic system, using hydrogen peroxide as the primary oxidant. The structures of these dendrimers are given in figure 10.

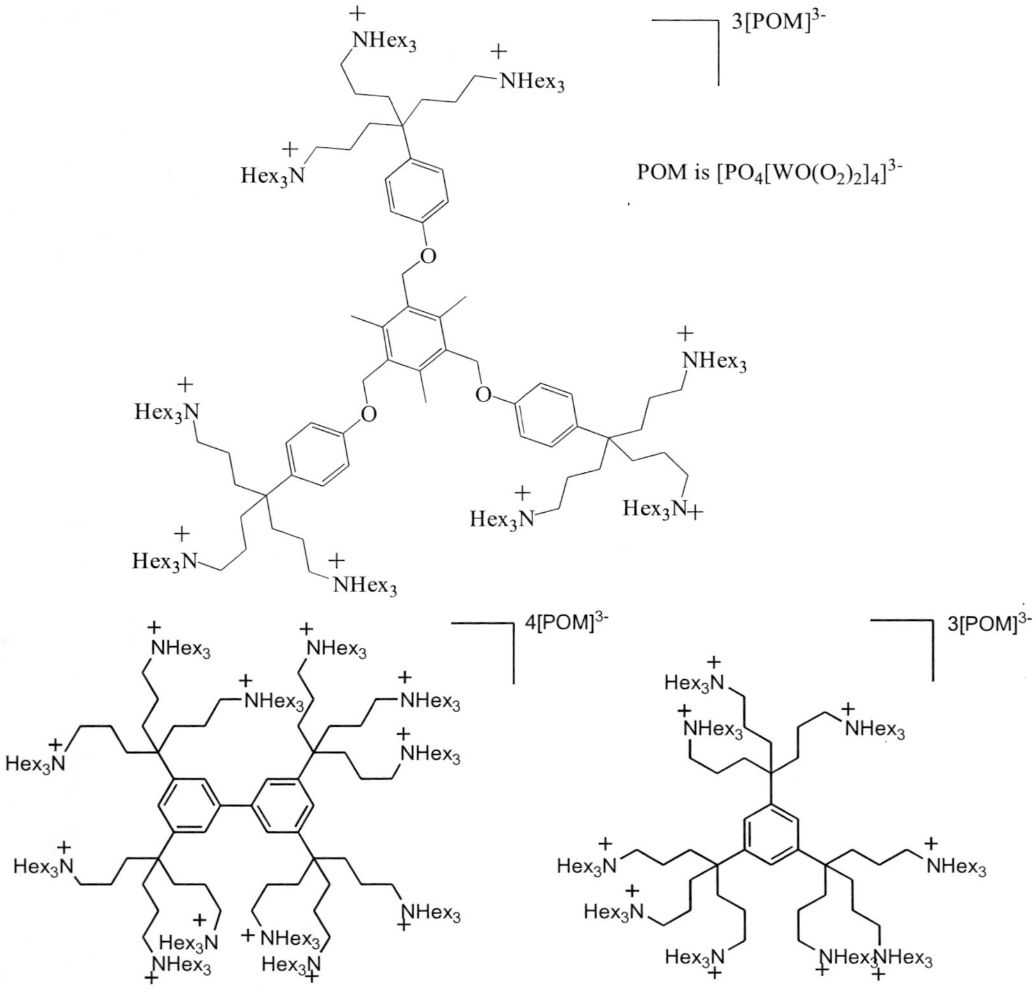

Figure 10. Dendrimers functionalized at the periphery with polyoxometalate (POM) units.

Rodriguez and coworkers explored the catalytic properties of a series of carbosilane dendrimer containing the P-stereogenic phosphine fragments [P(2- biphenylyl)$PhCH_2$] at the periphery in the hydrovinylation of styrene in $scCO_2$. [30] The reaction proceeded with good selectivity and enantiomeric excess. A negative dendrimer effect was observed and the monomeric species gave the better result followed by lower generation dendrimers.

Recently Niu et al. reported the preparation of few C3-symmetric 1,2,3-triazole linked dendrimers by Cu(I) catalyzed azide-alkyne click chemistry between arylether dendrimers carrying terminal alkynes on the periphery and ((2R,4S)-4-azidopyrrolidine-2-yl)diphenylmethanol (figure 11). [31] The dendrimer carrying chiral 1,2,3-triazole on the periphery were used as organocatalysts for the asymmetric borane reduction of prochiral ketones. The catalysts showed good activity and the products were isolated in good to excellent yields and good enantioselectivity.

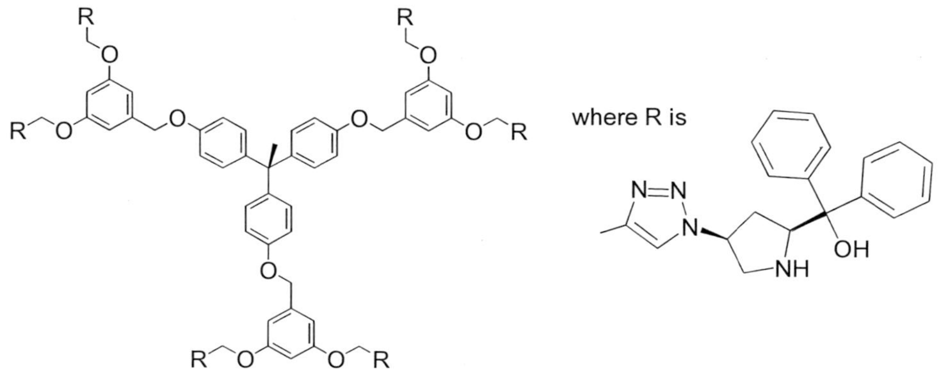

Figure 11. C3-symmetric 1,2,3-triazole linked dendrimers used as organocatalysts for the asymmetric borane reduction of prochiral ketones.

Simple dendrimers have also found application as macromolecular ligands in the synthesis of catalytically active metal complexes. Vassilev and co-workers used poly(propylene imine) dendrimers carrying eight [DAB-dendr-$(NH_2)_8$] and thirty two [DAB-dendr-$(NH_2)_{32}$] amino groups in the preparation of various metal complexes by interaction of the dendrimers with transition metal salts such as $FeCl_3.6H_2O$; $CoCl_2.6H_2O$; $CuCl_2.2H_2O$; $VOSO_4.5H_2O$; $Na_2MoO_4.2H_2O$ and $Na_2WO_4.2H_2O$ at room temperature in aqueous solutions. The content of metal ions in the complexes was found to be from 8.2 to 69.6 mg metal ion/g polymer carrier. It was found that the order of the catalytic activity of the complexes prepared using the higher generation dendrimer in the epoxidation of cyclohexene with organic hydroperoxides such as *tert*-butyl hydroperoxide (*t*-BHP), ethylbenzene hydroperoxide (EBHP) and cumene hydroperoxide (CHP) was as follows: D32-MoO_2^{2+}>D32-VO^{2+}>D32-WO_2^{2+} > D32-Co^{2+} > D32-Cu^{2+}>D32-Fe^{3+} where D32 is the dendrimer. The order of reactivity of organic hydroperoxides was found to follow the order *t*-BHP > EBHP > CHP. [32]

Similarly, bimetallic complexes of Pd and Cu and diaminobutane dendrimers carrying sixteen periphery nitrile groups were used as catalysts in the Wacker oxidation of terminal alkenes to methyl ketones. The peculiarity of this catalyst was that the yield of methyl ketones decreased with increase of generation of the dendrimer while selectivity for methyl ketones over isomeric ketones increased. Similar results were obtained when a dendrimer having diaminohexane core was used as ligand. These experiments showed that sterric crowding of catalytic sites in higher generation dendrimers introduced by the spherical structure of higher generation dendrimers has particular influence on catalytic activity and selectivity of dendrimer catalysts. Added advantages of the above mentioned catalysts are that they can be prepared insitu by simple mixing of dendrimer and metal salts and is reusable. [33]

Balieu etal. reacted first to fifth generations of poly(propyleneimine) dendrimers (PPI) with glycerol carbonate yielding a new family of glycerol-decorated PPIs. Owing to the presence of glyceryl units surrounding the PPI core, the high generation GD-PPI-5 can be successfully immobilized in a glycerol phase, thus offering a convenient route for possible utilization as a recyclable homogeneous catalyst. The authors showed that GD-PPI-5 can be used as a basic catalyst in the ring opening of epoxides with carboxylic acids in glycerol. The high affinity of GD-PPI-5 for the glycerol phase has allowed the selective extraction of the reaction products from the glycerol/GD-PPI-5 mixture and the GD-PPI-5 catalyst could be recycled. So this method offers a convenient method for recycling of dendrimer catalysts. [34]

4.2. Core or Focal Point Functionalized Dendrimers

In core or focal point functionalized dendrimers the catalytic species is placed either in the core of the dendrimer or in the focal point of the dendrimer. A schematic representation of these two conditions is given in figure 12.

In core or focal point functionalized dendrimers, the catalyst could benefit especially from the site isolation created by the dendritic environment. [4,5] For reactions that are deactivated by excess ligand or in cases in which bimetallic deactivation mechanism is operative, core-functionalized systems can specifically prevent such deactivation pathway, whereas periphery functionalized systems might suffer from relative low activity. Core functionalized dendrimers may benefit from the local catalyst environment created by the dendrimer. Effect of solvation of the substrate during the penetration of the dendrimer is another important factor. Another significant difference between core and periphery functionalized dendrimers is the molecular weight per catalytic site. Much higher costs will be involved in the application of core-functionalized systems and application can also be limited by the solubility of the system. On the other hand, solubility of the core-functionalized dendritic catalyst can be tuned by changing the end groups.

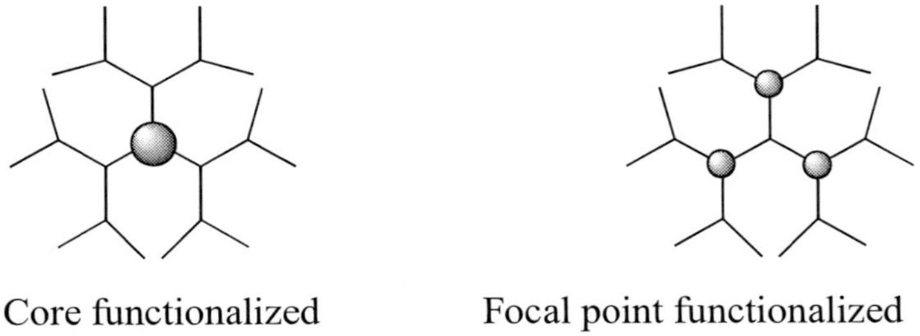

Figure 12. Schematic representation of core and focal point functionalized dendrimers.

The first example of a catalytic reaction at the core of the dendrimer was provided by Gitsov et al. using dendritic alcoholates as macro-initiatores for anionic ring opening polymerization of ε-caprolactone. [35] The structure of the dendrimer catalyst and reaction schemes are given in Figure 13 and 14.

The dendritic catalyst shown (figure 14 A) exhibited high catalytic activity compared to conventional alkali metal alcoholates used for the polymerization of ε-caprolactone. The G4 alcoholate acted as a highly effective catalyst producing high M_w polymers with a narrow distribution of 1.07. The catalysts showed a positive dendrimer effect and the best results were given by G4 dendrimer. G1 dendrimer gave only low molecular weight oligomers. The explanation for this behavior is that the large dendrimers prevent termination of the polymerization by shielding the growing tip from reaction with a chain of another growing tip.

Similarly, dendritic wedges were designed for use in the synthesis of living polymers and living block copolymers by controlled radical polymerization. [36] A nitroxyl radical was attached to the focal point of G1 to G3 aryl ether dendrimers (figure 14 B). In polymerization reaction, low polydispersities were obtained using the high generation dendrimers because of the irreversible release of the growing chains and slow recombination. Unfortunately, insolubility of the polymer –dendrimer complexes limited the growth of the chains.

Chow et al. reported the synthesis of a series of poly(alkyl aryl ether) dendrons (G0-G3) functionalized at the focal point with dendritic biz(oxazoline) ligands. [37, 38] Cu(II) complexes of these dendrimers catalyzed the Diels- Alder reaction between cyclopentadiene and N-2-butenoyl-2-oxazolidinone (figure 15). A detailed study revealed that the reaction followed enzyme like Michaelis-Menten kinetics. A reversible formation of the copper-dienophile complex is followed by the rate limiting conversion into the Diels-Alder adducts. The association constants of the catalyst-dienophile complex decreased slightly with the higher generation dendrimer. The G0-G2 systems showed almost similar activity, but the activity dropped dramatically in the case of the G3 dendrimer systems. Upon complexation of the dienophile at the focal point, it was observed that, the geometry of the metal center was changed. This resulted in an increase in steric repulsion between the dendritic wedges, which was more pronounced for the larger systems, and this resulted in the drop in activity.

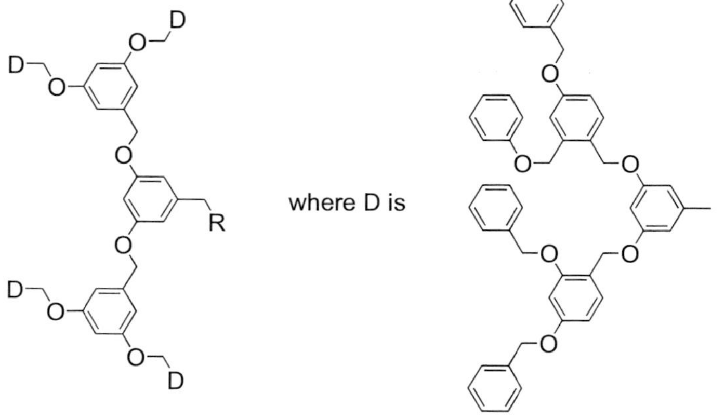

Figure 13. Dendritic alcoholate macro-initiators for anionic ring opening polymerization of ε-caprolactone.

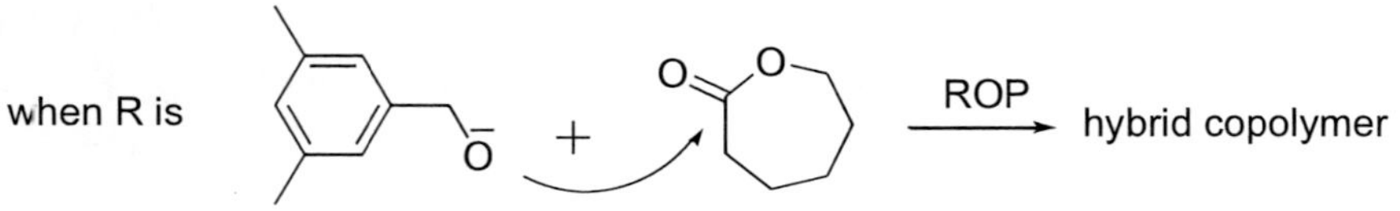

A

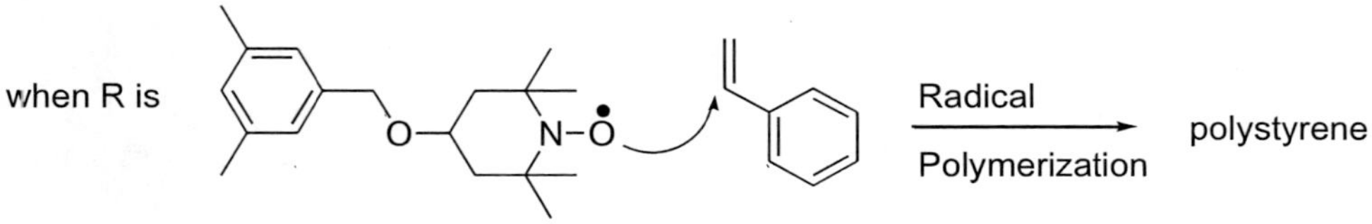

B

Figure 14. Ring opening polymerization of ε-caprolactone catalyzed by dendritic alcoholate macro-initiators

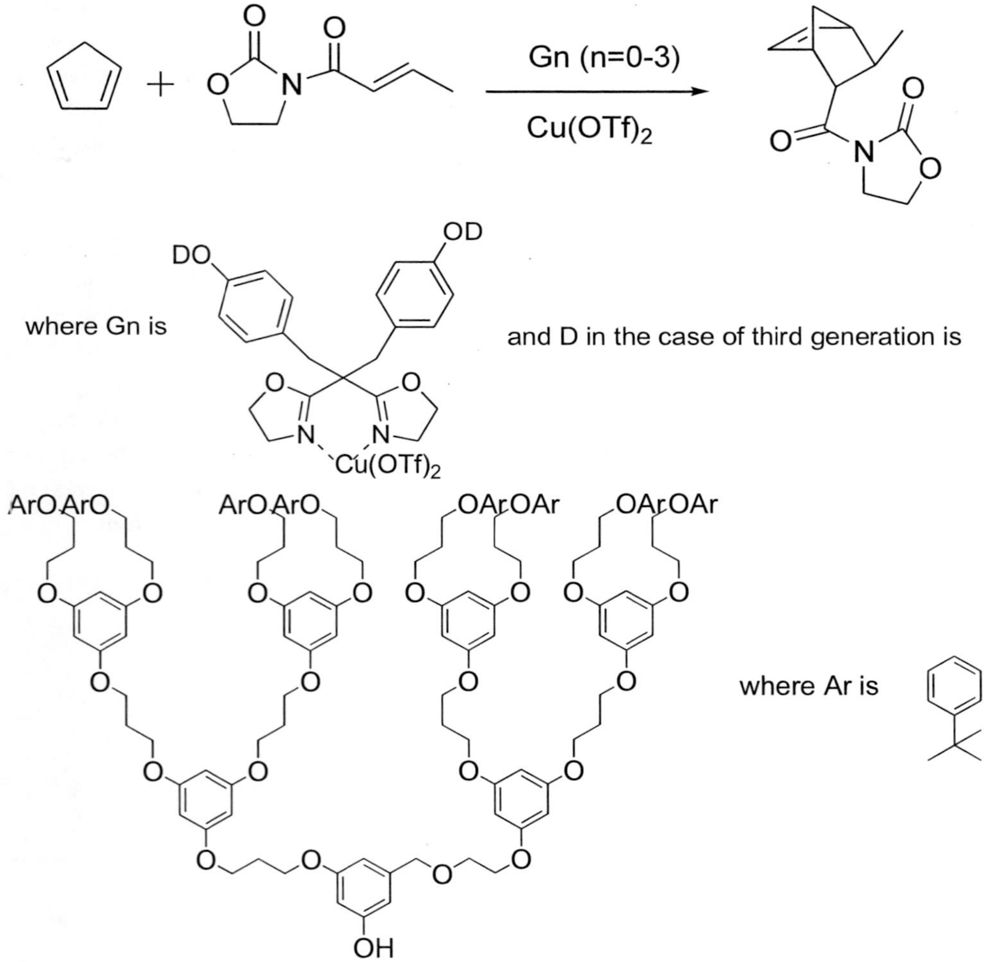

Figure 15. Diels- Alder reaction catalyzed by Cu(II) complexes of poly(alkyl aryl ether) dendron functionalized at the focal point with dendritic biz(oxazoline) ligands.

Introduction of dendritic wedges on metalloporphyrins resulted in core functionalized dendrimers in which the porphyrin unit was shielded from the bulk solution. Bhyrappa et al. used steric shielding of the porphyrin units to establish regio and shape selective catalysis. [39, 40] Eight bulky G1 and G2 poly(aryl ester) dendrons were connected to the porphyrin core to give the manganese(III) porphyrin complex (figure 16). These complexes were used as catalysts in epoxidation and it was observed that they showed greater preference for the least hindered double bond than less sterically hindered porphyrins do.

Another important core functionalized dendrimer catalyst to mention is the tertiary amine catalyst prepared by Zubia et al. [41] In their work, a single tertiary amine catalyst for the Henry reaction was encapsulated by different generations of aryl ether dendrimers (figure 17). It was shown that the catalytic ability of a single active site in this reaction decreased both with higher generation and with higher branching multiplicity.

Due to their strong bond with metals, N-heterocyclic carbene (NHC) ligands are very desirable and have been shown to be extremely versatile in their applicability to a variety of organic transformations. [42] Fujihara et al. have reported the use of dendritic NHC ligands in the Rh(I)-catalyzed hydrosilylation of acetophenone and cyclohexanone (figure 18). [43] A positive dendrimer effect resulting in an increase in the yield with increasing generation of the RhI complex was observed. This effect was attributed to the folding of the dendrimer around the active site that led to greater stability and hence greater overall turnover during the course of the reaction.

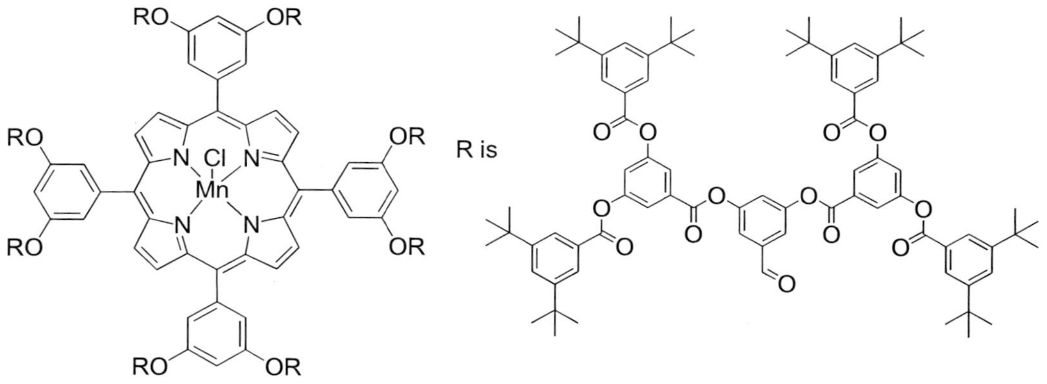

Figure 16. Dendritic metalloporphyrins.

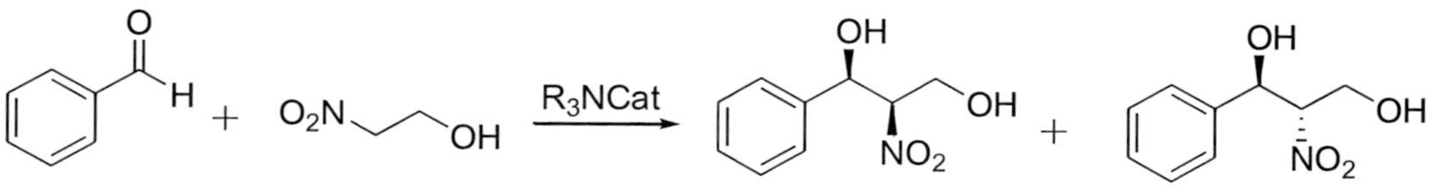

where catalysts are first(1), second(2) and third(3) generation aryl ether dendrimers

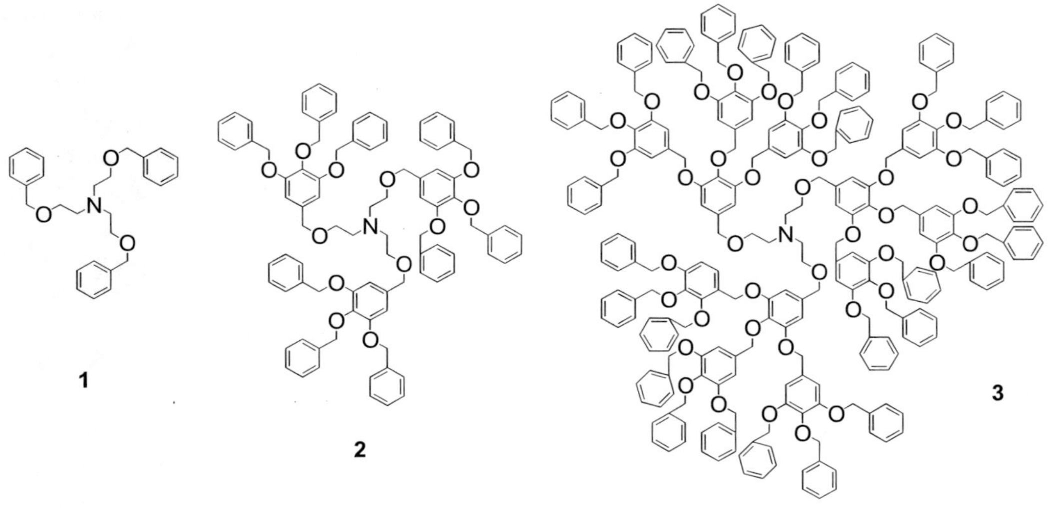

Figure 17. Henry reaction catalyzed by aryl ether dendrimers with tertiary amine core.

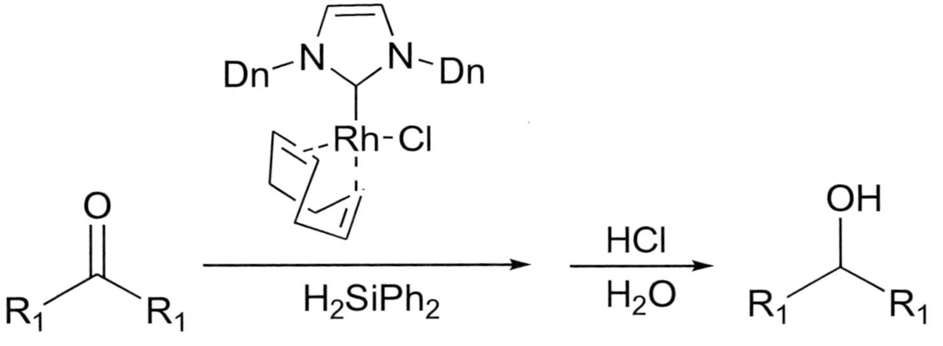

where Dn is aryl ether dendrimer

Figure 18. Hydrosilylation of ketones catalyzed by Rh complex of dendritic NHC ligand.

Nickel complexes bearing P,O ligands, such as o-phenylphosphinophenols, have been used widely in industry, for example, in the Shell higher olefin process. [44]

Reactions involving these nickel complexes are usually carried out in non-polar solvents such as toluene, although the use of more environmentally benign media has also been explored. However, in polar solvents like alcohols or water, there is a strong tendency for the formation of a bis-(P,O)nickel complex which is inactive. Muller and co-workers have constructed a second generation dendritic P, O ligand based on a carbosilane platform that was effective in preventing undesirable bis(P,O)nickel complexes in toluene (figure 19). [45] As a consequence of this site isolation, the dendritic ligand 2 (TOFavg=7700 h^{-1}) outperforms the parent ligand 1 (TOFavg=3600 h^{-1}) in ethylene oligomerization leading to higher yields of oligoethylene. The site isolation provided by the dendritic ligand 2 is mitigated in methanol as bis (P, O) nickel complexes are observed. However, the second P, O ligand appears to be labile at higher reaction temperatures in dendritic catalyst, which is not the case with the ligand and its bis (P, O) nickel complex of the monomric form. As a result, the nickel catalyst based on the dendritic ligand shows good activity (TOFavg=3242 h^{-1}) in methanol whereas the parent ligand itself does not produce any higher olefins. This is a classical example of site isolation provided by dendrimers in catalysis.

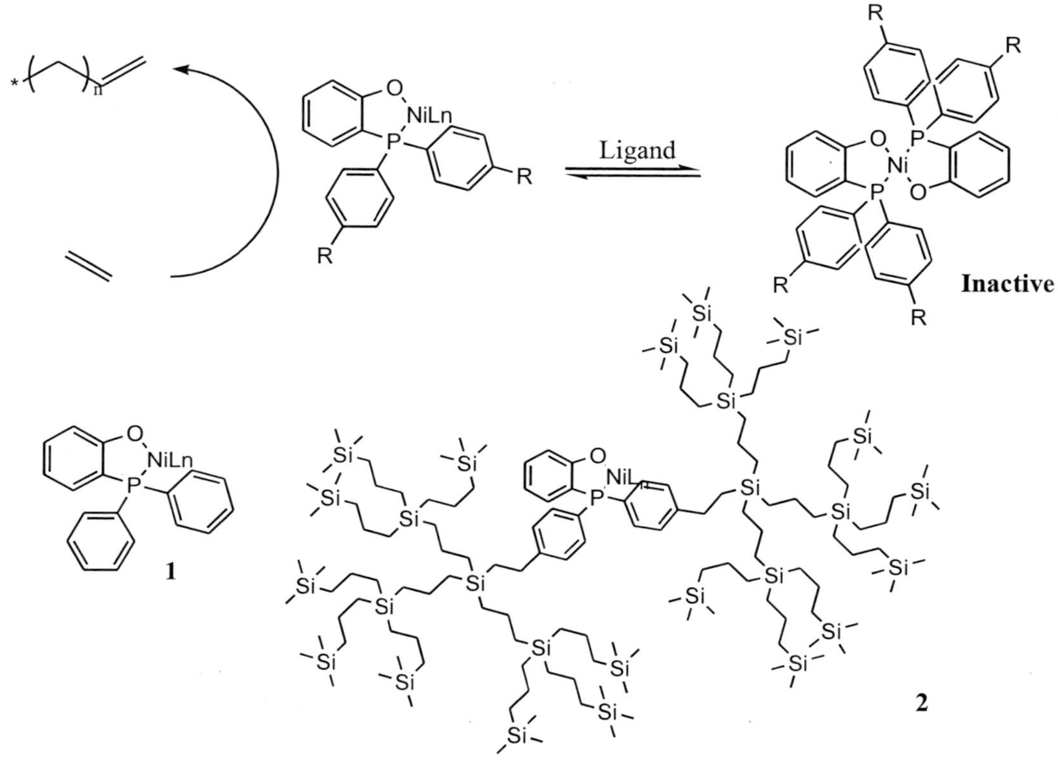

Figure 19. Shell higher olefin process catalyzed by Ni complex of dendritic ligand.

Apart from providing site isolation, the globular dendritic architecture around a catalytically active site also functions to bind guest molecules, another important factor for achieving efficient catalysis. This aspect of catalysis with dendrimers is reminiscent of molecular transformations achieved by enzymes in which substrate binding features prominently in determining the substrate selectivity and the extent of product inhibition. [46] The simple model describing the active site of an enzyme as a hydrophobic binding pocket for substrates held in close proximity to moieties that perform the transformation has often been applied, with mixed success, to hydrophobic, sometimes amphiphilic, dendrimers. Steric congestion resulting from the encapsulation of reactive groups within the dendrimer frequently leads to slower kinetics; however, there are notable exceptions: for instance, when substrate binding ability and nanoenvironment effects are paramount to achieving reactivity. In this respect, binding pockets with the desired characteristics will be found only in larger dendrimers. A nice example of this phenomenon was reported recently by Zhang et al. with a series of Frechet-type poly(benzyl ether) dendrimers (ref. figure 2) with a diselenide core that demonstrated generation-dependent glutathione peroxidase (GPx) activity (figure 20). [47]

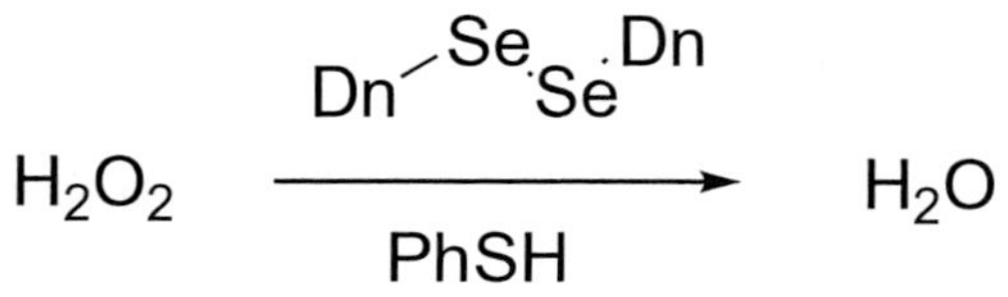

Figure 20. Reduction of hydrogen peroxide catalyzed by dendrimer with diselenium core.

The catalytic cycle involved the reduction of hydrogen peroxide by benzenethiol in the presence of dendrimer with diselenium core. Substrate binding was found to increase with dendrimer generation (Ka=16.4, 39.4 and 252.7M^{-1} for G1, G2 and G3, respectively) and the third generation catalyst performed much better than smaller molecules that catalyzed the same reaction.

Higher activities for the dendrimers were achieved upon the introduction of small amounts of water into the reaction medium, presumably as a result of stronger binding of the substrate to the dendrimer due to the hydrophobic effect. This study points to the importance of nanoenvironment effects in dendrimer catalysts where substrate binding requires larger dendrimers that are able to provide the more hydrophobic environment best suited for the reaction.

The dielectric nature of the nanoenvironment generated by the dendritic polymers also has a great role in catalysis. Piotti and co-workers reported an early example on the subject of nanoenvironment effects in catalysis with dendrimers. In this work, third and fourth generation unimolecular dendritic reverse micelles were prepared with a radial polarity gradient consisting of a nonpolar corona resembling the reaction solvent and more polar ester or alcohol functionalities at predetermined locations throughout the interior (figure 21). [48] These dendrimers were specifically designed to catalyze reactions in which a nascent positive charge was developed during the transition state: for example, the simple unimolecular dehydrohalogenation of 2-iodo-2-methylheptane, which proceeded via an E1 elimination mechanism. Dendrimers with polyol interiors outperformed those with polyester cores. In addition, yields were significantly improved using the larger G4 dendrimer. One of the most striking features of this catalyst system was its ability to achieve high turnover even at very low catalyst loadings. The authors attributed this result to a kinetic but thermodynamically driven "concentrator effect" whereby the polarity gradient inherent to the dendrimer had driven the relatively polar substrates into the macromolecule's core leading to its accumulation in a constrained nanoenvironment that was more suitable for the E1 elimination than the external reaction medium, hexane. Higher generation dendrimers showed better efficacy for the reaction due to a larger internal reaction volume that could accommodate a higher concentration of substrate molecules. In addition, the non-polar alkene product had a greater affinity for the hydrophobic surrounding medium thereby achieving the necessary turnover. The authors have described this latter phenomenon as a free-energy driven "catalytic pump."

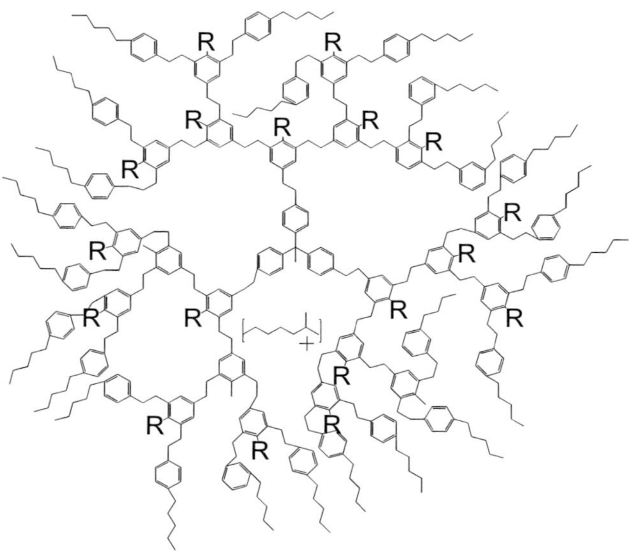

Figure 21. Dendrimer catalyst for E1 elimination reaction.

The concepts, of a "concentrator effect" to accumulate substrates near a catalytically active site and a "catalytic pump" to prevent product inhibition, has been broadly applied to other dendrimer catalysts with amphiphilic designs. Hecht and Frechet reported the first results on light-driven catalysis at the core of the dendrimers via photosensitization. [49] In their design, a $^1O_2$-sensitizing benzophenone core was incorporated into globular dendrimers having a radial polarity gradient consisting of a hydrophobic interior and a hydrophilic surface exposed to the polar solvent. The resulting nanoscale photoreactors were created and applied to the oxidation of cyclopentadiene (CP) with dioxygen. The non-polar core was designed both to favor substrate accumulation of the CP substrate and to increase the lifetime of the photogenerated $^1O_2$. The rate of the reaction and the stability of the catalyst increased with increase in generation of the catalyst. The scheme of the reaction and the structure of the third generation catalyst are given in figures 22 and 23.

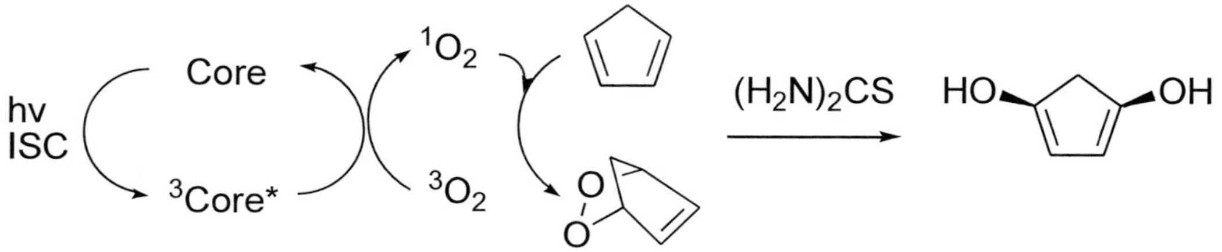

Figure 22. Oxidation of cyclopentadiene.

In order to study the effect of steric shielding on the kinetics of dendrimer catalysts, Mizugaki functionalized the periphery of a third generation PPI dendrimer with either C10 or C16 acyl chains. [50] Their interiors were quaternized with iodomethane to afford lipophilic tetraalkylammonium iodide dendrimers. These dendrimers were used as Lewis base catalysts (via iodide ions) for the Mukaiyama aldol reaction of 1-methoxy-2-methyl-1-

(trimethylsilyloxy)propene with various aldehydes in toluene. This study showed that dendritic iodides were remarkably more effective than other "small molecule" sources of iodide, such as tetrabutyl- or tetrahexylammonium salts. The authors argued that the higher activity of the dendrimers over the discrete small molecule iodides stemmed from the high polarity of the nanoenvironment encapsulating the iodide in the macromolecular catalysts. The reactions were promoted to a greater extent in polar solvents such as DMF as opposed to toluene. Thus, the concentration of multiple cationic charges within the nanoscopic confines of the dendritic interior appears uniquely capable of stabilizing the reactive anionic intermediate. In addition, a study of the steric effects showed that dendrimer with a more crowded periphery (i.e., with C16 chains) did not perform well as with C10 peripheral chains. This work clearly delineated the opportunistic design features of a dendritic catalyst concentrating reactive moieties within a small reaction volume at the core of the dendrimer and using the dielectric constant of this discrete nanoenvironment to enhance reactivity, while stressing that sufficient access to that environment is also important. The control of reaction selectivity through dielectric effects exerted by the dendrimer nanoenvironment has also been observed by Ooe and co-workers in Pd catalyzed Heck reactions and allylic aminations. [51]

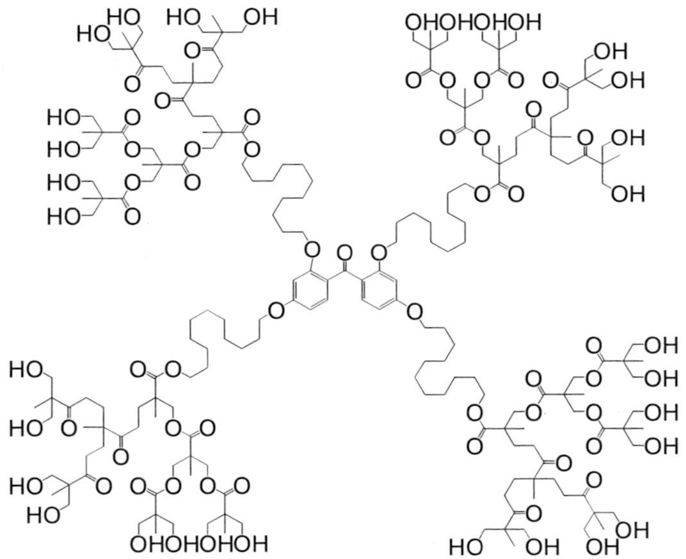

Figure 23. Dendritic photoreactor for the oxidation of cyclopentadiene with dioxygen.

Another salient feature of dendrimers bearing one or more catalytically active sites at their core are their ability to achieve substrate selectivity through steric restriction. In the literature, this has often been referred to as "shape selectivity." The pioneering work with shape selective catalysts was performed by Bhyrappa et al. with dendritic manganese porphyrins. [39,40]

The kinetic behavior of various guest molecules with different steric constraints was investigated using distance- dependent excited state quenching through photoinduced electron transfer. [52] Anthracene-cored dendrimers were prepared from Frechet-type dendrons up to the fourth generation. Core accessibility was probed using trialkylamines, which are known to quench the photoexcited state of the anthracene chromophore through electron transfer.

Various trialkylamines of different sizes and rotational degrees of freedom were used as quenchers. Each of these amine-based quenchers gave smaller Stern–Volmer bimolecular quenching rate constants, k_q, with increasing dendrimer generation. This is consistent with the greater site isolation achieved with the increasing larger dendritic shells at higher generations. In addition, geometrically constrained molecules were found to have better access to the dendrimer cores than their more flexible counterparts.

Core-modified dendrimers have also been used as biomimics. Liu and Breslow prepared a series of PAMAM dendrimers with pyridoxamine residues at their core. [53] The pyridoxamine PAMAM dendrimers, G1 to G6, were used as catalysts in the transamination of pyruvic acid and phenylpyruvic acid in aqueous buffer. The pyridoxamine PAMAM dendrimers displayed Michaelis–Menten kinetics and superior efficacy when compared to the simple pyridoxamine. Interesting positive dendrimer effects were drawn from the observed Michaelis–Menten parameters i.e., the second order rate constant k_2, the binding constant K_M, and the ratio k_2/K_M – as they changed with increasing generation. The larger dendrimers were much more efficient at facilitating general acid-base reactions along the catalytic pathway leading to a net increase of k_2.

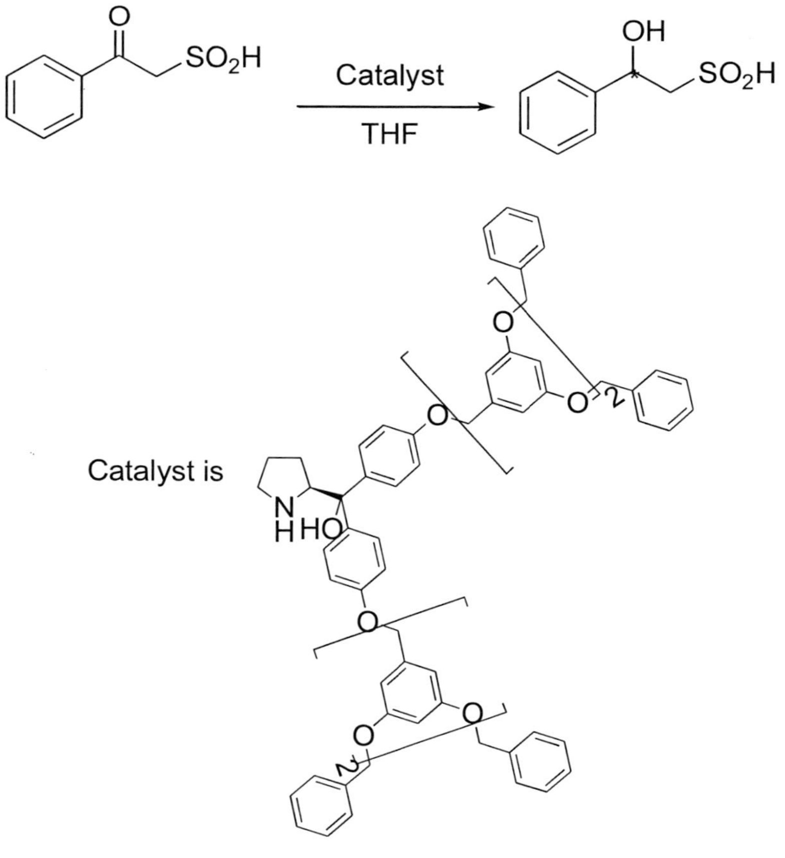

Figure 24. Reduction of ketones to secondary alcohols with borane catalyzed by aryl ether dendrimer with proline core.

Aryl ether dendrimers containing a proline core was used as a chiral catalyst for the reduction of ketones to secondary alcohols with borane (figure 24). Five different catalysts were prepared and screened for their catalytic properties. Alcohols were obtained in excellent yield with ee ranging from 49 to 98%. One of the catalysts was reused five times with out any loss of catalytic and stereochemical activity. [54]

The same catalyst with suitable modification on the proline core was used for asymmetric Michael addition also and in the enantioselective synthesis of (+)-sertraline. [55,56] The direct aldol reactions catalyzed by chiral dendritic catalysts derived from N-prolylsulfonamide and the same dendritic backbone mentioned above gave the corresponding products in high isolated yields (up to 99%) with excellent anti diastereoselectivities (up to >99:1) and enantioselectivities (up to >99% ee) in water. In addition, the catalyst may be recovered by precipitation and filtration and reused for at least five times without loss of activity. [57]

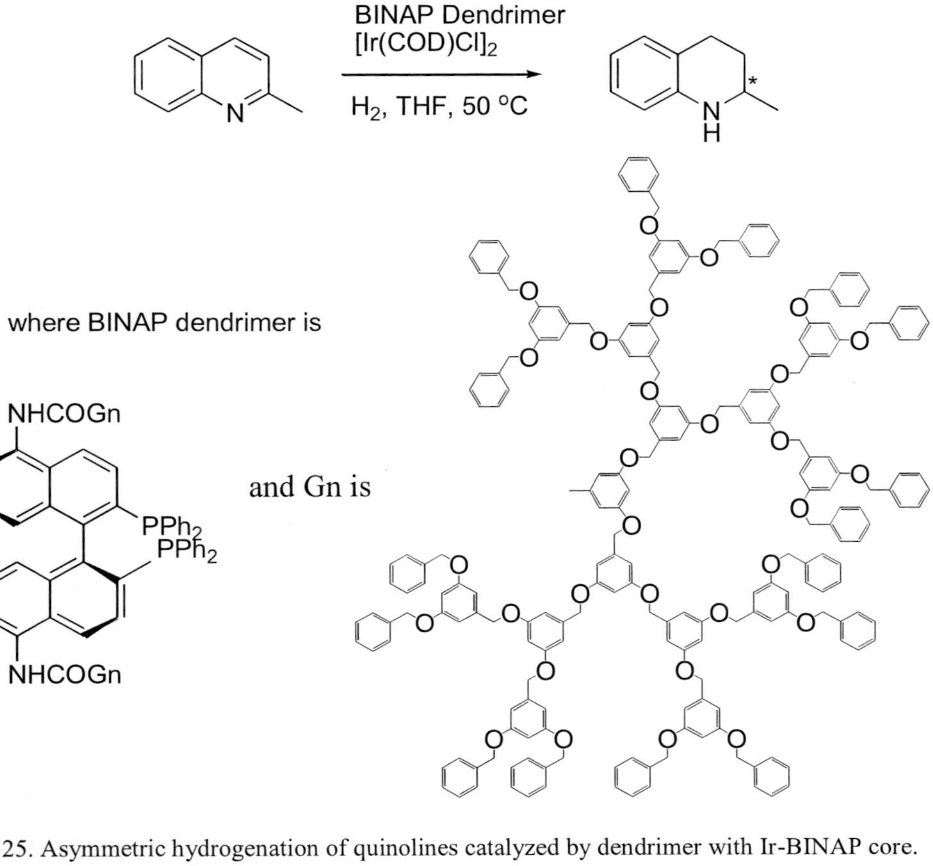

Figure 25. Asymmetric hydrogenation of quinolines catalyzed by dendrimer with Ir-BINAP core.

Muraki *et al.* reported a novel dendritic catalyst for three-component Mannich reaction in water. [58] The dendrimer contained a 2,2'-bipyridine core around which an aryl ether type dendrimer was constructed. The catalyst was prepared by treatment of $Cu(OTf)_2$ and the dendrimer in dichloromethane. The catalyst showed excellent activity compared to the $Cu(OTf)_2$ alone in the three component mannich reaction between aldehydes, amines and silyl enolates in water. A positive dendrimer effect was observed and the third generation catalyst gave better results. The yield of the product was low in all the cases because of the decomposition of the silyl enols in water.

Recently Wang and co-workers reported a new dendritic catalyst for the asymmetric hydrogenation of quinolines (Figure 25). [59]

The chiral dendrimer ligands were synthesized by condensation of the Frechet type aryl ether dendritic wedges carrying a COOH group at their focal point with (*S*)-5,5'-diamino BINAP in the presence of triphenylphosphite, pyridine, and calcium chloride in *N*-methyl-2-pyrrolidone (NMP). The catalyst generated in situ from the dendrimer and $[Ir(COD)Cl]_2$ was used in the hydrogenation of the quinolines in THF. More than 75% of conversion was obtained with all the substrates used and the ee was also high. A positive dendrimer effect was observed and the third generation dendrimer gave better results.

A series of polyether dendritic chiral phosphine Lewis bases were synthesized by Liu and Shi (figure 26). [60]

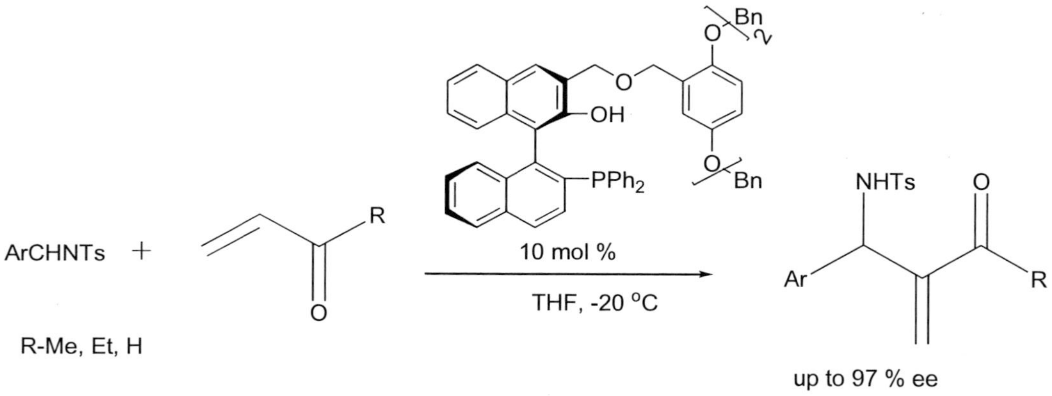

Figure 26. Aza-Morita-Baylis-Hillman reaction catalyzed by dendritic chiral phosphine Lewis bases.

These dendrimers carrying the chiral group at the core were successfully applied as reusable organocatalysts in the asymmetric aza-Morita-Baylis-Hillman reaction of *N*-sulfonated imines (*N*-arylmethylidene-4-methylbenzene sulfonamides) with methyl vinyl ketone, ethyl vinyl ketone, and acrolein to give the adducts in good to excellent yields along with up to 97 % ee. The enantioselectivity was higher for the dendritic catalyst compared to the original chiral phosphine Lewis bases.

Fujita and co-workers reported the synthesis of an aryl ether type dendrimer with an osmium tetroxide core. Initially a bis(ammonium bromide) dendrimer with aryl ether branches were synthesized followed by attaching OsO_4 to the core by an ion exchange reaction in which the bromide dendrimers were allowed to react with K_2OsO_4. Three generations of the dendrimer catalyst were synthesized. These catalysts were found to be active in cis-dihydroxylation of olefins. The catalyst can be precipitated after the reactions and can be reused upto five cycles. The catalytic activity remained almost same even after the fifth cycle. [61] The structure of the third generation catalyst is shown in figure 27.

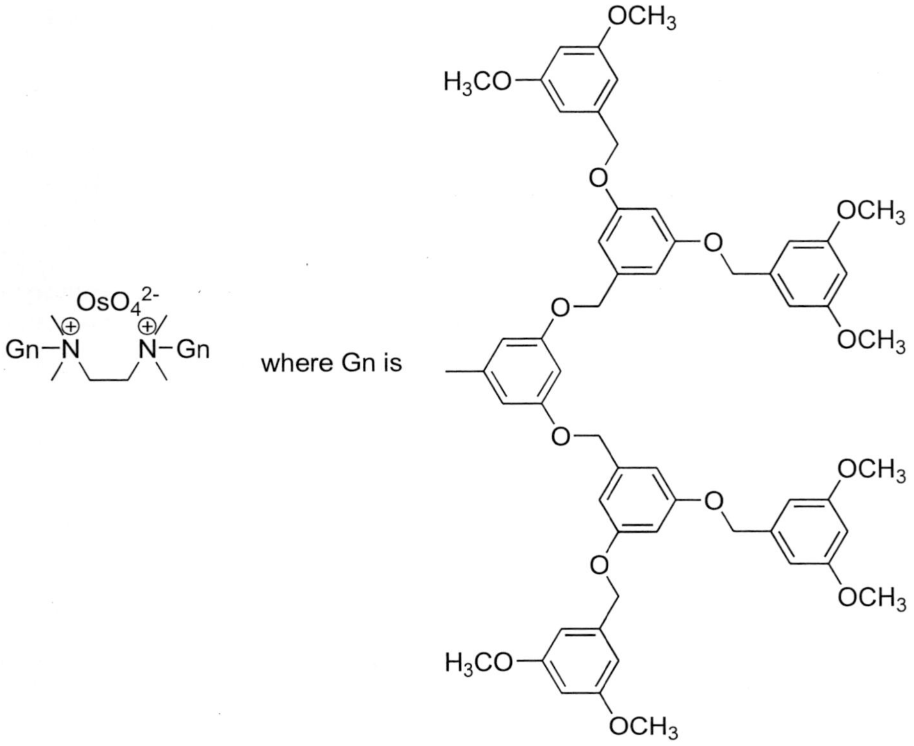

Figure 27. Aryl ether dendrimer with OsO_4 used for cis-dihydroxylation of olefins.

In short, both periphery and focal point functionalized dendrimers have found number of applications in homogeneous catalysis. The fine tuning of catalytic properties as well as selectivity obtained by placing the catalyst on suitable points of dendrimers aroused curiosities of both catalyst researchers and organic chemists.

5. Dendrimer Nanoparticle Composites in Catalysis

The current decade has witnessed the rise of nanotechnology and carrying out nanotechnology from laboratories to industry to some extent. The coming years are considered as the golden age of nanotechnology that will change the current world in a dramatic way in all fields from computers to medicine. [62] Catalysis is not an exception. Catalysis is the primary area that has used nanotechnology for many years and the classic examples are the transition metal colloids used for catalysis. [63] Recent developments in instrumentation methods help us to visualize and manipulate the nanoparticles and this helps to develop novel catalysts from nanoparticles.

There are two approaches in the application of nanoparticles in catalysis. In the first, nanoparticles themselves are used as catalysts. Nanoparticles show high catalytic activity compared to bulk materials of the same composition because of two reasons. As the bulk material is reduced to particles of nanodimension, the surface area get increased appreciably and as the size get smaller and smaller, the percentage of atoms present on the surface of the particle gets increased. It was estimated that for a 2 nm particle, more than 90% atoms reside on the surface. Since the basics of catalysis is interaction between the reactants and the surface atoms of the catalysts, the catalytic efficiency increases with more number of surface atoms, ie, with smaller particle size of the catalyst. Another factor which governs the catalytic activity of nanoparticles is the vacant sites present on surface atoms. When an atom is situated in the interior of a bulk material, all of its coordination sites are occupied. On the other hand, for the atoms situated on the surface, there are few vacant coordination sites for interaction with the approaching reactant species. If the atom is on an edge of a particle, there are more vacant coordination sites and the number increases again if it is on the corner. So the atoms staying at the corner of a nanoparticle is more catalytically active than a particle on the edge which in turn is more active than the particle on the surface. These facts help us to tune the catalytic activity of nanoparticles by controlling its size and shape. [63-67]

In a second approach, nanoparticles have no role in catalysis; but they merely act as carriers of catalysts. [68, 69] Nanoparticles stabilized with suitable ligands are able to carry a catalyst, most probably a transition metal complex, attached to the ligand. The ligands may have two different functional centers in which one has special affinity towards the nanoparticles. These groups get attached to the nanoparticle while the rest of the ligand gets complexed with the transition metal ion that acts as the catalyst. Generally, this kind of approach is used in the case of gold nanoparticles, which have special affinity towards thiol groups. Another method of using nanoparticles as carriers in catalysis is preparation of magnetic nanoparticle core polymer shell system. The polymer shell is functionalized with suitable functional groups followed by attachment of ligands. These ligands are complexed with transition metals and used in catalysis. The advantage of these magnetic systems is their easier separation by the application of an external magnetic field. But generally, these approaches are not used widely because of the difficulty and high cost in synthesizing nanoparticles that have no particular role in catalysis.

As described above, in the first approach, the nanoparticles directly take part in catalysis. So it becomes necessary to prepare small sized and monodisperse nanoparticle, which show good stability and activity with ease of separation. One common technique used for this is generating the nanoparticles on supports such as alumina, silica or carbon by impregnating these substrates with solutions of the corresponding metal salts followed by reduction of the salt into zero valent nanoparticle by an appropriate method. Even though these nanoparticles are robust and remain unaltered even after many reaction cycles, they are usually non homogeneous and show a wide distribution of particle size. Many particles are of very big size so that they may be little or not at all active in catalysis. Moreover, only a portion of these nanoparticles gets exposed to the reactants. Due to these reasons, only a portion of these metal particles contribute to catalysis. So it becomes necessary to synthesize nanoparticles with narrow range of particle size distribution that can be easily accessible to the substrates and easily separated from the reaction mixture. Many techniques are used for this purpose. In one approach, metal nanoparticles stabilized with organic ligands are used as catalysts in which the ligands worked as gates that control the rate of catalysis and product selectivity. Under strictly controlled condition the particle size distribution of nanoparticles can be restricted to a narrow distribution and attain excellent and selective catalysis. [70-72]

In the second method, nanoparticles are synthesized in a polymer matrix by reduction of metal complexes or metal salts.These nanoparticles trapped in the polymer matrix usually show a narrow distribution of particle size and good catalytic activity. [73-75]

Dendrimers are considered as organic nanoparticles and are expected to be the quantized building blocks for nanoscale synthetic organic chemistry. [76] They are used extensively in nanochemistry because of their particular shape, nanodimensional size and very narrow size and molecular weight distribution, which are generally not attained, by polymers or inorganic nanomaterials. [77] Dendrimers have internal voids and small molecules or clusters of suitable size can fit into these voids. [78-80] Crook and Tomalia demonstrated that metal nanoparticles could be synthesized in the internal voids of dendrimers. [81,82] Dendrimers are particularly attractive hosts for catalytically active metal nanoparticles for the following five reasons: (1) bearing fairly uniform composition and structure, the dendrimer templates themselves yield well-defined nanoparticle replica; (2) the nanoparticles are stabilized by encapsulation within the dendrimer, and therefore they do not agglomerate during catalytic reactions; (3) the nanoparticles are retained within the dendrimer primarily by steric effects and therefore a substantial fraction of their surface is unpassivated and available to participate in catalytic reactions; (4) the dendrimer branches can be used as selective gates to control the access of small molecules (substrates) to the encapsulated (catalytic) nanoparticles; (5) the dendrimer periphery can be tailored to control solubility of the hybrid nanocomposite and used as a handle to facilitate linking to surfaces and other polymers. [83] The synthesis of Dendrimer Encapsulated Nanoparticles (DEN) includes two simple steps. Initially, dendrimers of suitable generation with appropriate surface functional groups are allowed to complex with metal ions. The reduction of these dendrimer metal complexes, with a suitable reducing agent gives small nanoparticles of suitable dimensions that get trapped into the internal voids of dendrimers. Controlling the ratio of metal ions and dendrimer functional groups can regulate the size of these nanoparticles. [84] Moreover, the monomettalic, bimettalic, alloy type and core-shell nanoparticles can be synthesized by controlling various reaction parameters.[83,85-89] Figure 28 represents the general method of preparation of dendrimer nanoparticle conjugates. The red spots represent the corresponding metal ion and the dark mass represents the nanoparticle.

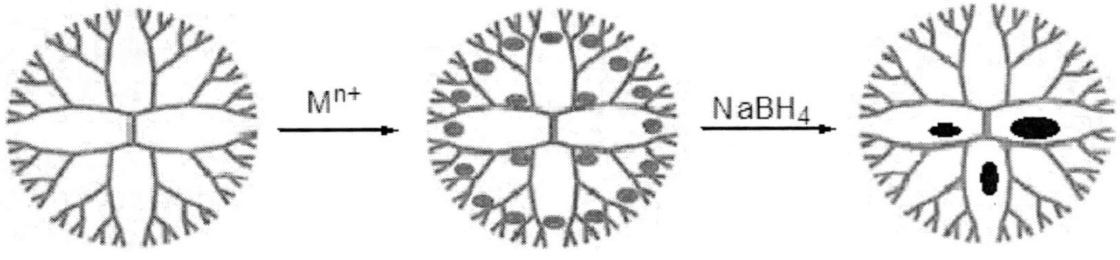

Figure 28. General method of preparation of dendrimer nanoparticle conjugates.

Any dendrimer-metal nanoparticle conjugate can be represented by the general structure given above and their method of preparation also may be the same. Only differences among the various such catalysts reported are the dendrimer and metal used for their preparation.

This dendrimer-metal nanoparticle conjugates showed appealing catalytic properties. One of the earliest applications of dendrimer-nanoparticle conjugates in catalysis was in the hydrogenation of olefins. Crooks and coworkers prepared nearly monodisperse palladium nanoparticles by using hydroxyl terminated PAMAM dendrimers as ligands. Three generations of dendrimers (4, 6 and 8) were used for this purpose. These dendrimer encapsulated nanoparticles were used as catalysts in the hydrogenation of allyl alcohol and α-substituted allyl alcohols. The results showed that catalysts prepared using higher generation dendrimers showed selectivity towards less substituted allyl alcohols. Catalysts based on lower generation dendrimers gave almost similar results in reduction of all olefins while those based on higher generation dendrimers preferably reduced substrates with minimum sterric factors. [83] The same group reported an efficient method for recycling the dendrimer-nanoparticle catalysts. They found that complexation of Pd/dendrimer composites with perfluorinated carboxylic acids renders the resulting nanocomposites preferentially soluble in fluorinated hydrocarbons. These new catalysts showed high activity and selectivity for biphasic hydrogenation of alkenes and conjugated dienes. On extraction of the reaction medium using some organic solvent leaves the catalysts in the fluorous phase and it can be recovered and reused for multiple reactions. [84]

Similar results were obtained by Kaneda *et.al.* Dendrimer-encapsulated Pd(0) nanoparticles inside poly(propylene imine) (PPI) dendrimers functionalized with triethoxybenzamide groups have been prepared by extraction of Pd^{2+} and subsequent chemical reduction. The nanoparticles were of 2-3 nm in diameter. The resulting dendrimer-Pd nanocomposites are unique catalysts for substrate-specific hydrogenation of polar olefins. The selectivity towards polar olefins was assumed to be due to the strong interaction between polar substrates and tertiary amino groups within the dendrimers. The dendrimers act not only as templates of Pd nanoparticles, but also as nanoreactors capable of molecular recognition.[85]

Dendrimer nanoparticle conjugates have found applications as catalysts in carbon-carbon bond forming reactions also. Christensen *et.al.* reported the catalytic activity of palladium nanoparticle conjugates prepared using fourth generation PAMAM dendrimer having surface hydroxyl groups in Suzuki coupling between aryl halides and benzene boronic acids. The catalyst followed the general trends in Suzuki coupling and higher yields were obtained for aryl iodides compared to other halides. With some substrates, a yield as high as 99% was reported.[86] Similar nanoparticle conjugates prepared from fourth generation hydroxyl terminated PAMAM dendrimers and palladium salts showed high activity in Stille coupling also. The advantages of this catalyst are that the reaction can be performed at room temperature and the amount of Pd required for efficient catalysis was as low as 0.1 atom%. [87]

Dendrimers were used for the immobilization of nanoparticles on various supports. The advantages of this method are two. First, due to the peculiar structure of dendrimers, it is possible to produce monodisperse nanoparticles with very small size. Second, the dendrimers can be removed from the system by some methods and so the nanoparticles obtained in such a way can be attached to some insoluble supports. A general diagram of dendrimer nanoparticle conjugates attached to some supports is shown in figure 29. Krishnan and Sreekumar used polystyrene supported PAMAM dendrimers for the immobilization of catalytically active Pd nanoparticles. The catalyst was active in Suzuki coupling between aryl halides and aryl boronic acids. As stated previously, this catalyst also followed the same general trends in Suzuki coupling. The catalyst can be recovered by simple filtration and was reusable up to six times without considerable loss of activity. [88]

In another approach, Murugan and Rangasamy reported the synthesis of polymer-supported poly(propyleneimine)-G2 dendrimer stabilized gold nanoparticle catalysts using crosslinked poly(4-vinylpyridine) matrix as support material. The resultant polymer-supported dendrimer stabilized Au nanoparticles were used as a heterogeneous catalyst for the reduction of 4-nitrophenol. The catalytic activity was found to be excellent and it could also be reused many times by simple filtration and activity remained unchanged. [89]

Caos *et.al.* synthesized PAMAM dendrimers on amine functionalized SBA-15 using standard solid phase synthesis strategy. A series of Pt nanoparticles (NPs) smaller than 3 nm were successfully encapsulated in these dendrimer/SBA-15 organic and inorganic hybrids. Charatcterization using TEM showed that the Pt nanoparticles with narrow size distribution are monodispersed in SBA-15 channels. Catalytic property of the supported Pt catalysts was investigated in both inorganic (ferricyanide to ferrocyanide by thiosulfate) and organic (*p*-nitrophenol to *p*-aminophenol by sodium borohydride) electron transfer reactions. In both cases, the reduction reactions followed smoothly and the catalysts showed excellent catalytic activity. Moreover, the catalysts can be easily separated and reused several times preserving good catalytic performance. [90]

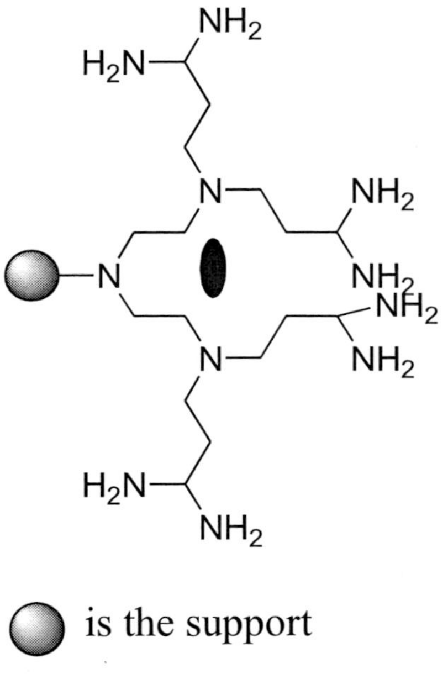

Figure 29. Dendrimer nanoparticle conjugates attached to a support.

In yet another approach, dendrimers were used as templates for growing nanoparticles which eventually incorporated into some supports followed by removal of the dendrimer part. This resulted in the synthesis of catalytically active nanoparticles attached to the supports. Here the role of dendrimer was only to assist the synthesis of nanoparticles and have no role in catalysis. It is really confusing to call such systems as dendrimer based catalysts. But they are also included here because dendrimers have some role in the preparation of such catalysts. Some important examples of such catalyst are given below.

Zaera et.al. used fourth generation PAMAM dendrimers as templates for the preparation of well-defined platinum-based catalysts. Nanoparticles with an average of 40 Pt atoms and an average diameter of 1.5 nm were produced this way, and dispersed on a sol–gel silica support. Different treatments were tested for the removal of dendrimer part and activation of the resulting catalyst, including heating in vacuum or in oxygen or hydrogen atmospheres. They found that heating in hydrogen led to fairly active catalysts with high selectivity for the conversion of trans olefins to their cis counter parts. The other heating methods did not produce the active catalyst. In this method, dendrimer played the role of templates and was eventually removed to obtain the active catalyst. [91] Another example of using dendrimers as templates was reported by Chandler and co-workers and they used PAMAM dendrimers as template of Pt, Au, and bimetallic Pt–Au dendrimer encapsulated nanoparticles (DENs) in solution. Adjusting the solution pH allowed for slow, spontaneous adsorption of the nanoparticles onto silica, alumina, and titania. After dendrimer removal, the catalysts were characterized with infrared spectroscopy of adsorbed CO and tested with CO oxidation catalysis. The bimetallic catalysts were found to be more active than the monometallic catalysts and had lower apparent activation energies. The titania supported Pt–Au catalyst was resistant to deactivation during an extended treatment at 300 °C. [92]

More details of synthesis, characterization and applications (not only catalysis) of dendrimer-nanoparticle composites can be found in the literature and excellent reviews on this subject are available. [93-95]

6. Conclusion

In conclusion, dendrimers have influenced the science of catalysis in a magnificent way. The possibility of tuning the activity and selectivity of catalysts by attaching them to properly designed dendrimers opened new routes in exploiting precious catalysts in a more fruitful manner. The area of catalyst recovery and recycling also changed tremendously after the appearance of dendrimers. In addition of acting as supports to catalysts, dendrimers found applications as templates in the synthesis of catalytically active metal nanoparticles. In short, the difficulty and cost in the synthesis of dendrimers complement the opportunities they offer in catalysis. A perfect picture of dendrimers, their synthesis, properties and applications can be obtained from a number of books and the interested reader may go through those available literature. [96-103]

References

[1] Majoral, P. J. *New. J. Chem.* 2007, *31*, 1039.

[2] Tomalia, D. A. *Porg. Polym. Sci.* 2005, *30*, 294.

[3] De Gennes, P. G.; Hervet, H. J. *Phys. Lett.* 1983, *44*, 351.

[4] Oosterom, G. E.; Reek, J. N. H.; Kamer, P. C. J.; Van Leeuwen, P. W. N. M. *Angew. Chem. Int. Ed. Eng.* 2001, *40*, 1828.

[5] Helms, B.; Frechet, J. M. J. *Adv. Synth. Catal.* 2006, *348*, 1125.

[6] Keijsper, J. J.; Van Leeuwen, P. W. N. M.; Van der Made, A. P.; EP 0456317, 1991; Shell Int. Research, US 5243079, 1993 (*Chem. Abstr.* 1992, 116, 129870).

[7] Knapen, J. W. J.; Van der Made, A. P.; De Wilde, J. C.; Van Leeuwen, P. W. N. M.; Wijkens, P.; Grove, D. M.; Van Koten, G. *Nature*, 1994, *372*, 659.

[8] Miedaner, A.; Curtis, C. J.; Barkley, R. M.; DuBois, D. L. *Inorg. Chem.* 1994, *33*, 5482.

[9] Reetz, M. T.; Lohmer, G.; Schwickardi, R. *Angew. Chem. Int. Ed. Engl.* 1997, *36*, 1526.

[10] Brinkmann, N.; Giebel, D.; Lohmer, G.; Reetz, M. T.; Kragl, U. *J. Catal.* 1999, *183*, 163.
[11] De Groot, D.; Eggeling, E. B.; De Wilde, J. C.; Kooijman, H.; Van Haaren, R. J.; Van der Made, A. W.; Spek A. L.; Vogt, D.; Van Leeuwen, P. W. N. M. *Chem. Commun.* 1999, 1623.
[12] Mery, D.; Heuze, K.; Astruc, D. *Chem. Commun.* 2003, 1934.
[13] Heuze, K.; Mery, D.; Astruc, D. *Chem. Commun.* 2003, 2274.
[14] Heuze, K.; Mery, D.; Gauss, D.; Blais, J. C.; Astruc, D. *Chem. Eur. J.* 2004, *10*, 3936.
[15] Sarkar, A.; Ilankumaran, P.; Kisanga, P.; Verkade, J. G. *Adv. Synth. Catal.* 2004, 346, 1093.
[16] Krishnan, G. R.; Thomas, J.; Sreekumar, K. *Arkivoc*, 2009, x, 106
[17] Krishnan, G. R.; Sreerekha, R.; Sreekumar, K. *Lett. Org. Chem.*, 2009, *6*, 17
[18] Mizugaki, T.; Ooe, M.; Ebitani, K.; Kaneda, K. *J. Mol. Catal. A.* 1999, *145*, 329.
[19] Findeis, R. A.; Gade, L. H. *Eur. J. Inorg. Chem*, 2003, *1*, 99.
[20] Seyferth, D.; Wyrwa, R.; Franz, U. W.; Becke, S. (PCT Int. Appl.) WO 9732908, 1997 [Chem. Abstr. 1997, 127, 263179p].
[21] Seyferth, D.; Wyrwa, R. WO 9732918, 1997 [Chem. Abstr. 1997, 127, 263180g].
[22] Mager, M.; Becke, S.; Windisch, H.; Cenninger, U. *Angew. Chem. Int. Ed. Engl.* 2001, 40, 1898.
[23] Zheng, Z. j.; Chen, J.; Li, Y. S. *J. Organomet. Chem.* 2004, 689, 3040.
[24] Benito, J. M.; de Jesus, E.; de La Mata, F. J.; Flores, J. C.; Gomez, R. *Chem. Commun.* 2005, 5217.
[25] Malgas-Enus, R.; Mapolie, S. F.; Smith, G. S. *J. Organomet. Chem.* 2008, *693*, 2279.
[26] Ropartz, L.; Morris, R. E.; Foster, D. F.; Cole-Hamilton, D. J. *Chem Commun.* 2001, 361.
[27] Ropartz, L.; Morris, R. E.; Foster, D. F.; Cole-Hamilton, D. *J. Mol. Catal. A: Chem.* 2002, *182-183*, 99.
[28] Delort, E.; Darbre, T.; Reymond, J. L. *J. Am. Chem. Soc.* 2004, *126*, 15642.
[29] Nlate, S.; Plault, L.; Astruc, D. *New J. Chem.* 2007, *31*, 1264.
[30] Rodriguez, L. I.; Rossell, O.; Seco, M.; Orejon, A.; Bulto, A. M. M. *J. Organometallic Chem.* 2008, *693*, 1857.
[31] Niu, Y. N.; Yan, Z. Y.; Li, G, Q.; Wei, H. L.; Gao, G. L.; Wu, L. Y.; Liang, Y. M. *Tetrahedron Asymmetry*, 2008, *19*, 912.
[32] K. Vassilev, S. Turmanova, M. Dimitrova, S. Boneva, *Eur. Polym. J.*2009, 45, 2269
[33] Karakhanov, E. A.; Maximov, A. L.; Tarasevich, B. N.; Skorkin. V. A. *J. Mol. Catal. A. Chem.* 2009, *297*, 73.
[34] Balieu, S.; Zein, A. E.; De Sousa, R.; Jérôme, F.; Tatibouët, A.; Gatard, S.; Pouilloux, Y.; Barrault, J.; Rollin, P.; Bouquillon, S. *Adv. Syn. Catal.* 2010, *352*, 1826
[35] Gitsov, I.; Ivanova, P. T.; Frechet, J. M. J. *Macromol. Rapid Commun.* 1994, *15*, 387.
[36] Matyjaszewski, K.; Shigemoto, T.; Frechet, J. M. J.; Leduc, M. *Macromolecules*, 1996, *29*, 4267.
[37] Mak, C. C.; Chow, H. F. *Macromolecules*, 1997, *30*, 1228.
[38] Chow, H. F.; Mak, C. C. *J. Org. Chem.* 1997, *62*, 5116.
[39] Bhyrappa, P.; Young, J. K.; Moore, J. S.; Suslick, K. S. *J. Am. Chem. Soc.* 1996, *118*, 5708.

[40] Bhyrappa, P.; Young, J. K.; Moore, J. S.; Suslick, K. S. *J. Mol. Catal. A. Chem.* 1996, *113*, 109.

[41] Zubia, A.; Cossio, F. P.; Morao, I.; Rieumont, M.; Lopez, X. *J. Am. Chem. Soc.* 2004, *126*, 5243.

[42] Herrmann, A. W. *Angew. Chem. It. Ed. Engl.* 2002, *41*, 1290.

[43] Fujihara, T.; Obora, Y.; Tokunaga, M.; Sato, H.; Tsuji, Y. *Chem. Commun.* 2005, 4526.

[44] Singleton, D. M. (Shell Oil Corp.), US Patent4, 472,522, 1985; *Chem. Abstr.* 1985, *102*, 46405.

[45] Muller, C.; Ackerman, L. J.; Reek, J. N. H.; Kamer, P. C. J.; van Leeuwen, P. W. N. M. *J. Am. Chem. Soc.* 2004, *126*, 14960.

[46] Liang, C.; Frechet, J. M. J. *Prog. Polym. Sci.* 2005, *30*, 385.

[47] Zhang, X.; Xu, H.; Dong, Z.; Wang, Y.; Liu, J.; Shen, J. *J. Am. Chem. Soc.* 2004, *126*, 10556.

[48] Piotti, M. E.; Rivera, Jr. F.; Bond, R.; Hawker, C. J.; Frechet, J. M. J. *J. Am. Chem. Soc.* 1999, *121*, 9471.

[49] Hecht, S.; Frechet, J. M. J. *J. Am. Chem. Soc.* 2001, *123*, 6959.

[50] Mizugaki, T.; Hetrick, C. E.; Murata, M.; Ebitani, K.; Amiridis, M. D.; Kaneda, K. *Chem. Lett.* 2005, *34*, 420.

[51] Ooe, M.; Murata, M.; Mizugaki, T.; Ebitani, K.; Kaneda, K. *J. Am. Chem. Soc.* 2004, *126*, 1604.

[52] Aathimanikandan, S. V.; Sandanaraj, B. S.; Arges, C. G.; Bardeen, C. J.; Thayumanavan, S. *Org. Lett.* 2005, *7*, 2809.

[53] Liu, L.; Breslow, R. *J. Am. Chem. Soc.* 2003, *125*, 12110.

[54] Wang, G.; Liu, X. Y; Zhao, G. *Synlett*, 2006, *8*, 1150.

[55] Li, Y.; Liu, X. Y; Zhao, G. *Tetrahedron: Asymmetry*, 2006, *17*, 2034.

[56] Wang, G.; Zheng, C.; Zhao, G. *Tetrahedron: Asymmetry*, 2006, *17*, 2074.

[57] Wu, Y.; Zhang, Y.; Yu, M.; Zhao, G.; Wang, S. *Org. Lett.* 2006, *8*, 4417.

[58] Muraki, T.; Fujita, K.; Terakado, D. *Synlett*, 2006, *16*, 2646.

[59] Wang, Z. J.; Deng, G. J.; Li, Y.; He, Y. M.; Tang, W. J.; Fan, Q. H. *Org. Lett.* 2007, *9*, 1243.

[60] Liu, Y. H.; Shi, M. *Adv. Synth. Catal.* 2008, *350*, 122.

[61] Fujita, K.; Ainoya, T.; Tsuchimoto, T.; Yasuda, H. *Tetrahedron Lett.* 2010, 51, 808

[62] Smalley, R. *Congressional Hearings, Summer*, 1999.

[63] Roucoux, A.; Schulz, J.; Patin, H.; *Chem. Rev.* 2002, *102*, 3757.

[64] Klabunde, K. J. In *Nanoscale Materials in Chemistry*; Klabunde, K. J., Ed.; Wiley Interscience, New York, 2001.

[65] Schmid, G. In *Nanoscale Materials in Chemistry*; Klabunde, K. J., Ed.; Wiley Interscience, New York, 2001.

[66] Hvolbaek, B.; Janssens, T. V. W.; Clausen, B. S.; Falsig, H.; Christensen, C. H.; Norskov, J. K. *Nano Today*, 2007, *2*, 14 and references there in.

[67] Klabunde, K. J.; Mulukutla, R. S. In *Nanoscale Materials in Chemistry*; Klabunde, K. J., Ed.; Wiley Interscience, New York, 2001.

[68] Fan, J.; Gao, Y. *J. Exp. Nanoscience*, 2006,*1*, 457.

[69] Zheng, Y.; Stevens, P. D.; Gao, Y. *J. Org. Chem.* 2006, *71*, 537.

[70] Oila, M. J.; Koskinen, A. M. P. *Arkivoc*, 2006, 76.

[71] Schmid, G.; Maihack, V.; Lantermann, F.; St. Peschel. *J. Chem. Soc. Dalton Trans.* 1996, 589.
[72] Schmid, G.; Emde, S.; Maihack, V.; Meyer-Zaika, W.; St. Peschel. *J. Mol. Catal. A.* 1996, *107*, 95.
[73] Bronstein, L. M.; Sidorov, S. N.; Valetsky, P. M. *Russ. Chem. Rev.*, 2004, *73*, 5, 501.
[74] Tabuani, D.; Monticelli, O.; Chincarini, A.; Bianchini, C.; Vizza, F.; Moneti, S.; Russo, S. *Macromolecules*, 2003, *36*, 4294.
[75] Li, Y.; Hong, X. M.; Collard, D. M.; El-Sayed, M. A. *Org. Lett.* 2000, *2*, 2385.
[76] Tomalia, D. A. *Aldrichimica Acta*, 2004, *37*, 39.
[77] Majoral, J. P. *New. J. Chem.* 2007, *31*, 1039.
[78] van Hest, J. C. M.; Delnoye, D. A. P.; Baars, M. W. P. L.; van Genderen, M. H. P.; Meijer, E. W. *Science* 1995, *268*, 1592.
[79] Johan, F. G. A.; Janson, J.; Meijer, E. W.; de Brabander-van den Berg, E. M. M. *J. Am.Chem. Soc.* 1995, *117*, 4417.
[80] Johan, F. G. A.; Janson, J.; de Brabander-van den Berg, E. M. M.; Meijer, E. W. *Science* 1995, *266*, 1266.
[81] Balogh, L.; Tomalia, D. A. *J. Am. Chem. Soc.* 1998, *120,* 7355.
[82] Zhao, M.; Sun, L.; Crooks, R. M. *J. Am. Chem. Soc.* 1998, *120*, 4877
[83] Niu, Y, Yeung, L. K.; Crooks, R. M. *J. Am. Chem. Soc.* 2001, *112*, 6840
[84] Ooe, M.; Murata, M.; Mizugaki, T.; Ebitani, K.; Kaneda, K. *Nano Lett.* 2002, *2*, 999
[85] Chechik, V.; Crooks, R. M. *J. Am. Chem. Soc.* 2000, *122*, 1243
[86] Pittelkow, M.; Moth-Poulsen, K.; Boas, U.; Christensen, J. B. *Langmuir*, 2003, *19*, 7682
[87] Garcia-Martinez, J. C.; Lezutekong, R.; Crooks, R. M. *J. Am. Chem. Soc.* 2005, *127*, 5097
[88] Krishnan, G. R.; Sreekumar, K. *Soft Mater.*2010, *8*, 114
[89] Murugan, E.; Rangasamy, R. *J. Polym. Sci. A. Polym. Chem.* 2010, *48*, 2525
[90] Li, H.; Lu, J.; Zheng, Z.; Cao, R. *J. Coll. Inter. Sci.* 2011, *353*, 149
[91] Zaera, F.; Morales, R.; Albiter, M. A.
[92] Auten, J. B.; Lang, H.; Chandler, B. D. *Appl. Catal. B. Eviron.* 2008, *81*, 225
[93] Niu, Y.; Crooks, R. M., *C. R. Chimie*, 2003, *6*, 1049.
[94] Crooks, R. M.; Zhao, M.; Sun, L.; Chechik, V.; Yeung, L. K. *Acc.Chem. Res.* 2001, *34*, 181.
[95] Scott, R. W. J.; Wilson, O.M.; Crooks, R. M. *J. Phy. Chem. B*, 2005, *109*, 692.
[96] *Dendrimers and Other Dendritic Polymers*; Frechet, J. M. J., Tomalia, D. A., Eds.; John Wiley & Sons, UK, 2001.
[97] *Dendrimers*; Vogtle, F., Ed.; Topics in Current Chemistry Vol. 197; Springer-Verlag: Berlin Heidelberg, 1998.
[98] *Dendrimers II: Architecture, Nanostructure and Supramolecular Chemistry*; Vogtle, F., Ed.; Topics in Current Chemistry Vol. 210; Springer-Verlag:Berlin Heidelberg, 2000.
[99] *Dendrimers III: Design, Dimension, Function*; Vogtle, F., Ed.; Topics in Current Chemistry Vol. 212; Springer-Verlag: Berlin Heidelberg, 2001.
[100] *Dendrimers IV: Metal Coordination, Self Assembly, Catalysis*; Vogtle, F., Schalley, C. A., Eds.; Topics in Current Chemistry Vol. 217; Springer-Verlag:Berlin Heidelberg, 2001.

[101] *Dendrimers V: Functional and Hyperbranched Building Blocks, Phtotophysical Properties, Applications in Materials and Life Sciences*; Vogtle, F., Schalley, C. A., Eds.; Topics in Current Chemistry Vol. 228; Springer-Verlag:Berlin Heidelberg, 2003.
[102] *Dendrimers in Medicine and Biotechnology: New Molecular Tools* Boas, U.; Christensen, J. B.; Heegaard, P. M. H.; The Royal Society of Chemistry: UK, 2006
[103] *Dendrimer Based Nanomedicine*; Majoros, I. J., Baker Jr., J. R., Eds. Pan Stanford Publishing Pte. Ltd. Singapore, 2008

In: Focus on Catalysis Research: New Developments
Editors: Minjae Ghang and Bjørn Ramel

ISBN: 978-1-62100-455-4

Chapter 14

LIQUID PHASE PARTIAL OXIDATIONS OVER ACTIVE TRANSITION METALS GRAFTED ON DIFFERENT HETEROGENEOUS SUPPORTS

Mahasweta Nandi[1,2] and Asim Bhaumik[1]
[1]Department of Materials Science, Indian Association for the Cultivation of Science, Jadavpur, Kolkata, India
[2]Integrated Science Education and Research Centre, Siksha Bhavana, Visva-Bharati, Santiniketan, India

1. INTRODUCTION

To start with, let us recall the definition of transition metals. According to the modern definition, given by the International Union of Pure and Applied Chemistry (IUPAC) a transition metal is 'an element whose atom has an incomplete *d* sub-shell, or which can give rise to cations with an incomplete *d* sub-shell.'[1] Electrons are fed into the *d*-orbitals starting from Group 3. For the first and second transition series, the Group 3 elements, scandium (Sc^{3+}) and yttrium (Y^{3+}), have a single *d* electron in their outermost shell but usually they are not considered as transition metals. In all of their compounds they exist as Sc^{3+} and Y^{3+} ions where there are no *d* electrons. Other elements with d^1 configuration are lanthanum and actinium, but they are classified under *lanthanoid* and *actinoid* series of elements, respectively. On the other hand, the Group 12 elements, namely zinc, cadmium, and mercury have an outer shell electronic configuration $d^{10}s^2$ with no incomplete *d* shell and hence they are not transition metals according to the above definition. In their +2 oxidation state they have a d^{10} electronic configuration while in the +1 oxidation state; there are no unpaired electrons because of the formation of dimer with a covalent bond between the two atoms. An interesting property of these transition metals is that they can exhibit two or more oxidation states which usually differ by one in their compounds. They have electrons of similar energy in both the *3d* and *4s* levels and thus a particular element can form ions of nearly the same stability by losing different numbers of electrons. The first row transition metal catalysts are of great utility in the oxidation chemistry because of their high reactivity and general utility

[2]. This is the reason for which the transition metals can be utilized as catalysts for redox reactions. In biological system, the transition metals play a very crucial role to perform different redox reactions which are otherwise difficult to carry other under normal conditions.

This basic knowledge about the transition metals motivated the scientists to design novel materials which contain these elements. These type of materials can be used as catalysts in various eco-friendly, selective and industrially important organic transformations. A catalyst is a compound or material that can affect the rate of a chemical reaction by providing an alternative and lower energy profile or pathway. That is, it only changes the cost of the activation energy. It is not related to the thermodynamics of the process and hence, the final product distribution. Complexation by transition metals affords access to a wide variety of oxidation states for the metal. This has the property of providing electrons or withdrawing electrons from the transition state of the reaction. Most industrially used catalysts are the transition metal in a bed, as a metal or bound structure.

In this context the development in the field of microporous materials started attracting widespread attention because of their exceptional surface areas and well-defined pore sizes. These are extremely desirable for the diffusion of bulky adsorbate or reactant molecules, which is one of the key requisites for being a good catalyst support. As early as 1983, Taramasso *et al.* invented a porous crystalline synthetic material, named as titanium silicalite 1 or TS-1 [3], where Ti atoms partially substitute Si of ZSM-5 structure. TS-1 found to be a very efficient and selective catalyst in a number of industrially important organic transformations involving small sixe molecules which could panitrate its medium size micropores of dimensions 5.4–5.6 Å. Later in 1990, Huybrechts *et al.* reported the oxidation of alkanes on a microporous crystalline titanium silicalite with hydrogen peroxide as oxidant. [4] But the major breakthrough came around two years later, when Kresge *et al.* first prepared the M41S family of mesoporous solids with regular arrays of uniform channels [5]. One of the members of this family, MCM-41 exhibited a hexagonal arrangement of uniform pores with dimensions tuneable from *ca.* 15 Å to more than 100 Å. The usefulness of these materials is attributed to their microstructure, which allows the molecules to access large surface area which enhances their catalytic activity and adsorption capacity. Followed by this discovery, Pinnavaia *et al.* reported a hexagonal mesoporous titanium containing silica, Ti-HMS [6], which worked as an efficient catalyst for the selective oxidation of alkanes, hydroxylation of phenol and epoxidation of alkenes in the presence of hydrogen peroxide. Titanium-containing mesoporous silica gradually became a substance of great interest both in academia and industry due to their potential to oxidize very bulky organic substrates which are otherwise difficult to oxidize over microporous TS-1 [3] under liquid-phase reaction conditions. This reflected the high potential of titanium containing catalysts in different industrially important reactions. Following this, numerous heterogeneous catalysts based on other transition metals started developing rapidly and they were also found to be very important catalysts for different chemical transformations [7].

Catalytic reactions can be performed either in homogeneous medium or heterogeneous. Both the media involved in the process have advantages and disadvantages over each other. In most of the cases, the heterogenized catalytic species worked well in the immobilized state, often better than expected. In homogeneous catalysis, all catalyst, substrate(s) and the reactant are in a single phase, mainly in liquid phase. In heterogeneous catalysis, the catalysts are insoluble in the liquid phase where reactants and substrates remain. Generally, homogeneous reaction requires low temperature, whereas heterogeneous catalytic reactions are carried out at high temperature. Immobilization or heterogenizing of transition metals on different solid supports, both organic and inorganic, offer the advantages of high catalytic activity and stability, and several other benefits like easy separation. In homogeneous catalysis, separation of catalysts from reaction mixture is a cumbersome task. Heterogeneous catalysts can be used for several times without appreciable loss of its activity in its next use.

There are many strategies for the design and the preparation of heterogeneous catalysts. Various types of supports can be used viz. encapsulation in zeolites [8], immobilization in porous alumina [9], immobilization in mesoporous silica [10-13], Y-zeolite [14], resin [15], grafting on polymers [16], dendritic [17, 18] and polymeric organic support [19] have been developed in terms of heterogenization of homogeneous catalysts. In this review article we shall confine our discussion particularly on three main types of solid supports: a) microporous and mesoporous silicas with ordered pore systems, b) non-porous silicas and c) polymers or resins. A general overview of the different types of catalysts based on active transition metals that can be used as catalyst liquid phase partial oxidation reactions will be discussed. These transition metals shall include titanium, vanadium, niobium, chromium, molybdenum, manganese, iron, cobalt, nickel and copper. In this book review we will limit our discussion up to those transition metal containing catalysts, which have been synthesized and studied by our group.

2. General Synthetic Techniques

2.1. Microporous and Mesoporous Silica with Ordered Pore Systems

The general synthetic strategy for the preparation of microporous and mesoporous [20] materials is based on templating mechanism [4]. Microporous materials are generally obtained by using single molecule templating pathways, where quaternary ammonium salts or primary, secondary or tertiary amine compounds are used as template along with the inorganic species under hydrothermal/solvothermal conditions. The inorganic phase is constructed around these organic moieties, which behave as 'single molecule templates'. The crystalline phases are then obtained by hydrothermal/solvothermal treatment of the reaction gel. In 1992, Mobil researchers extended this idea of template synthesis [5] to develop the MCM-41 family of mesoporous silicates and aluminosilicates with hexagonally packed pore system. They used liquid crystal templating mechanism, where surfactant micelles are used as the structure directing agents in a sol-gel assisted process. Self-assembly of amphiphilic surfactants forming micelles are encapsulated by an inorganic material by charge balance interaction to generate the mesostructured composites. Removal of the surfactants by calcination or extraction gives the mesoporous composite. With the advance in research in field of porous materials, many cationic [21, 22], anionic [23-25] and non-ionic [26] surfactants have been frequently employed as templates in various syntheses. Apart from that in some cases few amphoteric, gemini, bolaform, multi-headgroup and cationic fluorinated surfactants have also been used. Interactions between the surfactant head group and inorganic components mainly being *via* electrostatic and H-bonding. In addition to the purely inorganic frameworks, mesoporous hybrid silsesquioxane materials [27-32] have also been synthesized containing organic functionality as an integral part of the framework following similar approach. These basic ideas and concepts for the synthesis of mesoporous materials can be extended to design various transition metal-containing mesoporous solids which can show excellent catalytic properties [33]. The metal atoms are introduced into the framework structures either by *in situ* [33, 34] incorporation or by post-synthetic [35, 36] immobilization. In the following section a brief overview on the synthesis of different metal incorporated mesoporous materials based on purely inorganic as well as organic-inorganic hybrid frameworks will be discussed. For the immobilization of the metal complexes on to solid supports, we have mostly used MCM-41 type of 2D-hexagonal mesoporous silica. Thus in the preceding section we shall come across this material several times, for the synthesis of various metal containing catalyst. So it is worth mentioning here a brief outline for the synthesis of this 2D-hexagonal silica.

For the synthesis, a supramolecular self-assembly of cationic cetyltrimethyl ammonium bromide, CTAB) and non-ionic Brij-35 ($C_{12}H_{25}$–$(OC_2H_4)_{23}$–OH, a polyether and aliphatic hydrocarbon chain) mixed surfactant system is used as the structure directing agent (SDA) in the presence of tartaric acid (TA) as a mineralizer [37]. Though the use of a mixed surfactant system or addition of a mineralizer is not absolutely essential for the preparation of MCM-41 type of materials, but it has been found that their use produces mesoporous silica with highly ordered structure and high surface area. In the original report of MCM-41 quaternary ammonium bromide alone has been used as the template [5]. Organosilica, tetraethyl orthosilicate (TEOS) is used as the silica source for the syntheses. In a typical synthesis, 2.96 g of (CTAB) (8.14×10^{-3} mol) and 1.5 g of Brij-35 are dissolved in an acidic aqueous solution of 0.78 g TA in 60.00 g H_2O under vigorous stirring at room temperature for 30 min. Then 3.50 g (16.8×10^{-3} mol) of TEOS is added to the mixture under continuous stirring for 1 h. Tetramethylammonium hydroxide (TMAOH, 25% aqueous solution) or sodium hydroxide solution is then added drop wise to adjust the pH at *ca.* 11.0. The resulting mixture is aged overnight under stirring at room temperature and then heated at 353 K for 3 days without stirring. The solid product is then recovered by filtration, washed several times with water, and dried under vacuum in a lypholyzer. The resulting powder is calcined in the flow of air at 723 K for 8 h to remove all the organic surfactants.

2.2. Non-Porous Silica Supports

Transition metals can also be immobilized on non-porous supports to generate efficient heterogeneous catalysts [38, 39]. In this respect modified silica has been found to be highly efficient due to its chemical and thermal stability and its strong ability to bind to the metal complexes. Generally the silica surfaces are functionalized with active organic groups to generate some binding sites on their surface, which are then capable of efficiently binding metal ions. In typical procedures amorphous silica with an average particle size of 10 μm are taken and allowed to undergo reactions for surface functionalization. Then the metal ions are heterogenized over these modified silica materials through phase impregnation.

2.3. Polymer or Resin Supports

Immobilization of the active transition metal centres can also be carried out on organic supports which include organically modified polystyrene based materials, resins, polymers, etc [40-43]. Polystyrene frameworks are inexpensive hard plastic materials which are generally selected for their high chemical and thermal stability. Amino-functionalized polystyrene has a strong ability to bind to metal complexes and hence it can be used as a support for heterogenization. The polymer or resin supports are mostly available commercially; in addition to that different polymeric frameworks can be designed in order to provide adequate binding sites. Apart from that, the commercially available polymers or resins can be chemically modified to anchor chelating sites on to the frameworks which can facilitate the binding of the metal centers. In a general procedure, at first the precursor metal salt is dissolved in an organic solvent. The solid support is suspended in to the solution and the mixture is stirred for several hours, often under refluxing conditions. The solid metal supported catalyst is thereafter obtained by filtration, followed by repeated washing and drying. In the following section preparation of two different types of polymer supported catalysts has been given.

3. Heterogeneous Metal Catalysts

3.1. Titanium Containing Heterogeneous Catalysts

Titanium containing catalysts supported on porous silica frameworks are the mostly studied and pioneering materials in the field of transition metal supported heterogeneous catalysis. The titanium-containing porous silica, which is prepared by isomorphous substitution of Si^{4+} by Ti^{4+} ions, is of great interest from both academic and industrial viewpoints due to their versatile potential as oxidation catalysts in many organic reactions. They are generally prepared by *in situ* addition of the titanium precursor along with the organosilane precursors during the synthesis. Post-synthetic functionalization or grafting is not the usually practiced method for the preparation of titanium silicates, mainly associated with the reactive nature of the titanium precursors in air and moisture. In the following section we will discuss briefly about their synthesis, separating the materials under two heading *viz.* titanium silicates and hybrid titanium silicates.

3.1.1. Titanium Silicates

Microporous titanium silicates, TS-1 with different ratio of Si/Ti are synthesized using reported procedure. [3, 44, 45] In typical syntheses, tetraethyl orthosilane (TEOS) is added to tetrapropylammonium hydroxide and the solution is homogenized by stirring. Then titanium tetrabutoxide (TBOT) taken in dry isopropyl alcohol (IPA) is added slowly to the above reaction mixture. The solution thus obtained is autoclaved at 433 K for 6 h. The composition of the synthesis mixture is generally: 1 TEOS/ 0.4 TPAOH/ x TBOT/ 0.95 IPA/ 30 H_2O, where x determines the amount of titanium loading desired in the materials. The solid products are recovered by centrifugation followed by thorough washing with demineralized water, treatment with 1 wt % aqueous solution of ammonium acetate, washing thoroughly with a plentiful amount of demineralized water, and drying. Microporous TS-1 is then obtained by calcination in a flow of air. This concept can be extended to develop extra large pore titanium silicates by using diaminoalkanes [46] as the structure directing agent. In such syntheses, the diamines are dissolved in mixture of water and ethanol. Then the silica source, TEOS is added to it followed by the addition of tetraalkylammonium hydroxide to hydrolyze it. Finally titanium tetrabutoxide (Aldrich) taken in dry isopropyl alcohol is added as the titanium source and the mixture is aged for several hours at ambient temperature. The porous titanium silicate can be obtained by calcination of this as-synthesized material in a flow of air.

2D-hexagonal mesoporous titanium-silicates are also an important class of titanosilicates which are synthesized by using supramolecular assemblies of surfactant molecules. Their syntheses are almost similar to the synthesis of mesoporous silica frameworks, like MCM-41, described in Section 2.1. For the purely silicate frameworks, organosilane precursor tetraethylorthosilane is used whereas, in order to synthesize titanium silicates, titanium tetrabutoxide is added with the silane as the source of titanium. In such syntheses different types of molecules can be used as templates including mixture of surfactants. Two such templating systems used in the syntheses are mixture of cationic (cetyltrimethylammonium bromide, CTAB) and anionic surfactants (sodium dodecylsulphate, SDS) [47] and mixture of cationic (CTAB) and non-ionic surfactants (Brij-35) [48]. 2D-hexagonal ordered mesoporous titanium silicates are also synthesized employing non-ionic surfactant dodecylamine (DDA) [49] is the structure-directing agent (SDA) under moderately basic condition. In addition to that, titanium-rich mesoporous silica can also be prepared using unconventional Ti source $TiCl_4$ under strongly acidic pH condition in the presence of a cationic surfactant octadecyltrimethylammonium chloride as the SDA [50]. The basic synthetic strategy in all the cases involves the dissolution of the surfactant molecules in an aqueous medium followed by the addition of the organosilane precursor. After hydrolysis of the silane, titanium butoxide taken in dry isopropanol is added to the system. The pH is adjusted to a higher value by addition of a base in most of the cases, or an acid, e.g. in the synthesis of titanium rich silicates using $TiCl_4$. Finally the reaction gels are treated hydrothermally or under ambient condition e.g. for synthesis with dodecylamine, and filtered to get the solid products. The mesoporous titanium-silicates are obtained by calcination or extraction of the as-synthesized materials to remove the organic template molecules.

3.1.2. Hybrid Titanium Silicates

Surface modification by incorporating various organic functionalities [51–53] and silylation [54] induces hydrophobicity along with stability in the mesoporous materials as a result of which the catalytic activity of titanium silicates are improved considerably in the liquid phase oxidation reactions in the presence of water. One of the very well-known methods to prepare hybrid titanium silicates makes the use of different organosilica precursors containing various functionalities along with tetraalkylorthosilane. The synthesis procedure for these hybrid materials is similar to that used for the synthesis of the pure analogs, described in Section 3.1.1. The only difference lies in the use of different organic moieties attached to the silica precursor. These kind of precursors are either available commercially or the organic functional groups are introduced by some chemical reactions. Microporous titanium oxophenyl phosphonates can be prepared by using phenylphosphonic acid (PPA) as organophosphorus source without the aid of any structure directing agent [55]. Different organosilane precursors have been used for the syntheses of various mesoporous titanium silicates, *viz.* 3-chloro-propyltriethoxysilane [56], 3-chloropropyldimethoxymethylsilane [57] and polyvinyltrimethoxysilane (synthesized by the polymerization (Scheme 1) of vinyltrimethoxysilane in presence of azoisobutironitrile (AIBN) as the radical initiator) [58]. Apart from that, bridge bonded organic groups integrated into the silane presursors are also used for these syntheses. The bridging organosilanes include 1,2-bis-(trimethoxysilyl)ethane (BTME) [59, 60] and 1,2-bis(triethoxysilyl)-ethane (BTEE) [61]. The synthesis procedure being the same for all the materials, the templates are removed exclusively by solvent extraction method, since calcination will result in combustion of the organic groups.

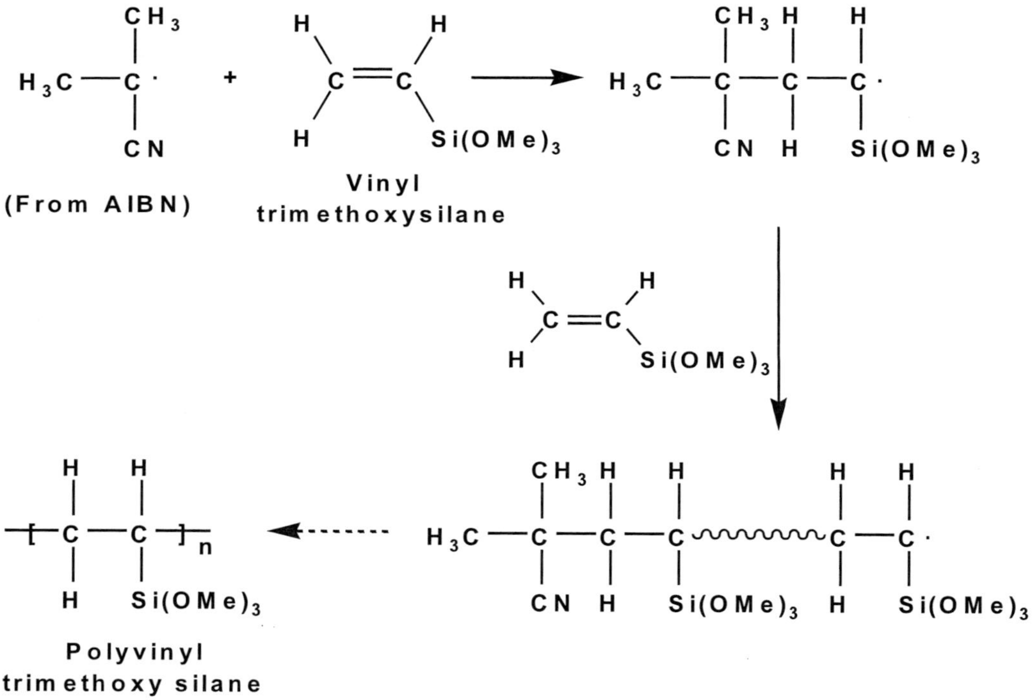

Scheme 1. Reaction scheme for the polymerization of vinyltrimethoxysilane to polyvinyltrimethoxysilane.

3.2. Vanadium and Niobium Containing Heterogeneous Catalysts

3.2.1. Microporous Vanadium Silicate Using 1,6-Hexamethylene-Bis (Benzyldimethylammonium Hydroxide) As Template

Microporous vanadium silicate is synthesized using a diquaternary ammonium cation, 1,6-hexamethylene-bis (benzyldimethylammonium hydroxide) as template [62]. The organic template is designed by the interaction of 1,6-dibromohexane with benzyldimethylamine in 1:2.2 molar ratio under refluxing condition using acetone as solvent. The white solid obtained is purified and the bromide salt thus obtained is electrochemically converted to the corresponding hydroxide. The structural formula of the organic template is given in Figure …

For the synthesis of the micropoorus vanadium silicate highly reactive fumed silica and V(O)SO4-3H20 are used as the silica and vanadium sources respectively. Fumed silica is added to an aqueous solution of the template under stirring followed by the addition of the required amount of V(O)SO4-3H20. Then the ph of the reaction mixture was adjusted to ca. 10 by the addition of an aqueous solution of NaOH. The mixture is treated hydrothermally at 433 K for 3 days and the product is recovered by filtration. The final solid catalyst is obtained by calcination of the as-synthesized material in a flow of air.

3.2.2. Amberlite-Anchored V (IV) Catalyst

Amberlite resins are gel-type ion-exchange resin with a styrenedivinylbenzene (DVB) framework composition [63]. It acts as good heterogeneous supports for different metal ions though it does not have a specific or true porosity. The intermolecular distances which give rise to an apparent porosity (ca. 4 nm) is sufficient for the ions/molecules to migrate through the gel structure to exchange sites [64] which can be responsible for high catalytic activity. The cationic sites of the amberlite resins can be exchanged with metal cations to generate various potential catalysts. In a typical procedure [65], the strongly acidic cation exchange resin is soaked in brine solution to convert the H-form of the resin is converted into Na-form, washed thoroughly with double distilled water and air dried. The vanadium resin is then obtained by exchanging Na^+ sites of Amberlite IR 120 resin with VO^{2+} obtained from an aqueous solution of vanadium (IV) oxide sulphate hydrate ($VOSO_4$).

3.2.3. Microporous Niobium Phosphate

For the synthesis H_3PO_4 and $NbCl_5$ are used as the phosphorous and niobium sources respectively [66]. For reproducibility of the data, the two sources are mixed together before adding to the surfactant solution. H_3PO_4 and $NbCl_5$ are mixed together in an aqueous solution and the pH of the solution is raised to ca. 4.9. The precipitate obtained is filtered and washed with water to remove excess chloride ions. To this precipitate, an aqueous solution of decylamine is added as the template under constant stirring. The pH is adjusted at 3.9 by addition of H_3PO_4 and the resultant gel is treated hydrothermally at 348 K for 1 day. The product is obtained filtration and drying and calcination of this material in a flow of air gives the microporous niobium phosphate.

3.3. Chromium and Molybdenum Containing Heterogeneous Catalysts

3.3.1. Mesoporous Cr-MCM-41

Cationic surfactant cetyltrimethylammonium bromide is used as the structure directing agent and chromium sulfate as the chromium source [67]. TEOS is allowed to mix with an aqueous solution of CTAB and the desired amount of $Cr_2(SO_4)_3$ dissolved in water is added to the silica sol in different mole ratio depending up on the desired loading of chromium. Then aqueous TMAOH solution is added into it until pH rose to *ca.* 11.0. After vigorous stirring for 1h the final mixture is autoclaved at 353K for 2–3 days and the product is obtained by filtration followed by drying. The final mesoporous Cr-MCM-41 is obtained by calcination in a flow of air.

3.3.2. Supermicroporous Chromium Oxophenylphosphate

For the synthesis, phenylphosphonic acid and chromium (III) chloride are used as the phosphorus and chromium sources, respectively [68]. Phenylphosphonic acid is at first dissolved in water to which an aqueous solution of chromium (III) chloride is added dropwise. Then resulting green mixture is stirred and the pH is adjusted to *ca.* 4-5 by the addition of tetramethylammonium hydroxide. The reaction gel is then treated hydrothermally at 443 K for 1 day and the porous metal oxophenylphosphate is obtained by fitration followed by drying.

3.3.3. Oxodiperoxomolybdenum(VI) Complexes Immobilized Over Mesoporous Silica

Highly ordered 2D-hexagonal MCM-41-type mesoporous silica is used for the heterogenization of molybdenum complexes [69]. Two complexes of molybdenum has been used for the preparation on the catalysts viz. a neutral Oxodiperoxo molybdenum(VI) adduct $[MoO(O_2)_2 \cdot 2QOH]$ [QOH = 8-quinilinol] and an anionic oxodiperoxo-8-quinolinolato-molybdate(VI) complex of tetraphenylphosphonium cation. Immobilization of the metal complexes is carried out by dispersing the mesoporous silica in a solution containing the metal complex dissolved in dry acetonitrile, followed by stirring at room temperature. The catalyst is finally obtained by filtration followed by drying.

3.4. Manganese Containing Heterogeneous Catalysts

Manganese containing heterogeneous catalysts are widely reported in the literature. They can be prepared either by direct incorporation using solvothermal technique or by post-synthesis modification of host materials by impregnation, ion exchange or encapsulation ("ship-in-a-bottle" approach). Mn-MCM-41 has been synthesized by post-treatment synthesis of MCM-41 with an aqueous solution of manganese salt [70]. Polymer-bound chiral Mn(III)-salen complex has been synthesized using polystyrene-divinylbenzene as the host and used for enantioselective epoxidation of unfunctionalised olefins [71]. Manganese containing catalysts prepared by immobilization of the manganese complexes on porous silica have interesting catalytic properties [72]. As stated earlier, we shall discuss only about those catalysts which has been synthesized by our group. Thus in this chapter we do not attempt to discuss in details about Mn-catalysts prepared and studied by other groups.

3.5. Iron Containing Heterogeneous Catalysts

3.5.1. Mesoporous Fe (III)-Borate

For the preparation of iron borate ferric chloride ($FeCl_3$, E-Merck) and boric acid are used as iron and borate sources respectively and cetyl trimethyl ammonium bromide as the template [73]. To an aqueous solution of CTAB required amount of H_3BO_3 dissolved in water is added and homogenized by stirring. Then $FeCl_3$ dissolved in water is added slowly to this and the pH is adjusted to *ca.* 6.0 with TMAOH. The resultant brown slurry is heated at 343 K for 1day. Surfactant is removed from the material by two consecutive HCl-EtOH extractions to obtain the mesoporous iron-borate.

3.6. Cobalt Containing Heterogeneous Catalysts

3.6.1. Co (III)-Containing Mesoporous Silica

Cobalt-containing mesoporous silica is synthesized under mild acidic condition by acid hydrolysis of tetraethyl orthosilicate precursor followed by the addition of base for precipitation and condensation [74]. $CoCl_3 \cdot 5H_2O$ is used as the Co(III) source and cationic surfactant cetyltrimethylammonium bromide (CTAB) as the structure directing agent. TEOS is mixed with an aqueous solution of CTAB and then required amount of Co(III) salt dissolved in water is added to the gel and then allowed to hydrolyze slowly under acidic pH. Finally tetramethylammonium hydroxide is added to increase the pH to 4.5–6.5, when a thick bluish homogeneous gel is formed. After stirring for the desired time the mixture is hydrothermally treated at 353K for 2–3 days to obtain the bluish gray solid product after filtration and drying. Calcination of the solid in air gives the grey colored Co(III) containing mesoporous silica.

3.7. Nickel Containing Heterogeneous Catalysts

3.7.1. Nickel Complexes with N_2O Donor Ligands Immobilized on Mesoporous Silica:

Terephthalato-bridged tetranuclear polymeric Ni(II) complexes, namely $[Ni_4L^1_4(\mu\text{-tp-}\kappa_4\text{-O})(H_2O)_2(\mu\text{-tp-}\kappa_2\text{-O})]\cdot 2C_2H_5OH\cdot CH_3OH\cdot 3H_2O$ (**1**) and $[Ni_4L^2_4(\mu\text{-tp-}\kappa_4\text{-O})(H_2O)_2(\mu\text{-tp-}\kappa_2\text{-O})]\cdot 3H_2O$ (**2**) [L^1 = *N*-(3-aminopropyl)-5-bromosalicylaldimine, L^2 = *N*-(3-aminopropyl) salicylaldimine, tp = terephthalato], and dicyanoargentate-bridged polymeric complexes $[Ni(L^1)(H_2O)\text{-}\{Ag(CN)_2\}]_\infty$ (**3**) and $[Ni(L^3)(MeOH)\{Ag(CN)_2\}]_\infty$ (**4**) [L^3 = *N*-(3-amino-2,2-dimethylpropyl)-5-bromosalicylaldimine] are heterogenized over MCM-41 type two dimensional hexagonal mesoporous silica [75]. For loading the Ni(II) complexes, the mesoporous silica is suspended in dimethyl formamide solution containing the complexes and stirred at room temperature.

3.8. Copper Containing Heterogeneous Catalysts

3.8.1. Copper Complexes with N_2O Donor Ligands Immobilized on Mesoporous Silica

Copper(II) complexes of the type $[Cu_4(O)(\mathbf{L''})_2(CH_3COO)_4]$ with N_2O-donor Schiff-base ligands are heterogenized over 2D-hexagonal mesoporous silica [76]. Four different types of complexes are synthesized where the ligands are $\mathbf{HL^1}$ = 4-methyl-2,6-bis(cyclohexylmethyl-iminomethyl)phenol, $\mathbf{HL^2}$ = 4-methyl-2,6-bis(phenylmethyliminomethyl)phenol, $\mathbf{HL^3}$ = 4-methyl-2,6-bis(((3-tri-fluoromethyl) phenyl) methyliminomethyl)phenol, $\mathbf{HL^4}$ = 4-methyl-2,6-bis(((4-tri-fluoromethyl) phenyl) methyliminomethyl)phenol. Immobilization of the metal complexes are carried out by dispersing the mesoporous silica in a solution containing the metal complex dissolved in dry acetonitrile, followed by stirring at room temperature. Then final material is obtained after filtration followed by drying.

3.8.2. Immobilized Schiff Base Complexes of Copper (II)on Non-Porous Silica

The immobilization of the copper complexes on non-porous silica surfaces are carried out in three steps at room temperature. Amorphous silica is ground and suspended in an ethanolic solution of 3-aminopropyl-trimethoxysilane to obtain the amino-propyl modified silica. The solid material is then stirred with an excess aqueous solution of cupric chloride to obtain the copper loaded silica. The copper-loaded samples are then reacted with solutions of ligands in methanol to obtain the immobilized copper complexes. Two types of ligands have been used in our works, viz. 8-hydroxyquinoline [39] and N-(hydroxyphenyl)salicyldimine [38]. In order to compare their reactivities before and after heterogenization, homogeneous complexes are synthesized by reacting the ligands with copper chloride in a methanolic solution.

3.8.3. Polystyrene-anchored Cu(II) catalyst

Macroporous beads of polystyrene are available commercially which can be functionalized to generate suitable binding sites which can coordinate with metal ions. For example, nitro group can be introduced into the polystyrene framework by nitration with HNO_3/CH_3COOH, which can be further reduced to an amino group. Amino groups are well known to undergo Schiff-base condensation reactions with aldehydes. When salicylaldehyde is allowed to react with amino-polystyrene a polymer anchored Schiff-base ligand is produced which can bind with metal ions to give heterogenized metal catalyst. In this present chapter we will discuss on a Cu(II) catalyst prepared by this method [42]. Polymer anchored Schiff base ligand and cupric chloride solution are taken in a methanolic solution and stirred under refluxing condition to get the polymer anchored Schiff-base catalyst (Scheme 2).

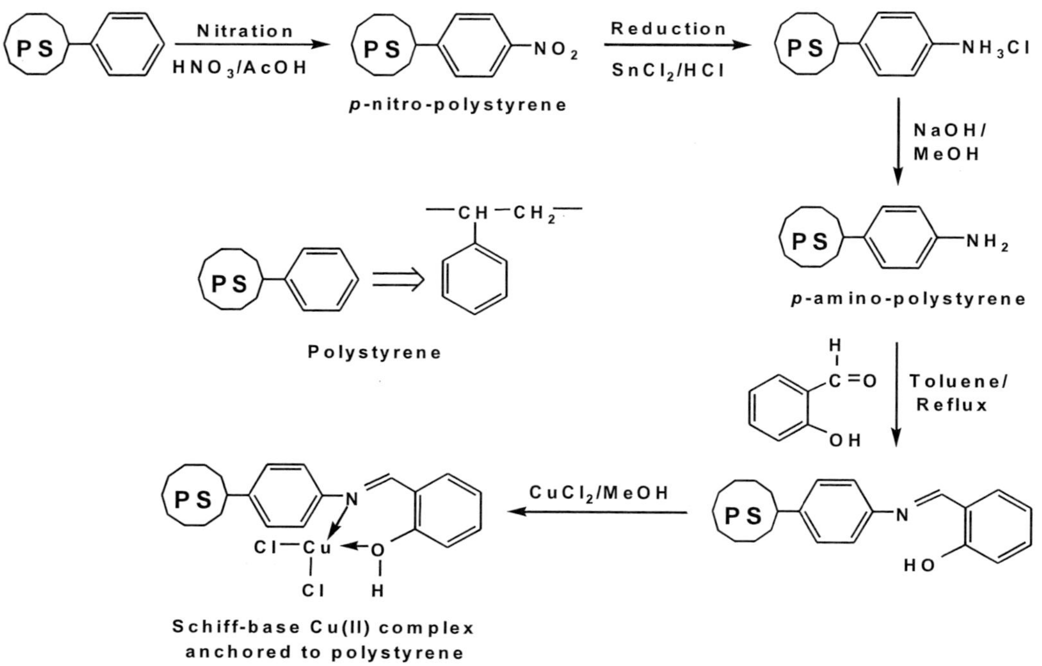

Scheme 2. .Synthesis of polystyrene anchored Cu(II) complex.

4. CHARACTERIZATION

4.1. Powder X-Ray Diffraction

Powder X-ray diffraction (PXRD) is one of the basic techniques to characterize nanoporous and nanostructured materials. PXRD patterns collected at low angles are important for the characterization of nanoporous materials and to have an understanding of the size as well as ordering of the pores. Small angle X-ray diffraction patterns of the titanium silicate samples synthesized by using a mixed surfactant system [48] of CTAB and Brij-35 (Section 3.1.1) are shown in Figure 1. Four peaks corresponding to the 100, 110, 200 and 210 planes of the 2D hexagonal mesophase [5, 6] are observed for the as-synthesized (Figure 1a) and calcined samples (Figure 1b). This type of XRD pattern indicates a highly ordered structure for the nanoporous materials with pore diameters *ca.* 2-3 nm. XRD patterns which exhibit a single peak in the low angle region are related to the presence of disordered mesophases. The position of the peak in the diffraction pattern (i.e. 2θ value) can be correlated to the size of the pores using Bragg's law. For microporous samples the position of the peak is shifted to higher values of 2θ for obvious reasons. The XRD pattern for microporous chromium phenylphosphonate (Section 3.3.2) is shown in Figure 2 [68]. This type of diffraction pattern corresponds to a lamellar type structure in the material. Thus by the help of PXRD it is possible to determine the nanostructure in a material.

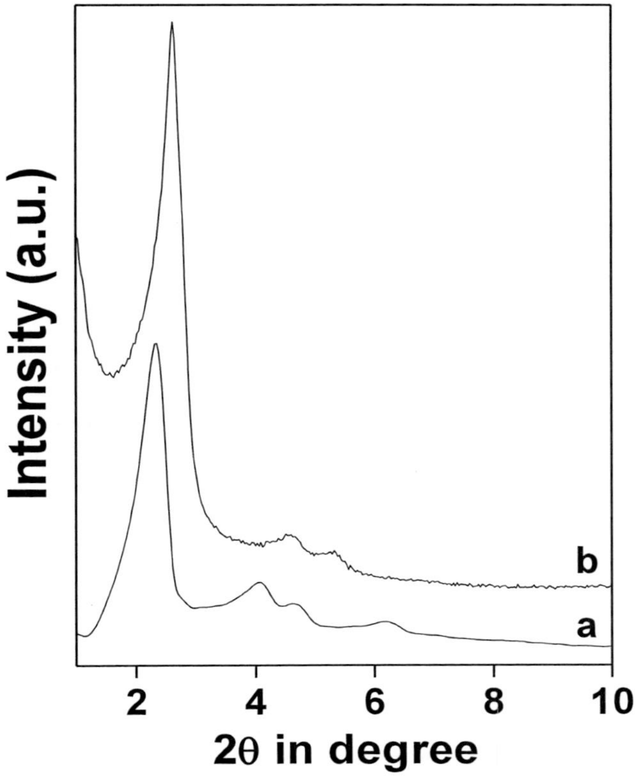

Figure 1. XRD patterns of as-synthesized (a) and (b) calcined mesoporous titanium silicate.

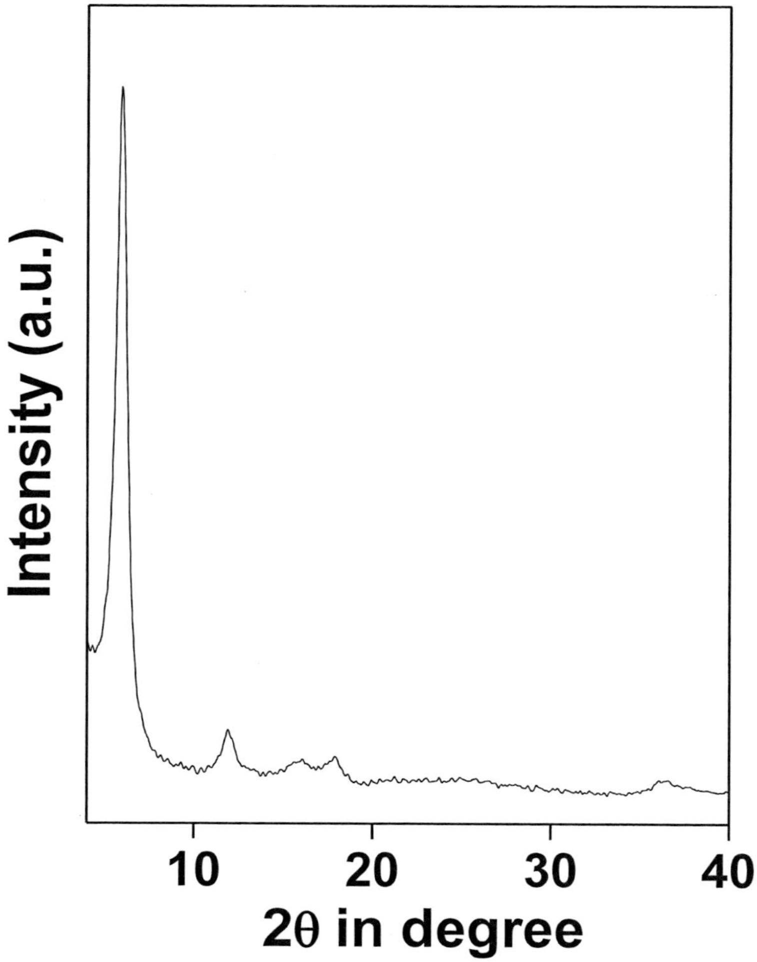

Figure 2. XRD pattern of super microporous chromium oxophenylphosphate.

4.2. FT-IR Spectroscopy

FT-IR spectroscopy is a common technique to determine the binding mode of various functionalities in a material. This technique has been successfully used for the characterization of heterogeneous catalysts. The FT-IR spectra for the template-extracted mesoporous organic-inorganic hybrid polyvinylsiloxane and its titanium analog (Section 3.1.2) are shown in Figure 3 as a representative case. For both the samples distinct broad band at 3600-3200 cm^{-1} are seen attributed to the –OH groups; and bands at 1090 and 900 cm^{-1} appear for Si–O–Si and Si–OH, respectively. The presence of organic hydrocarbon species as a bridging group in these hydrid materials can be confirmed from the FT-IR analysis. Bands 2850 and 810 cm^{-1} appear due to the aliphatic C-H stretching and C-H out-of-plane of the organic group. In addition to that, titanium silicates show strong bands at 960-965 cm^{-1} which is attributed to the Si-O-Ti stretching vibration [77]. In the FT-IR spectra (not shown here) of the Schiff base complexes of copper (II) immobilized on non-porous silica (Section 3.8.2), a band at 1573 cm^{-1} appears which can be assigned to the bending vibration of the primary amine and bands at 1458 and 1438 cm^{-1} are observed for C–H and $–CH_2$ bending vibrations, respectively. For the supported copper complex catalyst the band for the primary amine of aminopropyl group is shifted to 1560 cm^{-1}, suggesting the binding in the catalyst. Thus we see that FT-IR spectra can give inevitable information regarding the structural aspect of the catalysts.

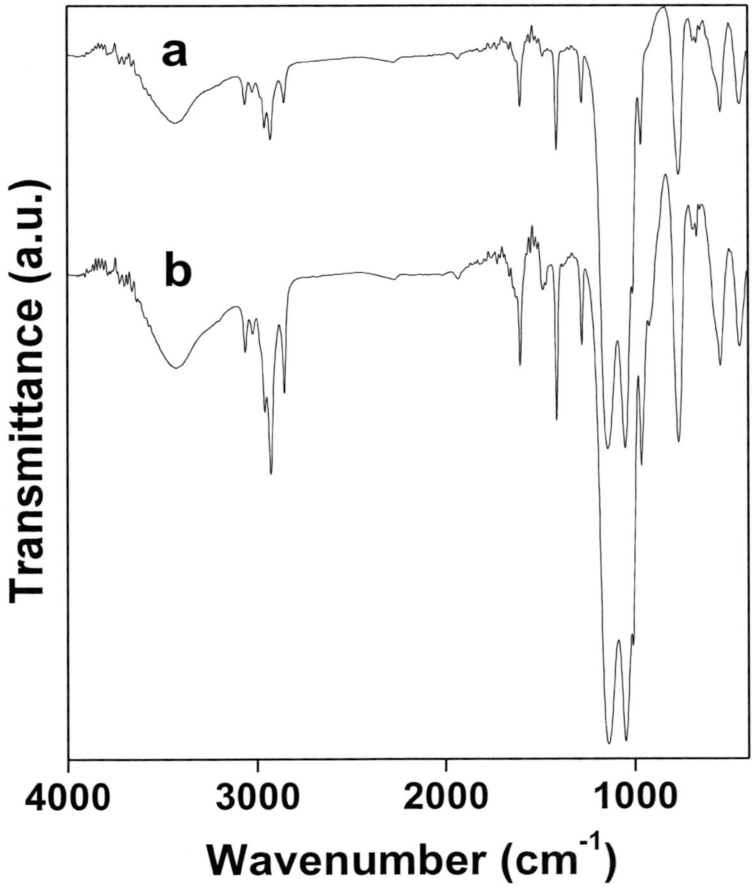

Figure 3. FT IR spectra of template-extracted mesoporous organic-inorganic hybrid polyvinylsiloxane (a) and its titanium analogue (b).

4.3. UV-Visible Spectroscopy

UV-Visible spectroscopy is a very effective tool to characterize transition metal containing materials because of the presence of d electrons within the metal atoms that can be excited from one electronic state to another. All transition metals have their characteristics electron transitions. For example, titanium silicate samples show strong high-energy absorption band at 200-290 nm due to the electronic transition from O^{2-} to Ti^{4+} which corresponds to a highly dispersed tetrahedral Ti (IV) in isolated silica environment. The UV–

Visible diffuse reflectance spectra of pure Amberlite resin (Figure 4a), vanadium loaded resin (Figure 4b) and V-resin in the presence of H_2O_2 (Figure 4c) show broad absorbance in the region of 700-800 nm which is assigned to the overlapping of d_{xy} to d_{xz} and d_{xy} to d_{x2-y2} transitions in vanadium species [78]. Strong bands observed at 236, 265 and 340 nm are characterized as π-π* charge transfer transitions originating from styrene-DVB moieties present in the resin. During the catalytic process, on addition of H_2O_2 the dark green vanadium loaded resin changes its color to orange. In the spectra two new bands at 390 and 450 nm appear which are attributed to the formation of hydroperoxo vanadium species and corresponding LMCT transition after the addition of H_2O_2 [79, 80]. For chromium containing samples characteristic absorption bands appear in the 200-620 nm wavelength regions with maxima *ca.* 430 and 600 nm and for molybdenum the maxima are obtained *ca.* 385 nm due to ligand-to-metal charge transfer transitions. For Mn, the absorptions take place ca. 270 and 500 nm attributed to the charge transfer transition from O^{2-} to Mn^{3+} in tetrahedral coordination and crystal field transitions of Mn^{2+}, respectively [70]. For octahedral chemical environment of Fe(III) in iron borate (Section 3.5.1) broad absorption band having maximums at *ca.* 365 nm and 435 nm are observed whereas for iron (III) silicates [81] the absorption occurs at much higher energy. Cobalt containing mesoporous silica shows a broad d–d transition in the 360–625 nm wavelength regions. For octahedral Ni in its d^8 configuration, bands in the 390–392 and 571–588 nm wavelength regions where as for Ni(II) ions in a square planar geometry band at *ca.* 460 nm appear [82].

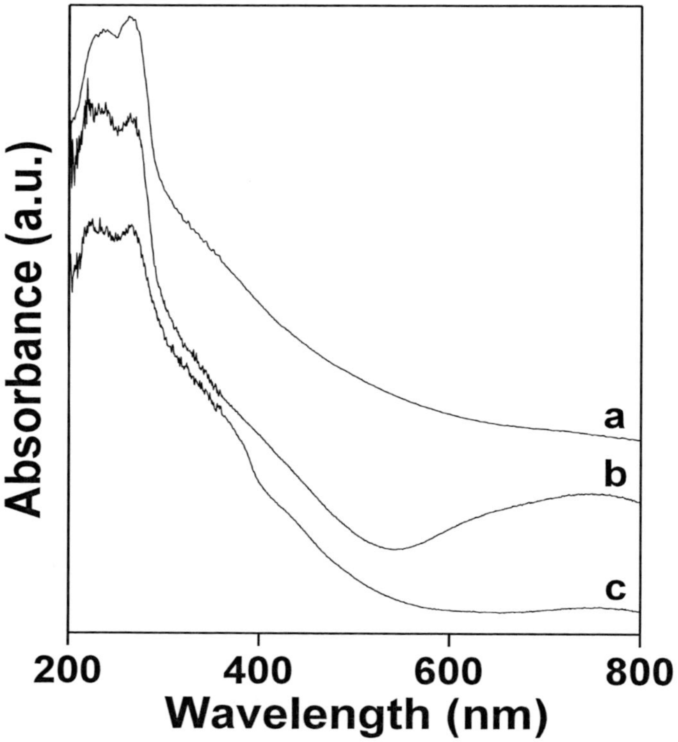

Figure 4. UV-visible diffuse reflectance spectra of amberlite resin (a), vanadium loaded amberlite resin (b) and vanadium loaded amberlite resin in presence of H_2O_2 (c).

In Figure 5, the UV-Vis spectra of Schiff base complex of copper (II) (Section 3.8.2) in acetonitrile (a), Cu(II)-complex immobilized on non-porous silica taken in acetonitrile and acetic acid (b) and Cu(II)-complex immobilized on non-porous silica taken in acetic acid-acetonitrile solution with H_2O_2 (c) are shown. Two absorption maximum at 436 and 250 nm correspond to intra-ligand and ligand to metal charge transfer transitions, respectively [83]. The catalytic reactions are carried out in acetic acid where two additional bands are formed at 326 and 503 nm, where the latter is assigned to $O_2^{2-}\sigma^* \rightarrow$ Cu(II) [84] transition. The Cu(II)-complex when immobilized on non-porous silica exhibits two distinct bands at *ca.* 250 and 700 nm, due to LMCT and d-d transition respectively.

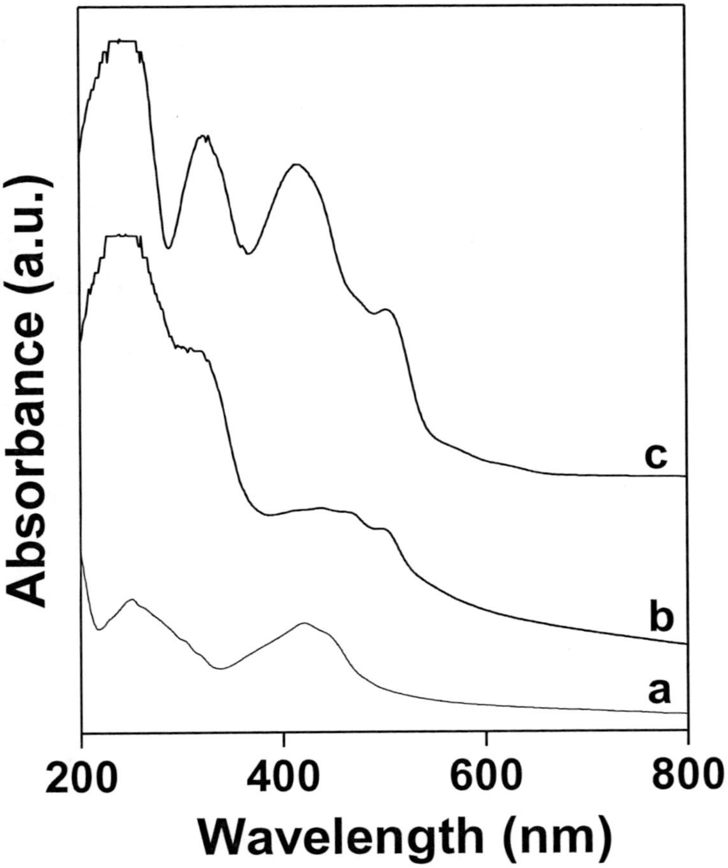

Figure 5. UV-visible diffuse reflectance spectra of Cu(II)-complex in acetonitrile (a), Cu(II)-complex immobilized on non-porous silica taken in acetonitrile and acetic acid (b) and Cu(II)-complex immobilized on non-porous silica taken in acetic acid-acetonitrile solution with H_2O_2 (c).

4.4. Surface Area Analysis

For the designing of a heterogeneous catalyst on a porous support the extent of porosity and the size of the pores in the host play a very crucial role in determining the activity of the catalyst. A highly porous catalytic bed is associated with higher efficiency for a heterogeneous catalyst. In order to find out the Brunauer-Emmett-Teller (BET) surface area and pore size of catalyst nitrogen sorption experiments are carried out. The nitrogen sorption isotherms for titanium silicate sample [47] synthesized using mixture of CTAB and SDS (Section 3.1.1) is shown in Figure 6. The isotherm resembles a typical type IV pattern with steep capillary condensation, characteristic of mesopores [85]. BET surface area and average pore diameter estimated using Barret-Joyner-Halenda (BJH) method are 1418 m2g-1 and 2.36 nm, respectively. In the low-pressure region of the isotherm a steep rise is observed along with an upward orientation towards the left side of the BJH pore size distribution which points towards the existence of microporosity in the material. The micropore size distribution plot obtained by using Horvath-Kawazoe (HK) method [86] showed a peak pore diameter of 0.58 nm. Thus we see that the titanium silicate is composed of a mixed micro- and meso-pore system [87]. These kinds of materials are of special interest for carrying out various selective oxidation reactions. Thus by using nitrogen sorption experiments it is possible to estimate the surface area of a supported heterogeneous catalysts as well as have an idea about the size of the pores.

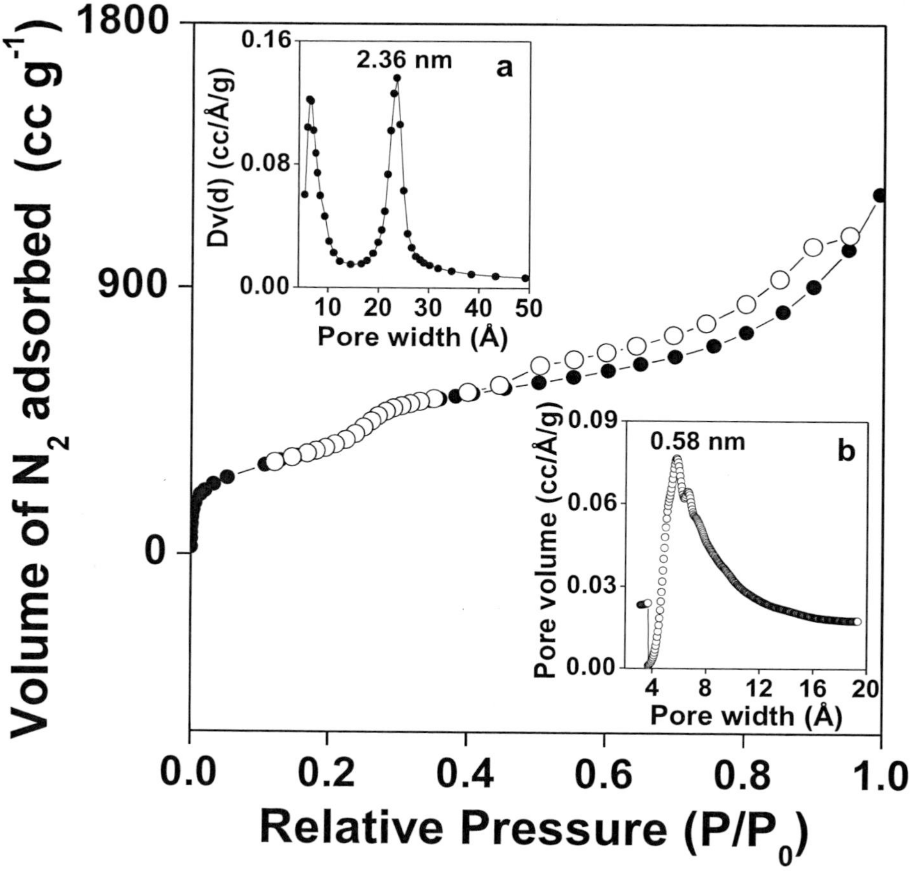

Figure 6. N_2 adsorption/desorption isotherms for mesoporous titanium silicate. Adsorption points are marked by filled cycles and desorption points are marked by open cycle. BJH (a) and HK (b) pore size distributions are given in the inset.

4.5. Transmission Electron Microscopy (TEM)

The nanoporous phases of the heterogeneous catalysts can be can be analyzed using transmission electron microscopy (TEM). The TEM image of microporous chromium oxophenylphosphate (Section 3.3.2) shown in Figure 7 depicts a multilamellar structural feature in this material. The low electron density uniform spots correspond to the presence micropores of *ca.* 1.38 nm diameters in this material. Figure 8 shows the TEM image of a highly ordered mesoporous titanium silicate [47] synthesized using mixed surfactant CTAB and SDS (Section 3.1.1). Hexagonal arrangement of the pores having different contrast than that of the pore walls can be clearly seen in the material. The selected area electron diffraction pattern for the sample is given in the inset of Figure 8 which also confirms the hexagonal ordering of the pores. Thus TEM can give useful information about the nanostructure of the metal containing catalysts.

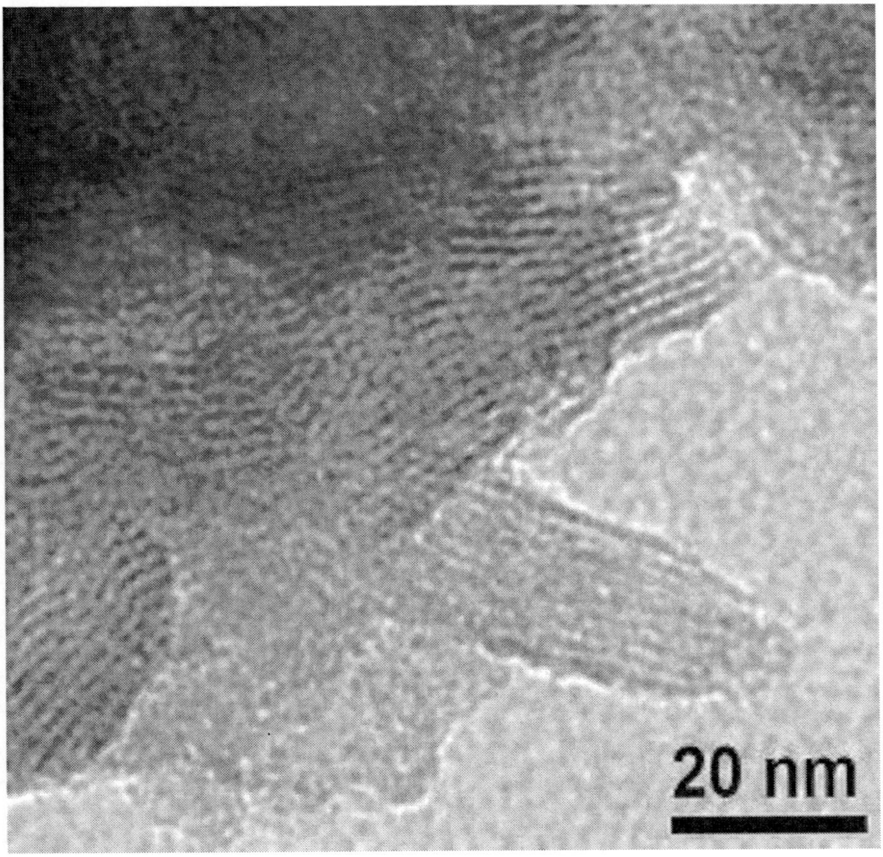

Figure 7. TEM image of supermicroporous chromium oxophenylphosphate.

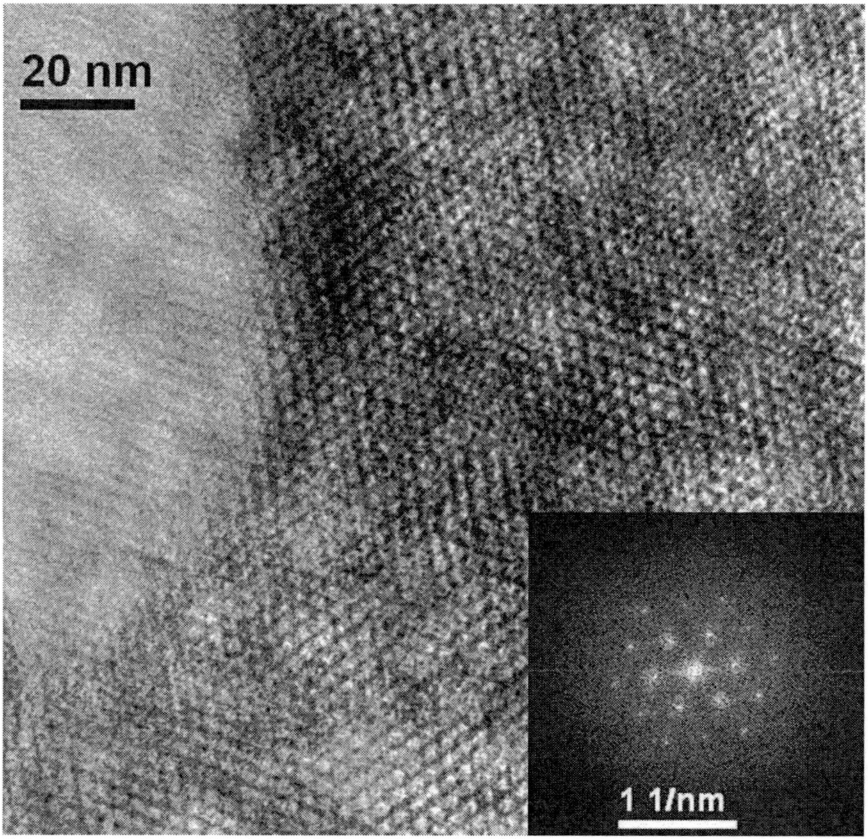

Figure 8. TEM image of highly ordered 2D-hexagonal mesoporous titanium silicate. Selected are electron diffraction pattern has been given in the inset.

4.6. Scanning Electron Microscopy and Energy-Dispersive X-Ray Spectroscopy (SEM-EDX)

Field emission scanning electron microscopy combined with an energy-dispersive X-ray (EDX) spectroscopy is a very useful analytical tool to analyze the surface morphology and chemical composition of heterogeneous catalysts. For example, TS-1 has a highly crystalline structure comprising of uniform cuboid crystals of 0.2–0.25 μm in size [88]. Figure 9 shows the SEM image of a typical TS-1 sample synthesized in the presence of H3PO4 promoter having uniform cubic small crystallites of dimensions 100–200 nm and a very narrow particle size distribution [89]. From EDX spectroscopy the composition of the materials can be determined and hence the amount of metal loading. In Figure 10, the result of the EDX analysis for microporous chromium oxophenylphosphate (Section 3.3.2) is shown as a representative case. From the data, the molar ratio of P/Cr, C/O and P/Cr are found to be 1.34, 1.51 and 1.2, respectively. Thus from EDX analysis it is possible to determine the amount of metal loading in a heterogeneous catalyst.

Figure 9. SEM image of TS-1.

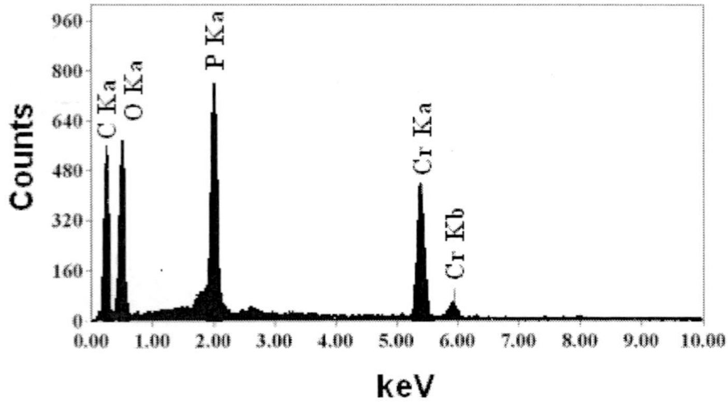

Element	keV	Mass %	Atomic %
C K	0.277	42.74	55.56
O K	0.525	37.70	36.80
P K	2.013	8.66	4.37
Cr K	5.411	10.89	3.27
Total		**100.00**	**100.00**

Figure 10. EDX data for supermicroporous chromium oxophenylphosphate.

4.7. Atomic Absorption Spectroscopy (AAS)

AAS can be used to analyze the concentration of over 62 different metals with concentrations as low as parts per billion of a gram in a sample. Thus for the transition metal containing catalysts it is a very good technique to determine the amount of active metal centres by wet chemical analysis. A known amount of the solid sample is taken and it is treated with strong acids like H_2SO_4, HNO_3, HCl and HF as required, to bring the sample in solution phase. Aqueous H_2O_2 often used in dissolving some insoluble solids. Then the solution is evaporated to dryness, extracted with water and analyzed by using AAS to find the amount of metal-loading. This technique is also performed to test whether the metal is leached out in the liquid phase during the oxidation reactions.

4.8. X-Ray Single Crystal Analysis

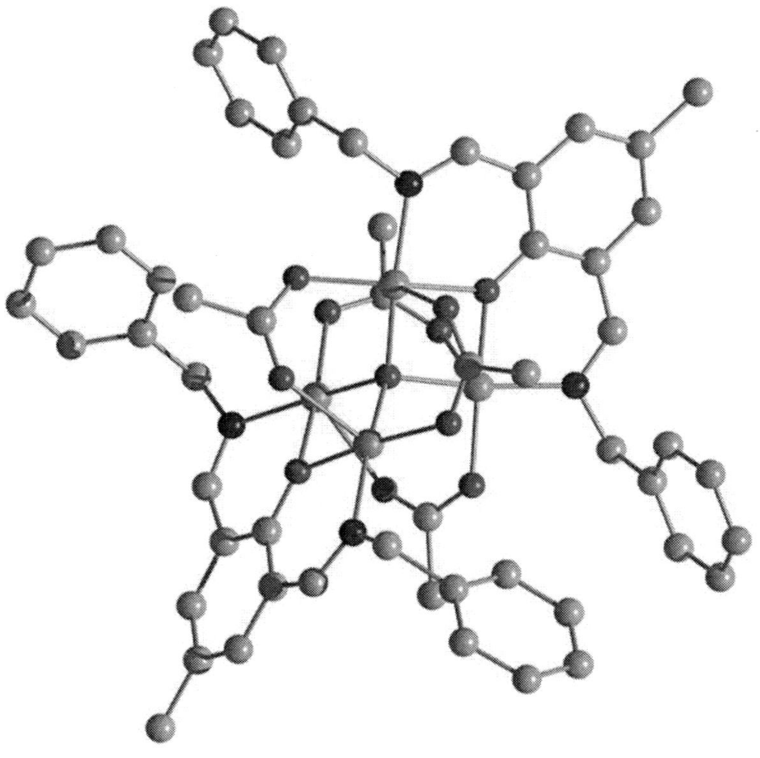

Figure 11. X-ray crystal structure of Cu(II) complex (Section 3.8.1). Hydrogen atoms are omitted for clarity. Color legend, Grey: carbon; Blue: nitrogen; Red: oxygen and Cyan: copper.

Single crystal X-ray diffraction analysis is often useful for characterization of a heterogeneous catalyst. This technique is used in cases where a metal-complex is synthesized at first and then it is immobilized on a solid support. Prior to heterogenization the structure of the complexes can be solved by this technique. In such cases the presence of metal ions in the particular material can be confirmed from the crystal structure itself. In our work we have synthesized a few metal-complexes using nickel, copper and molybdenum, as described in previous section. The crystal structures for the complexes have been determined by X-ray diffraction studies. In Figure 11, the crystal structure of $[Cu_4(O)(\mathbf{L''})_2(CH_3COO)_4]$ where $\mathbf{HL}^1$ = 4-methyl-2,6-bis(cyclohexylmethyliminomethyl)phenol, is given as a representative case. It reveals that the complex contains μ_4-oxo-bridged tetrameric copper(II) as the active center.

4.9. X-Ray Photoelectron Spectroscopy (XPS)

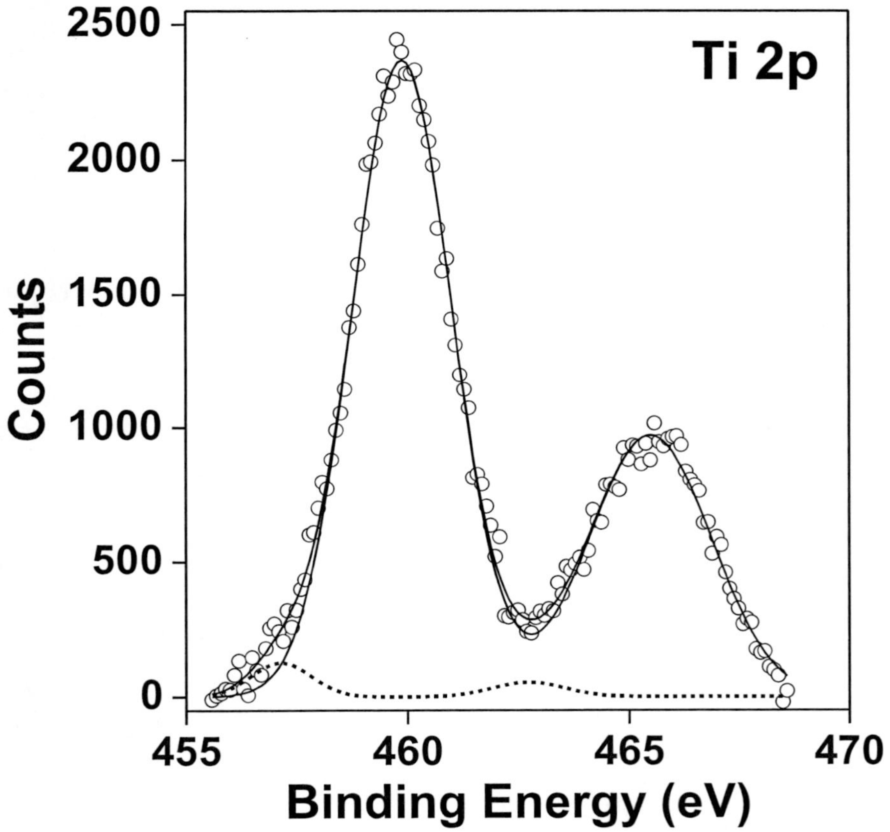

Figure 12. X-ray photoelectron spectra of mesoporous titanium silicate. Solid lines refer to the tetrahedrally coordinated framework Ti(IV) and broken lines to the octahedrally coordinated extraframework Ti(IV).

XPS is a quantitative spectroscopic technique which enables the estimation of the elemental composition, empirical formula, chemical state and electronic state of the elements present in a material. It is a surface chemical analysis technique which can be used to measure directly the loading of a metal in a heterogeneous catalyst. In Figure 12 the X-ray photoelectron spectrum of Ti2p electron in the titanium silicate sample [48] (Section 3.1.1) is shown as a representative case. The energy positions of the peaks are calibrated by fixing the position of the C1s peak at 285.0 eV. From the XPS measurement the atomic percentage ratios for Ti/Si and O/(Ti+Si) are found to be 4.1 and 2.79 %, respectively and the peak positions for Si2p and $Ti2p^{3/2}$ are 103.35 eV and 459.9 eV, respectively. For titanium silicalite framework, Ti(IV) located at the tetrahedral position have binding energies 460 and 466 eV for $Ti2p^{3/2}$ and $Ti2p^{1/2}$, respectively [90, 91], whereas the octahedrally coordinated extra-framework Ti(IV) have binding energies 458 and 464 eV for $Ti2p^{1/2}$ and $Ti2p^{3/2}$, respectively [90]. In this case, $Ti2p^{3/2}$ and $Ti2p^{1/2}$ peaks are present at 459.9 and 465.7 eV which confirms that most of the Ti(IV) species present in the sample are tetrahedrally coordinated with silica in the framework. The deconvoluted pattern (Figure 12) of the spectra also points to the fact that 97% of the total Ti(IV) are tetrahedrally coordinated (solid line) in the framework, whereas 3% are octahedrally coordinated (broken line) as extra-framework Ti(IV). In a similar way the framework composition for other transition metal containing catalysts can also be analyzed using XPS studies.

4.10. Magic Angle Spinning Nuclear Magnetic Resonance (MAS NMR) Spectroscopy

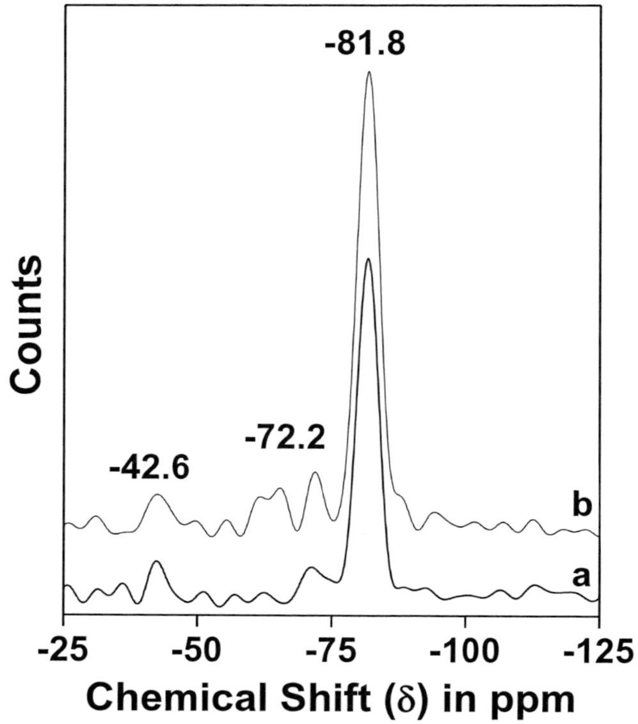

Figure 13. ^{29}Si MAS NMR spectra of template extracted organic-inorganic hybrid polyvinylsiloxane (a) and its titanium-analogue (b).

Solid state MAS NMR spectroscopy plays a crucial role and assists in the characterization and refinement of complex structures of pure and inorganic-organic hybrid materials. For silica samples the degree of functionalization can be quantitatively determined by using ^{29}Si MAS spectra, while ^{13}C and ^{31}P MAS NMR measurements can help in identifying the functional groups. The ^{29}Si MAS NMR spectra of organic-inorganic hybrid polyvinylsiloxane and its titanium analogue [58] are shown in Figure 13. For both the samples, downfield chemical shifts at –81.8, –72.2 and –42.6 ppm are observed. Comparison with hybrid mesoporous silica, where ^{29}Si peaks appear at chemical shifts values of *ca.* –80 ppm and –71 ppm corresponding to T^3 and T^2 species, respectively [92, 93], shows that incorporation of aliphatic hydrocarbon bridges takes place in these materials. The ^{13}C MAS NMR spectrum of titanium phenylphosphonate (Section 3.1.2) is shown in Figure 14. The material shows two strong signals, one at 127.4 ascribed to the C_2, C_3, C_5 and C_6 carbon atoms of the phenyl groups and the other at 131.6 ppm due to the C_4 atom of the phenyl group attached to the phosphonate moiety [55]. A downfield shift is observed in this material compared to phenylphosphonic acid due to the more confined environment of the phenyl group in the solid matrix. Thus from ^{13}C NMR data it can be concluded that the phenyl group in the phosphonate moiety remains intact in this material. The ^{31}P MAS NMR spectrum of the same material is given in Figure 15. It shows a very strong and sharp signal at 13.8 ppm attributed to the phosphorus moiety in the $C_6H_5\underline{P}O_3$ and a weak signal at 5.6 ppm due to the phosphorous in $C_6H_5\underline{P}(OH)O_2$ present in the framework. This result also confirms the presence of phenyl group in the material. Thus we see by using solid state NMR it is possible to identify the presence of organic functional groups, their structures and chemical environment in a given catalyst.

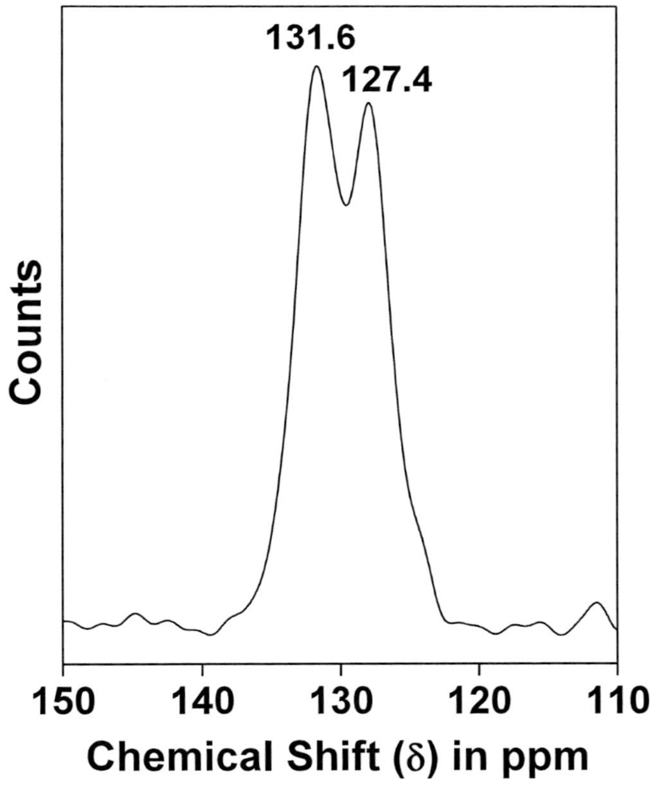

Figure 14. ^{13}C CP MAS NMR of microporous titanium oxophenylphosphonate.

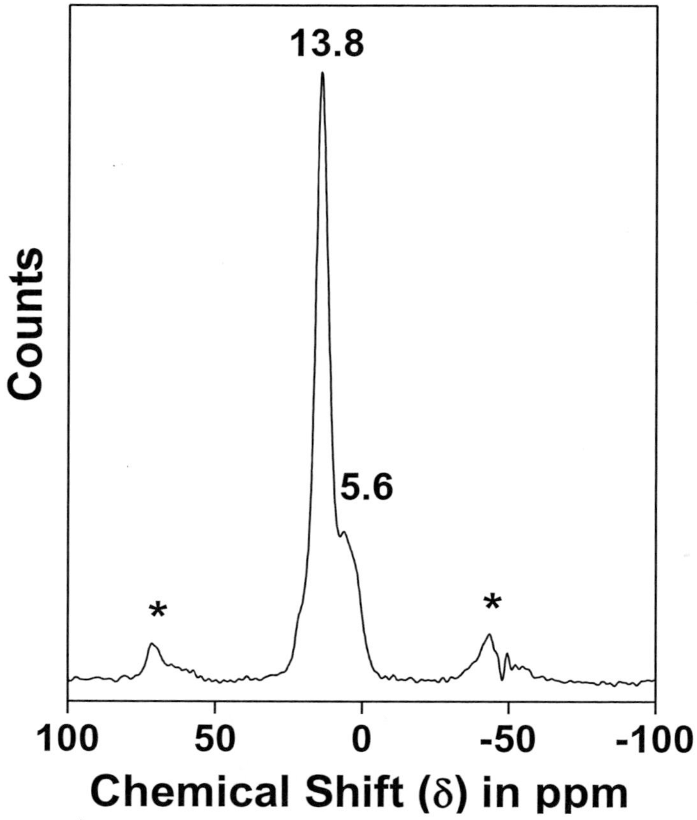

Figure 15. ^{31}P MAS NMR spectra of microporous titanium oxophenylphosphonate. Spinning side bands are marked in asterisks.

4.11. Thermo-Gravimetric and Differential Thermal Analysis (TG-DTA)

TG-DTA is one of the reliable techniques to determine the thermal stability of nanostructured materials which is an essential factor for their application in catalysis. The commonly investigated processes are thermal stability and decomposition, dehydration, oxidation, determination of volatile content and other compositional analysis. For hybrid organic-inorganic and polymer or resin supported materials it is very essential to know about their thermal stability before studying their application in catalysis. The TG and DTA plots of a representative sample of mesoporous organic-inorganic hybrid titanium polyvinylsiloxane [58] have been demonstrated in Figure 16. As seen from the plot the first weight loss takes place in the temperature range of 298-373 K which is due to desorption of physisorbed water (*ca.* 2.7 wt %). This is followed by three sharp decreases in the weight in the temperature range 373-1073K, attributed to bond breaking and removal of the hydrocarbon bridges present in hybrid titanium-siloxane framework. The total losses for these three steps are *ca.* 12.2, 7.5 and 10.0 wt%. Thus thermal analysis gives a clear picture regarding the stability of a material which can be used to characterize other heterogeneous catalysts in a similar way.

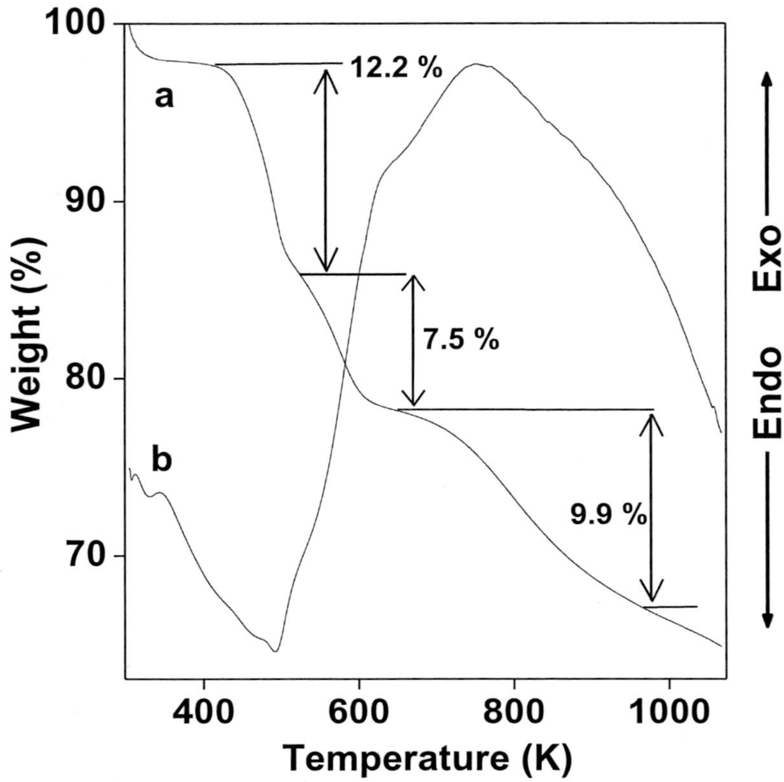

Figure 16. TG (a) and DTA (b) plots for extracted mesoporous organic-inorganic hybrid titanium polyvinylsiloxane.

4.12. Electron Paramagnetic Resonance Spectroscopy (EPR)

EPR spectroscopy is a powerful technique for studying the presence of chemical species at the catalyst surface that have one or more unpaired electrons. For vanadium and copper containing heterogeneous catalysts EPR spectroscopy is often used to confirm the loading of the metal ions. In Figure 17 the representative EPR spectra of vanadium containing amberlite resin [65] (Section 3.2.2) before and after the ammoxidation reactions are shown. The EPR spectra of the fresh catalyst and the oxidant treated catalyst display axial spectra at room temperature with well-resolved eight ^{51}V (I = 7/2) hyperfine lines [32-35] which indicates the presence of V^{4+} species in both the samples. The corresponding spectral data and various signal parameters are given in Table 1, which confirms the loading of the vanadium over the resin. It is seen that $g_{II} > g_{\perp}$ and $A_{II} > A_{\perp}$ suggesting the presence of axially compressed d_{xy} configuration. For copper (II) complex immobilized on non-porous silica [38, 39] (Section 3.8.2), the spectra shows four hyperfine lines in low and high field regions (not shown here). The hyperfine features are characteristic of copper nucleus (I = 3/2) interacting with the unpaired electron (S = ½) of the Cu^{2+} ion in its d^9 electronic configuration [94]. EPR spectra of the catalyst before and after the oxidation reactions are similar. Thus we see that for metal ions with unpaired electrons EPR spectra can give valuable information regarding its configuration.

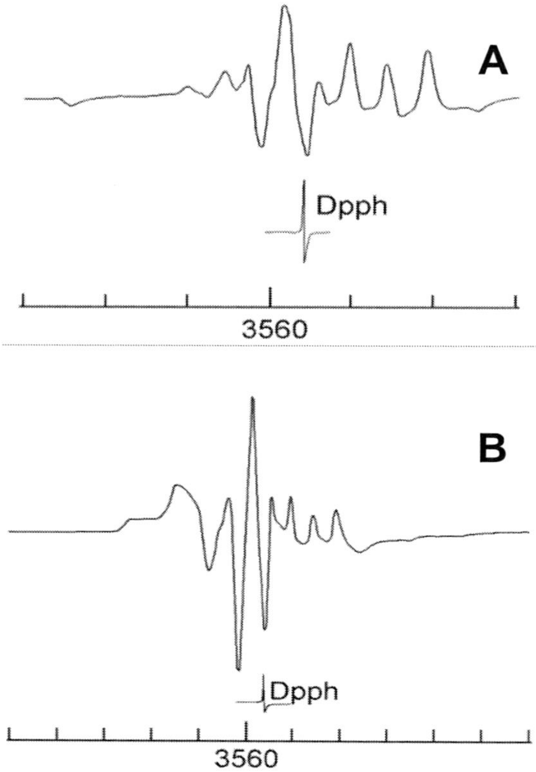

Figure 17. Room temperature EPR spectra of vanadium loaded amberlite resin (a) and and vanadium loaded amberlite resin in presence of H_2O_2 (b).

Table 1. EPR parameters for V-amberlite before and after the catalytic reactions

V-Amberlite	$g_{\parallel}$	$g_{\perp}$	$A_{\parallel}$ / 10^{-4} cm^{-1}	$A_{\perp}$ / 10^{-4} cm^{-1}	g_{av}	A_{av}
Before reaction	1.977	1.952	102.5	24.5	1.96	50.5
After reaction	1.962	1.431	130	22.3	1.61	58.2

Experimental errors: $\triangle g = \pm 0.002$ and $g_{av} = 1/3[2g_{\perp} + g_{\parallel}]$
$\triangle A = \pm 3\times10^{-4}$ cm^{-1} and $A_{av} = 1/3[2A_{\perp} + A_{\parallel}]$

5. CATALYSIS

We have discussed about synthetic routes of different materials containing transition metals as catalysts for different oxidation reactions. These materials demonstrated excellent catalytic activity in a wide range of oxidation reactions, e.g., oxidation of C–H, N–H and S–H bonds, epoxidations, hydroxylations, ammoximations etc. Generally, all catalytic reactions are carried out by liquid phase oxidation of substrate. Substrate, catalyst and oxidant are taken in a two necked round bottom flask. The reaction is carried out in an oil bath when elevated temperature is needed. Aliquots are collected at regular time interval. The reaction mixture is analyzed by gas chromatography and/or gas chromatography mass spectroscopy. In this section we will put light on the oxidation reactions we have studied with the heterogeneous catalysts prepared. These are epoxidation of alkene, oxidation of alkanes, ammoximation, Baeyer-Villiger oxidation of ketones, and benzylation and hydroxylation of activated aromatics.

5.1. Epoxidation

Epoxidation of alkenes, allylic alcohol etc. is very important reaction in synthetic organic chemistry as epoxides are versatile synthetic compounds. An epoxide is a cyclic compound constituting of two carbon and oxygen atoms. As it is in strained condition, so it is more reactive in nature than other ethers. These compounds can be used for constituting suitable building blocks for the synthesis of many products and fine chemicals [95]. Epoxides have been prepared by oxidation of olefins in the presence or absence of metals. In non-metal-catalyzed epoxidation reactions, the most familiar oxidizing agents are per acid, peroxide [96], and oxaziridines [97]. Several studies have been reported on the preparation of the epoxides employing transition metal compounds as the catalysts in the presence of different terminal oxidants such as NaOCl, peracid, *tert*-butyl hydroperoxide, hydrogen peroxide, molecular oxygen, iodosylbenzene *etc*. This reaction generally, becomes eco-friendly when hydrogen peroxide is used as the oxidant because water is the only waste product. Such oxidation reactions are performed following both homogeneous and heterogeneous media. We have generally carried out this liquid phase oxidation reaction in heterogeneous medium. The heterogeneous catalysts are titanium silicates, chromium oxophenylphosphate, nickel complexes immobilized on mesoporous silica, copper complexes immobilized on silica or anchored on polymer, oxodiperoxomolybdenum(VI) on mesoporous silica, etc. Highly porous organic-inorganic hybrid titanium silicate showed good catalytic activity towards oxidation of R-(–)-carvone to the corresponding epoxide using dilute aqueous hydrogen peroxide as the oxidant [58]. Microporous silicotitaniumphosphate and organic–inorganic hybrid silicotitanium phosphate showed high conversion of acrylic acid, cyclohexene, styrene and α-methyl styrene with 67.5, 95.5, 98.8 and 98.5% yield respectively to their corresponding epoxides with high selectivity [61]. Titanium containing inorganic–organic hybrid mesoporous materials exhibits an enhanced activity in epoxidation of α-pinene using hydrogen peroxide owing to the presence of tetrahedral active sites on the surface that are highly dispersed in nature. In this case the selectivity towards *α*-pinene oxide was over 99%. [60]. A generalized reaction mechanism for epoxidation of an olefin (e.g. styrene) considering a titanium catalyst as a representative case is shown in Scheme 3. The titanium-hydroperoxo species (I) generated *in situ* in the presence of the oxidant H_2O_2 is the reactive intermediate and it facilitates the formation of epoxide.

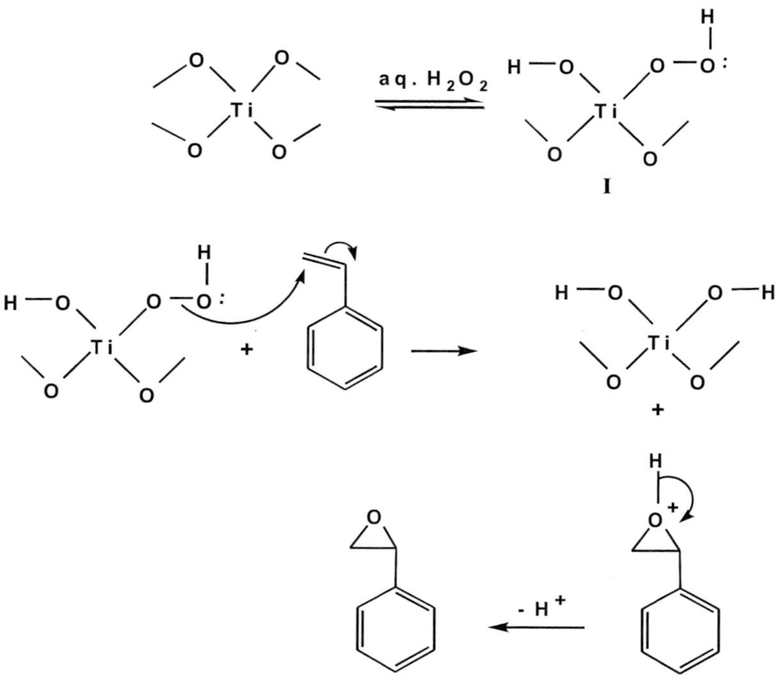

Scheme 3. General reaction mechanism for epoxidation of an olefin in presence of titanium catalyst.

Table 2. Epoxidation reactions over metal catalysts

Metal catalyst	Substrate	Oxidant	Reaction Time (h)	Conversion (%)	Product selectivity (%)	TOF
Ti Section 3.1.1	Styrene	H_2O_2	24	91.4	Epoxide = 70.5	3.55
Ti Section 3.1.2	R-(–)-carvone	H_2O_2	10	72.3	Epoxide = 68.2 Diketone = 13.3 Others = 17.5	5.9
Cr Section 3.3.2	Allyl alcohol	TBHP	24	31.1	Glycidol = 55.67 Glycerol = 44.33	1.8
Mo Section 3.3.3	Norbornene	H_2O_2	12	85	Epoxide = 85	150
Ni Section 3.7.1	α-methylstyrene	TBHP	24	56	Epoxide = 86	6.4
Cu Section 3.8.1	Cyclohexene	TBHP	10	79	Epoxide = 4 Cyclohexene-2-ol = 22 Cyclohexene -2-one = 74	18.2

Solvent = Acetonitile for all the above reactions
TOF = Turn over frequency, moles of substrate converted per mole of metal per hour

Supermicroporous chromium oxophenylphosphate material exhibited conversion of styrene, cyclohexene, allyl alcohol with moderate to good yield in the presence of *tert*-butylhydroperoxide as the oxidant [68]. Nickel complexes with N_2O donor Schiff-base ligands immobilized on mesoporous silica catalyzed the epoxidation of different olefins with moderate conversion [75]. Following similar procedure, some copper(II) complexes were also employed in epoxidation reactions [76]. Polystyrene bound copper(II) complex showed modest catalytic activity towards epoxidation [42]. Oxodiperoxomolybdenum(VI) complexes of 8-quinolinol immobilized over highly ordered 2D-hexagonal mesoporous silica showed very good catalytic activity and excellent recycling efficiencies for the epoxidation of different olefins in the presence of $H_2O_2/NaHCO_3$ system. High catalytic efficiency could be attributed to the heterogenization of soluble metal complexes on the high surface area mesoporous host [69]. In Table 2, a brief summary about some of the catalysts which can be used for epoxidation of various substrates and their activities in the different reactions are given.

5.2. Cycloalkane Oxidation

Study of oxidation of cycloalkanes is an important field of research because their products are essential raw materials for different chemical industries. For example, cyclohexane is oxidized to cyclohexanol and cyclohexanone. Cyclohexanol is mainly used in the manufacture of adipic acid, which is again a raw material of nylon, soaps and detergents, rubber chemicals, pesticides *etc.*, whereas cyclohexanone is utilized as an industrial solvent and activator in oxidation reactions. Copper(II) complexes have been used as the catalysts for such conversion [98-100]. We have shown that highly ordered mesoporous Cr-MCM-41 materials can act as active catalysts for the oxidation of cyclohexane and cyclooctane using dilute aqueous hydrogen peroxide or *tert*-butyl hydroperoxide as the oxidant under mild liquid phase reaction conditions. Cyclohexanone and cyclooctanone are the major product respectively, whereas cyclohexanol and cyclooctanol were produced in minor amount [67].

5.3. Ammoximation

Ammoximation of a ketone is carried out by using ammonia and dilute aqueous hydrogen peroxide as reagents to convert the ketone to respective oxime. Outstanding catalytic activity has been found for titanium containing ethane bridged hybrid mesoporous silsesquioxane in the ammoximation of bulky ketones using dilute aqueous hydrogen peroxide and ammonia under liquid phase conditions. Conversion of cyclohexanone can get up to 90% with more than 95% selectivity towards oxime formation. In some other conditions, the conversion is lower compared to the former but selectivity is 100% [30]. Microporous titanium silicate is also an extremely efficient catalyst for the liquid phase ammoximation of cyclohexanone and cyclododecanone to their corresponding oximes. They are found to be highly active with very high selectivity towards oxime formation [46]. The catalytic activity of some of the heterogeneous catalysts in ammoximation reactions is given in Table 3.

Vanadium resin also exhibits good catalytic activity for ammoximation of cyclohexanone, cyclooctanone and cyclododecanone [65] and shows moderate conversion and selectivity towards oxime formation. In the presence of H_2O_2 and ammonia, it either catalyzes the formation of hydroxylamine from ammonia through N–H bond oxidation, which can react with ketones to give the respective oximes, or it oxidizes the imine formed between the reaction of ketone and ammonia to its respective oxime. In either case one molecule each of H_2O_2 and NH_3 are necessary for the conversion of a molecule of the ketone to its corresponding oxime. It is pertinent to mention that high cation exchange ability of the resin matrix plays a crucial role as a catalyst support and helps to immobilize large amount of active V^{4+} species which provides a favourable environment to stabilize the hydroperoxo species in the catalyst [65].

Table 3. Ammoximation reactions over metal catalysts

Metal catalysts	Substrate	Substrate: NH_3:H_2O_2 Temperature	Reaction Time (h)	Conversion (%)	Product[a] selectivity (%)
Ti Section 3.1.1	Cyclohexanone	1 : 1.5 : 1.2 353 K	6	91.2	96.6
Ti Section 3.1.1	Cyclododecanone	-do-	6	75.6	100.0
Ti Section 3.1.2	Cyclohexanone	1 : 2.5 : 1.25 353 K	8	63.2	100
Ti Section 3.1.2	Cyclododecanone	-do-	8	42.0	100
V Section 3.2.2	Cyclohexanone	1 : 2.5 : 1.2 333 K	10	81.3	79.4
V Section 3.2.2	Cyclooctanone	-do-	10	56.1	75.3
V Section 3.2.2	Cyclododecanone	-do-	10	42.1	81.6

Solvent= *tert*-Butanol for all the reactions.
[a]Selectivity of corresponding oximes.

5.4. Baeyer-Villiger Oxidation

In Baeyer-Villiger oxidation reaction (BV reaction), a ketone is oxidized to an ester on treatment with peroxy acids or hydrogen peroxide. Normally used oxidants are *m*-chloroperbenzoic acid, peroxyacetic acid, peroxytrifluoroacetic acid, hydrogen peroxide, etc. Titanium silicate (TS-1) can be used for BV reaction in the presence of hydrogen peroxide to convert acetophenone to phenyl acetate [101] along with the formation of *o*- or *p*-hydroxyacetophenone. *p*-Hydroxyacetophenone is a raw material in pharmaceutical industry. Oxidation of cyclohexane using TS-1 gives ε-caprolactone, hydroxyketone and diketone under mild conditions. It is reported that TS-1 can exhibit Brønsted acidity in the presence of H_2O_2 due to formation of hydroperoxo species stabilized by a protic solvent e.g. water or alcohol (Scheme 4) [102]. Titanium hydroperoxo species stabilized by water attacks through nucleophilic oxygen at the *ortho* and *para*-positions to form respective hydroxyacetophenone. A plausible reaction path in the oxidation of acetophenone is shown in Scheme 5. The species II may act as a nucleophile, particularly in the presence of water (species III, Scheme 4). II may attack the electron deficient carbonium ion produced after the initial protonation step, hence forming 5 via a plausible intermediate 2 through the migration of the phenyl group. Similarly, the species II can also attack the partially electron deficient *o*- and *p*-positions of the deactivated aromatic ring of acetophenone, thus producing *o*- and *p*-hydroxyacetophenones.

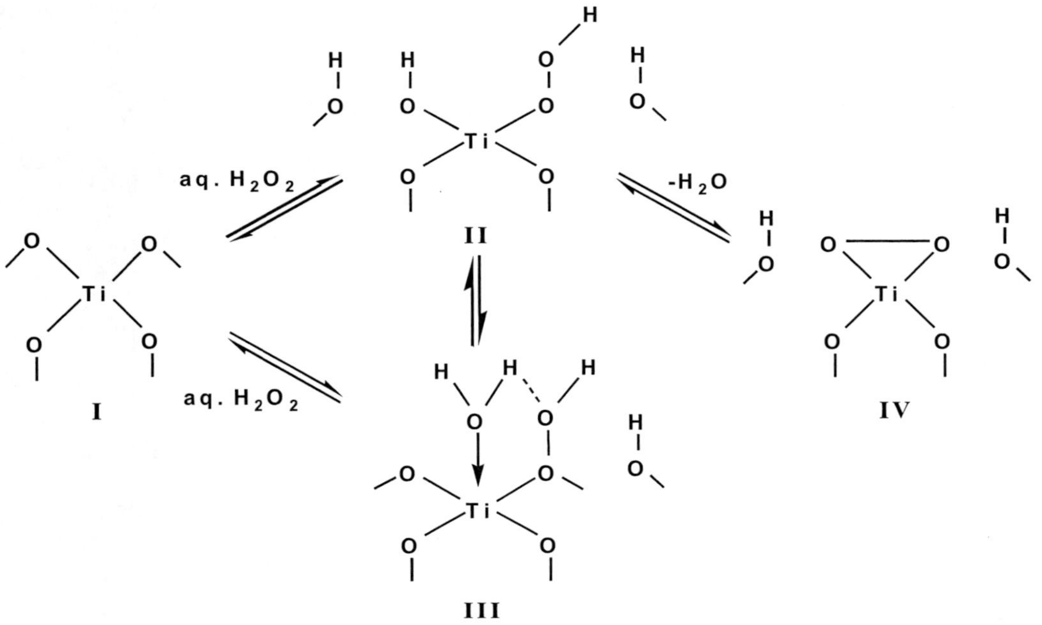

Scheme 4. Formation and reactions of titanium hydroperoxo species.

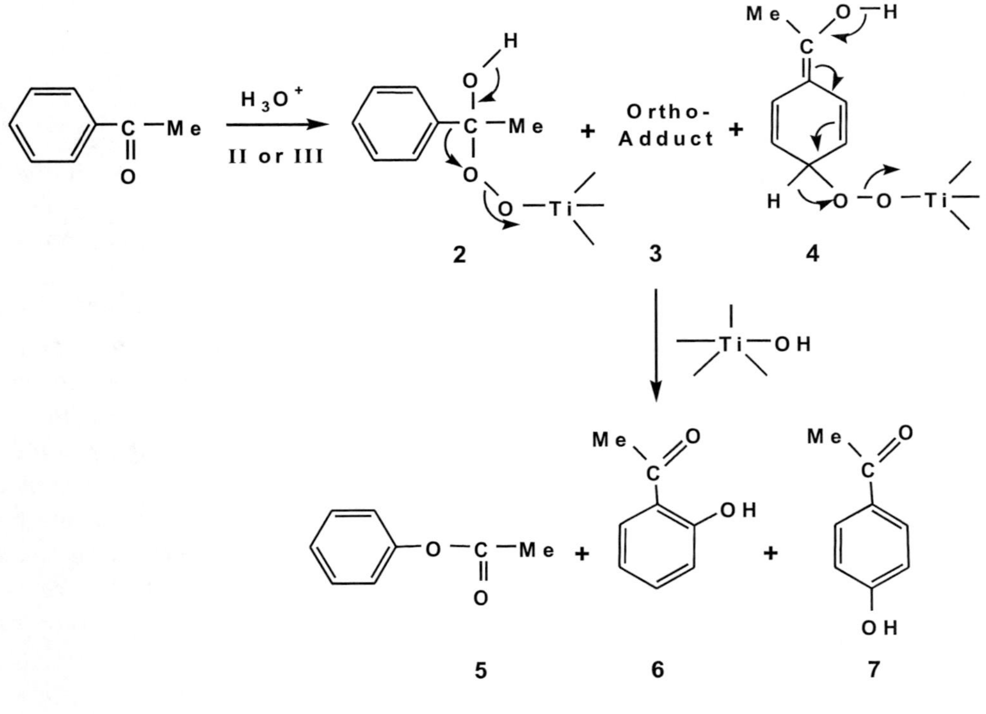

Scheme 5. Reaction mechanism for oxidation of acetophenone.

5.5. Benzylation Reaction

Friedel-Crafts alkylation is an important reaction for introducing alkyl groups into the aromatic ring systems. Solid acid catalysts having strong to moderate Lewis acid sites are usually used for benzylation reaction and benzyl chloride is used as alkylating agent. Benzylation of mesitylene and anisole are carried out over mesoporous $FeBO_3$ [73]. Results on different substrates and with different molar ratios of substrate to benzyl chloride are given in Table 4. From this result it is clear that very high yields of benzylated aromatics are obtained for both the substrates. In this benzylation reaction the desired product is the monoalkylated compound. For high substrate to benzyl chloride ratio, the product formed in the benzylation is mainly the mono-benzylated one and there was no formation of poly-condensation products.

Table 4. Benzylation reactions over mesoporous iron borate (Section 3.5.1)

Substrate	Substrate to benzyl chloride molar ratio	Conversion (%)	Product selectivity (%)	TON
Mesitylene	15:1	88.2	97.6	5.3
Mesitylene	10:1	95.6	100.0	8.7
Mesitylene	5:1	97.9	81.3	14.9
Anisole	10:1	99.2	52.6 (ortho) 47.4 (para)	9.0

Solvent= Acetonitrile for all the reactions, Temperature = 348 K
Reaction time = 1 h
TON = Turn over number = moles of substrate converted per mole of iron

5.6. Hydroxylation

Hydroxylation is a chemical reaction in which a hydroxyl (OH) group is introduced in an organic compound. Titanium silicate, vanadium silicate, cobalt silicate etc. have been used as active catalysts for the hydroxylation of different organic compounds e.g. toluene, anisole, cyclohexene, *m*-cresol etc. These reactions can be carried out under biphasic or triphasic reaction conditions. In biphase system, substrate, hydrogen peroxide and catalysts are taken in a common miscible solvent like acetonitrile (as co-solvent), whereas in triphase system, hydrogen peroxide, reactant and catalysts are dispersed in water. TS-1/H_2O_2 system is found to be efficient in oxidation of toluene, anisole, *m*-cresol, benzyl alcohol and cyclohexanol under biphasic (solid catalyst and reagents in a miscible co-solvent) and triphasic (solid catalyst, aqueous oxidant and organic reactant) conditions [103,104]. The results shown in Table 5 suggest that the conversion of each substrate is increased significantly under triphasic conditions. The reason could be that under biphasic conditions, the solvent may be competing with reactant for diffusion in the channels and adsorption at the active sites of TS-1 catalyst. This point is further supported by the fact that when *tert*-butanol, which is too bulky to enter in TS-1 (MFI topology) pores, is used as solvent in place of acetonitrile, very high *para* selectivity is obtained in the hydroxylation of anisole.

Table 5. Hydroxylation reactions over TS-1 (Section 3.1.1)

Substrate	Phase	Time (h)	Conversion (%)	Product selectivity (%)	TOF
Toluene	Biphase	12	5.5	*o*-Cresol = 69.7 *p*-Cresol = 28.4	0.42
Toluene	Triphase	6	14.8	*o*-Cresol = 41.4 *p*-Cresol = 55.9	2.28
Anisole	Biphase	16	42.2	2-Hydroxyanisole = 66.6 4-Hydroxyanisole = 30.3	2.07
Anisole	Triphase	8	66.5	2-Hydroxyanisole = 25.6 4-Hydroxyanisole = 72.3	6.54
Benzyl alcohol	Biphase	18	12.5	Benzaldehyde = 86.0 Benzoic acid = 14.0	0.55
Benzyl alcohol	Triphase	18	89.6	Benzaldehyde = 73.5 Benzoic acid = 25.6	3.93

Temperature = 353 K

TOF = Turn over frequency, moles of substrate converted per mole of metal per hour

TS-1 is also found to be a highly efficient and selective catalyst in dihydroxylation of bulky unsaturated alcohols or bulky unsaturated halides under triphase reaction conditions in the presence of dilute hydrogen peroxide [88]. Unsaturated alcohols and halides yield triol and diol as the major products, respectively. Under biphasic conditions, epoxide is the major product for these unsaturated substrates. When oxidation reactions of benzene, toluene, anisole and benzyl alcohol are carried out in the presence of TS-1/H_2O_2 under solvent free triphasic conditions, it is noted that there is a significant enhancement in the conversion for the oxidation reactions [89,104-106]. This enhancement under triphasic conditions may be attributed to (i) relative hydrophobic nature and restricted pore dimensions of titanium silicate; (ii) competition of diffusion of substrate with co-solvent; and (iii) Brønsted acidity of titanium silicate in the presence of hydrogen peroxide and water.

Vanadium silicate exhibits good catalytic activity for the hydroxylation of phenol, *n*-hexane, cyclohexane and ethylbenzene in the presence of hydrogen peroxide as the oxidant [62]. Microporous niobium phoshphate can be used as catalyst for hydroxylation of phenol with high selectivity for catechol formation *ca.* 95 %, in aqueous protic solvent (e.g. MeOH) with H_2O_2 as the oxidant [66]. Although yield of conversion is quite similar in protic and aprotic solvents, the selectivity for catechol formation in aprotic solvent (acetonitrile) is much lower compared to protic solvent. This unusual high catechol selectivity in protic solvent could be due to the formation of hydrogen bonds between the OH group of phenol and methanol in a transition state, thereby activating the *ortho* position of phenol to form catechol (Scheme 6), since the strong hydrophilic character of niobium phosphate has the tendency to coordinate with protic solvents and expand its coordination. For Co(III)-containing mesoporous silicate, excellent catalytic activity and high *trans*-selectivity in the liquid phase dihydroxylation of cyclohexene under mild liquid phase reaction conditions is observed using hydrogen peroxide, TBHP or molecular O_2, as oxidants and acetonitrile as solvent [74].

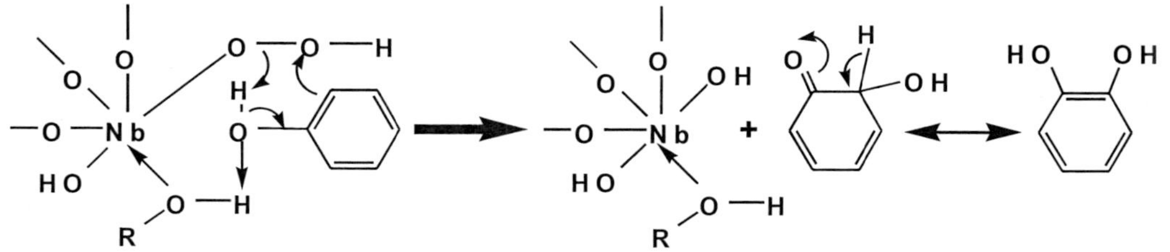

Scheme 6. Mechanism for the formation of catechol over microporous niobium phosphate in protic solvent (R= CH_3 or H).

6. CONCLUSIONS

In this book chapter we have discussed about different types of heterogeneous active catalysts containing active transition metals. For the preparation of heterogenized catalysts three main types of solid supports, *viz.* microporous and mesoporous silica with ordered pore systems, non-porous silica and polymer or resin, have been used. The transition metals used for these heterogeneous catalysts are titanium, vanadium, niobium, chromium, molybdenum, iron, cobalt, nickel and copper. The syntheses of these materials have been elaborated in details. Characterization of these materials using various instrumental techniques, like powder X-ray diffraction, nitrogen adsorption-desorption, transmission and scanning electron microscopy, UV-vis, FT IR and other spectral studies, have been presented along with the general implications that can drawn from these experimental observations. Finally, a generalized discussion has been done on the application of these active transition metal containing heterogeneous catalysts in different liquid phase oxidation reactions. Regarding the catalytic reactions, the discussion has been confine mainly up to liquid phase partial oxidation reactions. These reactions include oxidation of alkanes, alkenes and aromatic alcohols, ammoximation of cyclic ketones, dihydroxylation of alkenes, hydroxylation of aromatics, benzylation of alkanes, Baeyer-Villiger oxidation and benzylation reactions. From the above discussion we can see that all these heterogeneous metal catalysts plays crucial role in different types of organic transformation for the synthesis of diverse class of value added fine chemicals under eco-friendly conditions and hence these supported metal catalysts have huge potential in liquid phase partial oxidation reactions.

REFERENCES

[1] IUPAC *Compendium of Chemical Terminology* 2nd Edition (1997), R.B. 43

[2] D. T. Sawyer, A. Sobkowaik, F. T. Matshumita, *Acc. Chem. Res.* 29, 409 (1996).

[3] M. Taramasso, G. Perego and V. Notari, *US Patent No. 4410501* (1983).

[4] D. R. C. Huybrechts, L. de Bruycker and P. A. Jacobs *Nature* 345, 240 (1990).

[5] C. T. Kresge, M. E. Leonowicz, W. J. Roth, J. C. Vartuli and J. S. Beck, Nature 359, 710 (1992); J. S. Beck, J. C. Vartuli, W. J. Roth, M. E. Leonowicz, C. T. Kresge, K. D. Schmitt, C. T. W. Chu, D. H. Olson and E. W. Sheppard, *J. Am. Chem. Soc.* 114, 10834 (1992).

[6] P. T. Tanev, M. Chibwe and T. J. Pinnavaia, *Nature* 368, 321 (1994).

[7] A. Taguchi and F. Schüth, *Microporous Mesoporous Mater*. 77, 1 (2005).

[8] C. Jin, W. Fan, Y. Jia, B. Fan, J. Ma and R. Li, *J. Mol. Catal. A: Chem*. 249, 23 (2006).

[9] M. Salavati-Niasari, M. Hassani-Kabutarkhani and F. Davar, *Catal. Commun*. 7, 955 (2006).

[10] P. Roy, K. Dhara, M. Manassero and P. Banerjee, *Inorg. Chem. Commun*. 11, 265 (2008).

[11] V. Ayala, A. Corma, M. Iglesias and F. Sanchez, *J. Mol. Catal. A: Chem*. 221, 201 (2004).

[12] M. Nandi, P. Mondal, M. Islam and A. Bhaumik. *Eur. J. Inorg. Chem*. 221 (2011).

[13] C. E. Song and S. –G. Lee, *Chem. Rev*. 2002, *102*, 3495-3524.

[14] P. Chutia, S. Kato, T. Kojima and S. Satokawa, *Polyhedron* 28, 370 (2009).

[15] S. Pande, A. Saha, S. Jana, S. Sarkar, M. Basu, M. Pradhan, A. K. Sinha, S. Saha, A. Pal and T. Pal, *Org. Lett*. 10, 5179 (2008).

[16] A. R. Silva, C. Freire, B. de Castro, M. M. A. Freitas and J. L. Figueiredo, *Langmuir* 18, 8017 (2002).

[17] R. Kreiter, A. W. Kleij, R. J. M. Klein Gebbink and G. van Koten, In *Dendrimers IV: Metal Coordination, Self-Assembly, Catalysis;* F. Vögtle, C. A. Schalley, Eds.; Springer-Verlag: Berlin 217, 163 (2001).

[18] R. van Heerbeek, P. C. J. Kamer, P. W. N. M. van Leeuwen and J. N. H. Reek, *Chem. Rev*. 102, 3717 (2002).

[19] S. M. Islam, A. Bose, B. K. Palit and C. R. Saha, J. Catal. 173, 268 (1998).

[20] G. J. de A. A. Soler-Illia, C. Sanchez, B. Lebeau and J. Patarin, *Chem. Rev*. 102, 4093–4138 (2002).

[21] H.-P. Lin and C.-Y. Mou, *Acc. Chem. Res*. 35, 927 (2002).

[22] T. Czuryszkiewicz, F. Kleitz, F. Schüth and M. Lindén, *Chem. Mater*. 15, 3704 (2003).

[23] H. Lin, G. Zhu, J. Xing, B. Gao and S. Qiu, *Langmuir* 25, 10159 (2009)

[24] C. Gao, H. Qiu, W. Zeng, Y. Sakamoto, O. Terasaki, K. Sakamoto, Q. Chen and S. Che, *Chem. Mater*. 18, 3904 (2006).

[25] L. Han, Q. Chen, Y. Wang, C. Gao and S. Che, *Microporous Mesoporous Mater*. 139, 94 (2011).

[26] D. Zhao, Q. Huo, J. Feng, B. F. Chmelka and G. D. Stucky, *J. Am. Chem. Soc*. 120, 6024 (1998).

[27] S. Inagaki, S. Guan, Y. Fukushima, T. Ohsuna and O. Terasaki, *J. Am. Chem. Soc*. 121, 9611 (1999).

[28] T. Asefa, M. J. MacLachlan, N. Coombs and G. A. Ozin, Nature 402, 867 (1999).

[29] A. Sayari, S. Hamoudi, *Chem. Mater*. 13, 3151 (2001).

[30] A. Bhaumik, M. P. Kapoor and S. Inagaki, *Chem. Commun*. 470 (2003).

[31] K. Sarkar, S. C. Laha and A. Bhaumik, *J. Mater. Chem*. 16, 2439 (2006).

[32] D. Chandra, T. Yokoi, T. Tatsumi and A. Bhaumik, *Chem. Mater*. 19, 5347 (2007).

[33] A. Tuel, *Microporous Mesoporous Mater*. 27, 151 (1999).

[34] Á. Szegedi, M. Popova, V. Mavrodinova and C. Minchev, *Appl.Catal. A: Gen*. 338, 44 (2008).

[35] S. M. Rivera-Jiménez, S. Méndez-González and A. Hernández-Maldonado, *Microporous Mesoporous Mater*. 132, 470 (2010).

[36] T. A. Fernandes, C. D. Nunes, P. D. Vaz, M. J. Calhorda, P. Brandão, J. Rocha, I. S. Gonçalves, A. A. Valente, L. P. Ferreira, M. Godinho and P. Ferreira, *Microporous Mesoporous Mater*. 112, 14 (2008).
[37] D. Chandra, N. K. Mal, M. Mukherjee and A. Bhaumik, *J. Solid State Chem*. 179, 1802, (2006).
[38] S. Mukherjee, S. Samanta, B. C. Roy and A. Bhaumik *Appl. Catal. A: Gen*. 301, 79 (2006).
[39] S. Mukherjee, S. Samanta, A. Bhaumik and B. C. Ray, *Appl. Catal. B: Environ*. 68, 12 (2006).
[40] N. E. Leadbeater and M. Marco, *Chem. Rev*. 102, 3217 (2002).
[41] T. Punniyamurthy and L. Rout, *Coord. Chem. Rev*. 252, 134 (2008).
[42] C. G. Frost and L. Mutton, *Green Chem*. 12, 1687 (2010).
[43] S. Mukherjee, M. Nandi, K. Sarkar and A. Bhaumik, *J. Mol. Catal. A: Chem*. 301, 114 (2009).
[44] B. Notari, Stud. *Surf. Sci. Catal*. 37, 413 (1987).
[45] R. Kumar, A. Bhaumik, R. K. Ahedi and S. Ganapathy, *Nature* 381, 298 (1996).
[46] A. Bhaumik, S. Samanta and N. K. Mal, *Microporous Mesoporous Mater*. 68, 29 (2004).
[47] D. Chandra and A. Bhaumik, *Ind. Eng. Chem. Res*. 45, 4879 (2006).
[48] D. Chandra, N. K. Mal, M. Mukherjee and A. Bhaumik, *J. Solid State Chem*. 179, 1802 (2006).
[49] M, Nandi, K, Sarkar and A. Bhaumik, *Mater. Chem. Phys*. 107, 499 (2008).
[50] M. Nandi and A. Bhaumik, *Chem. Eng. Sci*. 61, 4373 (2006).
[51] S. L. Burkett, S. D. Simis and S. Mann, *J. Chem. Soc. Chem. Commun*. 1367 (1996).
[52] A. Corma, J. L. Jordá, M. T. Navarro and F. Rey, *J. Chem. Soc. Chem. Commun*. 1899 (1998).
[53] N. Igarashi, Y. Tanaka, S. Nakata and T. Tatsumi, *Chem. Lett*.1 (1999).
[54] T. Tatsumi, K. A. Koyano and N. Igarashi, *J. Chem. Soc. Chem. Commun*. 325 (1998).
[55] K. Sarkar, S. C. Laha, N. K. Mal and A. Bhaumik, *J. Solid State Chem*. 181, 2065 (2008).
[56] A. Bhaumik and T. Tatsumi, *J. Catal*. 189, 31 (2000).
[57] A. Bhaumik and T. Tatsumi, *Catal. Lett*. 66, 181 (2000).
[58] D. Chandra, S. C. Laha and A. Bhaumik, *Appl. Catal. A: Gen*. 342, 29 (2008).
[59] A. Bhaumik, M. P. Kapoor and S. Inagaki, *Chem. Commun*. 470 (2003).
[60] M. P. Kapoor, A. Bhaumik, S. Inagaki, K. Kuraoka and T. Yazawa, *J. Mater. Chem*. 12, 3078 (2002).
[61] K. Sarkar, M. Nandi and A. Bhaumik *Appl. Catal. A: Gen*. 343, 55 (2008).
[62] A. Bhaumik, M. K. Dongare, R. Kumar, *Microporous Mater*. 5, 173 (1995).
[63] L. A. Errede and G. V. D. Tiers, *J. Phys. Chem*. 100, 9918 (1996).
[64] K. Akagawa, S. Sakamoto, K. Kudo, *Tetrahedron Lett*. 48, 985 (2007).
[65] S. Mukherjee, M. Nandi, K. Sarkar and A. Bhaumik, *J. Mol. Catal. A: Chem*. 301, 114 (2009).
[66] N. K. Mal, A. Bhaumik, P. Kumarc and M. Fujiwara, *Chem. Commun*. 872 (2003).
[67] S. Samanta, N. K. Mal and A. Bhaumik, *J. Mol. Catal. A: Chem*. 236, 7 (2005).
[68] N. Pal, M. Paul and A. Bhaumik, *The Open Catal. J*. 2, 162 (2009).

[69] S. K. Maiti, S. Dinda, M. Nandi, A. Bhaumik and R. Bhattacharyya, *J. Mol. Catal. A: Chem.* 287, 135 (2008).
[70] Q. Zhang, Y. Wang, S. Itsuki, T. Shishido and K. Takehira, *J. Mol. Catal. A: Chem,* 188, 189 (2002).
[71] F. Minutolo, D. Pini and P. Salvadori, *Tetrahedron Lett.* 37, 3375 (1996).
[72] N. N. Tusăr, S. Jank and R. Gläser, *ChemCatChem*, DOI: 10.1002/cctc.201000311, (2011).
[73] S. K. Das, M. Nandi, S. Giri and A. Bhaumik, *Microporous Mesoporous Mater.* 117, 362 (2009).
[74] S. Samanta, S. C. Laha, N. K. Mal and A. Bhaumik, *J. Mol. Catal. A: Chem.* 222, 235 (2004).
[75] J. Chakraborty, M. Nandi, H. Mayer-Figge, W. S. Sheldrick, L. Sorace, A. Bhaumik and P. Banerjee, *Eur. J. Inorg. Chem.* 5033 (2007).
[76] P. Roy, M. Nandi, M. Manassero, M. Riccó, M. Mazzani, A. Bhaumik and P. Banerjee, *Dalton Trans.* 9543 (2009).
[77] A. Corma, H. Garcia, M. T. Navarro, E. J. Palomares and F. Rey, *Chem. Mater.* 12, 3068 (2000).
[78] C. J. Ballhausen and H. B. Gray, *Inorg. Chem.* 1, 11 (1962).
[79] M. R. Maurya, U. Kumar and P. Manikandan, *Eur. J. Inorg. Chem.* 2303 (2007).
[80] S. Shylesh and A. P. Singh, *Microporous Mesoporous Mater.* 94, 127 (2006).
[81] S. Samanta, S. Giri, P. U. Sastry, N. K. Mal, A. Manna and A. Bhaumik, *Ind. Eng. Chem. Res.* 42, 3012 (2003).
[82] M. Soibinet, I. Déchamps-Olivier, E. Guillon, J.-P. Barbier, M. Aplincourt, F. Chuburu, M. L.Baccon and H. Handel, *Polyhedron* 24, 143 (2005).
[83] P. L. Holland, K. R. Rodgers and W. B. Tolman, *Angew. Chem. Int. Ed.* 38, 1139 (1999).
[84] E. I. Solomon, U. M. Sundaram and T. E. Machonkin, *Chem. Rev.* 96, 2563 (1996).
[85] S. Inagaki, Y. Fukushima and K. Kuroda, *J. Chem. Soc., Chem. Commun.* 680 (1993).
[86] G. Horvath and K. Kawazoe, *J. Chem. Eng. Jpn.* 16, 470 (1983).
[87] S. M. Solberg, D. Kumar and C. C. D.; Landry, *J. Phys. Chem. B*, 109, 24331 (2005).
[88] A. Bhaumik and T. Tatsumi, *J. Catal.* 176, 305 (1998).
[89] A. Bhaumik, P. Mukherjee and R. Kumar, *J. Catal.* 178, 101 (1998).
[90] S. M. Mukhopadhyay and S. H. Garofalini, *J. Non-Cryst. Solids* 126, 202 (1990).
[91] Y. S. S. Wan, J. L. H. Chau, K. L. Yeung and A. Gavriilidis, *J. Catal.* 223, 241 (2004).
[92] I. Díaz, F. Mohino, T. Blasco, E. Sastre and J. Pérez-Pariente, *Microporous Mesoporous Mater.* 80, 33 (2005).
[93] M. H. Lim, C. F. Blanford and A. Stein, *Chem. Mater.* 10, 467 (1998).
[94] P. J. Carl and S. C. Larsen, *J. Phys. Chem. B* 104, 6568 (2000).
[95] K. A. Jorgensen, *Chem. Rev.* 89, 431 (1989).
[96] *The Chemistry of peroxides*; S. Patai, Ed.; Wiley: New York, (1983).
[97] F. A. Davis and A. C. Sheppard, *Tetrahedron* 45, 5703 (1989).
[98] C. M. Che and J. S. Huang, *Coord. Chem. Rev.* 242, 97 (2003).
[99] A. M. Kirillov, M. N. Kopylovich, M. V. Kirillova, M. Haukka, M. F. C. Guedes da Silva and A. J. L. Pombeiro, *Angew. Chem., Int. Ed.* 44, 4345 (2005).
[100] P. Roy and M. Manassero, *Dalton Trans.* 39, 1539 (2010).
[101] A. Bhaumik, P. Kumar and R. Kumar, *Catal. Lett.* 40, 47 (1996).

[102] G. Bellussi and V. Fattore, *Stud. Surf. Sci. Catal.* 69, 79 (1991).
[103] A. Bhaumik and R. Kumar, *J. Chem. Soc., Chem. Commun.* 349 (1995).
[104] R. Kumar and A. Bhaumik, *Microporous Mesoporous Mater.* 21, 497 (1998).
[105] R. Kumar, P. Mukherjee and A. Bhaumik *Catal. Today* 49, 185 (1999).
[106] P. Mukherjee, A. Bhaumik and R. Kumar, *Ind. Eng. Chem. Res.* 46, 8657 (2007).

INDEX

B

C

D

E

F

G

H

I

J

K

L

M

N

O

P

T

U

V

W